Lecture Notes in Computer Science　12391

More information about this series at http://www.springer.com/series/7409

Sven Hartmann · Josef Küng ·
Gabriele Kotsis · A Min Tjoa ·
Ismail Khalil (Eds.)

Database and Expert Systems Applications

31st International Conference, DEXA 2020
Bratislava, Slovakia, September 14–17, 2020
Proceedings, Part I

 Springer

Editors
Sven Hartmann
Clausthal University of Technology
Clausthal-Zellerfeld, Germany

Gabriele Kotsis
Johannes Kepler University of Linz
Linz, Austria

Ismail Khalil
Johannes Kepler University of Linz
Linz, Austria

Josef Küng
Johannes Kepler University of Linz
Linz, Austria

A Min Tjoa
IFS
Vienna University of Technology
Vienna, Wien, Austria

ISSN 0302-9743 ISSN 1611-3349 (electronic)
Lecture Notes in Computer Science
ISBN 978-3-030-59002-4 ISBN 978-3-030-59003-1 (eBook)
https://doi.org/10.1007/978-3-030-59003-1

LNCS Sublibrary: SL3 – Information Systems and Applications, incl. Internet/Web, and HCI

This Springer imprint is published by the registered company Springer Nature Switzerland AG
The registered company address is: Gewerbestrasse 11, 6330 Cham, Switzerland

Preface

This volume contains the papers presented at the 31st International Conference on Database and Expert Systems Applications (DEXA 2020). This year, DEXA was held as a virtual conference during September 14–17, 2020, instead of as it was originally planned to be held in Bratislava, Slovakia.

On behalf of the Program Committee we commend these papers to you and hope you find them useful.

Database, information, and knowledge systems have always been a core subject of computer science. The ever increasing need to distribute, exchange, and integrate data, information, and knowledge has added further importance to this subject. Advances in the field will help facilitate new avenues of communication, to proliferate interdisciplinary discovery, and to drive innovation and commercial opportunity.

DEXA is an international conference series which showcases state-of-the-art research activities in database, information, and knowledge systems. The conference and its associated workshops provide a premier annual forum to present original research results and to examine advanced applications in the field. The goal is to bring together developers, scientists, and users to extensively discuss requirements, challenges, and solutions in database, information, and knowledge systems.

DEXA 2020 solicited original contributions dealing with all aspects of database, information, and knowledge systems. Suggested topics included, but were not limited to:

- Acquisition, Modeling, Management and Processing of Knowledge
- Authenticity, Privacy, Security, and Trust
- Availability, Reliability and Fault Tolerance
- Big Data Management and Analytics
- Consistency, Integrity, Quality of Data
- Constraint Modeling and Processing
- Cloud Computing and Database-as-a-Service
- Database Federation and Integration, Interoperability, Multi-Databases
- Data and Information Networks
- Data and Information Semantics
- Data Integration, Metadata Management, and Interoperability
- Data Structures and Data Management Algorithms
- Database and Information System Architecture and Performance
- Data Streams and Sensor Data
- Data Warehousing
- Decision Support Systems and Their Applications
- Dependability, Reliability, and Fault Tolerance
- Digital Libraries and Multimedia Databases
- Distributed, Parallel, P2P, Grid, and Cloud Databases
- Graph Databases

- Incomplete and Uncertain Data
- Information Retrieval
- Information and Database Systems and Their Applications
- Mobile, Pervasive, and Ubiquitous Data
- Modeling, Automation, and Optimization of Processes
- NoSQL and NewSQL Databases
- Object, Object-Relational, and Deductive Databases
- Provenance of Data and Information
- Semantic Web and Ontologies
- Social Networks, Social Web, Graph, and Personal Information Management
- Statistical and Scientific Databases
- Temporal, Spatial, and High Dimensional Databases
- Query Processing and Transaction Management
- User Interfaces to Databases and Information Systems
- Visual Data Analytics, Data Mining, and Knowledge Discovery
- WWW, Databases and Web Services
- Workflow Management and Databases
- XML and Semi-structured Data

Following the call for papers which yielded 190 submissions, there was a rigorous refereeing process that saw each submission reviewed by three to five international experts. The 38 submissions judged best by the Program Committee were accepted as full research papers, yielding an acceptance rate of 20%. A further 20 submissions were accepted as short research papers.

As is the tradition of DEXA, all accepted papers are published by Springer. Authors of selected papers presented at the conference were invited to submit substantially extended versions of their conference papers for publication in special issues of international journals. The submitted extended versions underwent a further review process.

We wish to thank all authors who submitted papers and all conference participants for the fruitful discussions.

This year we have five keynote talks addressing emerging trends in the database and artificial intelligence community:

- "Knowledge Graphs for Drug Discovery" by Prof. Ying Ding (The University of Texas at Austin, USA)
- "Incremental Learning and Learning with Drift" by Prof. Barbara Hammer (CITEC Centre of Excellence, Bielefeld University, Germany)
- "From Sensors to Dempster-Shafer Theory and Back: the Axiom of Ambiguous Sensor Correctness and its Applications" by Prof. Dirk Draheim (Tallinn University of Technology, Estonia)
- "Knowledge Availability and Information Literacies" by Dr. Gerald Weber (The University of Auckland, New Zealand)
- "Explainable Fact Checking for Statistical and Property Claims" by Paolo Papotti (EURECOM, France)

In addition, we had a panel discussion on "The Age of Science-making Machines" led by Prof. Stéphane Bressan (National University of Singapore, Singapore).

This edition of DEXA features three international workshops covering a variety of specialized topics:

- BIOKDD 2020: 11th International Workshop on Biological Knowledge Discovery from Data
- IWCFS 2020: 4th International Workshop on Cyber-Security and Functional Safety in Cyber-Physical Systems
- MLKgraphs 2020: Second International Workshop on Machine Learning and Knowledge Graphs

The success of DEXA 2020 is a result of collegial teamwork from many individuals. We like to thank the members of the Program Committee and the external referees for their timely expertise in carefully reviewing the submissions.

Warm thanks to Ismail Khalil and the conference organizers as well as all workshop organizers.

We would also like to express our thanks to all institutions actively supporting this event, namely:

- Comenius University Bratislava (who was prepared to host the conference)
- Institute of Telekoopertion, Johannes Kepler University Linz (JKU)
- Software Competence Center Hagenberg (SCCH)
- International Organization for Information Integration and Web based Applications and Services (@WAS)

We hope you enjoyed the conference program.

September 2020

Sven Hartmann
Josef Küng

Organization

Program Committee Chairs

Sven Hartmann   Clausthal University of Technology, Germany
Josef Küng   Johannes Kepler University Linz, Austria

Steering Committee

Gabriele Kotsis   Johannes Kepler University Linz, Austria
A Min Tjoa   Vienna University of Technology, Austria
Ismail Khalil   Johannes Kepler University Linz, Austria

Program Committee and Reviewers

Javier Nieves Acedo	Azterlan, Spain
Sonali Agarwal	IIIT, India
Riccardo Albertoni	CNR-IMATI, Italy
Idir Amine Amarouche	USTHB, Algeria
Rachid Anane	Coventry University, UK
Mustafa Atay	Winston-Salem State University, USA
Faten Atigui	CEDRIC, CNAM, France
Ladjel Bellatreche	LIAS, ENSMA, France
Nadia Bennani	LIRIS, INSA de Lyon, France
Karim Benouaret	Université de Lyon, France
Djamal Benslimane	Université de Lyon, France
Morad Benyoucef	University of Ottawa, Canada
Catherine Berrut	LIG, Université Joseph Fourier Grenoble I, France
Vasudha Bhatnagar	University of Delhi, India
Athman Bouguettaya	The University of Sydney, Australia
Omar Boussai	ERIC Laboratory, France
Stephane Bressan	National University of Singapore, Singapore
Sharma Chakravarthy	The University of Texas at Arlington, USA
Cindy Chen	University of Massachusetts Lowell, USA
Gang Chen	Victoria University of Wellington, New Zealand
Max Chevalier	IRIT, France
Soon Ae Chun	City University of New York, USA
Alfredo Cuzzocrea	University of Calabria, Italy
Debora Dahl	Conversational Technologies, USA
Jérôme Darmont	Université de Lyon, France
Soumyava Das	Teradata Labs, USA
Vincenzo Deufemia	University of Salerno, Italy
Dejing Dou	University of Oregon, USA

Cedric Du Mouza	CNAM, France
Umut Durak	German Aerospace Center (DLR), Germany
Suzanne Embury	The University of Manchester, UK
Markus Endres	University of Passau, Germany
Nora Faci	Université de Lyon, France
Bettina Fazzinga	ICAR-CNR, Italy
Stefano Ferilli	Università di Bari, Italy
Flavio Ferrarotti	Software Competence Centre Hagenberg, Austria
Flavius Frasincar	Erasmus University Rotterdam, The Netherlands
Bernhard Freudenthaler	Software Competence Center Hagenberg, Austria
Steven Furnell	Plymouth University, UK
Aryya Gangopadhyay	University of Maryland Baltimore County, USA
Manolis Gergatsoulis	Ionian University, Greece
Javad Ghofrani	HTW Dresden University of Applied Sciences, Germany
Vikram Goyal	IIIT Delhi, India
Sven Groppe	University of Lübeck, Germany
William Grosky	University of Michigan, USA
Francesco Guerra	Università di Modena e Reggio Emilia, Italy
Giovanna Guerrini	DISI, University of Genova, Italy
Allel Hadjali	LIAS, ENSMA, France
Abdelkader Hameurlain	IRIT, Paul Sabatier University, France
Takahiro Hara	Osaka University, Japan
Ionut Iacob	Georgia Southern University, USA
Hamidah Ibrahim	Universiti Putra Malaysia, Malaysia
Sergio Ilarri	University of Zaragoza, Spain
Abdessamad Imine	Loria, France
Peiquan Jin	University of Science and Technology of China, China
Anne Kao	Boeing, USA
Anne Kayem	Hasso Plattner Institute, University of Potsdam, Germany
Uday Kiran	The University of Tokyo, Japan
Carsten Kleiner	University of Applied Science and Arts Hannover, Germany
Henning Koehler	Massey University, New Zealand
Michal Kratky	VSB-Technical University of Ostrava, Czech Republic
Petr Kremen	Czech Technical University in Prague, Czech Republic
Anne Laurent	LIRMM, University of Montpellier, France
Lenka Lhotska	Czech Technical University in Prague, Czech Republic
Wenxin Liang	Chongqing University of Posts and Telecommunications, China
Chuan-Ming Liu	National Taipei University of Technology, Taiwan
Hong-Cheu Liu	School of Information Technology and Mathematical Sciences, Australia
Jorge Lloret	University of Zaragoza, Spain
Hui Ma	Victoria University of Wellington, New Zealand

Qiang Ma	Kyoto University, Japan
Zakaria Maamar	Zayed University, UAE
Sanjay Madria	Missouri S&T, USA
Jorge Martinez-Gil	Software Competence Center Hagenberg, Austria
Elio Masciari	Federico II University, Italy
Atif Mashkoor	Software Competence Center Hagenberg, Austria
Jun Miyazaki	Tokyo Institute of Technology, Japan
Lars Moench	University of Hagen, Germany
Riad Mokadem	Paul Sabatier University, Pyramide Team, France
Anirban Mondal	The University of Tokyo, Japan
Yang-Sae Moon	Kangwon National University, South Korea
Franck Morvan	IRIT, Paul Sabatier University, France
Francesco D. Muñoz-Escoí	Universitat Politècnica de València, Spain
Ismael Navas-Delgado	University of Malaga, Spain
Wilfred Ng	The Hong Kong University of Science and Technology, Hong Kong
Teste Olivier	IRIT, France
Carlos Ordonez	University of Houston, USA
Marcin Paprzycki	Systems Research Institute, Polish Academy of Sciences, Poland
Oscar Pastor Lopez	Universitat Politècnica de València, Spain
Clara Pizzuti	CNR-ICAR, Italy
Elaheh Pourabbas	CNR, Italy
Birgit Proell	Johannes Kepler University Linz, Austria
Rodolfo Resende	Federal University of Minas Gerais, Brazil
Werner Retschitzegger	Johannes Kepler University Linz, Austria
Claudia Roncancio	Grenoble Alps University, France
Viera Rozinajova	Slovak University of Technology in Bratislava, Slovakia
Massimo Ruffolo	ICAR-CNR, Italy
Shelly Sachdeva	National Institute of Technology Delhi, India
Marinette Savonnet	University of Burgundy, France
Stefanie Scherzinger	Universität Passau, German
Wieland Schwinger	Johannes Kepler University Linz, Austria
Florence Sedes	IRIT, Paul Sabatier University, France
Nazha Selmaoui-Folcher	University of New Caledonia, New Caledonia
Michael Sheng	Macquarie University, Australia
Patrick Siarry	Université de Paris 12, France
Gheorghe Cosmin Silaghi	Babes-Bolyai University, Romania
Hala Skaf-Molli	University of Nantes, LS2N, France
Bala Srinivasan	Monash University, Australia
Stephanie Teufel	University of Fribourg, iimt, Switzerland
Jean-Marc Thévenin	IRIT, Université Toulouse I, France
Vicenc Torra	University Skövde, Sweden
Traian Marius Truta	Northern Kentucky University, USA
Lucia Vaira	University of Salento, Italy

Ismini Vasileiou	De Montfort University, UK
Krishnamurthy Vidyasankar	Memorial University, Canada
Marco Vieira	University of Coimbra, Portugal
Piotr Wisniewski	Nicolaus Copernicus University, Poland
Ming Hour Yang	Chung Yuan Chritian University, Taiwan
Xiaochun Yang	Northeastern University, China
Haruo Yokota	Tokyo Institute of Technology, Japan
Qiang Zhu	University of Michigan, USA
Yan Zhu	Southwest Jiaotong University, China
Marcin Zimniak	Leipzig University, Germany
Ester Zumpano	University of Calabria, Italy

Organizers

Institute for
Telecooperation

International Organization for
www.iiwas.org

Information Integration and
Web-based Applications & Services

Abstracts of Keynote Talks

Knowledge Graph for Drug Discovery

Ying Ding

The University of Texas at Austin, USA

Abstract. A critical barrier in current drug discovery is the inability to utilize public datasets in an integrated fashion to fully understand the actions of drugs and chemical compounds on biological systems. There is a need to intelligently integrate heterogeneous datasets pertaining to compounds, drugs, targets, genes, diseases, and drug side effects now available to enable effective network data mining algorithms to extract important biological relationships. In this talk, we demonstrate the semantic integration of 25 different databases and showcase the cutting-edge machine learning and deep learning algorithms to mine knowledge graphs for deep insights, especially the latest graph embedding algorithm that outperforms baseline methods for drug and protein binding predictions.

Incremental Learning and Learning with Drift

Barbara Hammer

CITEC Centre of Excellence, Bielefeld University, Germany

Abstract. Neural networks have revolutionized domains such as computer vision or language processing, and learning technology is included in everyday consumer products. Yet, practical problems often render learning surprisingly difficult, since some of the fundamental assumptions of the success of deep learning are violated. As an example, only few data might be available for tasks such as model personalization, hence few shot learning is required. Learning might take place in non-stationary environments such that models face the stability-plasticity dilemma. In such cases, applicants might be tempted to use models for settings they are not intended for, such that invalid results are unavoidable.

Within the talk, I will address three challenges of machine learning when dealing with incremental learning tasks, addressing the questions: how to learn reliably given few examples only, how to learn incrementally in non-stationary environments where drift might occur, and how to enhance machine learning models by an explicit reject option, such that they can abstain from classification if the decision is unclear

From Sensors to Dempster-Shafer Theory and Back: The Axiom of Ambiguous Sensor Correctness and Its Applications

Dirk Draheim

Tallinn University of Technology, Estonia
dirk.draheim@taltech.ee

Abstract. Since its introduction in the 1960s, Dempster-Shafer theory became one of the leading strands of research in artificial intelligence with a wide range of applications in business, finance, engineering, and medical diagnosis. In this paper, we aim to grasp the essence of Dempster-Shafer theory by distinguishing between ambiguous-and-questionable and ambiguous-but-correct perceptions. Throughout the paper, we reflect our analysis in terms of signals and sensors as a natural field of application. We model ambiguous-and-questionable perceptions as a probability space with a quantity random variable and an additional perception random variable (Dempster model). We introduce a correctness property for perceptions. We use this property as an axiom for ambiguous-but-correct perceptions. In our axiomatization, Dempster's lower and upper probabilities do not have to be postulated: they are consequences of the perception correctness property. Furthermore, we outline how Dempster's lower and upper probabilities can be understood as best possible estimates of quantity probabilities. Finally, we define a natural knowledge fusion operator for perceptions and compare it with Dempster's rule of combination.

Knowledge Availability and Information Literacies

Gerald Weber

The University of Auckland, New Zealand

Abstract. At least since Tim Berners-Lee's call for 'Raw Data Now' in 2009, which he combined with a push for linked data as well, the question has been raised how to make the wealth of data and knowledge available to the citizens of the world. We will set out to explore the many facets and multiple layers of this problem, leading up to the question of how we as users will access and utilize the knowledge that should be available to us.

Explainable Fact Checking for Statistical and Property Claims

Paolo Papotti

EURECOM, France

Abstract. Misinformation is an important problem but fact checkers are overwhelmed by the amount of false content that is produced online every day. To support fact checkers in their efforts, we are creating data-driven verification methods that use structured datasets to assess claims and explain their decisions. For statistical claims, we translate text claims into SQL queries on relational databases. We exploit text classifiers to propose validation queries to the users and rely on tentative execution of query candidates to narrow down the set of alternatives. The verification process is controlled by a cost-based optimizer that considers expected verification overheads and the expected claim utility as training samples. For property claims, we use the rich semantics in knowledge graphs (KGs) to verify claims and produce explanations. As information in a KG is inevitably incomplete, we rely on rule discovery and on text mining to gather the evidence to assess claims. Uncertain rules and facts are turned into logical programs and the checking task is modeled as a probabilistic inference problem. Experiments show that both methods enable the efficient and effective labeling of claims with interpretable explanations, both in simulations and in real world user studies with 50% decrease in verification time. Our algorithms are demonstrated in a fact checking website (https://coronacheck.eurecom.fr), which has been used by more than twelve thousand users to verify claims related to the coronavirus disease (COVID-19) spread and effect.

"How Many Apples?" or the Age of Science-Making Machines (Panel)

Stéphane Bressan (Panel Chair)

National University of Singapore, Singapore

Abstract. Isaac Newton most likely did not spend much time observing apples falling from trees. Galileo Galilei probably never threw anything from the tower of Pisa. They conducted thought experiments.

What can big data, data science, and artificial intelligence contribute to the creation of scientific knowledge? How can advances in computing, communication, and control improve or positively disrupt the scientific method?

Richard Feynman once explained the scientific method as follows. "In general, we look for a new law by the following process. First, we guess it; don't laugh that is really true. Then we compute the consequences of the guess to see what, if this is right, if this law that we guessed is right, we see what it would imply. And then we compare those computation results to nature, or, we say, compare to experiment or experience, compare directly with observations to see if it works. If it disagrees with experiment, it's wrong and that simple statement is the key to science." He added euphemistically that "It is therefore not unscientific to take a guess."

Can machines help create science?

The numerous advances of the many omics constitute an undeniable body of evidence that computing, communication, and control technologies, in the form of high-performance computing hardware, programming frameworks, algorithms, communication networks, as well as storage, sensing, and actuating devices, help scientists and make the scientific process significantly more efficient and more effective. Everyone acknowledges the unmatched ability of machines to streamline measurements and to process large volumes of results, to facilitate complex modeling, and to run complex computations and extensive simulations. The only remaining question seems to be the extent of their unexplored potential.

Furthermore, the media routinely report new spectacular successes of big data analytics and artificial intelligence that suggest new opportunities. Scientists are discussing physics-inspired machine learning. We are even contemplating the prospect of breaking combinatorial barriers with quantum computers. However, except, possibly for the latter, one way or another, it all seems about heavy-duty muscle-flexing without much subtlety nor finesse.

Can machines take a guess?

Although the thought processes leading to the guesses from which theories are built are laden with ontological, epistemological, and antecedent theoretical assumptions, and the very formulation of the guesses assumes certain conceptual views, scientists seem to have been able to break through those glass ceilings again and again and invent entirely new concepts. Surely parallel computing, optimization algorithms, reinforcement learning, or genetic algorithms can assist in the exploration of the space of combinatorial compositions of existing concepts. In the words of Feynman again: "We set up a machine, a great computing machine, which has a random wheel and if it makes a succession of guesses and each time it guesses a hypothesis about how nature should work computes immediately the consequences and makes a comparison to a list of experimental results that it has at the other hand. In other words, guessing is a dumb man's job. Actually, it is quite the opposite and I will try to explain why." He continues: "The problem is not to make, to change or to say that something might be wrong but to replace it with something and that is not so easy."

Can machines create new concepts?

The panelists are asked to share illustrative professional experiences, anecdotes, and thoughts, as well as their enthusiasm and concerns, regarding the actuality and potential of advances in computing, communication, and control in improving and positively disrupting the scientific process.

Contents – Part I

Data Mining

Databases and Data Management

Information Retrieval

Prediction and Decision Support

Contents – Part II

Knowledge Discovery

Machine Learning

Semantic Web and Ontologies

Stream Data Processing

Temporal, Spatial, and High Dimensional Databases

Keynote Paper

From Sensors to Dempster-Shafer Theory and Back: The Axiom of Ambiguous Sensor Correctness and Its Applications
Keynote at DEXA'2020 – The 31st International Conference on Database and Expert Systems Applications

Dirk Draheim[1(✉)] and Tanel Tammet[2]

[1] Information Systems Group, Tallinn University of Technology, Akadeemia tee 15a, 12618 Tallinn, Estonia
dirk.draheim@ttu.ee
[2] Applied Artificial Intelligence Group, Tallinn University of Technology, Akadeemia tee 15a, 12618 Tallinn, Estonia
tanel.tammet@ttu.ee

Abstract. Since its introduction in the 1960s, Dempster-Shafer theory became one of the leading strands of research in artificial intelligence with a wide range of applications in business, finance, engineering and medical diagnosis. In this paper, we aim to grasp the essence of Dempster-Shafer theory by distinguishing between *ambiguous-and-questionable* and *ambiguous-but-correct* perceptions. Throughout the paper, we reflect our analysis in terms of signals and sensors as a natural field of application. We model *ambiguous-and-questionable* perceptions as a probability space with a quantity random variable and an additional perception random variable (Dempster model). We introduce a correctness property for perceptions. We use this property as an axiom for *ambiguous-but-correct* perceptions. In our axiomatization, Dempster's lower and upper probabilities do not have to be postulated: they are consequences of the perception correctness property. Even more, we outline how Dempster's lower and upper probabilities can be understood as best possible estimates of quantity probabilities. Finally, we define a *natural* knowledge fusion operator for perceptions and compare it with Dempster's rule of combination.

Keywords: Dempster-Shafer theory · Dempster's rule of combination · Knowledge fusion · Sensor fusion · Artificial intelligence · Machine learning · Probabilistic reasoning · Bayesian inference · Uncertainty

1 Introduction

Reasoning under uncertainty has been thoroughly investigated for at least a century, leading to a proliferation of different theories and mechanisms, each of

© Springer Nature Switzerland AG 2020
S. Hartmann et al. (Eds.): DEXA 2020, LNCS 12391, pp. 3–19, 2020.
https://doi.org/10.1007/978-3-030-59003-1_1

which is well suited for certain kinds of problems and ill-suited for other kinds. Dempster-Shafer theory [1–6] is one of the major frameworks for reasoning under uncertainty. Since its introduction in the 1960s, Dempster-Shafer theory became one of the leading strands of research in artificial intelligence [7] with a wide range of applications in business [8,9], finance [10,11], engineering [12,13] and medical diagnosis [14,15].

In this paper, we aim to grasp the essence of Dempster-Shafer theory by distinguishing between *ambiguous-and-questionable* and *ambiguous-but-correct* perceptions. We model *ambiguous-and-questionable* perceptions as a probability space with a quantity random variable and an additional perception random variable (Dempster model). On the basis of this, this paper features the following contributions:

(i) We introduce a correctness property for perceptions. We use this property as an axiom for *ambiguous-but-correct* perceptions.
(ii) In our axiomatization, Dempster's lower and upper probabilities do not have to be postulated: they are consequences of the perception correctness property.
(iii) We outline how Dempster's lower and upper probabilities can be understood as best possible estimates of quantity probabilities.
(iv) We define a *natural* knowledge fusion operator of perceptions and compare it with Dempster's rule of combination.

Throughout the paper we reflect our analysis in terms of signals and sensors. Signals are a natural and particularly successful field of application of Dempster-Shafer theory [16–18]; and arguments are often particularly easy to follow when presented against the *ad hoc* understanding of technical signals/sensors. A kind of tension comes in through this approach due to the fact that signal/sensor examples lay rather in the realm of frequentist interpretation of probability theory, whereas Dempster clearly localizes Dempster-Shafer theory in the strand of Bayesian inference [4] and subjective/epistemic probabilities [19] (although his original motivating example in [3,4], i.e., a partially covered map with regions of land and sea, is, in our humble opinion, rather an example from the realm of frequentist probability). For us, this is not a problem, as we are rather agnostic with respect to the debate of Bayesian/subjective probability (Carnap's probability-1 [20,21]) *versus* frequentist/objective probability (Carnap's probability-2) [20,21], see [22], pp. 5–7, for a discussion, compare also with [23]. Actually, the understanding of signals/sensors can be (quite) easily generalized from technical signals/sensors to more conceptual perceptions: at least for the purpose of this paper, we see all this rather as a matter of presentation and instruction (in our role as working data scientists, we anyhow rather agree with Jerzy Neyman [24]).

In Sect. 2, we set the stage by informally describing the scenarios of ambiguous-and-correct perception through a simple signals/sensors example. In Sect. 3, we formalize ambiguous-and-questionable perceptions as a probability space with two random variables, which actually meets the original model of Dempster as introduced in the 1960s. In Sect. 4 we introduce the axiom of ambiguous sensor correctness and introduce a central Lemma, which states that Dempster's lower and upper probabilities are actually lower and upper bounds for probabilities of signal events – as a consequence of the sensor correctness axiom. In Sect. 5, we deal with the exact relationship between signals and their perceptions. The corresponding lemma also provides the proof for the lower- and upper-bound lemma of the previous section. In Sect. 6, we explain in how far Dempster's lower and upper probabilities are best possible estimates (and not only lower and upper bounds for probabilities of signal events). In Sect. 7, we introduce a natural knowledge fusion operator and compare it with Dempster's rule of combination. We finish the paper with a conclusion in Sect. 8.

2 Ambiguous-but-Correct Perception

Let us start – *in medias res* – with a very small example. The purpose of this example is to explain the scenario with which we deal in this paper: *ambiguous-but-correct perceptions*. Please note, that all essential ingredients (such as the fact that a certain sensor message is correct with respect to certain signals) are meant to be basic assumptions of the scenario and not just properties of that particular example.

2.1 A Simple Signal/Sensor Example

Let us assume that we have a signal source that periodically sends one of the following light signals:

- 'red', 'green', 'blue'   (1)

The signal can have exactly one of the colors in (1), i.e., the signal color is *unambiguous* and there are no more than three signal colors. Next, there is a sensor that detects the light signal and reports its signal detection to a decision maker. The crucial point is that the sensor is *sometimes not sure* about the exact color of the signal. Actually, the sensor might report any of the following information to the decision maker:

- "*red*", "*green*", "*blue*"
- "*either red or green*", "*either red or blue*", "*either green or blue*"   (2)
- "*either red or green or blue*"

For example, if the sensor reports "*red*", then the decision maker can be absolutely sure that the signal has been 'red' (similarly for "*green*" and "*blue*"). Now, if the sensor reports, e.g., "*either red or green*", the decision maker cannot be sure any more, whether the signal has been 'red' or 'green'; however (and this is crucial for the described scenario), he can be sure that the signal has not been 'blue'! So, the sensor report "*either red or green*" is ambiguous but still carries some information, i.e., that the signal must have been *either* 'red' *or* 'green' (similarly for *"either red or blue"* and *"either green or blue"*). The least information is carried by the sensor message *"either red or green or blue"*, here, the decision maker does not know at all anymore, which color the signal actually had. We can summarize what we just explained as a *correctness property*, i.e.:

> *The sensor messages might be ambiguous, but they are correct; i.e., a sensor might not be sure which signal has occurred exactly; however, with respect to the potential signals that the sensor flags, the sensor is always correct.* (3)

We are not ready with the description of the scenario yet. First, each of the signal colors in (1) occurs with a certain probability. Next, also the sensor messages in (2) occur with certain probabilities. Now, the probabilities of the sensor messages are all known to the decision maker, whereas the exact probabilities of the signal colors are not known to him! Still, even if the exact probabilities of the signal colors are not know to him, he at least knows certain guaranteed lower and upper bounds for them, because he can infer those from the probabilities of the sensor messages; and those lower and upper bounds is exactly what Dempster-Shafer theory is about: in Dempster's writing [2–4] they are simply called *lower probability* and *upper probability*, in Shafer's writing *belief* and again *upper probability* [6], in further literature [7,25] they might be called *belief* and *plausibility* and so on; whereas we will coin the terms *certainty* and *possibility* for them, for reasons to become clear in due course – compare with Table 2.

Sensors can greatly vary with respect to their ambiguity – it is all up to the distribution of probabilities that the several sensor messages have. A *perfectly unambiguous* sensor would always only show either *"red"*, *"green"* or *"blue"*, i.e., all other "ambiguous" sensor messages have a probability of 0% (i.e., never occur). For a perfectly unambiguous sensor, the probabilities of *"red"*, *"green"* and *"blue"* are exactly the same as of the corresponding signals 'red', 'green' resp. 'blue'. This follows immediately from the correctness requirement that we have formulated for sensors in our scenario. The most "ambiguous" sensor always shows *"either red or green or blue"* (i.e., with probability 100%) and is actually useless as a sensor. Dempster-Shafer theory deals with scenarios that lay somewhere in between the two extremes of perfectly unambiguous and completely ambiguous (useless) sensors.

The described scenario can be naturally modelled as a probability space [26–28] with signals as outcomes and two random variables: one modeling the sent signals as would be perceived by the decision maker directly, and one for the information as detected by the sensor (and handed over to the decision maker as message). This is what we will do in Sect. 3 and also, what Dempster did in [2–4] up to some technical and terminological details, as we will see in due course. As a preparation of what comes in Sect. 3, let us delve a little bit deeper into our current example. We have said that the the signal source sends one out of the three signals listed in (1). That was good in order to start the discussion, but now, after the explanation of the scenario so far, we can (and need to) understand, that the structure of the sent signals is actually more fine-grained.

The signals in (1), i.e., 'red', 'green' and 'blue' are actually the signals that can be perceived by the decision maker, if he could look at the signals directly, i.e., without the intermediate sensor. But signals of the same color (as perceived by the decision maker), might be different with respect to the message they trigger in the sensor, so the actual structure of the signals is more fine-grained which could be modeled artificially (and explained in due course), for example, as follows:

- $\langle$'red',"*red*"$\rangle$, $\langle$'red',"*green*"$\rangle$, $\langle$'red',"*blue*"$\rangle$,
  $\langle$'red',"*either red or green*"$\rangle$, $\langle$'red',"*either red or blue*"$\rangle$,
  $\langle$'red',"*either green or blue*"$\rangle$, $\langle$'red',"*either red or green or blue*"$\rangle$
- $\langle$'green',"*red*"$\rangle$, $\langle$'green',"*green*"$\rangle$, $\langle$'green',"*blue*"$\rangle$,
  $\langle$'green',"*either red or green*"$\rangle$, $\langle$'green',"*either red or blue*"$\rangle$,   (4)
  $\langle$'green',"*either green or blue*"$\rangle$, $\langle$'green',"*either red or green or blue*"$\rangle$
- $\langle$'blue',"*red*"$\rangle$, $\langle$'blue',"*green*"$\rangle$, $\langle$'blue',"*blue*"$\rangle$,
  $\langle$'blue',"*either red or green*"$\rangle$, $\langle$'blue',"*either red or blue*"$\rangle$,
  $\langle$'blue',"*either green or blue*"$\rangle$, $\langle$'blue',"*either red or green or blue*"$\rangle$

If the decision maker would look at a signal from (4) directly, he would perceive signal colors from (1), modeled as left-hand components of the signals in (4). For the sensor, some further conditions to the signals (or properties of the signals) are relevant that make the sensor to detect the signals ambiguously (at least sometimes). Such conditions/properties are modeled as right-hand components of the signals in (4). Note, that we conduct the explicit construction of the outcome signals in (4) only for explanatory purposes. It is something that we usually would never do in a mathematical model. In probability theory, we usually keep the outcome space Ω abstract and model the probability of a concrete phenomena Y that interests us only via a random variable $Y : \Omega \longrightarrow I$. Again, that is exactly, what Dempster did and what we will do in Sect. 3.

2.2   Generalizing from Signals/Sensors

The example in Sect. 2.1 heavily relies upon the reader's *ad-hoc* intuition of signals and sensors. Henceforth, we often stay in the picture of signal/sensors even when it comes to more abstract concepts (all in service of readability). However, the concepts that we discuss in this paper are not bound to the domain of signal/sensors. Also, signal/sensors can be interpreted themselves differently from an *ad hoc* understanding of technical signals and their detection, compare with Table 1.

Table 1 is a first attempt to grasp some possible, tentative terminology for concepts of ambiguous-but-correct perceptions in different (both technical and cognitive) domains. Throughout the paper we use *signal* as synonym for *signal source*, if clear from the context. Similarly, we equally use *quantity* for *signals* (and *signal sources*), as well as *perception* for *sensor messages* and *sensors* themselves.

Table 1. Concepts of ambiguous-but-correct perceptions: some possible, tentative terminology in different technical and cognitive domains (structured by the components of Dempster models V, ϕ, and Ψ according to Definition 1).

	V	ϕ	Ψ	$\mathsf{P}(\Psi \subseteq T)$	$\mathsf{P}(\Psi \cap T = \emptyset)$
This paper	values	quantity	perception	certainty	possibility
Sensor	signals	signal	sensor	must	may
Human	signals	signal	sense	is convinced	can imagine
Kolmogorov	outcomes	experiment	observation	must	may
Transmission	messages	sender	receiver	clearly	noisy
Production	goods	producer	consumer	demand	potential
Accounting	assets	asset	utility	debitable	credible
Epistemology	facts	truth	knowledge	trustable	speculative
Estimating	facts	knowledge	guess	reliable	possible
News	facts	fact	news	believable	probable
Opinion	facts	fact	opinion	insisted	debatable
Experience	happenings	happening	experience	worst case	best case

We say that ambiguous-but-correct perception is essentially what Dempster-Shafer theory is about. Actually, we do not claim that Dempster have stated it in that way or that the model that he introduced is restricted to what we will explain and formalize as *correct perception*. However, the concepts of lower probability (called *belief* by Shafer) and upper probability, which are at the core of Dempster-Shafer theory, are valid (and, in general, *only* valid) in scenarios of ambiguous-but-correct perception.

3 Dempster Models and Dempster-Shafer Theory

In this section we formalize the scenario as described in Sect. 2, however, yet without formalizing the notion of perception correctness as outlined in (3). This means, we only formalize the part of *ambiguous-and-questionable perception*; the formalization of ambiguous-but-correct perception is deferred until Sect. 4.

We immediately give our own formalization in terms of well-established, standard measure-theoretic terminology in Definition 1. Next, we will explain it by one-to-one comparison with the basic example from Sect. 2 – again: this is for explanatory reasons only, the actual, conceptual formalization of ambiguous-and-questionable perception is only in Definition 1. Only after that explanation, we will compare our definition with the original contribution and formalization of Dempster [2–4] and, furthermore, its reformulation by Shafer.

Definition 1 (Ambiguous-and-Questionable Perception)   An *ambiguous-and-questionable perception*, also called *Dempster model* for short, is a tuple $((\Omega, \Sigma, \mathsf{P}), V, \phi, \Psi)$ that consists of a probability space $(\Omega, \Sigma, \mathsf{P})$, a finite set of *(quantity) values* V, a *quantity* random variable $\phi : \Omega \longrightarrow V$, and a *perception* random variable $\Psi : \Omega \longrightarrow \mathbb{P}(V) \backslash \emptyset$.

We now model our signal/sensor example from Sect. 2 as a Dempster model according to Definition 1. We assume that signals are sent and that we draw them from the set of outcomes Ω of the model. Signals are subject to probability, as determined via the σ-algebra Σ and the probability measure P of the model. As we have said earlier, we can keep the set Ω abstract; there is no need to construct the signal outcomes in Ω in a concrete manner and it would be rather unusual to do so. Still, if the reader finds it instructive, he might think of Ω as the set of *artificially constructed* signal outcomes in (4).

The set of signals 'red', 'green' and 'blue' from (1) becomes the set of *quantity values* V. Now, the event that a signal is sent that is either 'red', 'green' or 'blue' is modeled via the *quantity* random variable ϕ.

Next, the event that a sensor reports a certain message back to the decision maker is modeled by the *quantity* random variable Ψ. Consequentially, the range of Ψ is the power set $\mathbb{P}(\{\text{'red', 'green', 'blue'}\})$ of the quantity values V, which perfectly corresponds to the sensor messages listed in (2). We are done. Henceforth, we can express the probabilities of arbitrary events. For example, we could express the probability that the sensor detects that the sent signal is *either* 'red' *or* 'green', although the sent signal was actually 'blue' as

$$\mathsf{P}\big((\Psi = \text{'red'} \vee \Psi = \text{'green'}) \wedge \phi = \text{'blue'}\big) \tag{5}$$

Now: is it possible that the probability of (5) is greater than 0%, i.e., that actually 'blue' has been sent, although the sensor does not report it as a potentially detected signal? Yes and no. Formally: yes, it is possible, because we have not yet forbidden it in our formal model. Informally: no, it should not be possible according to the informal correctness property (3) of our scenario description. It will be exactly the task of Definition 2 in Sect. 4 to turn the informal constraint (3) into a formal axiom.

The model provided in Definition 1 formalizes the notion of ambiguous perception as described in our scenario. Up to technical details and terminology, it is the same model that has been introduced by Dempster in [2–4]. In [3,4], Dempster refers to his model as the tuple (X, S, μ, Γ). X and S are two spaces, μ is a probability measure carried by X, and Γ is a multivalued mapping from X to S, i.e., in symbols: $\Gamma : X \longrightarrow \mathbb{P}(S)$. The model is slightly richer. First, as a minor, necessary technical detail, there is also the σ-algebra $\mathfrak{F}$ over X [2], i.e., $(X, \mathfrak{F}, \mu)$ forms a probability space. Next, there is also a random variable $\psi : X \longrightarrow S$. This random variable ψ is there, but Dempster keeps is *anonymous*, i.e., he does not talk about it explicitly as a function. But is is surely there and is it very important, e.g., Dempster talks about the probability $\mathsf{P}(\psi = s)$ as "*probability judgement [...] about [...] $s \in S$*" [2]. Given that $\Gamma : X \longrightarrow \mathbb{P}(S)$ and $(X, \mathfrak{F}, \mu)$ is a probability space, we have that Γ is a random variable, although Dempster does not call it a random variable and also does not use the established notation usually used when working with random variables. To summarize, we have that Dempster's model $((X, \mathfrak{F}, \mu), S, \psi, \Gamma)$ (almost) exactly corresponds to our model $((\Omega, \Sigma, \mathsf{P}), V, \phi, \Psi)$ in Definition 1; actually, the only difference between the two models is in the target domains of Γ respectively Ψ. Γ ranges over the full power set $\mathbb{P}(S)$, whereas Ψ ranges over $\mathbb{P}(V)\backslash\emptyset$, which conceptually makes no difference: actually, $\emptyset$ needs to be excluded. Dempster maintains the exclusion of $\emptyset$ in all his subsequent definitions; whereas, we exclude it from the beginning (Similarly, Shafer does not exclude $\emptyset$ from the beginning [6]; therefore, he needs to care for the special case of $\emptyset$ explicitly by setting his probability mass function m to zero for $\emptyset$, i.e., $m(\emptyset) = 0$.).

In [6], Shafer reformulates Dempster's model. He gets rid of Dempster's underlying structure $((X, \mathfrak{F}, \mu), S, \psi, \Gamma)$ and introduces the probability mass function $\mathsf{P}(\Gamma = \_) : \mathbb{P}(S) \longrightarrow [0,1]$ directly as a function called m, i.e., without reference to the random variable Γ, by explicitly axiomatizing its properties that otherwise (in Dempster's original contribution) follow from the Kolmogorov axioms. In Table 2, we summarize common terminology in existing literature for concepts of ambiguous-but-correct perceptions and Dempster-Shafer theory, in particular.

Table 2. Concepts of ambiguous-but-correct perceptions: some common terminology in existing literature.

	V	ϕ	Ψ	$\mathsf{P}(\Psi \subseteq A)$	$\mathsf{P}(\Psi \cap A = \emptyset)$
This paper	*values*	*quantity*	*perception*	*certainty*	*possibility*
Dempster [2–4]	space S	outcome of $s \in S$	multivalued mapping Γ	lower probability $\mu(A_\star)$	upper probability $\mu(A^\star)$
Shafer [6]	true values, Θ, *frame of discernment*	quantity	basic probability, m	belief, $Bel(A)$	upper probability, $P^\star(A)$
Dempster-Shafer as in, e.g., [7, 25]	answers [7], *frame of discernment* [25]	question [7]	basic probability[7], m, basic belief [25]	belief, $Bel(A)$	plausibity, $Pl(A)$

4   The Axiom of Ambiguous Sensor Correctness

4.1 γ-Perception Models

We now step from ambiguous-and-questionable perception to ambiguous-but-correct perception: we formalize (3) by the so-called γ-property in (6) and restrict the Dempster models from Definition 1 to those that fulfill this property.

Definition 2 (γ-Perception)   An *ambiguous-but-correct perception*, also called *γ-perception model* for short, is a Dempster model $((\Omega, \Sigma, \mathsf{P}), V, \phi, \Psi)$ so that the following *perception correctness* (also called *Ψ-correctness of γ-property* for short) holds for all $A \subseteq V$ for which $\mathsf{P}(\Psi = A) \neq 0$:

$$\mathsf{P}(\phi \in A \mid \Psi = A) = 1 \qquad (6)$$

The γ-property connects the behavior of the sensor Ψ to the behavior of the signal ϕ. Equation (6) simply states, that one of the signals that has been detected the sensor as potential signal *must* have actually occurred (been sent by the signal source); or, to express it differently, no other signal than those that have been detected as potential must have occurred:

$$P(\phi \notin A \mid \Psi = A) = 0 \tag{7}$$

Actually, we have that (6) and (7) are equivalent due to the law of total probabilities. It is also instructive to look at the γ-property from the perspective of signal outcomes. Lemma 1 rephrases (6) only in terms of signal outcomes and the random variables ϕ and Ψ.

Lemma 1 (Ambiguous-but-Correct Perception)   *Given a γ-perception model $((\Omega, \Sigma, P), V, \phi, \Psi)$ we have the following for all $\omega \in \Omega$:*

$$\phi(\omega) \in \Psi(\omega) \tag{8}$$

Corollary 1 lists some consequences of the γ-property (6) from different instructive angles.[1]

Corollary 1 (Perception Correctness)   *Given a γ-perception model $((\Omega, \Sigma, P), V, \phi, \Psi)$ and sets $A, B \subseteq V$, we have that:*

$$P(\phi \in A, \Psi = A) = P(\Psi = A) \tag{9}$$
$$A \subseteq B \implies P(\phi \in A, \Psi = B) \leqslant P(\Psi = B) \tag{10}$$
$$B \subseteq A \implies P(\phi \in A, \Psi = B) = P(\Psi = B) \tag{11}$$
$$P(\phi \in A, \Psi \subseteq A) = P(\Psi \subseteq A) \tag{12}$$

$$P(\phi \notin A, \Psi = A) = 0 \tag{13}$$
$$a \notin B \implies P(\phi = a, \Psi = B) = 0 \tag{14}$$
$$A \cap B = \emptyset \implies P(\phi \in A, \Psi = B) = 0 \tag{15}$$
$$P(\phi \in A, \Psi \cap A = \emptyset) = 0 \tag{16}$$

4.2   Certainties and Possibilities

Given an ambiguous-but-correct perception, it is common to assume that the probabilities of the perceptions (sensor messages) are known, but the probabilities of the signals are not. Still, the probabilities of the perceptions reveal (more or less vague) information about the probabilities of the signals. Given the probabilities of the perceptions, we at least know lower bounds and upper bounds for the probabilities of *signal events*, as expressed (see Footnote 1) by Lemma 2.

---

[1] Note, that standard the notation for random variables applies throughout all the paper, i.e., given a random variables $X : \omega \longrightarrow I$, we have that $Expr(X)$ denotes the event $\{\omega \mid Expr(X(\omega))\}$ for each common mathematical expression $Expr(\_)$. For example, $(X = y)$ stands for $X^{-1}(y) = \{\omega \mid X(\omega) = y\}$ as usual; $X \subseteq A$ stands for $\{\omega \mid X(\omega) \subseteq A\}$; $A \subset X$ stands for $\{\omega \mid A \subset X(\omega)\}$ etc.

Lemma 2 (Lower and Upper Probabilities in γ-Perception Models)
Given a γ-perception model $((\Omega, \Sigma, \mathsf{P}), V, \phi, \Psi)$ and a set $A \subseteq V$, we have that:

$$\mathsf{P}(\Psi \subseteq A) \quad \leqslant \quad \mathsf{P}(\phi \in A) \quad \leqslant \quad \mathsf{P}(\Psi \cap A \neq \emptyset) \qquad (17)$$

Proof. Immediate corollary from Lemma 3.   □

We call $\phi \in A$ a signal event. If clear from the context, we call also $A \subseteq V$ itself a signal event. Similarly, if clear from the context, we call A a perception (i.e., depending on the context, A sometimes plays the role of a signal event, sometimes it plays the role of a perception). We call the lower probability bound $\mathsf{P}(\Psi \subseteq A)$ of a signal event $\phi \in A$ the certainty of A – it is *certain* that $\mathsf{P}(\phi \in A)$ is *at least* $\mathsf{P}(\Psi \subseteq A)$. Similarly, we call the upper probability bound $\mathsf{P}(\Psi \cap A \neq \emptyset)$ of a signal event $\phi \in A$ the possibility of A – *at most* it is *possible* that $\mathsf{P}(\phi \in A)$ is $\mathsf{P}(\Psi \subseteq A)$. Lemma 2 only gives us known lower and upper probability bounds; but it is not yet clear whether those are tightest bounds, i.e., in a sense greatest lower bounds and least upper bounds. It is not as straightforward to come up with appropriate notions of best possible estimates; this will be the subject of Sect. 6.

Definition 3 (Certainties and Possibilities)   Given a γ-perception model $((\Omega, \Sigma, \mathsf{P}), V, \phi, \Psi)$ and a set of signals $A \subseteq V$, we call $\mathsf{P}(\Psi \subseteq A)$ the *certainty* of the signal event A, whereas we call $\mathsf{P}(\Psi \cap A \neq \emptyset)$ its *possibility*.

Lemma 2 provides known lower and upper bounds for signal events. It is instructive to see the lower and upper bounds for single signals. For a single signal $a \in V$, Lemma 2 shows as follows:

$$\mathsf{P}(\Psi = \{a\}) \quad \leqslant \quad \mathsf{P}(\phi = a) \quad \leqslant \quad \mathsf{P}(a \in \Psi) \qquad (18)$$

Table 2 summarizes different terminology and notation for the probabilities $\mathsf{P}(\Psi \subseteq A)$ and $\mathsf{P}(\Psi \cap A \neq \emptyset)$. In [2–4], Dempster introduces the notation $A_\star$ for the event $(\Psi \subseteq A)$ and the notation $A^\star$ for the event $(\Psi \cap A \neq \emptyset)$. He calls the probability $\mathsf{P}(A_\star)$ *lower probability* and the probability $\mathsf{P}(A^\star)$ *upper probability*. In [6], Shafer explicitly defines the *belief of* A (which he denotes as $Bel(A)$) and the *upper probability of* A (which he denotes as $P^\star(A)$) via the probability mass function $m : \mathbb{P}(V) \longrightarrow [0, 1]$ as follows:

$$Bel(A) = \sum_{B \subseteq A} m(B) \qquad P^\star(A) = \sum_{B \cap A \neq \emptyset} m(B) \qquad (19)$$

Now, by the law of total probabilities, we have that (17) is equivalent to:

$$\underbrace{\sum_{B \subseteq A} \mathsf{P}(\Psi = B)}_{Bel(A)[6],\, \mathsf{P}(A_\star)[2]} \quad \leqslant \quad \mathsf{P}(\phi \in A) \quad \leqslant \quad \underbrace{\sum_{B \cap A \neq \emptyset} \mathsf{P}(\Psi = B)}_{P^\star(A)[6],\, \mathsf{P}(A^\star)[2]} \qquad (20)$$

In our framework, the probability mass function m shows as $\mathsf{P}(\Psi = \_)$, i.e., $m(A)$ equals $\mathsf{P}(\Psi = A)$ for each A. Against this background, please note the direct correspondences between (20) and (19). Again, note that in Dempster-Shafer theory, $Bel(A)$ and $P^\star(A)$ (and $\mathsf{P}(A_\star)$ and $\mathsf{P}(A^\star)$ likewise) are introduced explicitly (postulated); whereas (20) is a consequence of the γ-property of γ-perception models (via Lemma 2).

5   The Exact Relationship of Quantities and Perceptions

With Lemma 2, we have determined lower and upper bounds for the probability of signal events. In this section, we turn the inequations (17) into equations of the following form:

$$\mathsf{P}(\Psi \subseteq A) + \Delta = \mathsf{P}(\phi \in A) = \mathsf{P}(\Psi \cap A \neq \emptyset) - \Delta' \tag{21}$$

With equations of the form (21), we exactly characterize the relationship between signal probabilities and perception probabilities on the basis of the *certainties* and *possibilities* as provided by Lemma 2. The characterization is provided by Lemma 3. Note, that the differences Δ and Δ' as provided by Lemma 3 involve knowledge about the signal random variable, and, therefore, are not known to the decision maker in our described scenario. Still, Lemma 3 is instructive for considerations such as conducted in Sect. 6 and, furthermore, yields us the yet missing proof of Lemma 2 (Lemma 2 follows from Lemma 3 as an immediate corollary).

Lemma 3 (Exact Relationship of ϕ- and Ψ-Probabilities)   *Given a γ-perception model $((\Omega, \Sigma, \mathsf{P}), V, \phi, \Psi)$ and a set of values $A \subseteq V$, we have that:*

$$\mathsf{P}(\phi \in A) = \mathsf{P}(\Psi \subseteq A) \quad\ \ + \mathsf{P}(\phi \in A,\ A \subset \Psi) \tag{22}$$
$$\mathsf{P}(\phi \in A) = \mathsf{P}(\Psi \cap A \neq \emptyset) - \mathsf{P}(\phi \notin A, A \subset \Psi) \tag{23}$$

Proof. We start with proving (22). By the law of total probabilities (case distinction), we have that $\mathsf{P}(\phi \in A)$ equals

$$\mathsf{P}(\phi \in A,\ \Psi \subseteq A) + \mathsf{P}(\phi \in A,\ \Psi \not\subseteq A) \tag{24}$$

By a further case distinction, we have that (24) equals

$$\mathsf{P}(\phi \in A,\ \Psi \subseteq A) + \mathsf{P}(\phi \in A, A \subset \Psi) + \mathsf{P}(\phi \in A,\ \Psi \cap A = \emptyset) \tag{25}$$

Due to (12), we have $\mathsf{P}(\phi \in A,\ \Psi \subseteq A)$ equals $\mathsf{P}(\Psi \subseteq A)$. Furthermore, due to (16), we have that $\mathsf{P}(\phi \in A,\ \Psi \cap A = \emptyset)$ equals zero. Therefore, we have that (25) equals

$$\mathsf{P}(\Psi \subseteq A) + \mathsf{P}(\phi \in A, A \subset \Psi) \tag{26}$$

which proofs (22).

We proceed with (23). Actually, we can step further from (26). By (22), we have that $\mathsf{P}(\phi \in A)$ equals (26). Due to the fact that $\Psi : \Omega \longrightarrow \mathbb{P}(V)\backslash\emptyset$, we have that (26) equals

$$\mathsf{P}(\Psi \cap A \neq \emptyset,\ \Psi \subseteq A) + \mathsf{P}(\phi \in A, A \subset \Psi) \tag{27}$$

Due to the law of total probabilities, we have that (27) equals

$$\mathsf{P}(\Psi \cap A \neq \emptyset, \Psi \subseteq A) + \mathsf{P}(A \subset \Psi) - \mathsf{P}(\phi \notin A, A \subset \Psi) \tag{28}$$

Again, due to $\Psi : \Omega \longrightarrow \mathbb{P}(V) \setminus \emptyset$, we have that (28)

$$\mathsf{P}(\Psi \cap A \neq \emptyset, \Psi \subseteq A) + \mathsf{P}(\Psi \cap A \neq \emptyset, A \subset \Psi) - \mathsf{P}(\phi \notin A, A \subset \Psi) \tag{29}$$

Next, by a simple transformation, we have that (29)

$$\mathsf{P}(\Psi \cap A \neq \emptyset, \Psi \subseteq A) + \mathsf{P}(\Psi \cap A \neq \emptyset, \Psi \not\subseteq A) - \mathsf{P}(\phi \notin A, A \subset \Psi) \tag{30}$$

Finally, due to the law of total probabilities, we have that (30) equals

$$\mathsf{P}(\Psi \cap A \neq \emptyset) - \mathsf{P}(\phi \notin A, A \subset \Psi) \tag{31}$$

□

6 Certainty and Possibility as Optimal Estimates

In how far are the lower and upper probability bounds of a signal event as provided by Lemma 2 best possible estimates? In this section, we provide an answer to this question. The target is to characterize the lower bound as a greatest lower bound, and the upper bound as a least upper bound. The key is to look at systems of sensor messages (and *only* sensor messages, as information about probabilities is only available for sensor messages in our prescribed scenario) in regards of all possible (unknown) probability measures ranging over *potential* γ-perception models. Technically, we introduce the notion of a *Dempster pre-model* for this purpose, which is a Dempster model *without* a probability measure plus an operator that turns a Dempster pre-model D and a *compatible* probability measure μ (i.e., a probability measure that fulfills the γ-probability) into a γ-perception model, denoted as $D \cdot \mu$.

Definition 4 (Dempster Pre-Model)   A *Dempster pre-model* is a tuple $(\Omega, \Sigma, V, \phi, \Psi)$ that consists of a set of outcomes Ω, a σ-algebra Σ over Ω, a finite set V, a function $\phi : \Omega \longrightarrow V$, and a function $\Psi : \Omega \longrightarrow \mathbb{P}(V) \setminus \emptyset$. Given a Dempster pre-model $D = (\Omega, \Sigma, V, \phi, \Psi)$ and a probability measure $\mu : \Omega \longrightarrow V$, we denote the tuple $((\Omega, \Sigma, \mu), V, \phi, \Psi)$ as $D \cdot \mu$, if and only if $((\Omega, \Sigma, \mu), V, \phi, \Psi)$ forms a γ-perception model.

For the best possible lower bound, we look at all of those systems of sensor messages that have a combined probability that is smaller than the probability of a specified signal event A with respect to all possible *compatible* probability measures. Out of those, the system of sensor messages $\{B | B \subseteq A\}$ has always the largest combined probability (which is $\mathsf{P}(\Psi \subseteq A)$). This characterization of $\mathsf{P}(\Psi \subseteq A)$ as best possible (lower) estimate is formalized by Lemma 4 (similarly for the upper probability bounds as provided by Lemma 2).

Corollary 2 (Lower and Upper Probabilities are Optimal Estimates)
Given a Dempster pre-model $D = (\Omega, \Sigma, V, \phi, \Psi)$ we have that the following holds for each probability measure $\mu : \Sigma \longrightarrow [0,1]$, each system $S \subseteq \mathbb{P}(V)$ of subsets of V and each subset $A \subseteq V$:

$$\Big(\forall D \cdot \mu \,.\, \mu(\Psi \in S) \leqslant \mu(\phi \in A) \Big) \implies \Big(\forall D \cdot \mu \,.\, \mu(\Psi \in S) \leqslant \mu(\Psi \subseteq A) \Big) \quad (32)$$

$$\Big(\forall D \cdot \mu \,.\, \mu(\Psi \in S) \geqslant \mu(\phi \in A) \Big) \implies \Big(\forall D \cdot \mu \,.\, \mu(\Psi \in S) \geqslant \mu(\Psi \cap A \neq \emptyset) \Big) \quad (33)$$

7 Knowledge Fusion

In this section, we define a *natural* knowledge fusion operator for perceptions. The combinator only works for perceptions that are founded in the same probability space. This is natural, as we deal with two different sensors, but the signals are the same (independent of which sensor they are detected). We do not know exactly the probabilities of the signals, i.e., we only know the probabilities up to the bounds provided by Lemma 2. However, we know that the signals and their probabilities are the same for the two sensors. The decision maker can exploit that fact. For example, image that one sensor reports the message *"either red or green"* upon detection of a signal. Due to the γ-property the decision maker knows that the signal must have been either 'red' or 'green'. Now, imagine that the decision maker has a second, different sensor available that reports back (simultaneously to the first sensor, i.e., with respect to the same signal) the sensor message *"either green or blue"*. In this situation, the decision maker can be sure, that the signal must have been 'green', as he can combine the ambiguous-but-correct information of both sensors!

For a given signal, and sensor messages from different sensors the signal must be among the cut of potential signals of all involved sensors. This is exactly, how we formalize our fusion operator in Definition 5.

Definition 5 (Knowledge Fusion)   Given two γ-perception models $((\Omega, \Sigma, \mathsf{P}), V, M, \phi, \Psi)$ and $((\Omega, \Sigma, \mathsf{P}), V, M, \phi, \Psi')$ we define the knowledge fusion $\Psi \cdot \Psi'$ as follows:

$$\Psi \cdot \Psi' = \big(\omega \in \Omega \mapsto \Psi(\omega) \cap \Psi'(\omega) \big) \quad (34)$$

Note that the knowledge fusion $\Psi \cdot \Psi'$ is always well-defined. In particular, the event $\Psi(\omega) \cap \Psi'(\omega)$ can never by empty, because neither $\Psi(\omega)$ nor $\Psi'(\omega)$ can be empty; and, furthermore, due to the γ-property, the events $\Psi(\omega)$ and $\Psi'(\omega)$ must at least overlap on the actual signal $\phi(\omega)$.

The definition of the fusion operator in Definition 5 is intuitive, and it is useful, because the combined perception is always less ambiguous than both of the perceptions individually (more precisely: usually less ambiguous, but never more ambiguous). Yet, the fusion operator is not very useful in decision scenarios.

We need to know, how the fusion operator works on the known probabilities of sensor messages. This is exactly, what is provided next in Lemma 4, i.e., it tells us the probability of a *combined perception* in terms of probabilities of *simultaneously occurring* perceptions.

Lemma 4 (Knowledge Fusion)   Given two γ-perception models $((\Omega, \Sigma, \mathsf{P}), V, \phi, \Psi)$ and $((\Omega, \Sigma, \mathsf{P}), V, \phi, \Psi')$ we have that the following holds for all $A \subseteq V$:

$$\mathsf{P}(\Psi \cdot \Psi' = A) = \sum_{B \cap C = A} \mathsf{P}(\Psi = B, \Psi' = C) \qquad (35)$$

Proof. Due to random variable notation, we have that $\mathsf{P}(\Psi \cdot \Psi' = A)$ equals

$$\mathsf{P}(\{\, \omega \mid \Psi \cdot \Psi'(\omega) = A\}) \qquad (36)$$

Due to Definition 5, we have that (36) equals

$$\mathsf{P}(\{\, \omega \mid \Psi(\omega) \cap \Psi'(\omega) = A\}) \qquad (37)$$

Next, we can rewrite (37) as

$$\mathsf{P}\Big(\underbrace{\bigcup_{B \cap C = A} \{\, \omega \mid \Psi(\omega) = B \wedge \Psi'(\omega) = C\}}_{(i)} \Big) \qquad (38)$$

Now, all sets of the form (i) in (38) are disjoint for different B, B' and/or different C, C'. Therefore, we have that (38) equals

$$\sum_{B \cap C = A} \mathsf{P}(\{\, \omega \mid \Psi(\omega) = B \wedge \Psi'(\omega) = C\}) \qquad (39)$$

Finally, we have that (39) equals

$$\sum_{B \cap C = A} \mathsf{P}(\Psi = B, \Psi' = C) \qquad (40)$$

$\square$

In order to apply the knowledge fusion $\Psi \cdot \Psi'$ along the lines of Lemma 4, it is not sufficient to know the probabilities of perceptions $\mathsf{P}(\Psi = A)$ and $\mathsf{P}(\Psi' = A)$ for all $A \subseteq V$ individually. Instead we need the probabilities of *simultaneous perceptions* $\mathsf{P}(\Psi = B, \Psi' = C)$ for all possible combinations of perceptions $B \subseteq V$ and $C \subseteq V$. This might be considered a limitation, e.g., in scenarios where the perceptions take the form of *judgments* provided by individuals separately. We could argue, that we should always have enough *empirical* data to know about the probabilities of simultaneously occurring perceptions. We do not want to delve into this discussion here, but want to leave it for further writings instead.

Next, we compare the knowledge fusion $\Psi \cdot \Psi'$ with the knowledge fusion as provided by Dempster's rule of combinations that we denote as $\Psi \oplus \Psi'$ in this paper. In order to distinguish our knowledge fusion $\Psi \cdot \Psi'$ from other fusion operators, we sometimes also call it *sensor fusion*.

Actually, $\Psi \cdot \Psi'$ and $\Psi \oplus \Psi'$ behave completely differently, as we will see in due course with a simple example. Technically, the operators are different, as $\Psi \cdot \Psi'$ combines two random variables and $\Psi \oplus \Psi'$ combines two mass probability functions. However, conceptually, this is a minor detail that can be neglected; what counts is the different behavior with respect to the probabilities resulting from the different knowledge fusions. Dempster's rule of combination is defined in Definition 6 along the lines of [6], compare also with [7].

Definition 6 (Dempster's Rule of Combination [6])   Given two probability mass functions $\Psi : \mathbb{P}(V)\backslash\emptyset \longrightarrow [0,1]$ and $\Psi' : \mathbb{P}(V)\backslash\emptyset \longrightarrow [0,1]$, *Dempster's rule of combination* [6] defines the new combined probability mass function $\Psi \oplus \Psi' : \mathbb{P}(V)\backslash\emptyset \longrightarrow [0,1]$ as follows (compare also with [7], p. 5) for all $A \subseteq V$ as follows:

$$\Psi \oplus \Psi'(A) = \frac{\sum\limits_{B \cap C = A} \Big(\Psi(B) \times \Psi'(C) \Big)}{\sum\limits_{B \cap C \neq \emptyset} \Big(\Psi(B) \times \Psi'(C) \Big)} \tag{41}$$

Now, let us explore the different behavior of $\Psi \cdot \Psi'$ and $\Psi \oplus \Psi'$. Let us look into a most reductionist example for this purpose, i.e., a signal source that can send either of exactly two values 'red' and 'green' plus two *completely unambiguous* sensors Ψ and Ψ', i.e., the probability of the perception "*either red or green*" is zero, and the probabilities of the perceptions "*red*" and "*green*" are exactly the probabilities of the signals 'red' and 'green', please compare with Table 3. Note, that for the example, it is important that the probabilities of 'red' and 'green' are not equal.

Now, the sensor fusion $\Psi \cdot \Psi'$ behaves neutral. It does not change any of the probabilities, as one might expect, given that Ψ and Ψ' are exactly the same. $\Psi \oplus \Psi'$ behaves differently; it further increases the higher probability of the perception "*red*" and further decreases the lower probability of the perception "*green*", again, please compare with Table 3. At this point, we want to stop the discussion with this important observation and leave a deeper discussion of knowledge fusion for further writings.

Table 3. Sensor fusion $\Psi \cdot \Psi'$ vs. Dempster's rule of combination $\Psi \oplus \Psi'$.

$A \in \mathbb{P}(V)$	$\mathrm{P}(\Psi = A)$ $\Psi(A)$	$\mathrm{P}(\Psi' = A)$ $\Psi'(A)$	$\mathrm{P}(\Psi \cdot \Psi' = A)$	$\Psi \oplus \Psi'(A)$
"*red*"	0.6	0.6	0.6	≈ 0.69
"*green*"	0.4	0.4	0.4	≈ 0.31
"*either red or green*"	0	0	0	0

8   Conclusion

We have characterized ambiguous-and-correct perception through an informal description of a simple signals/sensors example. We claim that ambiguous-but-correct perceptions can be considered as the universe of discourse of Dempster-Shafer theory. The reader might not want to agree with that. Still, we claim that ambiguous-but-correct perceptions are worth considering and that we have introduced them as a well-defined notion. We have formalized ambiguous-and-questionable perceptions as a probability space with two random variables, one for the signal (quantity) and one for the perception. We have explained that this is actually, what Dempster did with his original model as introduced in the 1960s. We have formalized ambiguous-and-correct perceptions via the introduction of a sensor correctness axiom. We have proven that Dempster's lower and upper probabilities are actually lower and upper bounds for probabilities of signal events – as a consequence of the sensor correctness axiom. We have stepped beyond that and have explained, in how far Dempster's lower and upper probabilities are best possible estimates. Finally, we have introduced a natural knowledge fusion operator and started comparing it with Dempster's rule of combination.

References

1. Dempster, A.P.: New methods for reasoning towards posterior distributions based on sample data. Ann. Math. Stat. **37**(2), 355–374 (1966)
2. Dempster, A.P.: Upper and lower probabilities induced by a multivalued mapping. Ann. Math. Stat. **38**(2), 325–339 (1967)
3. Dempster, A.P.: A generalization of Bayesian inference. Technical report AD 664 659, Harvard University, November 1967
4. Dempster, A.P.: A generalization of Bayesian inference. J. Roy. Stat. Soc. B **30**(2), 205–247 (1968)
5. Dempster, A.P.: Upper and lower probability inferences based on a sample from a finite univariate population. Biometrika **45**(3), 515–528 (1967)
6. Shafer, G.: A Mathematical Theory of Evidence. Princeton University Press, Princeton (1976)
7. Liu, L., Yager, R.R.: Classic works of the Dempster-Shafer theory of belief functions: an introduction. In: Yager, R.R., Liu, L. (eds.) Classic Works of the Dempster-Shafer Theory of Belief Functions. STUDFUZZ, vol. 219, pp. 1–35. Springer, Heidelberg (2008). https://doi.org/10.1007/978-3-540-44792-4_1
8. Wu, C., Barnes, D.: Formulating partner selection criteria for agile supply chains: a Dempster-Shafer belief acceptability optimisation approach. Int. J. Prod. Econ. **125**(2), 284–293 (2010)
9. Müller, J., Piché, R.: Mixture surrogate models based on Dempster-Shafer theory for global optimization problems. J. Global Optim. **51**(1), 79–104 (2011). https://doi.org/10.1007/s10898-010-9620-y
10. Xiao, Z., Yang, X., Pang, Y., Dang, X.: The prediction for listed companies' financial distress by using multiple prediction methods with rough set and Dempster-Shafer evidence theory. Knowl. Based Syst. **26**, 196–206 (2012)

11. Sevastianov, P., Dymova, L.: Synthesis of fuzzy logic and Dempster-Shafer theory for the simulation of the decision-making process in stock trading systems. Math. Comput. Simul. **80**(3), 506–521 (2009)

12. Hong, S., Lim, W., Cheong, T., May, G.: Fault detection and classification in plasma etch equipment for semiconductor manufacturing e-diagnostics. IEEE Trans. Semicond. Manuf. **25**(1), 83–93 (2012)

13. Lonea, A., Popescu, D., Tianfield, H.: Detecting DDoS attacks in cloud computing environment. Int. J. Comput. Commun. Control **8**(1), 70–78 (2013)

14. Straszecka, E.: Combining uncertainty and imprecision in models of medical diagnosis. Inf. Sci. **176**(20), 3026–3059 (2006)

15. Bloch, I.: Some aspects of Dempster-Shafer evidence theory for classification of multi-modality medical images taking partial volume effect into account. Pattern Recogn. Lett. **17**(8), 905–919 (1996)

16. Murphy, R.R.: Dempster-Shafer theory for sensor fusion in autonomous mobile robots. IEEE Trans. Robot. Autom. **14**(2), 197–206 (1998)

17. Wu, H., Siegel, M., Stiefelhagen, R., Yang, J.: Sensor fusion using Dempster-Shafer theory. In: Proceedings of IMTC 2002 - The 19th IEEE Instrumentation and Measurement Technology Conference, vol. 1, pp. 7–12. IEEE (2002)

18. Basir, O., Yuan, X.: Engine fault diagnosis based on multi-sensor information fusion using Dempster-Shafer evidence theory. Inf. Fusion **8**(4), 379–386 (2007)

19. Dempster, A.P.: Foreward. In: Shafer, G. (ed.) A Mathematical Theory of Evidence. Princeton University Press, Princeton (1976)

20. Carnap, R.: The two concepts of probability – the problem of probability. Philos. Phenomenol. Res. **5**(4), 513–532 (1945)

21. Carnap, R.: On inductive logic. Philos. Sci. **12**(2), 72–97 (1945)

22. Draheim, D.: Generalized Jeffrey Conditionalization – A Frequentist Semantics of Partial Conditionalization (with a Foreword by Bruno Buchberger). Springer, Cham (2017). https://doi.org/10.1007/978-3-319-69868-7

23. Draheim, D.: Semantics of the Probabilistic Typed Lambda Calculus – Markov Chain Semantics, Termination Behavior, and Denotational Semantics. Springer, Heidelberg (2017). https://doi.org/10.1007/978-3-642-55198-7

24. Neyman, J.: Frequentist probability and frequentist statistics. Synthese **36**, 97–131 (1977)

25. Smets, P.: Belief functions: the disjunctive rule of combination and the generalized Bayesian theorem. In: Yager, R.R., Liu, L. (eds.) Classic Works of the Dempster-Shafer Theory of Belief Functions. Studies in Fuzziness and Soft Computing, vol. 219, pp. 633–664. Springer, Heidelberg (2008). https://doi.org/10.1007/978-3-540-44792-4_25

26. Kolmogorov, A.: Grundbegriffe der Wahrscheinlichkeitsrechnung. Springer, Heidelberg (1933). https://doi.org/10.1007/978-3-642-49888-6

27. Kolmogorov, A.: Foundations of the Theory of Probability. Chelsea, New York (1956)

28. Kolmogorov, A.: On logical foundation of probability theory. In: Itô, K., Prokhorov, J.V. (eds.) Probability Theory and Mathematical Statistics. Lecture Notes in Mathematics, vol. 1021, pp. 1–5. Springer, Dordrecht (1982). https://doi.org/10.1007/BFb0072897

Big Data Management and Analytics

A Graph Partitioning Algorithm for Edge or Vertex Balance

Adnan El Moussawi[✉], Nacéra Bennacer Seghouani, and Francesca Bugiotti

Laboratoire de Recherche en Informatique LRI, 91190 Gif-sur-Yvette, France
{adnan.moussawi,nacera.seghouani,francesca.bugiotti}@lri.fr
http://www.lri.fr

Abstract. The definition of effective strategies for graph partitioning is a major challenge in distributed environments since an effective graph partitioning allows to considerably improve the performance of large graph data analytics computations. In this paper, we propose a multi-objective and scalable Balanced GRAph Partitioning (B-GRAP) algorithm to produce balanced graph partitions. B-GRAP is based on Label Propagation (LP) approach and defines different objective functions to deal with either vertex or edge balance constraints while considering edge direction in graphs. The experiments are performed on various graphs while varying the number of partitions. We evaluate B-GRAP using several quality measures and the computation time. The results show that B-GRAP (i) provides a good balance while reducing the cuts between the different computed partitions (ii) reduces the global computation time, compared to Spinner algorithm.

Keywords: Large graph partitioning · Vertex balance · Edge balance · Parallel processing.

1 Introduction

In recent years, large-scale graph analytics and mining have been widely used in various domains such as communication network, urban transportation, biological data and social networks. In this context the efficient processing of large graphs becomes a new challenging task. Many research works focused on graph-based parallel computation algorithms in distributed systems [7,19,29]. The distribution of the workloads on several machines helps to reduce the overhead computation time. However, this distribution requires multiple exchanges of messages between the machines with a typically high cost.

Graph Partitioning (GP) algorithms have taken a lot of attention in recent decade as a key prerequisite for an efficient processing and many works focused on this problem [3,6,17,22]. An efficient partitioning algorithm allows to minimize the total computation cost while a good balanced load makes better leverage of the entire system. The GP problem aims to divide the graph into a given number of partitions while minimizing the number of their inter-connecting edges (called

© Springer Nature Switzerland AG 2020
S. Hartmann et al. (Eds.): DEXA 2020, LNCS 12391, pp. 23–37, 2020.
https://doi.org/10.1007/978-3-030-59003-1_2

cuts) and balancing their sizes w.r.t. the number of vertices or the number of edges. An *edge-balanced* GP divides the edges of the graph into nearly equal sized partitions. In contrast, a *vertex-balanced* GP divides the vertices of the graph into equisized partitions. Each objective has its own advantage. For a graph analysis task that needs few communications between vertices, the *vertex-balanced* GP is more beneficial. On the contrary, for task where the vertices exchange messages frequently, balancing the number of edges has more advantage.

Different GP partitioning strategies approaches have been studied in the literature. Multilevel GP approaches [12] defined a well *vertex-balanced* partitioning algorithm which shown high quality in terms of *cuts*. But it requires high resource usage and computation time and it does not scale with large graphs. Streaming GP methods [11,27] used an online graph partitioning, by considering *vertex-balance* or *edge-balance* constraints which reduces the overhead computation comparing to multilevel approaches. But on the other hand, the results of such methods are of less quality and depend on the order of vertex or edge processing. Moreover the partitioning is not adaptive to the graph's changes. Recent works have taken advantage of the lightweight mechanism of Label Propagation (LP) approach to improve the partitioning process [20,21,28]. In [21,26], the authors used LP to coarsen the graph in a multilevel partitioning approach while balancing the vertices. [20] extended LP to compute the entire partitioning basing on Giraph [7] programming model, while considering only *edge-balance* constraint.

In this paper we propose a new multi-objective and scalable *B*alanced *GRA*ph *P*artitioning algorithm (B-GRAP), based on LP approach, to produce balanced graph partitions. Our main contributions are:

- An optimized partitioning initialization that helps to improve the propagation of labels and to reduce the computational overhead comparing to similar approaches.
- A new scalable and parallel partitioning algorithms B-GRAP$_{VB}$ and B-GRAP$_{EB}$ that respectively address the vertex balance and the edge balance problems on both directed and undirected graphs.
- We implement our algorithm on top of the open source distributed graph processing system Giraph [7]. This allows us to take advantage from the parallel processing architecture in order to effectively parallelize B-GRAP.
- The evaluation of B-GRAP, using different measures (quality and time) on heterogeneous real-worlds and synthetic graphs, shows good performance while scaling with the number of partitions and size of a graph.

This paper is structured as follows. In Sect. 2, we detail some approaches related to graph partitioning problem. In Sect. 3, we present LP approach and B-GRAP main notations. In Sect. 4, we define B-GRAP, its initialization, the propagation functions for vertex or edge balance, then the measures we use to evaluate the quality of the partitioning are presented in Sect. 5. In Sect. 6, we provide the experimental study conducted in order to evaluate B-GRAP. Finally, in Sect. 7, we give our conclusions and future perspectives.

2   Related Work

During the last decade, research communities working on graph datasets have given a lot of interest to the definition of new strategies for large graph parallel computing and analytics in a distributed environment. This context opened up new challenges to define efficient graph partitioning algorithms [2,3,10,21,22]. One of the main challenges consists in defining graph partitioning algorithms that allow to balance the workload among the nodes of a distributed computing environment and to reduce, at the same time, the communication load over the network.

A common strategy in large graph partitioning is to use multilevel approaches [3]. The idea is to generate a first partition on the basis of a reduced view of the graph in which a vertex represents many vertices of the original graph. For example a triangle of three vertices can be reduced to one. The algorithms then expands the graph taking into account the whole initial graph. This family of approaches alternates three main phases: (i) coarsen the graph by collapsing adjacent vertices satisfying some matching criteria, (ii) partition the coarsened graph using any partitioning algorithm, (iii) the un-coarsening or refinement, which means generalizing the partition from last phase by mapping back the results to the original graph. METIS [12] is one of the multilevel graph partitioning algorithm family. This algorithm is known for its ability to produce partitioning with high quality w.r.t. the number of cuts, but with the disadvantage of the high computation time to obtain several intermediate results. Another known multilevel graph partitionner is Scotch [23] which deals with the graph changes and does not require to start the partitioning from scratch, in contrast to METIS. The parallel version of both algorithms, ParMETIS [13] and Pt-Scotch [5], show good cuts quality but their performance scales poorly with respect to the number of processors as shown in [21].

During the last years stream graph partitioning has been proposed in order to reduce the complexity [27] of multilevel approaches, since they take into account the entire input graph during the whole computation. These algorithms assign edges and vertices to various partitions by running a single pass through the whole graph. The goal, of the most part of these algorithms, is to guarantee the edge balance [8,27] and to find a partitioning that reduces the usage of the resources and the computation overhead. These methods are faster than multilevel algorithms but they build partitioning with lower quality, in term of cuts, due to the sensitivity to the stream order. Moreover, it's generally difficult to parallelize streaming algorithms.

Other works have used the label propagation approach (LP) [24] to partition large graphs. LP was mainly used for community detection in social networks [4,9]. Making use of LP for the graph partitioning problem was motivated by the lightweight mechanism that uses the network structure to guide its progress. LP partitioning methods generate less intermediary results than multilevel approaches, which need to store many intermediate results such as the coarser graph, and run with a lower complexity. Furthermore, LP method is semantic-aware, given the existence of local closely connected substructures, a label tends to propagate within such structures.

The authors of [21] propose ParHIP, a distributed memory parallel partitioning algorithm, that takes advantage of both multilevel and LP approaches. The authors adapt and parallelize LP technique for both coarsening and refinement step, using the Message Passing Interface (MPI), while considering the vertex-balance constraint. Their experimental results show that ParHIP is more scalable and achieves higher quality than existing state of the art methods like ParMETIS and PT-Scotch.

Finally, in [20] the authors define a distributed partitioning algorithm called Spinner that considers only edge balance. Spinner is based on LP approach and runs on the top of Giraph API [7]. Compared to the previous work, Spinner supports the parallelism and can adapt an existing partitioning to consider graph updates by adding or removing vertices and edges and changing the number of partitions. The algorithm divides N vertices across K partitions, while trying to keep similar the number of edges in each partition.

In this paper we present a new algorithm for balanced graph partitioning based on LP approach and using Giraph programming model. Compared to the literature, our algorithm B-GRAP deals with edge-based or vertex-based balanced partitioning while decreasing the number of cuts and computation time. Moreover, B-GRAP defines an initialization heuristic which allows to improve the propagation of labels across the graph and to accelerate the convergence of the algorithm on large graphs.

3   Preliminaries

Given a number of partitions K, a directed graph $G = \langle V, E, \omega \rangle$, where V is a set of vertices and E a set of weighted edges with $\omega : E \rightarrow \mathbb{R}^+$. Let $L = \{l\}_{l=1}^K$ be a set of partition labels defined by a labeling function $\phi : V \rightarrow L$ such that $\phi(v) = l$ means that v belongs to the partition with label l. The naïve LP algorithm proceeds as follows. Initially, a unique label l_v is assigned to each vertex v. Then, the label of each $v \in V$ is propagated and updated iteratively to its neighborhood $N(v) = \{u \in V | (v, u) \vee (u, v) \in E\}$ and is updated until a given convergence criteria is reached. The label updating is done by taking into account the most frequent label among $N(v)$ labels. More formally, let $\mathcal{F}_{\mathrm{LP}}(v, l)$ be the frequency of a label l in the neighborhood of v, defined by:

$$\mathcal{F}_{\mathrm{LP}}(v, l) = \sum_{u \in N(v)} \omega(v, u)\delta\big(\phi(u), l\big) \tag{1}$$

where $\phi(u)$ gives the current label of u and δ is the Kronecker delta function, which equals 1 if $\phi(u) = l$, and 0 otherwise. The label of vertex v is replaced by the label that maximizes the frequency function:

$$l_v = \underset{l}{\mathrm{argmax}} \ \mathcal{F}_{\mathrm{LP}}(v, l) \tag{2}$$

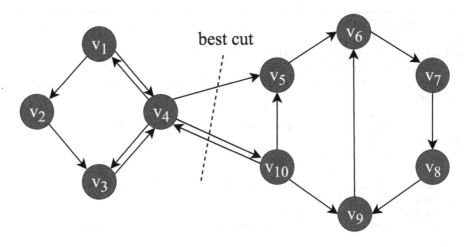

Fig. 1. Vertex-balanced and edge-balanced 2-partitioning graph example

If many maximal labels exist and do not include the current label of v, one of them is randomly chosen. LP algorithm stops if $\sum_{v \in V} \sum_{l \in L} \mathcal{F}_{LP}(v, l)$ converges according to a given threshold ϵ.

We note that naïve LP algorithm does not take into account the directions of edges. To consider directed graphs, virtual edges are added such that: $\forall (v, u) \in E, \omega(v, u) = 2$ and if $(u, v) \notin E, (u, v)$ is added with $\omega(u, v) = 1$ which we call *virtual edge*. Note that the Giraph data model is a distributed directed graph, where every vertex is aware of its outgoing edges only, but not of the incoming ones. Adding virtual edges in this case, allows to a vertex to discover its entire neighborhood $N(v)$, while the weight allows to consider the direction as well as to distinguish these virtual edges added.

4 B-GRAP Algorithm

Our goal is to define a K-balanced and LP-based partitioning algorithm that decreases the total cuts while considering the vertex balance or the edge balance constraints in directed graphs.

To illustrate our objectives we consider the example provided in Fig. 1. This example presents a small graph of 10 vertices and 16 directed edges. We would divide the graph into two balanced partitions, using either a vertex-balanced partitioning or the edge-balanced partitioning. First, we note that the best 2-partitioning that minimizes the edge cuts is $P_1 = \{v_1, v_2, v_3, v_4\}$ and $P_2 = \{v_5, v_6, v_7, v_8, v_9, v_{10}\}$, where the cuts $(v_4, v_5), (v_4, v_{10}), (v_{10}, v_4)$. To achieve a vertex-balanced partitioning, a vertex from P_2 should be moved to P_1, while caring about the cuts. In this case, moving v_{10} to P_1 is the most advantageous because it introduces less edge cuts. For the edge-balanced partitioning, no change is needed. Indeed, both partitions holds $|E|/2$ *directed edges* and the number of edge cuts is minimized. Note that if the directness of edges is ignored,

---

Algorithm 1. B-GRAP

---

input $G = \langle V, E, w \rangle$, K, τ, ϵ
output a partitioned graph $G = \langle V, E, w, L \rangle$
 1: Initialize the set of labels $L = \{l\}_{l=1}^{K}$
 2: **for** $(v \in V, d^+(v) > \tau)$ initialize $\phi(v)$ randomly from L and propagate to $N(v)$
 3: **repeat**
 4:    {*Search frequent labels*}
 5:    **for** $(v \in V, l \in L)$ get the set of frequent labels w.r.t an update function
 6:    {*Update and propagate*}
 7:    Update and propagate $\phi(v)$ to $N(v)$
 8:    {*Check unassigned vertices*}
 9:    **for** $(v \in V, \phi(v) = \emptyset)$ initialize $\phi(v)$ randomly from L and propagate to $N(v)$
10: **until** $\Delta\big(\mathcal{F}_{LP}(G, L)\big) \leq \epsilon$
11: **return** $G = \langle V, E, w, L = \{\phi(v)\}_{v \in V} \rangle$

---

this partitioning remains unbalanced. In fact, in an undirected graph, each edge is considered to be bidirectional, as a result the number of edges in P_1 is 12 and 16 in P_2.

In the following, we present in detail B-GRAP algorithm. First, we present the initialization strategy, then we define the update functions $\mathcal{F}$ to build vertex (or edge) balanced partitions, and finally we present the measures used for the evaluation of partitioning quality.

4.1   Initialization Optimization

To improve the performance of propagation approach in our algorithm, we define an initialization strategy, called B-GRAP$_{init}$, which considers only *hub* vertices having a high outgoing degree $d^+(.)$. The intuition behind this choice is that the higher $d^+(v)$, the more $\phi(v)$ will be propagated and considered as frequent label. This differs from LP approach that considers all the vertices. As a result, the candidates to be considered as frequent labels (i.e. the labels with high probability to be the most frequent) are propagated faster at the first *partial propagation* iteration. The initialization we defined should guarantee a faster label propagation and smaller number of exchanges between vertices given the fact that the nodes having the higher probability to be selected to propagate their label have been already initialized.

B-GRAP is described in Algorithm 1. Let τ be a given minimum out degree threshold to consider that a vertex v as a *hub* vertex. The algorithm proceeds as follows. First, we initialize the set of labels L (Line 1). Then, each $v \in V$, such $d^+(v) > \tau$ is assigned a random label $\in L$ and those labels are propagated to neighbors (Line 2). Then, the label of these neighbors are updated and propagated iteratively using an update function (Lines 4–7). The vertices are then checked and those not reached by the update/propagation step are initialized randomly, to ensure that all vertices are assigned a label (Line 9). The algorithm repeats the update/propagate step until convergence (Line 10).

4.2   Balanced Partitioning

In the basic LP partitioning, the label update is done without caring about the size of the partitions. Consequently, this can lead to an unbalanced partitioning. Moreover the update function of LP (Eq. 1) has a trivial optimal solution that consists of assigning all vertices to a single label, i.e. to a single partition. A standard resolution approach to deal with such a problem is to integrate the balance constraints to the update function via a penalty function. The LP update function becomes:

$$\mathcal{F} = \mathcal{F}_{\text{LP}} + \lambda \mathcal{P} \tag{3}$$

where $\mathcal{P}$ represents penalty terms and λ is a weight parameter. In B-GRAP algorithm, we define two update functions $\mathcal{F}_{\text{VB}}$ and $\mathcal{F}_{\text{EB}}$ which respectively deal with vertex and edge balance constraints.

Vertex Balance: Given a directed graph $G = \langle V, E, \omega \rangle$, a vertex-balanced partitioning divides the vertices into disjoint partitions of nearly equal size, while minimizing the number of edge cuts between partition. Let $size(V, l)$ be number of vertices having l as label, $size(V, l) = |\{v \in V \mid \phi(v) = l\}|$.

In a perfect balanced partitioning, the size of each partition should be equal to $|V|/K$. In other words, the distribution of vertices in the partitions should be close to a uniform distribution $\mathbf{U} = \langle 1/K, \ldots, 1/K \rangle$, where $1/K$ is called the balance factor. To handle the balance between the partitions, we define vertex-balance $\mathcal{P}_{\text{VB}}$ penalty function that penalizes $\mathcal{F}$ when trying to assign a vertex to a partition violating the balance constraints as follows:

$$\mathcal{P}_{\text{VB}}(l) = \frac{1}{K} - \frac{size(V, l)}{|V|} \tag{4}$$

This function measures the difference between the balance factor $1/K$ and the ratio of vertices assigned to l label. The larger the ratio of vertices with label l is, the higher the penalty to update the vertex label with l is.

At this stage, the number of edge cuts between the partitions is not considered. Thus, a vertex could move to a partition that increases the edge cuts. Given a vertex v and label l, we define a second penalty function as follows:

$$\mathcal{P}_{\text{EC}}(v, l) = \frac{|cut(v, l)|}{d^{+}(v)} \tag{5}$$

where $cut(v, l) = \{(v, u) \in E \mid \phi(u) = l\}$ is the set of edges outgoing from v to vertices in a partition with label l. This function measures the ratio of cuts which penalizes a vertex v to move to a partition with l label if the number of its outgoing edges to this partition is low (normalized to the out degree of v). Thus, when a vertex has more connections to a partition than to the others, the penalty gives more advantage to move to this partition and vice versa. By considering the penalty functions defined in Eq. 4 and Eq. 5, the vertex balance update function is defined in the following equation:

$$\mathcal{F}_{VB}(v,l) = n\mathcal{F}_{LP} + \lambda\Big(\kappa\mathcal{P}_{EC}(v,l) + (1-\kappa)\mathcal{P}_{VB}(v,l)\Big) \tag{6}$$

where n is a normalization constant equal to $\frac{1}{\sum_{u \in N(v)} \omega(v,u)}$. The balance factor $\frac{1}{K}$ could be omitted as it is constant, in this case $\mathcal{P}_{VB}$ variate $\in [0..1]$. The parameter κ is a weight ranging between 0 and 1 which gives more or less importance to balance penalty against the edge cuts penalty. We set κ to 0.5 by default.

Edge Balance: An edge-balanced partitioning divides the graph into disjoint partitions holding nearly equal number of edges, while minimizing the number of edge cuts between partition. Let $size(E,l)$ be the number of outgoing edges from a partition with label l, $size(E,l) = \sum_{v \in V, \phi(v)=l} |d^+(v)|$. Similarly to the vertex-balance partitioning, we define the following edge-balance penalty function:

$$\mathcal{P}_{EB}(l) = \frac{1}{K} - \frac{size(E,l)}{|E|} \tag{7}$$

This function discourages a vertex move to a partition with l label, when the ratio of edges in the partition l is closer or larger than the balance factor. Comparing to vertex balance, edge balance maximizes the edge locality in each partition, which contributes to minimizing the edge cuts. Thus, there is no need to add additional penalty to the update function as defined in Eq. 6. The edge-balance update function is formulated as follows:

$$\mathcal{F}_{EB}(v,l) = n\mathcal{F}_{LP} + \lambda\mathcal{P}_{EB}(l) \tag{8}$$

We note that Spinner algorithm [20] (see Sect. 3) uses the normalized unbalance as penalty function. Comparing to Eq. 7, the edge-size of a partition is normalized by the size of a perfect balanced partition, i.e. $\frac{|E|}{K}$. Moreover, their penalty function that measures the edge balance for each partition, considers both virtual and real edges. The function we defined in Eq. 8 considers only real edges.

5   Partitioning Evaluation Measures

To evaluate the quality of the partitioning produced by our algorithm B-GRAP, we use two standard measures: the ratio of edge cuts **EC** and the Jensen Shannon divergence (**JSD**) [18].

The edge cuts ratio is the ratio of edges connecting each two vertices in two different partitions w.r.t the total number of edges.

$$\mathbf{EC} = \frac{\sum_{v \in V} \sum_{l=1}^{K} |cut(v,l)|}{|E|} \tag{9}$$

The Jensen Shannon divergence (JSD) is the symmetric version of the Kullback–Leibler divergence known as a standard measure to compute the divergence between two distributions. This is a symmetric measure varying in the interval $[0\ldots1]$, where a value close to 0 indicates that the distributions are similar. Let $\mathbf{P} = \langle p_1,\ldots,p_K\rangle$ and $\mathbf{Q} = \langle q_1,\ldots,q_K\rangle$ two distributions with the same size. The **JSD** divergence is computed as follows:

$$\mathbf{JSD}(\mathbf{P}\|\mathbf{Q}) = \frac{1}{2}\Big(D_{\mathrm{KL}}(\mathbf{P}\|M) + D_{\mathrm{KL}}(\mathbf{Q}\|M)\Big) \tag{10}$$

$$\text{with } D_{\mathrm{KL}}(\mathbf{P}\|\mathbf{Q}) = \sum_{l=1}^{K} p_l \log(\frac{p_l}{q_l}) \quad \text{and } M = \frac{1}{2}(\mathbf{P}+\mathbf{Q})$$

In our case, $\mathbf{P}$ represents the distribution of vertices (or edges) on the partitions, where p_l is the ratio of vertices (or edges) in the partition with label l, and $\mathbf{Q}$ equals to the uniform distribution $\mathbf{U}$. The **JSD** considers the balance of the whole partitioning, comparing to other measures such the maximum normalized unbalance metric (**MNU**) [20]. This last used to measure unbalance and represents the percentage-wise difference of only the largest partition from a perfectly balanced partition.

$$\mathbf{MNU}_{\mathrm{VB}} = \frac{\max(|V_l|)}{|V|/K}, \tag{11}$$

$$\mathbf{MNU}_{\mathrm{EB}} = \frac{\max(|E_l|)}{|E|/K}, \quad \text{with } l \in L.$$

Finally, it is important to notice that for **EC**, **JSD**, and **MNU** we consider the directed edges in the original input graph. The virtual edges added for neighborhood discovery (see Sect. 3) are note taken into account.

6   Experiments

We achieve different experiments on different graph data sets in order to evaluate the quality of edge and vertex-balanced partitioning using **EC**, **JSD** and **MNU** measures defined previously. We compare our approach to Spinner [20] because it has shown better results comparing to some existing algorithms. Moreover, since both B-GRAP and Spinner are developed using Apache Giraph environment, we can also provide an evaluation in the same system conditions.

In the following, we first describe the data sets and the experiment settings. Then, we present in detail the results of B-GRAP$_{\mathrm{VB}}$ and B-GRAP$_{\mathrm{EB}}$, compared against Spinner, and achieved on nine graphs.

6.1   Data Sets Description and Experiment Settings

All the experiments are done on a Hadoop cluster of 8 machines, with 64 GB RAM and 8 compute cores. B-GRAP algorithm is implemented in Java using

Table 1. Data sets description

Graph	WikiTalk (W)	BerkeleyStanf (B)	Flixster (F)	DelaunaySC (F)	Pokec (P)	LiveJournal (L)	Orkut (O)	Graph500 (G)	SK-2005 (S)
Directed	yes	yes	yes	yes	yes	yes	yes	no	yes
$|V|$	2.4M	0.7M	2.5M	8.4M	16M	4.8M	2.7M	4.6M	50.6M
$|E|$	5M	7.6M	7.9M	25.2M	30.1M	69M	117.2M	258.5M	1.9B
Source	[14]	[16]	[25,30]	[2,25]	[15]	[1]	[25]	[25]	[21]

Apache Giraph environment [7]. Giraph is an open source implementation of distributed programming framework Pregel [19], designed for Google cluster architecture, with several performance improvement like multi-threading and memory usage optimization. It's built on Hadoop infrastructure to make distributed graph processing and can work with many data storage system supporting graph data (Neo4j, DEX, RDBMS, etc.). In Giraph, the graph is randomly partitioned on several workers (machines) after a complete in-memory load. As in Pergel, Giraph uses a vertex-centric approach to deal with large scale graph processing. In their approach, the computation of the user defined function is done locally, i.e. on each vertex, and in parallel. A vertex contains information about itself and its outgoing edges, it can change its state and the state of these edges by exchanging messages with other vertices at the same iteration, called *super-step*.

In our experiments, we use nine graph data sets of different degree distributions and different sizes in terms of edge and vertex number as summarized in Table 1. Wikitalk (W), Pockec (P), Flixster (F), LiveJournal (L) and Orkut (O) are social online networks graphs. BerkeleyStanf (B) is the berkely.edu and stanford.edu web graph, SK-2005 (S) is hyperlinks on '.sk' web. DelaunaySC (D) and Graph500 (G) are synthetic graphs. Notice that only (G) is an undirected graph.

Experimental setting: We evaluate our algorithm over all the graphs presented in Table 1, by varying the number of partitions K from 2 to 32. More precisely, we execute 10 runs of B-GRAP$_{VB}$ and B-GRAP$_{EB}$ for each graph and each value of K to ensure the significance of the results. For all experiments, we compute the average variation of the following measures with respect to the number of partitions K and over the runs:

- The maximum normalized unbalance of vertices (**MNU**$_{VB}$) and of edges (**MNU**$_{EB}$).
- The divergence between the distribution of vertices (respectively of edges) and the uniform distribution **JSD**$_{VB}$ (respectively **JSD**$_{EB}$).
- The edge-cuts ratio (**EC**).
- The computation time saving ratio (**ΔTime**) of B-GRAP w.r.t Spinner[1]. This ratio is computed using the total CPU time in seconds spent to execute the algorithm, from the initialization until the convergence.

---

[1] Δ**Time** $= \frac{Time(Spinner) - Time(\text{B-GRAP})}{Time(Spinner)}$.

Note that Δ**Time** > 0 means a better performance of our algorithm and a value close to 0 means similar performances with Spinner.

For all experiments, we set $\epsilon = 10^{-3}$ as a threshold stop value and we set τ average out degree $\bar{d}^+ = \frac{|E|}{|V|}$. The penalty term weight parameter λ in the update function $\mathcal{F}$ is set to 1. This gives an equal importance to the penalty term $\mathcal{P}$ and to $\mathcal{F}_{LP}$ according to the update functions defined in Sect. 4.2.

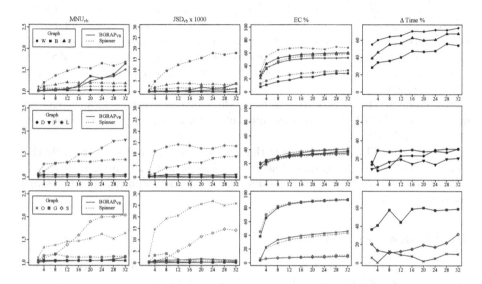

Fig. 2. Variation of the average scores of **MNU**$_{VB}$, **JSD**$_{VB}$, **EC**, and Δ**Time** for the partitioning obtained with B-GRAP$_{VB}$ and Spinner, w.r.t. K

6.2   B-GRAP Vertex Balance

The experiments presented in this section consider the vertex balance constraints to partition a graph. The main objective is to evaluate the ability of our algorithm B-GRAP$_{VB}$ to produce balanced partitions with respect to the number of vertices, while improving the quality of cuts, using the vertex-balance update function defined in Eq. 6.

For this aim, using the experimental protocol described in Sect. 6.1, we compare the balance and the cuts quality of B-GRAP$_{VB}$ partitioning with Spinner partitioning.

Results: The results are presented in Fig. 2. This figure shows, for each graph and algorithm, the average variation of the **MNU**$_{VB}$, **JSD**$_{VB}$, **EC**, and Δ**Time** according the number of partitions K.

We analyze first the variation of the unbalance degree **MNU**$_{VB}$ and the total balance of the partitioning **JSD**$_{VB}$. As shown on the Fig. 2, B-GRAP$_{VB}$ produces generally a low unbalance degree for the most part of graphs (seven over nine w.r.t. $\mathbf{MNU}_{VB}$) while varying the number of partitions K. On the other side, the results of Spinner show a high unbalance degree $\mathbf{MNU}_{VB}$ (> 1.1) when scaling with K, in particular for $K \geq 4$, except for (D) graph. We notice only two exceptions for B-GRAP$_{VB}$ on (B) and (W) graphs, when $K \geq 24$. However, the $\mathbf{MNU}_{VB}$ of B-GRAP$_{VB}$ is still lower then Spinner for (B) graph.

Similarly, the results of $\mathbf{JSD}_{VB}$ show that B-GRAP$_{VB}$ performs generally better than Spinner. The value of $\mathbf{JSD}_{VB}$ is very close to 0 over all graphs and for all K. This means that B-GRAP$_{VB}$ produces high balanced partitions. B-GRAP$_{VB}$ gives better results for 6 graphs over 9 (with 5 significant differences for (B), (D), (P), (O), and (S) and similar results for the others). We note that for the exceptions on (W) and (B) noticed previously for $\mathbf{MNU}_{VB}$, the $\mathbf{JSD}_{VB}$ values are very close to 0 which means that the partitioning has a high global balance degree.

We compare the quality of cuts for both algorithms. Figure 2 shows similar quality of $\mathbf{EC}$. B-GRAP$_{VB}$ shows significant better results on *BerkeleyStanf* and *WikiTalk* graphs.

Finally, the $\Delta\mathbf{Time}$ curves show that B-GRAP$_{VB}$ improves significantly the computation time on all graphs. The time saving percent $\Delta\mathbf{Time}$ is higher than 10% for all the graphs and all K values, except of (O) graph, where the results are better but less significant.

6.3   B-GRAP Edge Balance

Now we compare the performance of our algorithm B-GRAP$_{EB}$ with Spinner, using the edge-balance update function defined in Eq. 8.

Results: We present the results of this experiment in Fig. 3. For each graph and algorithm we show the average variation of the following measures w.r.t. K: $\mathbf{MNU}_{EB}$, $\mathbf{JSD}_{EB}$, $\mathbf{EC}$, and $\Delta\mathbf{Time}$.

Figure 3 shows that the partitioning produced by B-GRAP$_{EB}$ has a low edge unbalance degree for all graphs under analysis. In fact, the average $\mathbf{MNU}_{EB}$ is generally less than 1.05, except in the case of *WikiTalk* for $K = 28$ and $K = 32$ where the average $\mathbf{MNU}_{EB}$ is equal to 1.12 and 1.13, respectively. However, if we analyze the results obtained from running Spinner, we see that we obtain an unbalance degree $\mathbf{MNU}_{EB}$ generally higher than 1.05 and $\mathbf{MNU}_{EB}$ shows bad values while increasing the number of partitions K. On the contrary, the variation of $\mathbf{MNU}_{EB}$ for B-GRAP$_{EB}$ shows that it scales with K with a stable balance quality.

The behaviour of $\mathbf{JSD}_{EB}$ shows that B-GRAP$_{EB}$ generally scales up with K while maintaining a good global balance, with few exceptions. Furthermore, B-GRAP$_{EB}$ obtains better performance than Spinner over five graphs and gives similar $\mathbf{JSD}_{EB}$ scores for the others.

The quality of cuts is generally close for both algorithms (Fig. 3), with only one significant better result on *WikiTalk*.

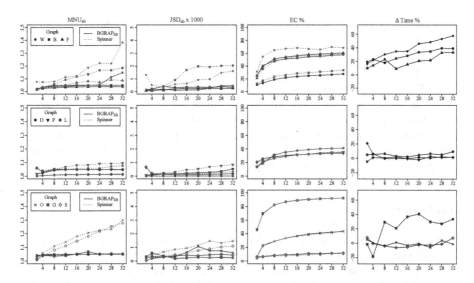

Fig. 3. Variation of the average scores of MNU_{EB}, JSD_{EB}, EC and $\Delta Time$ for the partitioning obtained with B-GRAP$_{EB}$ and Spinner, w.r.t. K

Finally, Δ**Time** variation shows significant better results for B-GRAP$_{EB}$ on (W), (B), (F) and (G) graphs. The computation time is slightly better for other graphs. In fact, this is only with few exceptions on (G) for $K \leq 4$ and for (S).

Summary: B-GRAP$_{EB}$ algorithm computes higher edge balanced partitioning without impacting the quality of cuts and while showing generally faster computation time. Moreover, the results on the undirected graph *Graph500* show that the initialization step is efficient for the time execution of the algorithm. Finally, the results given for the JSD_{EB} and MNU_{EB} show that a better balance can be obtained if we consider the directness of edges for a directed graph.

7   Conclusion and Perspectives

In this paper we proposed two scalable and parallel partitioning algorithms B-GRAP$_{VB}$ and B-GRAP$_{EB}$, based on LP, that address the vertex balance and the edge balance problems respectively on both directed and undirected graphs. We defined the initialization strategy of our algorithm that allows to speed up the convergence and two update functions to produce either vertex balanced or edge balanced partitioning.

Our results show good performances of B-GRAP on various graphs and with different scales. We show that B-GRAP produces high vertex balanced and high edge balanced partitioning with a good cuts quality comparing to Spinner algorithm (significant values for 5 graphs and slightly better values for others), on either directed and undirected graphs. Moreover, the computation time of B-GRAP is better than Spinner, with few exception for B-GRAP$_{EB}$ on two graphs.

The additional experiments we conducted to study the initialization step B-GRAP$_{init}$ show that the selection of the seed vertices has an impact on the quality of the partitioning and the computation time. We would study more deeply this step in order to optimize our method.

We would also study the impact of the partitioning on algorithms of graph analytics with respect to the balance strategy, such as Shortest Path Computation, PageRank, and Community Detection.

References

1. Backstrom, L., Huttenlocher, D., Kleinberg, J., Lan, X.: Group formation in large social networks. In: Proceedings of the 12th ACM SIGKDD International Conference on Knowledge discovery and data mining, p. 44. ACM Press (2006)
2. Bader, D., Meyerhenke, H., Wagner, D.: Graph Partitioning and Graph Clustering, Contemporary Mathematics, vol. 588 (2013)
3. Buluç, A., Meyerhenke, H., Safro, I., Sanders, P., Schulz, C.: Recent advances in graph partitioning. In: Kliemann, L., Sanders, P. (eds.) Algorithm Engineering. LNCS, vol. 9220, pp. 117–158. Springer, Cham (2016). https://doi.org/10.1007/978-3-319-49487-6_4
4. Chakraborty, T., Dalmia, A., Mukherjee, A., Ganguly, N.: Metrics for community analysis: a survey **50**(4), 1–37 (2016)
5. Chevalier, C., Pellegrini, F.: PT-Scotch: a tool for efficient parallel graph ordering. Technical report, pp. 6-8 (2008)
6. Das, H., Kumar, S.: A parallel TSP-based algorithm for balanced graph partitioning. In: 2017 46th International Conference on Parallel Processing (ICPP), pp. 563–570. IEEE (2017)
7. Giraph, A.: Giraph : Large-scale graph processing in Hadoop (2012)
8. Gonzalez, J.E., Low, Y., Gu, H., Bickson, D., Guestrin, C.: PowerGraph: distributed graph-parallel computation on natural graphs. In: Proceedings of the 10th USENIX Conference on Operating Systems Design and Implementation, pp. 17–30 (2012)
9. Gregory, S.: Finding overlapping communities in networks by label propagation **12**(10), 103018 (2010)
10. Heidari, S., Simmhan, Y., Calheiros, N., Buyya, R.: Scalable graph processing frameworks: a taxonomy and open challenges **51**, 1–53 (2018)
11. Henzinger, A., Noe, A., Schulz, C.: ILP-based local search for graph partitioning (2018)
12. Karypis, G., Kumar, V.: Multilevel graph partitioning schemes. In: Proceedings of the 24th International Conference on Parallel Processing (ICPP) 1955, vol. 3, pp. 113–122 (1995)
13. Karypis, G., Kumar, V.: A parallel algorithm for multilevel graph partitioning and sparse matrix ordering, **48**(1), 71–95 (1998)
14. Leskovec, J., Huttenlocher, D., Kleinberg, J.: Signed networks in social media. In: Proceedings of the 28th International Conference on Human factors in computing systems, p. 1361 (2010)

15. Leskovec, J., Krevl, A.: SNAP datasets: stanford large network dataset collection (2014). https://snap.stanford.edu/data/index.html
16. Leskovec, J., Lang, K.J., Dasgupta, A., Mahoney, M.W.: Community structure in large networks: natural cluster sizes and the absence of large well-defined clusters 6(1), 29–123 (2009)
17. Li, Y., Constantin, C., du Mouza, C.: A block-based edge partitioning for random walks algorithms over large social graphs. In: Cellary, W., Mokbel, M.F., Wang, J., Wang, H., Zhou, R., Zhang, Y. (eds.) WISE 2016. LNCS, vol. 10042, pp. 275–289. Springer, Cham (2016). https://doi.org/10.1007/978-3-319-48743-4_22
18. Lin, J.: Divergence measures based on the shannon entropy. IEEE Trans. Inf. Theory 37(1), 145–151 (1991)
19. Malewicz, G., et al.: Pregel: a system for large-scale graph processing. In: Proceedings of the 2010 ACM SIGMOD International Conference on Management of Data, pp. 135–146 (2010)
20. Martella, C., Logothetis, D., Loukas, A., Siganos, G.: Spinner: scalable graph partitioning in the cloud. In: Proceedings - International Conference Data Engineering (2017)
21. Meyerhenke, H., Sanders, P., Schulz, C.: Parallel graph partitioning for complex networks, 28, 2625–2638 (2017)
22. Nguyen, D.: Graph Partitioning. ISTE (2011)
23. Pellegrini, F., Roman, J.: Scotch: a software package for static mapping by dual recursive bipartitioning of process and architecture graphs. In: Proceedings of the International Conference and Exhibition on High-Performance Computing and Networking, pp. 493–498. HPCN Europe 1996 (1996)
24. Raghavan, U.N., Albert, R., Kumara, S.: Near linear time algorithm to detect community structures in large-scale networks, p. 036106 (2007)
25. Rossi, R.A., Ahmed, N.K.: The network data repository with interactive graph analytics and visualization. In: Proceedings of the 29 AAAI (2015)
26. Sanders, P., Schulz, C.: Think locally, act globally: highly balanced graph partitioning. In: Bonifaci, V., Demetrescu, C., Marchetti-Spaccamela, A. (eds.) SEA 2013. LNCS, vol. 7933, pp. 164–175. Springer, Heidelberg (2013). https://doi.org/10.1007/978-3-642-38527-8_16
27. Tsourakakis, C., Gkantsidis, C., Radunovic, B., Vojnovic, M.: FENNEL: streaming graph partitioning for massive scale graphs. In: Proceedings of the 7th ACM International Conference on Web search and data mining, pp. 333–342 (2014)
28. Ugander, J., Backstrom, L.: Balanced label propagation for partitioning massive graphs. In: Proceedings of the 6th ACM International Conference on Web Search and Data Mining, pp. 507–516. WSDM 2013, ACM (2013)
29. Xin, R.S., Gonzalez, J.E., Franklin, M.J., Stoica, I.: GraphX: a resilient distributed graph system on spark. In: First International Workshop on Graph Data Management Experiences and Systems, pp. 1–6. ACM Press (2013)
30. Zafarani, R., Liu, H.: Users joining multiple sites: distributions and patterns (2014)

DSCAN: Distributed Structural Graph Clustering for Billion-Edge Graphs

Hiroaki Shiokawa[1(✉)] and Tomokatsu Takahashi[2]

[1] Center for Computational Sciences, University of Tsukuba, Tsukuba, Japan
shiokawa@cs.tsukuba.ac.jp
[2] Graduate School of SIE, University of Tsukuba, Tsukuba, Japan
shihakata@kde.cs.tsukuba.ac.jp

Abstract. The structural graph clustering algorithm (SCAN) is an essential graph mining tool that reveals clusters, hubs, and outliers included in a given graph. Although SCAN is used in various applications, it has two serious drawbacks when handling large graphs. First, SCAN is computationally expensive since it requires iterative computations for all nodes and edges. Second, SCAN is not designed to handle large graphs that cannot fit in the main memory. This paper presents a distributed structural graph clustering algorithm, DSCAN, to address the aforementioned problems on a cluster of computers. DSCAN employs edge pruning techniques to reduce the communication and computation overheads of the distributed algorithm. Our extensive experiments on real-world billion-edge graphs demonstrate that DSCAN outperforms state-of-the-art algorithms in terms of running time even though DSCAN outputs the same clusters as SCAN.

Keywords: Graph · Clustering · Distributed algorithm · Community detection

1 Introduction

Graph clustering is a fundamental data mining tool that reveals community structures hidden in complex networks. The structural graph clustering algorithm (SCAN) [23] is one of the most successful graph clustering methods. The main idea underlying SCAN is it places nodes into the same cluster only if the nodes have dense internal connections. SCAN excludes sparsely connected nodes from the clusters, and instead classifies them as hubs or outliers. Unlike conventional graph clustering algorithms [1, 7, 13], SCAN finds clusters, hubs, and outliers in a graph simultaneously.

Although SCAN helps find accurate clusters, it has two serious drawbacks when handling large-scale graphs with millions or even billions of edges. First, SCAN is computationally expensive because it must find all clusters included

This work was done when T. Takahashi was a student of University of Tsukuba. He is currently a member of Nippon Telegraph and Telphone Corporation.

© Springer Nature Switzerland AG 2020
S. Hartmann et al. (Eds.): DEXA 2020, LNCS 12391, pp. 38–54, 2020.
https://doi.org/10.1007/978-3-030-59003-1_3

in a given graph before classifying hubs and outliers. In the worst case, this computation entails $O(m^{1.5})$ time [3], where m is the number of edges in the graph. Second, SCAN cannot handle a large graph whose memory footprint exceeds the main memory size. Although graph sizes have recently exceeded 600 GiB (Table 1), SCAN does not deal with such large memory footprints. Thus, SCAN suffers from large I/O overheads between the main memory and storage, which significantly degrades the clustering efficiency.

1.1 Existing Works and Challenges

Recently, many studies have strived to overcome the above problems. One major approach is to reduce the number of computed nodes and edges by skipping unnecessary computations. *SCAN++* [17] and *pSCAN* [3] are the most successful ones to date. SCAN computes the *structural similarity* for all edges to evaluate how adjacent nodes are densely connected. However, it is more reasonable to compute the structural similarity only for adjacent nodes that yield a dense connection. By incrementally removing nodes that cannot be in any clusters, SCAN++ and pSCAN successfully improve the clustering efficiency. Seo and Kim, however, recently pointed out that SCAN++ and pSCAN are still computationally expensive for billion-edge graphs [15]. That is, they require a large running time for large graphs.

Instead of the above pruning-based approaches, thread-parallel algorithms, e.g., *ppSCAN* [4] and *ScaleSCAN* [19,22], have been proposed. Modern CPUs are generally equipped with multiple physical cores that share a main memory. Thus, the algorithms load all nodes and edges into the main memory, and then compute the structural similarity in a thread parallel manner. Although they certainly reduce the running time of SCAN, it is not trivial to compute large graphs whose memory footprint exceeds the main memory size. To handle such large memory footprints, several distributed algorithms [8,25] have also been proposed in a recent few years. For instance, *PSCAN* [25] and *CASS* [8] implemented SCAN algorithm on distributed frameworks, Apache Hadoop [20] and Apache Spark [24], respectively. However, those distributed algorithms incur expensive I/O costs and communication overheads among distributed machines resulting in large computation time for billion-edge graphs. Hence, designing the structural graph clustering algorithm to efficiently compute massive graphs remains a challenging task.

1.2 Our Approaches and Contributions

This paper addresses the problem of speeding up SCAN for billion-edge graphs that do not fit on a main memory. We present a novel distributed parallel SCAN algorithm, namely *DSCAN*, which efficiently performs structural graph clustering on distributed memories. Given a graph, DSCAN first deploys disjointed subgraphs of the graph to the memories. Then it performs the structural graph clustering in distributed and parallel ways. By distributing the disjointed sub-graphs to multiple memories, DSCAN deals with a large memory footprint that does not fit a single main memory.

To further improve the clustering efficiency, DSCAN employs *skewness-aware edge-pruning*. As we briefly described in Sect. 1.1, distributed graph algorithms [8,25] typically suffer from large communication overheads among the distributed memories since many edges are spread across the distributed machines [11,25]. To reduce the overheads, distributed frameworks generally apply graph partitioning algorithms [11] before the distributed computation. However, this approach is impractical for billion-edge graphs because graph partitioning itself consumes a larger memory footprint than the main memory, even though we utilize parallel partitioning approaches [12]. That is, DSCAN needs to deploy subgraphs so that they reduce the overhead *without* using graph partitioning algorithms. By employing skewness-aware edge-pruning, DSCAN tries to reduce edges that involve unnecessary communications even if the subgraphs are randomly generated. Consequently, our proposed algorithm DSCAN has the following attractive properties:

Algorithm 1. SCAN

Procedure SCAN(G, ϵ, μ)
1: **for each** edge $(u, v) \in E$ **do**
2:     Compute $\sigma(u, v)$ by Definition 1;
3: $\mathbb{C} = \emptyset$;
4: **for each** non-visited node $u \in V$ **do**
5:     $C_u = \{u\}$;
6:     **for each** non-visited node $v \in C_u$ **do**
7:         **if** $|N_v^\epsilon| \geq \mu$ **then**
8:             $C_u = C_u \cup N_v^\epsilon$;
9:             Mark v as visited;
10:     **if** $|C_u| \geq 2$ **then**
11:         $\mathbb{C} = \mathbb{C} \cup C_u$;
12: Detect $\mathbb{H}$ and $\mathbb{O}$;
13: **return** $\mathbb{C}$, $\mathbb{H}$, and $\mathbb{O}$;

---

1. **Efficiency:** Compared with state-of-the-art sequential and distributed algorithms, DSCAN is superior in terms of running time on billion-edge graphs.
2. **Scalability:** DSCAN has good scalability. As the numbers of threads and machines increase, the speed-up is almost linear.
3. **Exactness:** Although we employ pruning techniques to improve efficiency, DSCAN outputs the same clusters as those of the original algorithm SCAN.

Extensive evaluations clarified that DSCAN runs up to 763.4 times faster than state-of-the-art distributed methods while keeping its clustering qualities. Specifically, DSCAN computes a graph with 5.5 billion edges within 10 s, while most of the state-of-the-art methods could not handle the graph because they ran out of memory. Although the structural graph clustering is essential in many applications, it suffers from performance limitations and cannot handle billion-edge graphs. By introducing our approaches, DSCAN helps improve the effectiveness of future applications.

Organization: The rest of this paper is organized as follows: In Sect. 2, we briefly review the baseline algorithm SCAN. Section 3 presents our proposed algorithm DSCAN. After that we report our experimental analysis in Sect. 4, and we briefly review related work in Sect. 5. Finally, we conclude this paper in Sect. 6.

2   Baseline Method: SCAN

In this section, we briefly introduce the baseline algorithm SCAN. Let $G = (V, E)$ be an unweighted and undirected graph, where V and E denote a set of nodes and edges, respectively. Given a density threshold $\epsilon \in [0, 1]$ and a minimum size of a cluster $\mu \in \mathcal{N}$, the structural graph clustering SCAN [23] returns sets of clusters $\mathbb{C}$, hubs $\mathbb{H}$, and outliers $\mathbb{O}$ simultaneously (Algorithm 1). SCAN initially extracts $\mathbb{C}$ as groups of densely connected nodes before detecting sparsely connected nodes as $\mathbb{H}$ or $\mathbb{O}$. SCAN places the nodes into the same cluster only if two nodes have a dense connection. To measure the density, SCAN evaluates the *structural similarity*, which is defined as:

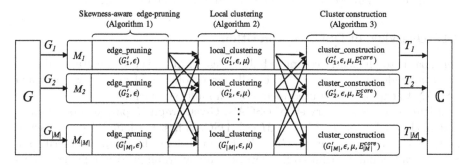

Fig. 1. Overview of DSCAN

Definition 1 (Structural similarity). *Let N_u be a structural neighborhood of node u such that $N_u = \{v \in V | (u, v) \in E\} \cup \{u\}$, the structural similarity $\sigma(u, v)$ between node u and v is defined as $\sigma(u, v) = |N_u \cap N_v| / \sqrt{d_u d_v}$, where $d_i = |N_i|$.*

We denote nodes u and v are *similar* if $\sigma(u, v) \geq \epsilon$; otherwise, *dissimilar*.

SCAN first computes $\sigma(u, v)$ for all edges (lines 1–2 in Algorithm 1). Then it finds a seed of a cluster, called *core node*, defined as follows:

Definition 2 (Core node). *Given user-specified parameters, ϵ and μ, node u is a core node iff $|N_u^\epsilon| \geq \mu$, where $N_u^\epsilon = \{v \in N_u | \sigma(u,v) \geq \epsilon\}$.*

Once SCAN finds a core node, it expands to a cluster from the core node (lines 7–9 in Algorithm 1). Let node u be a core node, SCAN places all nodes in N_u^ϵ into the same cluster of node u, which is denoted by C_u (line 8). If C_u has non-visited core nodes, SCAN recursively expands the cluster until no non-visited nodes can be found in C_u (lines 6–9). Formally, cluster $C_u \in \mathbb{C}$ is defined as follows:

Definition 3 (Cluster). *Let node u be a core node and C_u be a cluster that is initially set to $C_u = \{u\}$. SCAN finds a cluster C_u defined as $C_u = \{w \in N_v^\epsilon | v \in C_u\}$.*

By following Definition 3, SCAN finds all clusters in G.

Finally, SCAN identifies hubs and outliers from non-clustered nodes (line 12). SCAN regards node u as a hub if it bridges multiple clusters in $\mathbb{C}$; otherwise, it is an outlier. Once $\mathbb{C}$ is obtained, hubs and outliers can be detected in a straightforward manner in $O(n + m)$ time, where n and m are the number of nodes and edges, respectively. Hereafter, we thus focus on only extracting $\mathbb{C}$ in G.

3   Proposed Method: DSCAN

DSCAN can efficiently compute billion-edge graphs that cannot fit on a single main memory. Here we overview DSCAN and describe our algorithm.

3.1   Overview of DSCAN Algorithm

Our goal is to efficiently compute billion-edge graphs even if the graphs do not fit on a single main memory. To handle large volumes of graphs, we designed DSCAN to exploit distributed memories on multiple machines.

Figure 1 shows an overview of DSCAN. Given graph G, parameters, ϵ and μ, and machines $M = \{M_1, M_2, \ldots, M_{|M|}\}$, DSCAN first randomly deploys a set of nodes V_i resulting in subgraph $G_i = (V_i, E_i)$ for each machine M_i. As described in Sect. 1, traditional distributed frameworks employ graph partitioning algorithms to obtain disjointed subgraphs that reduce the communication overheads among machines. However, partitioning algorithms are not applicable to billion-edge graphs since (1) billion-edge graphs have larger memory footprints than a single main memory and (2) graph partitioning itself is computationally expensive. Hence, to achieve a low communication overhead for billion-edge graphs, DSCAN employs *skewness-aware edge-pruning* (Sect. 3.2) before performing distributed graph clustering. First, DSCAN randomly partitions V into equally sized partition, i.e., $|V_1| \approx |V_2| \approx \ldots |V_{|M|}|$, each of which yields a subgraph $G_i = (V_i, E_i)$. Then, DSCAN assigns the subgraphs to a machine to balance the loads. Last, DSCAN drops unnecessary edges that entail extraneous communication overheads among the machines by skewness-aware edge-pruning.

Afterwards DSCAN invokes *local clustering* (Sect. 3.3) in each machine to find all core nodes included in each subgraph. As discussed in Sect. 2, structural similarity computations require $O(m^{1.5})$ time, where m is the total number of edges in G. Consequently, each machine M_i still requires $O(|E_i|^{1.5})$ time to extract all core nodes in G_i. To reduce the computational costs, DSCAN performs thread-parallel and data-parallel algorithms for the structural similarity computations.

Finally, DSCAN constructs clusters from the core nodes over distributed machines (Sect. 3.4). To maintain clustering results over distributed machines, DSCAN employs a union-find tree [5] in distributed and parallel ways, and returns $\mathbb{C}$ by merging the results obtained from each machine.

3.2   Skewness-Aware Edge-Pruning

We propose *skewness-aware edge-pruning* to reduce the communication costs entailed by distributed computing. DSCAN prunes unnecessary edges that do not contribute to the clusters prior to distributed clustering.

Suppose that edge (u, v) spans across two machines M_i and M_j, i.e., $u \in V_i$ on M_i and $v \in V_j$ on M_j. In this case, we need to perform communications between M_i and M_j to compute structural similarity $\sigma(u, v)$ since $N_u \not\subseteq V_j$ and $N_v \not\subseteq V_i$. However, we can skip the communications for edge (u, v) if the edge is dissimilar since $v \notin N_u^\epsilon$ and $u \notin N_v^\epsilon$ when $\sigma(u, v) < \epsilon$. That is, edge (u, v) does not need to be computed because it is not used to construct clusters by Definition 3.

To eliminate unnecessary communication overhead, DSCAN prunes dissimilar edges (u, v), i.e., $\sigma(u, v) < \epsilon$, *without* computing the structural similarity. To find dissimilar edges without structural similarity computations, we introduce a simple criterion that evaluates the degree skewness of adjacent nodes:

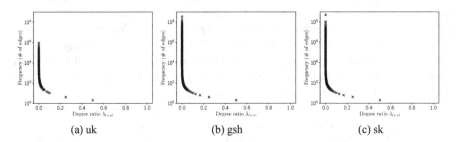

(a) uk   (b) gsh   (c) sk

Fig. 2. Distribution of degree ratio $\lambda_{(u,v)}$

Definition 4 (Degree-skewness $\lambda_{(u,v)}$). *Let edge (u, v) be in E, degree-skewness $\lambda_{(u,v)}$ is defined as $\lambda_{(u,v)} = \min\left\{\frac{d_u}{d_v}, \frac{d_v}{d_u}\right\}$, where $d_i = |N_i|$.*

Algorithm 2. Skewness-aware edge-pruning

Procedure edge_pruning(G_i, ϵ)
1: $E_i^{prune} = \emptyset$;
2: **for each** $(u, v) \in E_i$ **do in thread-parallel**
3:     **if** $\lambda_{(u,v)} < \epsilon^2$ **then** $E_i^{prune} \cup \{(u, v)\}$;
4: $E_i' = E_i \backslash E_i^{prune}$;
5: $E_i^{send} = \{(u, v) \in E' | u \in V_i, v \in V_j \text{ for } i \neq j\}$
6: **for each** $(u, v) \in E_i^{send}$ **do in thread-parallel**
7:     **send** N_u to M_j, **receive** N_v from M_j, and $V_i \cup N_v$;
8: **return** $G_i' = (V_i, E_i')$;

Definition 4 implies that $\lambda_{(u,v)}$ becomes small if nodes u and v have imbalanced degrees; otherwise, $\lambda_{(u,v)}$ approaches 1.

From Definition 1 and 4, we have the following lemma:

Lemma 1 (Prunable-edges). *If $\lambda_{(u,v)} < \epsilon^2$, the structural similarity between u and v is always smaller than ϵ, i.e.,$\sigma(u, v) < \epsilon$.*

Proof. We suppose $d_u < d_v$ without loss of generality. Since $\lambda_{(u,v)} = d_u/d_v$, we have $d_u/\epsilon^2 < d_v$ if $\lambda_{(u,v)} < \epsilon^2$. Thus, from Definition 1, $\sigma(u, v) = |N_u \cap N_v|/\sqrt{d_u d_v} < \epsilon|N_u \cap N_v|/d_u < \epsilon$, which completes the proof. □

Lemma 1 indicates that DSCAN regards edge (u, v) as *dissimilar* if $\lambda_{(u,v)} < \epsilon^2$. From Definitions 2 and 3, dissimilar edges are not included in any clusters. Thus, edges whose degree ratio is less than ϵ are pruned prior to distributed clustering.

We experimentally observed the impact of the skewness-aware edge-pruning on real-world graphs. Figure 2 shows the distributions of $\lambda_{(u,v)}$ in the real-world billion-edge graphs shown in Table 1. Each plotted point indicates the number of edges that yield $\lambda_{(u,v)}$ value in the corresponding graph. As shown in Fig. 2, the three graphs have almost the same distributions. Most edges have quite small $\lambda_{(u,v)}$ values. These observations imply that edges in billion-edge graphs prefer to connect nodes whose degrees are significantly different. Consequently, DSCAN effectively excludes a large part of E_i by skewness-aware edge-pruning. For example, existing studies [17,23] have reported that $\epsilon \in [0.5, 0.8]$ is a reasonable choice to obtain accurate clustering results. That is, if $\epsilon = 0.5$, DSCAN can prune up to 91% of the edges from the graphs.

Finally, Algorithm 2 shows the pseudo codes for skewness-aware pruning. To reduce the communication overheads, DSCAN prunes edges using Lemma 1 in a thread-parallel manner (line 2–3). DSCAN can check if an edge is dissimilar in $O(1)$ by Lemma 1. Thus, by letting T be the number of threads invoked in M_i, DSCAN finds all dissimilar edges of E_i in $O(\frac{|E_i|}{T}) \approx O(\frac{|E|}{|M|T})$ time.

DSCAN then shares structural neighborhoods of *similar* edges if they span across different machines (lines 4–7). First, a set of similar edges E_i^{send} are extracted (line 5). Then the structural neighborhoods N_u and N_v for each edge (u, v) are exchanged (lines 6–7). Suppose that u and v are located in M_i and M_j, respectively. DSCAN sends N_u to M_j, and it receives N_v from M_j.

Algorithm 3. Local clustering

Procedure local_clustering(G_i', ϵ, μ)
1: **for each** $(u, v) \in E_i'$ **do in thread-parallel**
2:     **if** $sd[u] < \mu$ and $ed[u] \geq \mu$ **then**
3:         Compute $\sigma(u, v)$ **in data-parallel**;
4: **return** $E_i^{core} = \{(u, v) \in E_i' | sd[u] \geq \mu \text{ and } sd[v] \geq \mu\}$;

3.3 Local Clustering

In each machine, DSCAN computes the structural similarities and detects all core nodes in G_i in a thread-parallel manner. Algorithm 3 shows the pseudo codes. To reduce redundant similarity computations, a node-pruning technique is performed [3] (line 2). DSCAN maintains two integers for each node in V_i: *similar degree sd* and *effective degree ed*. Let similar and effective degrees of node u be $sd[u]$ and $ed[u]$, respectively. $sd[u]$ is the number of nodes in N_u that have already been computed as similar, *i.e.*, $\sigma(u, v) \geq \epsilon$, while $ed[u]$ is the number of non-computed nodes in N_u. From Definition 2, if $sd[u] \geq \mu$, then node u is a core node. By contrast, node u is not a core node if $ed[u] < \mu$. DSCAN compares $sd[u]$ and $ed[u]$ in the thread-parallel manner (lines 1–3). $\sigma(u, v)$ is not computed only if $sd[u] \geq \mu$ and $ed[u] < \mu$ (or $sd[v] \geq \mu$ and $ed[v] < \mu$); otherwise, DSCAN computes $\sigma(u, v)$.

To further improve efficiency, DSCAN computes the structural similarities in a data-parallel manner (line 3). As shown in Definition 1, each structural similarity computation $\sigma(u, v)$ requires a set intersection between N_u and N_v, this is, however, a time-consuming task. Thus, we employed SIMD-wise set intersection method, proposed by Inoue *et al.* [6], to compute $N_u \cap N_v$ in Definition 1. We omit the details of the SIMD-wise set intersection due to the space limitation.

3.4 Cluster Construction

Finally, DSCAN runs Algorithm 4 to construct clusters over distributed machines. To efficiently find clusters, DSCAN employs a union-find tree [5] in each machine. The union-find tree efficiently maintains a set of nodes partitioned into disjoint clusters by two fundamental operators, called find(u) and union(u, v). find(u) looks up a cluster C_u, and union(u, v) merges two clusters C_u and C_v into the same cluster. Both operators run at most $\Omega(A(n))$ times, where A is Ackermann function.

In Algorithm 4, DSCAN first constructs local clusters in each machine by a thread-parallel manner. DSCAN initializes a union-find tree T_i in each machine M_i (line 1), and DSCAN finds out all clusters from core nodes by checking edges included in E_i^{core} obtained by Algorithm 3 (lines 2–5). Given $(u, v) \in E_i^{core}$, DSCAN looks up C_u and C_v by find(u) and find(v) operations (line 3). If $C_u \neq C_v$, C_u and C_v are merged by union(u, v) operation only if (u, v) is similar, *i.e.*, $\sigma(u, v) \geq \epsilon$ (lines 3–5). To avoid write-write conflicts of union operations among multiple threads, DSCAN employs CAS atomic operations

Algorithm 4. Cluster construction

Procedure cluster_construction$(G_i', \epsilon, \mu, E_i^{core})$
1: **initialze** *union-find tree* T_i for all $u \in V_i$;
2: **for each** $(u, v) \in E_i^{core}$ **do in thread-parallel**
3:     **if** $T_i.\text{find}(u) \neq T_i.\text{find}(v)$ **then**
4:        **if** (u, v) has not been computed **then** compute $\sigma(u, v)$;
5:        **if** (u, v) is *similar* **then** $T_i.\text{union}(u, v)$;
6: **execute** (lines 2-5) for $E_i' \backslash E_i^{core}$ instead of E^{core};
7: **for each** $(u, v) \in E_i'$ **do in thread-parallel**
8:     **if** $u \in G_i$ and $v \in G_j$ **then**
9:        **send** $\langle\langle u, v \rangle, T_i.\text{find}(u)\rangle$ to M_j;
10:       **receive** $\langle\langle u, v \rangle, T_j.\text{find}(v)\rangle$ from M_j;
11:       $T_i.\text{union}(u, v)$;

Table 1. Statistics of real-world datasets

Dataset name	# of nodes	# of edges	Memory footprint	Data source
uk	39,459,925	936,364,282	16 GiB	uk-2005 [2]
gsh	68,660,142	1,802,747,600	30 GiB	gsh-2015-host [2]
sk	50,636,154	1,949,412,601	33 GiB	sk-2005 [2]
union	133,633,040	5,507,679,822	86 GiB	uk-union-2006-06-2007-05 [2]
clueweb	978,408,098	42,574,107,469	691 GiB	clueweb12 [2]

before merging the clusters. DSCAN also clusters non-core nodes whose edges included in $E_i \backslash E_i^{core}$ in the same way (line 6).

After finding all local clusters, DSCAN tries to merge local clusters over distributed machines (lines 7–11). Each machine sends its local clustering results as pairs of $\langle\langle u, v \rangle, T_i.\text{find}(u)\rangle$ only if edge $(u, v) \in E_i'$ spans two machines. As shown in Algorithm 2, E_i' is a set of edges with $\lambda_{(u,v)} \geq \epsilon^2$. Hence, DSCAN merges C_u and C_v by $\text{union}(u, v)$ once it receives $\langle\langle u, v \rangle, T_j.\text{find}(v)\rangle$ from the other machines (lines 10–11).

4   Experimental Analysis

We conducted extensive experiments to evaluate the effectiveness of our proposed algorithm. We designed our experiments to demonstrate that:

- **Efficiency:** DSCAN is faster than the state-of-the-art algorithms on billion-edge graphs. Our proposal computes a graph having 5.5 billion edges within 10 s.
- **Scalability:** DSCAN shows better scalability than the state-of-the art methods. It linearly increases performances as increasing numbers of threads and machines.
- **Exactness:** Although DSCAN drastically reduces clustering time, it always returns exactly same clustering results as those of the original algorithm.

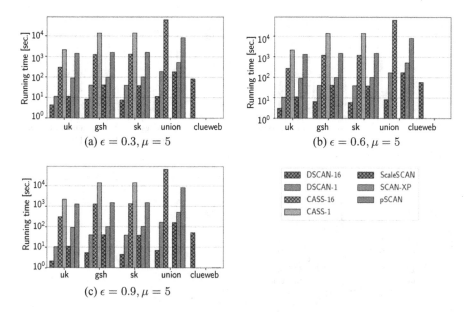

Fig. 3. Running time by varying ϵ.

4.1   Experimental Setup

Datasets: We evaluated our proposed algorithm DSCAN on five real-world graphs, which were from the Laboratory of Web Algorithmics (LAW) [2]. Table 1 summarizes the statistics of real-world datasets. In addition, we also used synthetic graphs generated by LFR-benchmark [9], which is considered as the *de facto* standard model for generating graphs. The settings will be detailed later.

Methods: In our experiments, we compared DSCAN with the state-of-the-art sequential algorithm pSCAN [3], the state-of-the-art thread-parallel algorithms SCAN-XP [22] and ScaleSCAN [19], and the state-of-the-art distributed algorithm CASS [8]. We also tested an another recent distributed algorithm proposed by Zhao *et al.* [25], which is implemented on distributed computation frameworks. However, we omitted the results from this paper since the algorithm could not return any results within 24 h on all datasets shown in Table 1.

We implemented the above algorithms, except for CASS, in C++ and compiled them by gcc compiler with -O3 compile option. Since CASS requires Apache Spark framework [24], we implemented CASS in Java by following the original paper [8]. For the thread-parallel and data-parallel implementations (SCAN-XP, ScaleSCAN, and DSCAN), we used OpenMP and AVX512 instructions, respectively. Additionally, we used MPI for the distributed processing in DSCAN.

All experiments were conducted on a computer cluster composed of 16 machines that are inter-connected by Intel Omni Path (12 GiB/s). Each machine was equipped with one Intel Xeon Phi 7250 processor (64 physical cores with 1.40 GHz default frequency) and 96 GiB DDR4 RAM. Unless otherwise stated, we used 64 threads for SCAN-XP, ScaleSCAN, and DSCAN, which is the same

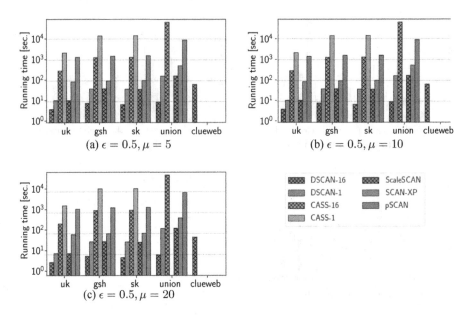

Fig. 4. Running time by varying μ.

number as the physical cores. Additionally, we run pSCAN, SCAN-XP, and ScaleSCAN on one machine since they are not distributed algorithms. We report the average score of 20 tests for each method.

4.2 Efficiency

We assessed the running time of DSCAN through wall clock time. In this evaluation, we tested two types of settings for DSCAN and CASS. DSCAN-1 and CASS-1 were performed on one machine, and DSCAN-16 and CASS-16 were on 16 machines. Figure 3 shows the running times of the algorithms for different ϵ settings with a fixed μ value, *i.e.*, $\mu = 5$. Similarly, Fig. 4 shows the running time with a fixed $\epsilon = 0.5$ and different μ values such as 5, 10, and 20. Note that we omitted several results (1) if the algorithms crashed due to the out of memory problem or (2) if they could not return any results within 24 h.

Overall, DSCAN-16 outperforms the state-of-the-art algorithms CASS, ScaleSCAN, SCAN-XP, and pSCAN, although DSCAN-1 is competitive with ScaleSCAN. In specific, DSCAN-16 is up to 763.4, 5.94, 17.1, and 234.9 times faster than CASS, ScaleSCAN, SCAN-XP, and pSCAN, respectively. Furthermore, only DSCAN-16 returns clustering results for clueweb. Actually, DSCAN-16 computes union and clueweb within 9.55 s and 68.2 s on average, respectively. This is because we designed DSCAN-16 so that (1) it effectively drops off unnecessary communications among distributed machines and (2) it utilizes distributed memories even if the graph volumes do not fit on a single main memory. Consequently, DSCAN-16 reduces the running time for billion-edge graphs.

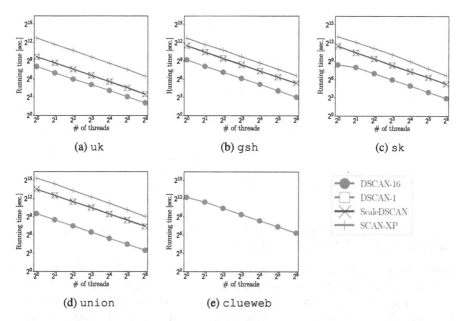

Fig. 5. Running time by varying # of threads ($\epsilon = 0.5, \mu = 5$).

We observed from Fig. 3 that DSCAN-16 very slightly decreases its running time as the size of ϵ increases, but is nearly independent of μ. This is due to the skewness-aware edge pruning shown in Sect. 3.2. From Lemma 1, DSCAN-16 excludes edges that require communication overheads among machines if the edges have a degree skewness smaller than ϵ^2, *i.e.,* $\lambda_{(u,v)} < \epsilon^2$. As observed in Fig. 1, most of the edges can be pruned by Lemma 1 for large ϵ settings in the real-world graphs. That is, DSCAN-16 can remove a large quantity of communication overheads if ϵ is large. As a result, DSCAN-16 can improve its efficiency by increasing the size of ϵ. By contrast, the running time for DSCAN-16 is almost constant for all μ settings (Fig. 4). However, as discussed in the literature [23], large μ settings are not suitable to detect clusters with high accuracy; In practice, $\mu = 2$ is recommended for real-world graphs.

4.3 Scalability

Thread-Parallel Scalability: To assess the scalability of DSCAN, we first compared the running times of the parallel algorithms by varying the number of threads invoked in each machine. Figure 5 shows the running times of the parallel algorithms. Similar to the previous section, we evaluated two types: DSCAN-16 and DSCAN-1, which perform structural graph clustering on 16 machines and a single machine, respectively. Since DSCAN-1, ScaleSCAN, and SCAN-XP cannot handle clueweb on a single main memory, the results are omitted from Fig. 5. As we can see from Fig. 5, all the algorithms linearly improve their running time as the number of threads increases. Especially, DSCAN-16 shows a better

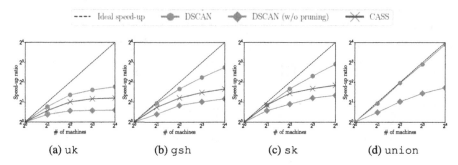

Fig. 6. Strong-scaling test by varying # of machines ($\epsilon = 0.5, \mu = 5$).

scalability than the other parallel algorithms. These results clarify that DSCAN is scalable against the number of threads.

Strong-Scaling Test: We then assessed the strong-scalability of distributed algorithms including DSCAN against the number of machines increases. To evaluate the impact of our pruning approach, we also tested *DSCAN (w/o pruning)* that lacks the skewness-aware edge-pruning. Figure 6 shows the speed-up ratio against the running time on one machine by varying the number of machines (1, 2, 4, 8, or 16). Note that results of CASS are omitted from union since CASS on one machine could not finish within 24 h. Similarly, we do not report the scalability on clueweb because all methods on one machine did not return results within 24 h.

As shown in Fig. 6, the performance of DSCAN increases almost linearly as the number of machines increases if the graphs are sufficiently large. By contrast, the performance of DSCAN (w/o pruning) and CASS peak earlier. By comparing the speed-up ratio between DSCAN and DSCAN (w/o pruning), DSCAN is approximately twice as efficient as DSCAN (w/o pruning). These results imply that our skewness-aware edge pruning successfully moderates the communication overheads, which degrades the scalability of distributed algorithms. As discussed in Sect. 3.2, DSCAN removes a large subset of given edges that cause the communication overheads by checking the degree skewness (Lemma 1). Hence, DSCAN shows a better scalability than DSCAN (w/o pruning) and a strong-scaling property for large graphs.

Weak-Scaling Test: For weak-scaling test, we generated synthetic graphs by using LFR benchmark with 16, 32, 64, 128, and 256 million nodes with an average degree 30. We compute those synthetic graphs by DSCAN and CASS on 1, 2, 4, 8, 16, and 32 machines, respectively. We also examined the running time of DSCAN w/o edge-pruning by using the same settings. Figure 7 shows the running time of each experimental setting. In Fig. 7, DSCAN keeps its running times almost constant even if we increase the number of nodes and machines. In contrast, DSCAN w/o edge-pruning and CASS gradually increase the running time as the number of machines and the graph size increased. As discussed in Sect. 3.2, DSCAN reduces the communication costs by using the skewness-aware edge-

Table 2. NMI varying ϵ

Parameters	uk	gsh	sk	union
$\epsilon = 0.3, \mu = 5$	1.0	1.0	1.0	1.0
$\epsilon = 0.6, \mu = 5$	1.0	1.0	1.0	1.0
$\epsilon = 0.9, \mu = 5$	1.0	1.0	1.0	1.0

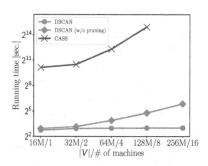

Fig. 7. Weak scaling test.

Table 3. NMI varying μ

Parameters	uk	gsh	sk	union
$\epsilon = 0.5, \mu = 5$	1.0	1.0	1.0	1.0
$\epsilon = 0.5, \mu = 10$	1.0	1.0	1.0	1.0
$\epsilon = 0.5, \mu = 20$	1.0	1.0	1.0	1.0

pruning. Hence, DSCAN shows better weak-scaling performances than DSCAN w/o edge-pruning and CASS.

4.4 Exactness of Clustering Results

Finally, we experimentally verified the exactness of the clustering results obtained by DSCAN. We used the information-theoretic metric, Normalized Mutual Information (NMI) [10], which outputs 1 if two clustering results are the same. As reported in [22], SCAN-XP returns the same clustering results as the original algorithm SCAN [23]. Thus, we measured the NMI scores between DSCAN and SCAN-XP.

Tables 2 and 3 show the NMI scores of DSCAN on various ϵ and μ settings. Since SCAN-XP did not finish clustering on `clueweb` due to the out of memory problem, the results are omitted. As shown in Table 2 and 3, DSCAN shows 1 for all conditions we examined. These results indicate that the clusters obtained by DSCAN are the same as the original algorithm SCAN even though DSCAN removes edges by skewness-aware edge pruning. As discussed in Lemma 1, our edge-pruning approach does not exclude edges that yield $\sigma(u,v) \geq \epsilon$. Thus, DSCAN does not sacrifice the clustering quality compared with the original algorithm.

5 Related Work

The structural graph clustering [23] is a fundamental tool to analyze complex data structures obtained from various applications [14,18]. Unlike traditional graph clustering algorithms [16], it can reveal not only densely connected clusters but also hubs and outliers from a given graph. As we described in Sect. 1, the original algorithm SCAN incurs $O(m^{1.5})$ time, which is known as the worst-case optimal cost. To overcome this expensive cost, several distributed and parallel algorithms have been proposed in a recent few years. Here, we briefly review several successful algorithms.

Parallel Algorithms: With the development of the many-core processors, parallel algorithms are one of the popular ways to efficiently perform large-scale structural graph clustering. SCAN-XP [22] and GPUSCAN [21] are the first parallel algorithms that provide scalable and exact algorithm for the structural graph clustering. The key ideas underlying those algorithms are to parallelize the structural similarity computation by using the many physical cores equipped on Intel Xeon Phi co-processors and GPUs. Shiokawa *et al.* [19] and Che *et al.* [4] further extended those algorithms by introducing several edge-pruning and optimization techniques. As a result, they succeeded in reducing the clustering time of SCAN for the real-world graphs. However, as we described in Sect. 1, the computation time for billion-edge graphs are still large, and they can not handle large graphs whose memory footprint exceeds the main memory size.

Distributed Algorithms: To address the large memory foot prints, the distributed algorithms are natural choices for improving the running time of SCAN. To the best of our knowledge, PSCAN [25] is the first solution that performs SCAN on distributed frameworks [20,24]; PSCAN also leads several extension works [8,11] in a recent few years. The common strategy among those distributed algorithms is to partition a given graph into disjoint subgraphs so that those subgraphs mitigate the communication overheads among distributed machines. However, this strategy is impractical for billion-edge graphs since graph partitioning [7] generally requires a large time- and space-complexity than those of SCAN. By contrast, DSCAN can reduce the communication overheads by the skewness-aware edge-pruning even though it does not use graph partitioning algorithms. Consequently, as we experimentally confirmed in Sect. 4, DSCAN shows better efficiency and scalability than those distributed algorithms.

6   Conclusion

We developed a novel distributed algorithm DSCAN for the structural graph clustering problem. DSCAN employs skewness-aware edge-pruning to reduce the communication overheads of the distributed algorithm. Our extensive experiments clarified that DSCAN outperforms the state-of-the-art methods without sacrificing clustering quality. Of particular interest, DSCAN successfully computes a graph whose memory footprint is 691 GiB within 1.5 min on 16 machines. The structural graph clustering is a fundamental graph data mining tool for current and prospective applications in various disciplines. By providing our efficient distributed method, DSCAN will help to improve the effectiveness of future applications.

Acknowledgement. This work was supported by JSPS KAKENHI Early-Career Scientists Grant Number JP18K18057, and JST ACT-I.

References

1. Blondel, V.D., Guillaume, J.L., Lambiotte, R., Mech, E.L.J.S.: Fast unfolding of communities in large networks. J. Stat. Mech.: Theory Experiment **2008**(10), P10008 (2008)
2. Boldi, P., Vigna, S.: The WebGraph framework I: compression techniques. In: Proceedings of the 13th International Conference on World Wide Web, pp. 595–601 (2004)
3. Chang, L., Li, W., Qin, L., Zhang, W., Yang, S.: pSCAN: fast and exact structural graph clustering. IEEE Trans. Knowl. Data Eng. **29**(2), 387–401 (2017)
4. Che, Y., Sun, S., Luo, Q.: Parallelizing pruning-based graph structural clustering. In: Proceedings of the 47th International Conference on Parallel Processing, pp. 77:1–77:10. ICPP (2018)
5. Cormen, T.H., Leiserson, C.E., Rivest, R.L., Stein, C.: Introduction to Algorithms. The MIT Press, Cambridge (2009)
6. Inoue, H., Ohara, M., Taura, K.: Faster Set Intersection with SIMD instructions by Reducing Branch Mispredictions. Proc. Very Learge Data Bases (PVLDB) **8**(3), 293–304 (2015)
7. Karypis, G., Kumar, V.: A fast and high quality multilevel scheme for partitioning irregular graphs. SIAM J. Sci. Comput. **20**(1), 359–392 (1998)
8. Kim, J., et al.: CASS: a distributed network clustering algorithm based on structure similarity for large-scale network. PLOS ONE 13(10), 1–22 (2018)
9. Lancichinetti, A., Fortunato, S., Radicchi, F.: Benchmark graphs for testing community detection algorithms. Phys. Rev. E **78**, 046110 (2008)
10. Manning, C.D., Raghavan, P., Schütze, H.: Introduction to Information Retrieval. Cambridge University Press, New York (2008)
11. Onizuka, M., Fujimori, T., Shiokawa, H.: Graph partitioning for distributed graph processing. Data Sci. Eng. **2**(1), 94–105 (2017)
12. ParMETIS – Parallel Graph Partitioning and Fill-reducing Matrix Ordering. http://glaros.dtc.umn.edu/gkhome/metis/parmetis/overview (2006–2008)
13. Rosvall, M., Axelsson, D., Bergstrom, C.T.: The map equation. The European Physical Journal Special Topics 178(1), 13–23 (2009)
14. Sato, T., Shiokawa, H., Yamaguchi, Y., Kitagawa, H.: FORank: fast objectrank for large heterogeneous graphs. Companion Proc. Web Conf. **2018**, 103–104 (2018)
15. Seo, J.H., Kim, M.H.: pm-SCAN: an I/O efficient structural clustering algorithm for large-scale graphs. In: Proceedings of the 2017 ACM on Conference on Information and Knowledge Management (CIKM 2017), pp. 2295–2298 (2017)
16. Shiokawa, H., Amagasa, T., Kitagawa, H.: Scaling Fine-grained modularity clustering for massive graphs. In: Proceedings of the Twenty-Eighth International Joint Conference on Artificial Intelligence, IJCAI-19. pp. 4597–4604 (2019)
17. Shiokawa, H., Fujiwara, Y., Onizuka, M.: SCAN++: efficient algorithm for finding clusters, hubs and outliers on large-scale graphs. Proc. Very Learge Data Bases **8**(11), 1178–1189 (2015)
18. Shiokawa, H., Onizuka, M.: Scalable graph clustering and its applications. Encyclopedia of Social Network Analysis and Mining, pp. 2290–2299 (2018)
19. Shiokawa, H., Takahashi, T., Kitagawa, H.: ScaleSCAN: scalable density-based graph clustering. In: Proceedings of the 29th International Conference on Database and Expert Systems Applications, pp. 18–34. DEXA (2018)
20. Shvachko, K., Kuang, H., Radia, S., Chansler, R.: The Hadoop distributed file system. In: IEEE 26th Symposium on Mass Storage Systems and Technologies (MSST 2010), pp. 1–10 (2010)

21. Stovall, T.R., Kockara, S., Avci, R.: GPUSCAN: GPU-based parallel structural clustering algorithm for networks. IEEE Trans. Parallel Distrib. Syst. **26**(12), 3381–3393 (2015)
22. Takahashi, T., Shiokawa, H., Kitagawa, H.: SCAN-XP: parallel structural graph clustering algorithm on intel xeon phi coprocessors. In: Proceedings of the 2nd International Workshop on Network Data Analytics, pp. 6:1–6:7 (2017)
23. Xu, X., Yuruk, N., Feng, Z., Schweiger, T.A.J.: SCAN: a structural clustering algorithm for networks. In: Proceedings of the 13th ACM SIGKDD International Conference on Knowledge Discovery and Data Mining, pp. 824–833 (2007)
24. Zaharia, M., Chowdhury, M., Franklin, M.J., Shenker, S., Stoica, I.: Spark: cluster computing with working sets. In: Proceedings of the 2nd USENIX Conference on Hot Topics in Cloud Computing, p. 10. HotCloud 2010, USENIX Association, USA (2010)
25. Zhao, W., Martha, V., Xu, X.: PSCAN: a parallel structural clustering algorithm for big network in MapReduce. In: Proceedings of the 2013 IEEE 27th International Conference on Advanced Information Networking and Applications (2013)

Accelerating All 5-Vertex Subgraphs Counting Using GPUs

Shuya Suganami[1]([✉]), Toshiyuki Amagasa[2], and Hiroyuki Kitagawa[2]

[1] Graduate School of SIE, University of Tsukuba, Tsukuba, Japan
suganami@kde.cs.tsukuba.ac.jp
[2] Center for Computational Sciences, University of Tsukuba, Tsukuba, Japan
{amagasa,kitagawa}@cs.tsukuba.ac.jp

Abstract. The subgraph counting problem is the problem of counting the number of occurrences of graph patterns in the target graph and is widely used as a fundamental technique for network analyses in different domains. The computational cost of subgraph counting grows drastically as the size of the pattern increases; it takes much time even with the state-of-the-art algorithms when counting 5-vertex patterns. To this problem, this paper proposes a subgraph counting method using GPUs. More precisely, we employ one of the state-of-the-art algorithms for 5-vertex subgraph counting and extend it so that counting is executed in parallel using massive threads. We conducted experiments for evaluating the performance of our proposed method by using real-world datasets, and the results demonstrate that our proposed method is about 4x to 10x and about 3× to 5× times faster than the original method in computing 5-vertex and 4-vertex subgraphs, respectively.

Keywords: Subgraph counting · GPU computing · Graphlet counting

1 Introduction

Given a set of patterns and a target graph, the subgraph counting is to count the number of occurrences of the pattern in the target graph. As opposed to the global features of a graph, the subgraph counts of a graph show a local feature and have been widely used in different problems, such as community detection [25], analysis of biological networks [10,24], and social networks [6,23], and others [7,21].

One of the biggest problems of subgraph counting is that it is computationally demanding. More precisely, the cost drastically increases as the size of patterns increases, especially when the subgraph pattern has more than five vertices. This is due to the combinatorial explosion; i.e., as described in [15], the frequencies of most 5-vertex subgraphs are more than billions in the graph, even with graphs with only millions of edges. Besides, to count the occurrences of a pattern, we need to manage the candidate patterns whose occurrences are far more than the target pattern, which deteriorates the performance.

© Springer Nature Switzerland AG 2020
S. Hartmann et al. (Eds.): DEXA 2020, LNCS 12391, pp. 55–70, 2020.
https://doi.org/10.1007/978-3-030-59003-1_4

For this reason, most of the existing work for subgraph counting dealt with up to 4-vertex subgraphs for exact cases [3, 13] and approximated cases [16, 26]. Recently, ESCAPE [15] addressed the problem of a 5-vertex subgraph counting by adopting the following techniques: (1) dividing patterns into smaller patterns and (2) conversion of edges to directed edges for reducing the search space.

Nevertheless, the performance of ESCAPE is not sufficient in particular when dealing with huge graphs, because of the long execution time. One promising way to address this problem is to apply parallelization. Especially, GPU (graphics processing unit) has shown remarkable progress in its performance and has been applied in a wide range of data-intensive workloads, including graph analysis. In this paper, we propose parallel subgraph counting methods for a GPU. We basically use the algorithm proposed in ESCAPE and parallelize it for GPU. To this end, we divide the algorithm into two parts; i.e., the part for extracting candidate patterns from the target graph and the part of computing aggregations based on the extracted candidates. The first part is executed on CPU, while the second part is executed in parallel on GPU. To our knowledge, among the GPU-based method for subgraph counting, this is the only method that can compute 5-vertex subgraph countings for large graphs. Our experiments using real-world datasets showed that our GPU-based methods could count all 5-vertex patterns up to 10× faster than ESCAPE. Besides, our methods can compute all 4-vertex patterns 3× to 5× faster than ESCAPE.

2   GPU Computing

"GPU computing" means using GPUs (graphics processing units) for general-purpose computing. Although GPUs have been originally designed for graphic processing, they have been used in different problem domains for accelerating tasks such as machine learning and scientific computation by its high parallelism.

In general, designing a parallel algorithm or converting non-parallel algorithms to parallel one is not easy. Moreover, it is even harder for us to make the best use of the GPUs' performance due to their characteristics, i.e., independent memory space, hierarchical memory structure, different programming environments, such as CUDA and OpenCL, etc.

In this work, we use NVIDIA's GPU and OpenACC [2] for GPU computing. Compared to CUDA, which has been widely used so far, OpenACC is a relatively new programming model and has been gathering attention due to its portability, maintainability, and productivity. The following describes the structure of NVIDIA's GPU and a program using OpenACC.

2.1   NVIDIA GPU

The GPU computing model consist of multiple SMs (Streaming Multiprocessors) which is composed of many SPs (Scalar Processors). SP is also called "CUDA core" or simply "core." For instance, the NVIDIA Tesla V100 GPU has 80 SMs, and each of them has 64 FP32 cores and 32 FP64 cores.

GPUs have roughly three types of memories: global memory, shared memory, and registers. Global memory is the largest memory on the GPU, and all SMs can access to it, but the latency is high. Meanwhile, shared memory is a memory with a capacity smaller than the global memory and can be accessed only from SPs on the same SM. The feature is that it presents much higher bandwidth and lower latency than global memory. Registers are the fastest memory among others, but the size is limited. So, to improve the performance, it is important to use registers and shared memory as much as possible, while reducing accesses to global memory.

2.2 OpenACC

OpenACC is a parallel programming standard for many-core accelerators [2]. Similar to OpenMP, OpenACC takes the directive-based programming model; programmers use directives to annotate their C/C++ or Fortran codes for parallelization. An OpenACC compatible compiler, like the PGI compiler, interprets the directives and generates code that can be executed in parallel by accelerators like GPU.

The OpenACC directives can be classified into three types: compute directives, data management directives, and loop directives. Compute directives specify a block of code to be executed in parallel on the accelerator. Data management directives instruct the program to transfer data from the host memory to the device memory on the accelerator and vice versa. Note that data transfer between the host and the device memory prone to be the bottleneck of the performance. So, we need to design the data transfer strategy on a parallel program carefully. Loop directives indicate how loop-iterations are distributed among different hardware components. This is another point to achieve high performance in an OpenACC program.

3 Preliminaries

In this paper, we focus on a connected, unweighted, and undirected graph $G = (V(G), E(G))$, where $V(G)$ and $E(G)$ are a set of vertices and edges of G, respectively. We assume that the graph does not contain any multiple edges or self-loops. We denote by $G^{\rightarrow}$ the directed graph. The set of out-neighbors and in-neighbors of $v \in V(G)$ of $G^{\rightarrow}$ is denoted by $N^+(v)$ and $N^-(v)$, respectively.

Figure 1 shows all possible graphs containing up to five vertices. We call them *patterns* and denote by H_i the i-th pattern; e.g., in Fig. 1, we refer to the 29th pattern (5-clique) by H_{29}. Besides, Table 1 shows the main notations used in the sequel discussion. To discuss subgraph counting, we must differentiate the induced subgraph from the subgraph. The following gives the concrete definitions.

Definition 1. (SUBGRAPH) *Given a graph $G = (V(G), E(G))$, a graph $G' = (V(G'), E(G'))$ is subgraph of G if $V(G') \subseteq V(G)$ and $E(G') \subseteq E(G)$.*

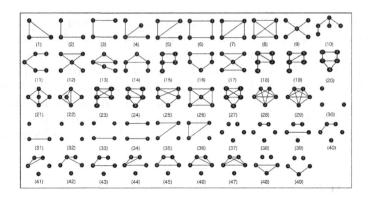

Fig. 1. All graphs with at most five vertices

Definition 2. (INDUCED SUBGRAPH) *Let $G' = (V(G'), E(G'))$ be a subgraph of $G = (V(G), E(G))$. G' is a induced subgraph of G if G' contains all edges $(u, v) \in E(G)$ with $u, v \in V(G')$. We also denote G' by $G[V(G')]$.*

We may call a non-induced subgraph a *subgraph* when there is no ambiguity.

In the following, we define the terminology used in this paper.

Definition 3. (ISOMORPHISM) *A graph $G' = (V(G'), E(G'))$ is said to be isomorphic to $G = (V(G), E(G))$ if there exists a bijection $\phi : V(G) \to V(G')$ such that $(u, v) \in E(G)$ iff $(\phi(u), \phi(v)) \in E(G')$ and such a map ϕ is called isomorphism.*

Definition 4. (AUTOMORPHISM) *Let $G = (V(G), E(G))$ be a graph. A automorphism is an isomorphism with itself, i.e., there exists a bijection $\phi : V(G) \to V(G)$ such that $(u, v) \in E(G)$ iff $(\phi(u), \phi(v)) \in E(G)$. The set of automorphisms of G is denoted by $AUT(G)$.*

Now, we introduce the subgraph counting and graphlet counting.

Definition 5. (SUBGRAPH COUNTING and GRAPHLET COUNTING) *Given a (potentially large) graph G and a set of mutually non-isomorphic graphs $\mathcal{G}$, the subgraph (graphlet) counting of $\mathcal{G}$ over G is to count for each $g \in \mathcal{G}$ the number of non-induced (induced) subgraphs G' in G that are isomorphic to g.*

3.1   Problem Statement

Our goal in this paper is, given a graph G and a set $\mathcal{H}$ of all 5-vertex patterns shown in Fig. 1, to compute the subgraph or graphlet counting as quick as possible

The result of the subgraph and graphlet counting of H_i is denoted by F_i and F_i^{IND}, respectively. As shown in [15], the subgraph counts of the disconnected patterns can be derived from the subgraph counts of connected patterns by a simple conversion. Besides, it demonstrates that we can convert all F_i into F_i^{IND} by a linear transformation. (We omit this matrix due to the space limitation.) Hence, it is sufficient to compute the subgraph counts for a graph G and a set of $\mathcal{H} = \{H_i | i \in \mathbb{N}, 1 \leqq i \leqq 29\}$.

Table 1. Notations

Symbol	Definition
G	Undirected graph
$G^\rightarrow$	Directed graph
$V(G)$	Set of vertices in G
$E(G)$	Set of edges in G
$N(v)$	Set of neighbors of vertex v
$d(v)$	Degree of vertex v
$G[S]$	Subgraph of G induced by S
$N^+(v)$	Set of out-neighbors of vertex v
$N^-(v)$	Set of in-neighbors of vertex v
$W(u,v)$	The number of wedges between vertices u, v
$W_{++}(u,v)$	The number of out-wedges between vertices u, v
$W_{+-}(u,v)$	The number of in-wedges from vertex v to vertex u
$T(u,v)$	Set of triangles incident to edge (u,v)
$T^+(u,v)$	Set of vertices w such that (u,v,w) is a triangle and $u,v \prec w$
$k_4^+(u,v,w)$	Set of vertex k such that (u,v,w) is a triangle and (k,u,v,w) is a 4-clique and $u,v,w \prec k$

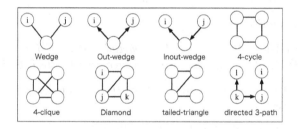

Fig. 2. Fundamental patterns

3.2 ESCAPE

We briefly introduce the algorithm of ESCAPE proposed by Pinar et al. [15], which is the base of our proposed method. ESCAPE is one of the state-of-the-art algorithms for counting all 5-vertex patterns based on the two main ideas for preventing the combinatorial explosion, which is one of the biggest problems in counting algorithms.

The first idea is cutting (or decomposing) patterns into smaller patterns; all connected k-vertex patterns except for k-clique can be split into smaller disconnected patterns with size less than k by removing some vertices (called cutset), thereby making it possible to count the frequency as a combination of smaller patterns. More details can be found below.

The second idea is to convert undirected graph G into the DAG $G^{\rightarrow}$ to reduce the search space. This idea has been employed in many triangle counting algorithms [5,22]. More precisely, we get $G^{\rightarrow}$ from G as follows. Note here that the degree ordering is denoted by $\prec$. For vertex u and v, we order $u \prec v$ if either $d(u) < d(v)$ or $d(u) = d(v)$ and $u < v$ (i.e., order by vertex ID). Thus, undirected graph G covert into directed graph $G^{\rightarrow}$ by orienting all edges following the degree ordering $\prec$. Since degree ordering is a total order, $G^{\rightarrow}$ is DAG.

The Cutting Framework. We describe the cutting framework. Let H be a pattern we want to count and C be a cutset of H. For convenience, we assume that G and H are labeled, although they are actually unlabeled. Actually, at the final step, we compensate the difference between labeled and unlabeled graphs, thereby allowing us to compute the correct answer.

We define match[1] on follows.

Definition 6. (MATCH) *A match of H in G is a bijection $\pi : S \rightarrow V(H)$ where $S \subseteq V(G)$ and $\forall u, v \in S$ is an edge of G if $(\pi(u), \pi(v)) \in E(H)$. The set of disjoint match of H in G is denoted by match(H).*

If π is only an injection (i.e., $|S| < |V(H)|$), we call π a partial match.

Definition 7. (EXTEND) *A match $\pi : S \rightarrow V(H)$ extends a partial match $\sigma : S' \rightarrow V(H)$ if $S' \subset S$ and $\forall u \in S', \pi(u) = \sigma(u)$.*

Next, we define H-degree.

Definition 8. (H-DEGREE) *Let σ be a partial match of H in G. We call the number of matches of H that extend σ H-degree and denote it by $Deg_H(\sigma)$.*

The fragment of G is obtained by splitting H into smaller patterns, which is defined as follows:

Definition 9. (FRAGMENT) *Let H be divided into connected components $S_1, S_2, \ldots$ by removing the vertices of C. The C-fragments of H are the subgraphs of H induced by $C \cup S_1, C \cup S_2, \ldots$. The set of C-fragments of H is denoted by $Frag_C(H)$.*

Let us consider a match σ of $H[C]$. If σ can extend to all elements of $Frag_C(H)$ and these elements are disjoint with each other, we can extend σ to H by merging these elements. When these elements of $Frag_C(H)$ are not disjoint, merging these elements result in patterns that are different from H, i.e., H', which we call a shrinkage.

Definition 10. (C-SHRINKAGE) *Let H' be a pattern different from H and $Frag_C(H) = \{F_1, F_2, \ldots F_{|Frag_C(H)|}\}$. A C-shrinkage of H into H' is a set of*

---

[1] We have slightly changed the definition from the original one in the ESCAPE paper to maintain the consistency of the theorem.

maps $\{\sigma, \pi_1, \pi_2, \ldots, \pi_{|Frag_C(H)|}\}$ such that $\sigma : H[C] \to H'$ is a partial match of H' and $\forall F_i \in Frag_C(H)$, $\pi_i : F_i \to H'$ is a partial match of H' extends σ and $\forall (u', v') \in V(H')$, there are some index $i \in |Frag_C(H)|$ and $u, v \in F_i$ such that $\pi_i(u) = u'$ and $\pi_i(v) = v'$. The $Shrink_C(H)$ denotes the set of patterns H' ($\neq H$) such that there exists at least one C-shrinkage of H in H'. For $H' \in Shrink_C(H)$, the number of disjoint C-shrinkage is denoted by $numSh_C(H, H')$.

Now, we are ready to introduce main lemma:

Lemma 1.

$$|match(H)| = \sum_{\sigma \in match(H[C])} \prod_{F \in Frag_C(H)} deg_{F(\sigma)}$$

$$- \sum_{H' \in Shrink_C(H)} numSh_C(H, H') \cdot match(H')$$

This lemma implies that we can compute $|match(H)|$ if we find the following three things: $deg_F(\sigma)$ for every copy of $H[C]$, for every C-fragment, the counts of every possible shrinkage.

Conversion from a Match of H to a Subgraph Count of H. As mentioned above, we consider H and G to be labeled graphs. In practice, however, H and G are unlabeled graphs. Hence, in order to obtain subgraph counts from $|match(H)|$, we have to divide $|match(H)|$ by $AUT(H)$.

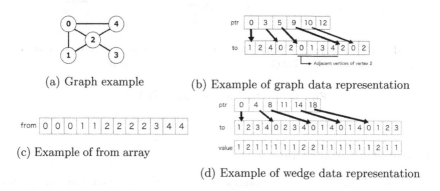

(a) Graph example

(b) Example of graph data representation

(c) Example of from array

(d) Example of wedge data representation

Fig. 3. Example of data representation

4   Proposed Method

In this section, we introduce our proposed method. We aim to accelerate the subgraph counting by parallelizing the counting part of ESCAPE using GPU. To this end, we reorganize the method of ESCAPE by introducing two stages. The first stage is to extract candidate patterns from the target graph and construct a dedicated data structure on CPU. Then, in the second stage is to compute aggregations based on the extracted candidates on GPU in parallel. This design is due to the separated memory space between CPU and GPU.

4.1 Data Structure

We introduce a data structure used in this paper. To store graph data, we adopt a data structure based on CSR (compressed sparse row), which is a common data structure to store graph data. A graph is stored using *ptr* array and *to* array when the graph is unweighted. *to* array records the adjacent vertices of all vertex, and each element in *ptr* array points the border of the list of adjacent vertices belonging to the neighbor vertices in *to* array. In Fig. 3, for example, the graph in Fig. 3a is represented as CSR format (Fig. 3b).

In this work, in addition to the arrays mentioned above, we use *from* array that stores reverse edge information. This is necessary for us to process the graph edge-centric manner rather than a node-centric manner, thereby allowing us to achieve better load balancing among different threads. This feature is important when dealing with power-law graphs where degree distribution follows a power-law distribution.

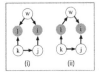
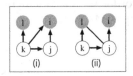

(a) All directed acyclic of ths 5-cycle

(b) All shrinkage of 5-cycle DAGs

(c) DAG version of 5-clique

Fig. 4. Directed patterns and shrinkages

4.2 Counting Algorithm

This section describes how to compute 5-vertex subgraph counting, and hence we assume that the result of up to 4-vertex subgraph counting is ready.

First Step: Extracting Candidate Patterns. The first step is to extract from the graph three candidate patterns, namely, triangle, wedge, and diamond. More precisely, we maintain the following information for each candidate. Regarding triangles, for each edge, we maintain all vertices that form triangles. Regarding wedges, for each vertex, we maintain all vertices corresponding to in- and out-wedges. Regarding diamonds, for each wedge, we maintain all vertices that form a diamond. For example, in Fig. 2, we can see a vertex that forms an out-wedge

---

Algorithm 1. Counting H_{16} (5-cycle) by GPU

```
1: F₁₆ ← 0
2: for each (i, j) ∈ E do in parallel // i ≺ j
3:     for all k ∈ N⁻(j) do // k ≺ j
4:         for all l ∈ N⁺(k) do // k ≺ l
5:             if vertex i, l is out-wedge or inout-wedge then
6:                 F₁₆ ← F₁₆ + W⁺⁺(i, l) + W⁺⁻(i, l)
7:             end if
8:             if i ∈ N(k) then // i, k is an edge.
9:                 F₁₆ ← F₁₆ - 1
10:            end if
11:            if j ∈ N(l) then // j, l is an edge.
12:                F₁₆ ← F₁₆ - 1
13:            end if
14:        end for
15:    end for
16: end for
```

---

for vertex i is the ones that correspond to vertex j. Notice that we do not care whether i and j is connected or not.

We maintain the above information using CSR format. For example, let us assume wedges. For counting the number of candidates, we need to maintain not only the connectivity between two vertices but also the number of possible wedges between them. This information cannot be stored only using *ptr* and *to* arrays. In this work, we additionally use *value* array to store the count information. For example, the wedges in the graph in Fig. 3a can be represented as shown in Fig. 3d; e.g., vertex 0 and 2 form a couple of wedges, $(0, 1, 2)$ and $(0, 4, 2)$. So, the value in the *value* array is 2. Similarly, we store information related to diamonds using CSR format, but it requires *value* array only.

Second Step: Computing Aggregations. Having constructed the data structure, we transfer it to GPU and process it in parallel on GPU.

In this work, we carefully designed algorithms for all 5-vertex patterns, but we cannot describe all of them due to the page limitation. So, we will describe algorithms for computationally heavy patterns, i.e., H_{16}, H_{25}, H_{28}, and 5-clique (H_{29}).

H_{16} *(5-cycle):* If we ignore the symmetric cases, there are two patterns of DAGs, as shown in Fig. 4a. If we let i, l be the cutset, then we get directed 3-path and a wedge as the fragments where there are two different directions in the wedge. Thus, we find these patterns from $G^{\rightarrow}$. The algorithm is shown in Algorithm 1.

H_{25} *(Diamond-wedge):* The fragments of diamond-wedge is a diamond and a wedge. Consequently, we search for them from the graph. Algorithm 2 shows the algorithms. Line 2 to 4 is to find i, j, and k that form a diamond, followed by checking whether i and j form a wedge in Line 5. If it forms a wedge, then we count up the counter by $W(i, l)$.

H_{28} *(Almost-5clique):* Algorithm 3 shows the algorithm for counting H_{28} (almost-5clique). It first identifies the vertices k that form triangles with the edge (i, j) in Line 3. In Line 5, identifies vertices l that form triangles with the

Algorithm 2. Counting H_{25} (diamond-wedge) by GPU

```
1: F_25 ← 0
2: for each (i, j) ∈ E do in parallel
3:     for all k ∈ T(i, j) do
4:         for all l ∈ T(j, k) do
5:             if vertex i, l is a wedge then
6:                 F_25 ← F_25 + W(i, l)
7:             end if
8:         end for
9:     end for
10: end for
```

Algorithm 3. Counting H_{28} (almost-5clique) by GPU

```
1: F_28 ← 0
2: for each (i, j) ∈ E do in parallel
3:     for all k ∈ T(i, j) do
4:         fourClique ← 0
5:         for all l ∈ T(i, j) do
6:             if k ≺ j and i ≺ k and k ∈ N(l) then
7:                 fourClique ← fourClique + 1
8:             end if
9:         end for
10:        F_28 ← F_28 + (fourClique choose 2)
11:    end for
12: end for
```

Algorithm 4. Counting H_{29} (5-clique) by GPU

```
1: F_29 ← 0
2: for each (i, j) ∈ E do in parallel
3:     for all k ∈ T^+(i, j) do
4:         // we assume that k_4^+(i, j, k) has already sorted by degree ordering
5:         for all l_u ∈ k_4^+(i, j, k) do
6:             for all l_v ∈ {l_{u+1}, ... l_{|k_4^+(i,j,k)|}} do
7:                 if l_v ∈ N(l_u) then
8:                     F_29 ← F_29 + 1
9:                 end if
10:            end for
11:        end for
12:    end for
13: end for
```

edge (i, j), and Line 6 checks the condition; if it passes, variable fourClique is incremented. Line 10 increments F_{28} according to the formula.

H_{29} *(5-clique):* Since we cannot cut a clique into smaller fragments, we only exploit edge direction as the clue to count the occurrences. If we ignore symmetric cases, there exists only one pattern (Fig. 4c). So, we find this from $G^{\rightarrow}$. Algorithm 4 shows the algorithms. We assume that 4-clique $k_4^+(i, j, k)$ are already identified. Line 2 to 4 search for triangles (i, j, k) such that $i \prec j \prec k$. Line 5 and 6 choose two vertices from $k_4^+(i, j, k)$ such that $l_u \prec l_v$. In Line 5, if there exists an edge between $l_u \prec l_v$, then $\{i, j, k, l_u, l_v\}$ forms a 5 clique.

4.3 Handling Huge Graphs

Although CSR format is a space-efficient data structure, when processing huge graphs containing massive wedges or diamonds, the data cannot be loaded on the GPU's device memory. For such graphs, we partition the array for wedges (or diamonds) and send each partition to GPU one by one. GPU process partial data and the final part of the program collects the partial results and aggregates them for generating the final result.

Table 2. Elapsed time for subgraph counting (Second).

Dataset	$|V|$	$|E|$	$|T|$	ESCAPE-4	Proposal-4	ESCAPE-5	Proposal-5
soc-brightkite	56.7K	426K	494K	0.13	0.03	7.16	1.79
soc-pokec	1.63M	22.3M	32.6M	36.05	11.2	1.79K	174.1
tech-as-skitter	1.69M	28.8M	28.8M	11.3	2.68	1.27K	321.5
web-wiki-ch-internal	1.93M	8.5M	18.2M	12.8	3.87	1.73K	210.4
web-hudong	1.98M	14.43M	21.6M	22.2	5.37	2.55K	396.7
web-baidu-baike	2.14M	17.01M	25.2M	27.1	9.41	3.61K	596.7
tech-ip	2.25M	21.6M	2.3M	60.8	11.67	-	18.1K

5 Experiments

To test the performance of the proposed scheme, we have conducted a set of experiments. We have implemented our proposed method using C++ compiled by pgc++ 18.5-0.

5.1 Experimental Setup

As we will discuss later in related work, there are some subgraph counting algorithms up to five vertices, such as ORCA [9] and ESCAPE. However, ORCA takes plenty of time to execute 5-vertex subgraph counting. Therefore, we chose ESCAPE, the state-of-the-art method for 4- and 5-vertex subgraph counting, as the comparative method. The implementation of ESCAPE is based on the code provided by the authors of the original paper [1], which runs on a CPU with single thread. ESCAPE is compiled with g++ (GCC) 4.8.5. Notice that we also tested pgc++ compiler to compile ESCAPE, which turned out to be slower than g++.

For running the codes, we used a Linux server with Intel(R) Xeon(R) CPU E5-2660 v4 (2.00 GHz) and 64 GB memory running Red Hat Enterprise Linux Workstation release 7.7. For GPU, we used NVIDIA Tesla V100 with 32 GB device memory. Besides, the version of OpenACC is 2.6.

5.2 Dataset

We used a couple of graph datasets, namely, Citation Network Dataset [19] and SNAP [12]. For the dataset with directed edges, we ignored the edge direction and removed self-loops. Table 2 summarizes the characteristics of the datasets.

5.3   Experimental Results

Execution Time. Table 2 shows the comparison of the execution time of ESCAPE and the proposed scheme. The excution time is average of three runs. In the table, the suffix denotes the size of the pattern, e.g.,"proposal-4" means the execution time for processing 4-vertex subgraph counting. Notice that the result of ESCAPE-5 for tech-ip is not shown due to the long-running time.

As can be seen from the results, the proposed scheme significantly outper-formed ESCAPE for both 4- and 5-vertex subgraph counting. The proposed scheme was about 5× faster than ESCAPE for 4-vertex subgraph counting on tech-ip dataset with 2.25M vertices and 21.6M edges.

Interestingly, the proposed scheme took almost the same time (11 s) for soc-pokec and tech-ip datasets while ESCAPE took about 36 s and 60 s soc-pokec and tech-ip datasets, respectively. This result implies that, for the proposed scheme, most of the time was spent on finding wedges by CPU. More precisely, it took almost the same time to find wedges in soc-pokec and tech-ip. Mean-while, the proposed scheme successfully reduced the time required for subsequent counting phase by GPU-based parallelization (less than one second), while the counting phase took longer in ESCAPE, resulting in the different execution times between the datasets.

As for 5-vertex subgraph counting, the proposed scheme was about 10× faster than ESCAPE for soc-pokec with 1.63M vertices and 22.3M edges, while the proposed scheme showed the least speed-up rate (about 4×) with tech-as-skitter dataset with 1.69M vertices and 28.8M edges. One of the reasons for the different speed-up rates is the number of substructures, such as wedges and diamonds, differs depending on the datasets; i.e., a large number of substructures lead to long execution time.

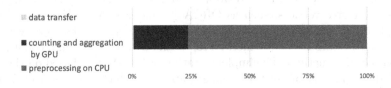

Fig. 5. Time breakdown of the proposed scheme for 5-vertex subgraph counting (tech-as-skitter: 1.69M vertices, 28.8M edges).

5.4   Bottleneck Analysis

To further discuss the bottleneck of the proposed scheme, we have measured the time required for different processes in the algorithm. Figure 5 shows the result. We measured the time for 5-vertex subgraph counting on tech-as-skitter dataset and measured the time for (1) preprocessing on CPU for finding substructures, such as wedges and diamonds, (2) counting and aggregation by GPU, and (3) data transfer between CPU and GPU memory. As we can see from the figure, the preprocessing on CPU took the longest time. So, to further improve the performance, we need to speed up the process of finding substructures, which is a part of our future work.

6   Related Work

The problem of subgraph counting, which is to count the number of (small) subgraphs called patterns in the target graph, has been well studied [18] and applied in various problem domains, e.g., computer science and bioinformatics. Due to the page limitation, we only review the ones that are most relevant to this work, i.e., we discuss the methods that deal with 4- or 5-vertex subgraph counting, while we do not mention triangle counting.

4-vertex Subgraph Counting. For the method of 4-vertex subgraph counting, there have been many works, such as RAGE by Marcus et al. [13], Ortman et al. [14], and PGD by Ahmed et al. [3]. PGD is relatively faster than others by edge-based parallelization, which requires less than 1 h for graphs with 10M edges.

There have been methods that exploit GPUs. Rossi [20] proposed a method that computes not only the number of occurrences of 4-vertex subgraph patterns but also the number of occurrences of 4-vertex patterns for each edge.

5-vertex Subgraph Counting. There are some works for 5-vertex subgraph counting, such as ORCA by Hočeva et al. [9] and ESCAPE by Pinar et al. [15]. Note that ORCA not only computes the number of occurrences of each 5-vertex pattern but also computes, for each vertex, the occurrences of each subgraph pattern. However, it is too slow to deal with huge graphs. In the meantime, ESCAPE is regarded as the state-of-the-art method. Besides, ESCAPE is faster than PGD for counting 4-vertex subgraph patterns. Our proposed method is based on ESCAPE.

6.1   Approximate Methods

The above methods output exact counts of subgraph patterns and are called exact methods. On the other hand, there is another line of researches on appropriate subgraph counting that output approximate counts of subgraph patterns. Such methods are needed due to the high computational cost of exact methods, and, in some applications, an exact number of counts is not necessary.

There have been several works on approximate 4-vertex subgraph counting, such as Elenberg et al. [8] and Madhav et al. [11]. In particular, the method by Elenberg et al. can compute more accurate results than GUISE and GRAFT, which we will mention below, and is faster than them by tens to hundred times.

For approximate 5-vertex subgraph counting, there have been works by GUISE by Bhuiyan et al. [4], GR AFT by Rahman et al. [17], and MOSS-5 by Wang et al. [26]. Among this category, MOSS-5 is the fastest and is hundreds to thousand times faster than GRAFT and GUISE and more accurate as well.

7   Conclusion

In this paper, we have proposed a parallel subgraph counting methods up to five vertices for a GPU. Our proposed methods consist of two parts. First part is a precomputation on CPU and second part is a counting and aggregation in parallel by GPU. Experiments show that our proposed methods can count all 5-vertex subgraphs up to 10 times faster and all 4-vertex subgraphs 3 to 5 times faster than one of the state-of-the-art algorithm for 4- and 5-vertex subgraph counting. In the future, we will improve first part which is the current bottleneck by executing in parallel, and also we will compare the results with other GPUs like GTX.

Acknowledgement. We thank to Prof. Hiroaki Shiokawa and Prof. Ryohei Kobayashi at the Center for Computational Sciences, University of Tsukuba for their useful discussions and support. This research was supported (in part) by Multidisciplinary Cooperative Research Program in CCS, University of Tsukuba.

Appendix

A   Counting Other Subgraphs

Table 3. Additional Notation

Symbol	Definition
$t(v)$	The number of triangles incident to vertex v
$C_4(G)$	The number of 4-cycle in graph G
$C_4(v)$, $C_4(u,v)$	The number of 4-cycle incident to vertex v, and edge (u,v)
$K_4(u)$, $K_4(u,v)$	The number of 4-clique incident to vertex v, edge (u,v)
$D(G)$	The number of diamond in graph$G$
$TT(G)$	The number of tailed-triangle in graph G

For other subgraphs, we can use simple formulas as described below. We can easily parallelize them.

$$F_3 = \sum_{(i,j \in E)} (d(i)-)(d(j)-1) - 3F_1, \quad F_4 = \sum_{i \in V} \binom{d(i)}{3}, \quad F_5 = \sum_{i \in V} t(i)(d(i)-2),$$

$$F_6 = \sum_{i \in V} \sum_{j \prec i} \binom{W_{++}(i,j) + W_{+-}(i,j)}{2}, \quad F_7 = \sum_{(i,n) \in E} \binom{|T(i,j)|}{2}, \quad F_9 = \sum_{i \in V} \binom{d(i)}{4}$$

$$F_{10} = \sum_{(i,j) \in E} [(d(i)-1)\binom{d(j)-1}{2} + (d(j)-1)\binom{d(i)-1}{2}] - 2F_5$$

$$F_{11} = \sum_{i \in V} \sum_{(i,j) \in E} (d(j)-1) - 4F_6 - F_5 - 3F_1, \quad F_{12} = \sum_{i \in V} t(i)(d(i)-2)$$

$$F_{13} = \sum_{(i,j) \in E} ((d(i)-1)t(j) + (d(j)-1)t(i)) - 4F_7 - 6F_1 - 2F_5$$

$$F_{14} = \sum_{(i,j) \in E} |T(i,j)|(d(i)-2)(d(j)-2) - 2F_7, \quad F_{15} = \sum_{i \in V} C_4(i)(d(i)-2) - 2F_7$$

$$F_{17} = \sum_{i \in V} \binom{t_i}{2} - 2F_7, F_{18} = \sum_{(i,j) \in E} \sum_{k \in T(i,j)} (d(k)-2)(|T(i,j)|-1) - 12F_8$$

$$F_{19} = \sum_{(i,j) \in E} \binom{|T(i,j)|}{2}(d(i)+d(j)-6), \quad F_{20} = \sum_{(i,j) \in E} C_4(i,j)|T(i,j)| - 4F_7$$

$$F_{21} = \sum_{i \in V} \sum_{i \prec j} \binom{W(i,j)}{3}, \quad F_{22} = \sum_{(i,j) \in E} \binom{|T(i,j)|}{3}, \quad F_{23} = \sum_{i \in V} K_4(i)(d(i)-3)$$

$$F_{24} = \sum_{(i,j) \in E} \sum_{k \in T(i,j)} ((|T(i,j)|-1)(|T(i,k)|-1)$$
$$+ (|T(i,j)|-1)(|T(j,k)|-1) + (|T(j,k)|-1)(|T(i,k)|-1))$$

$$F_{26} = \sum_{u < |diamondValue[u]|} \binom{diamonValue[u]}{2}, \quad F_{27} = \sum_{(i,j) \in E} K_4(i,j)(|T(i,j)|-2)$$

References

1. Escape. https://bitbucket.org/seshadhri/escape
2. Openacc-standard.org. https://www.openacc.org/sites/default/files/inline-images/Specification/OpenACC.3.0.pdf
3. Ahmed, N.K., Neville, J., Rossi, R.A., Duffield, N.: Efficient graphlet counting for large networks. In: 2015 IEEE International Conference on Data Mining, pp. 1–10. IEEE (2015)
4. Bhuiyan, M.A., Rahman, M., Rahman, M., Al Hasan, M.: Guise: uniform sampling of graphlets for large graph analysis. In: 2012 IEEE 12th International Conference on Data Mining, pp. 91–100. IEEE (2012)
5. Chiba, N., Nishizeki, T.: Arboricity and subgraph listing algorithms. SIAM J. Comput. **14**(1), 210–223 (1985)
6. Choobdar, S., Ribeiro, P., Bugla, S., Silva, F.: Comparison of co-authorship networks across scientific fields using motifs. In: 2012 IEEE/ACM International Conference on Advances in Social Networks Analysis and Mining, pp. 147–152. IEEE (2012)
7. Durak, N., Pinar, A., Kolda, T.G., Seshadhri, C.: Degree relations of triangles in real-world networks and graph models. In: Proceedings of the 21st ACM International Conference on Information and Knowledge Management, pp. 1712–1716 (2012)
8. Elenberg, E.R., Shanmugam, K., Borokhovich, M., Dimakis, A.G.: Distributed estimation of graph 4-profiles. In: Proceedings of the 25th International Conference on World Wide Web, pp. 483–493 (2016)
9. Ho č evar, T.ž., Dem š ar, J.: A combinatorial approach to graphlet counting. Bioinformatics **30**(4), 559–565 (2014)

10. Hayes, W., Sun, K., Pr ž ulj, N.š.a.: Graphlet-based measures are suitable for biological network comparison. Bioinformatics **29**(4), 483–491 (2013)
11. Jha, M., Seshadhri, C., Pinar, A.: Path sampling: a fast and provable method for estimating 4-vertex subgraph counts. In: Proceedings of the 24th International Conference on World Wide Web, pp. 495–505 (2015)
12. Leskovec, J., Krevl, A.: SNAP Datasets : Stanford large network dataset collection. http://snap.stanford.edu/data June 2014
13. Marcus, D., Shavitt, Y.: Rage-a rapid graphlet enumerator for large networks. Comput. Netw. **56**(2), 810–819 (2012)
14. Ortmann, Mark, Brandes, Ulrik: Quad census computation: simple, efficient, and orbit-aware. In: Wierzbicki, Adam, Brandes, Ulrik, Schweitzer, Frank, Pedreschi, Dino (eds.) NetSci-X 2016. LNCS, vol. 9564, pp. 1–13. Springer, Cham (2016). https://doi.org/10.1007/978-3-319-28361-6_1
15. Pinar, A., Seshadhri, C., Vishal, V.: Escape: efficiently counting all 5-vertex subgraphs. In: Proceedings of the 26th International Conference on World Wide Web, pp. 1431–1440. International World Wide Web Conferences Steering Committee (2017)
16. Rahman, M., Bhuiyan, M., Hasan, M.A.: Graft: an approximate graphlet counting algorithm for large graph analysis. In: Proceedings of the 21st ACM International Conference on Information and Knowledge Management, pp. 1467–1471 (2012)
17. Rahman, M., Bhuiyan, M.A., Al Hasan, M.: Graft: an efficient graphlet counting method for large graph analysis. IEEE Trans. Knowl. Data Eng. **26**(10), 2466–2478 (2014)
18. Ribeiro, P., Paredes, P., Silva, M.E., Aparicio, D., Silva, F.: A survey on subgraph counting: concepts, algorithms and applications to network motifs and graphlets. arXiv preprint arXiv:1910.13011 (2019)
19. Rossi, R.A., Ahmed, N.K.: The network data repository with interactive graph analytics and visualization. In: AAAI (2015). http://networkrepository.com
20. Rossi, R.A., Zhou, R.: Leveraging multiple GPUS and CPUS for graphlet counting in large networks. In: Proceedings of the 25th ACM International on Conference on Information and Knowledge Management, pp. 1783–1792. ACM (2016)
21. Rupp, M., Schneider, G.: Graph kernels for molecular similarity. Molecular Inform. **29**(4), 266–273 (2010)
22. Schank, Thomas, Wagner, Dorothea: Finding, counting and listing all triangles in large graphs, an experimental study. In: Nikoletseas, Sotiris E. (ed.) WEA 2005. LNCS, vol. 3503, pp. 606–609. Springer, Heidelberg (2005). https://doi.org/10.1007/11427186_54
23. Shizuka, D., McDonald, D.B.: A social network perspective on measurements of dominance hierarchies. Animal Behav. **83**(4), 925–934 (2012)
24. Sporns, O., Kötter, R.: Motifs in brain networks. PLoS Biology **2**(11) 56–62 (2004)
25. Tsourakakis, C.E., Pachocki, J., Mitzenmacher, M.: Scalable motif-aware graph clustering. In: Proceedings of the 26th International Conference on World Wide Web, pp. 1451–1460 (2017)
26. Wang, P., Zhao, J., Zhang, X., Li, Z., Cheng, J., Lui, J.C., Towsley, D., Tao, J., Guan, X.: Moss-5: a fast method of approximating counts of 5-node graphlets in large graphs. IEEE Trans. Knowl. Data Eng. **30**(1), 73–86 (2017)

PhoeniQ: Failure-Tolerant Query Processing in Multi-node Environments

Yutaro Bessho[1]([✉]) [iD], Yuto Hayamizu[1] [iD], Kazuo Goda[1] [iD],
and Masaru Kitsuregawa[1,2] [iD]

[1] The University of Tokyo, 7-3-1 Hongo, Bunkyo-ku, Tokyo, Japan
{bessho,haya,kgoda,kitsure}@tkl.iis.u-tokyo.ac.jp
[2] National Institute of Informatics, 2-1-2 Hitotsubashi, Chiyoda-ku, Tokyo, Japan

Abstract. Parallel processing is a flagship approach for answering analytical queries on large-scale database. As the database scale increases, a larger number of processing nodes are likely to be incorporated to increase the degree of parallelism. However, this solution results in an increased probability of node failure. If such a failure happens during query processing, the processing often has to restart from scratch. This temporal cost may not be acceptable for the user. In this paper, we propose PhoeniQ, a fault-tolerant query processing mechanism for analytical parallel database systems. PhoeniQ takes a package-level checkpoint for every operator pipeline and replicates the output of stateful operators among different processing nodes. If a single processing node fails during processing, another node is enabled to resume the execution state of the failed node, so that the query can continue to run. This paper presents our intensive experiments based on our prototype, which demonstrate that PhoeniQ can continue the query processing in the face of node failures with significantly smaller cost than the conventional approach.

Keywords: Parallel database system · Fault tolerance · Query processing

1 Introduction

A wide spectrum of big data applications have spurred the growth of database capacity. Petabyte-scale databases are no longer uncommon, especially in cloud-scale companies [6,11,20,23]. The trend of utilizing IoT sensor data is likely to boost the growth further [2].

Parallel query processing is a standard tactic to service analytical queries on large database [5,13,15]. A parallel database system is composed of multiple processing nodes, each of which executes the processing of an assigned part of a given query in parallel. This approach has been actively studied in academia and widely deployed in industry.

Y. Bessho—Currently, he works for NTT.

© Springer Nature Switzerland AG 2020
S. Hartmann et al. (Eds.): DEXA 2020, LNCS 12391, pp. 71–85, 2020.
https://doi.org/10.1007/978-3-030-59003-1_5

A major drawback of such parallel processing is that query processing becomes vulnerable to node failures [17]. As database accommodates larger data, an increased number of processing nodes are often incorporated into the database system. This approach performs well to increase the parallelism. At the same time, it causes a higher aggregate probability of node failure. A relational query is often composed of one or more pipelined operators, each of which can hold an internal execution state. For example, an aggregation operator keeps its in-process data in the memory buffer. If a processing node fails during query processing, the database system loses such an execution state held in the failed node. The system needs to restart the query from ground zero, no matter how far the process has progressed at the point of failure. This time penalty is likely to be unacceptable, particularly for users who run hour-long or day-long analytical queries.

This paper proposes PhoeniQ, a novel fault-tolerant query processing mechanism for analytical parallel database systems. The technical points of PhoeniQ are two-fold. First, PhoeniQ takes a package-level checkpoint for every operator pipeline. Second, it replicates the output of stateful operators among different processing nodes. If a single processing node fails during query processing, another node is enabled to resume the execution state of the failed node, so that the query can continue to run. This paper presents an intensive experiment that we performed with our prototype in a public cloud infrastructure. The experimental result demonstrates that PhoeniQ can continue the query processing in the face of node failure with a significantly smaller time penalty than the conventional approach.

The rest of this paper is structured as follows. Section 2 presents a design overview of PhoeniQ, and Sect. 3 offers a technical deep-dive. Section 4 provides prototype-based experiments in a public cloud environment. Section 5 reviews related work and Sect. 6 concludes the paper.

2   Overview of PhoeniQ

First of all, we present a design overview of PhoeniQ, a novel fault-tolerant query processing mechanism for analytical parallel database systems. Figure 1 highlights PhoeniQ by comparing it with the conventional execution mechanism. In this paper, we assume a shared-storage architecture [14] for simplicity[1]. As Fig. 1(a) illustrates, a parallel database system is composed of a single storage node that stores the entire database and multiple processing nodes that process query operators by fetching data from the storage node. According to the query execution plan generated from a given query, a set of pipelined operators are assigned to each node. The first operator in the pipeline is mostly a scan operator that fetches tuples from the storage node, processes them, and passes output tuples to its next operator. The next operator similarly processes received tuples and passes its output tuples to its next operator. Such data flow may travel over

---

[1] The idea of PhoeniQ can be easily extended to a shared-nothing architecture [26]. Due to the space limitation, we will present further discussion in a separate paper.

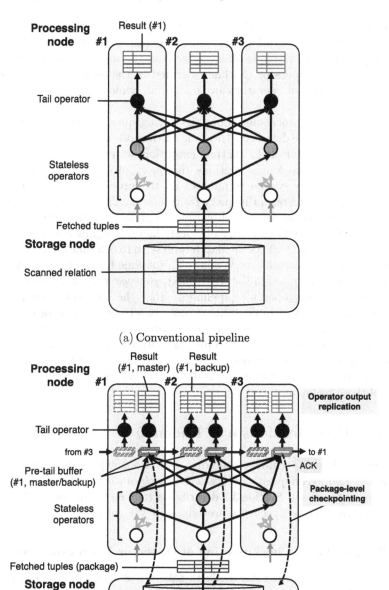

(a) Conventional pipeline

(b) PhoeniQ pipeline

Fig. 1. Execution comparison. PhoeniQ takes a package-level checkpoint for every operator pipeline and replicates the output of tail (sometimes stateful) operators among different processing nodes.

multiple processing nodes via the network connection. Finally, the last operator (which we refer to as the *tail operator* in this paper) generates the result of the operator pipeline; the result is often buffered in the memory to be shipped to the user or another operator pipeline or stored in the memory to be accessed later. Note that the tail operator can be stateful, while the other operators are stateless.

Assume that one processing node fails during query processing. The failed node loses intermediate data flows traveling on the node at the moment of failure. Worse, if some operators hold runtime execution states, such data is also lost and the system can no longer continue the query processing. In conventional practice, the database system terminates all the processing whenever a node failure happens, and restarts the query from scratch. This naive solution works, but all the work done so far gets discarded no matter how far the process has progressed. This restarting strategy obviously incurs a significant time penalty.

In contrast, PhoeniQ allows the database system to continue query processing even when a single processing node fails[2]. This unique feature is enabled by two novel techniques illustrated in Fig. 1(b). First, *package-level checkpointing* takes a checkpoint for every operator pipeline, so that the database system can be ready to identify the lost intermediate data flow at the moment of failure and restart merely the affected processing that is necessary to recover the lost data. Second, *operator output replication* copies the output of tail operators to another processing node, allowing the node to resume the execution states of the failed node even when the tail operator is stateful. Section 3 focuses on the technical details of these techniques.

3   Execution Mechanism of PhoeniQ

This section gives a technical deep-dive into PhoeniQ. Sections 3.1 and 3.2 explain the two techniques: *package-level checkpointing* and *operator output replication*, respectively. Section 3.3 describes a tagging technique behind them. Finally, Sect. 3.4 shows a recovery procedure for PhoeniQ.

3.1   Package-Level Checkpointing of Operator Execution States

PhoeniQ allows each processing node to take a checkpoint for every operator pipeline to the storage node. Thanks to this unique capability, whenever a processing node fails, the remaining processing nodes can identify the lost intermediate data flow and restart the only processing that is necessary to recover the lost data. In this subsection, we firstly explain how the storage node manages the execution states of every operator pipeline. The execution state is managed for each tuple of relations scanned by the first operator of the pipeline. However,

---

[2] For simplicity and due to the space limitation, this paper merely presumes a single-node crash failure of processing nodes. The same idea can be easily applied to other cases, such as a double-node failure. Another exploration is necessary to protect against a failure of the storage node.

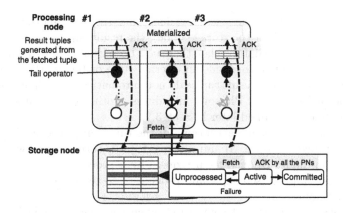

Fig. 2. The execution state management and checkpointing by PhoeniQ. *In this figure, the tail operator is stateless and result tuples are immediately materialized.*

a naive implementation leads to significant performance overhead. We secondly introduce a package-level state management technique to mitigate the overhead.

Execution State Management at the Storage Node. As mentioned earlier, we assume the shared-storage architecture. The basic function of the storage node is to store the database and to deliver a tuple upon a fetch request from a processing node. In processing a given query, each processing node requests tuple fetches to the storage node (mostly for executing a scan operator). In response to each request, the storage node feeds back a tuple to the requesting processing node in the on-demand manner [16].

Our novel idea is to let the storage node additionally manage an execution state of the operator pipeline for each tuple of scanned relations. After delivering a tuple to a processing node, the storage node tracks the execution state of the pipeline for the concerned tuple (Fig. 2). This mechanism enables the storage node to identify if the pipeline execution for each tuple has been completed or not. Thanks to this capability, the storage node can identify the lost part of the query processing in the face of failure, offering the recoverability for the failed query processing. We call this technique *checkpointing* in this paper.

PhoeniQ defines three states for a tuple in the scanned relation as follows.

Unprocessed. The initial state. An unprocessed tuple is not being processed by any of the processing nodes.

Active. An active tuple has been fetched and is being processed by the processing nodes. The tuples generated from an active tuple are not all *fault-safe* (defined later).

Committed. The tuples generated from a committed tuple are all fault-safe.

For a tuple in the pipeline, to be fault-safe means that it would not be lost in the face of a failure. If the tail operator is stateless and immediately materializes the result tuples (e.g. by delivering them to another processing node, a client terminal, or the storage node), the result tuples become fault-safe automatically and immediately when they have been generated and materialized. In contrast, if a tail operator holds an internal state or buffers the result tuples in the memory buffer, the result tuples do not automatically become fault-safe in the conventional approach. PhoeniQ introduces another technique for handling this case, which is explained later in Sect. 3.2.

At the start of a pipeline execution, all the tuples managed in the database are in the unprocessed state. An unprocessed tuple is turned active when fetched by a processing node. Upon a fetch request by a processing node, the storage node selects an unprocessed tuple and provides it to the requesting node in the on-demand manner.

An active tuple is turned committed when the storage node has received from every processing node a message called *ACK* associated with the tuple. Each processing node sends an ACK for a managed (active) tuple when it has finished its assigned part of the process to make the in-flight tuples generated from the managed tuple all fault-safe. As illustrated in Fig. 2, if the tail operator is stateless and result tuples are immediately materialized, each processing node sends an ACK for the tuple when all the corresponding result tuples generated in the node have been materialized. At the end of the pipeline execution, all the managed tuples have reached the committed state.

On a node failure, only the in-flight tuples corresponding to active tuples are lost and reprocessed after recovery. The recovery procedure is explained in Sect. 3.4.

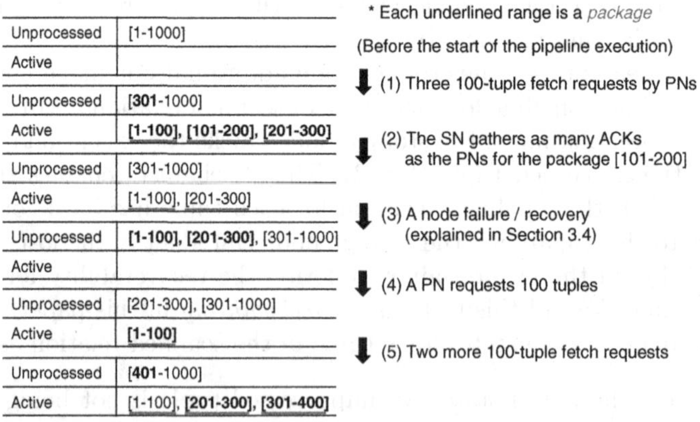

Fig. 3. An example of tuple processing states management at the storage node. This shows the tuple states changing over a pipeline execution where a 1000-tupled relation is scanned. *[x, y] denotes a range of tuples whose ID is between x and y (containing both ends).*

Package-Based State Management. A naive implementation of the check-pointing scheme would be to take checkpoints at the granularity of tuple. This is impractical, however, because it would incur significant memory footprint and considerable performance overhead. PhoeniQ instead takes a *range-based* app-roach to reduce the managed information and the processing overhead. This is achieved with the following three techniques.

- The processing nodes fetch multiple tuples from the storage node in one fetch request, instead of one tuple in one request.
- The storage node manages ranges of tuples for each state in a bulky manner, instead of managing a state of each tuple.
- The storage node explicitly manages information of unprocessed or active tuples only and omits that of committed ones. Tuples not present in the unprocessed nor active range list are regarded as committed.

Figure 3 illustrates an example of the package-based state management. Before the pipeline execution, the managed information consists of one range of tuples in the unprocessed state containing all the managed tuples. On receiving a fetch request from a processing node, the storage node cuts out a subset range from the unprocessed range, reads and sends the selected range of tuples to the processing node, and finally turns the tuples active. We call such a group of tuples read out and turned active in response to a fetch request a *package*.

This package-level checkpointing method poses the requirement to process tuples so that the tuples in every package are committed at once, i.e., the tuples generated from a package become fault-safe at once. When the tail operator is stateless and result tuples are materialized as soon as generated, every group of result tuples generated from a package are buffered, gathered, and then mate-rialized at once. How PhoeniQ meets the requirement when result tuples are buffered in the memory during the pipeline execution is explained in Sect. 3.2.

3.2 Operator Output Replication Among Different Processing Nodes

In addition to the checkpointing method, PhoeniQ employs *operator output repli-cation* technique, which copies the output of the tail operator to another pro-cessing node. This technique is motivated by situations where the pipeline result needs to be buffered until the end of the pipeline processing (e.g., when the tail operator is stateful). Such accumulated execution states are lost in a failure.

As shown in Fig. 1(b), when operator output replication is enabled, each processing node replicates the computation of the tail operator and its result partition in the logically neighboring node. Thanks to this redundancy, a spare node can restore the lost result as the failed node had at the moment of failure, by receiving it from the neighbors (explained in Sect. 3.4).

Tuples are not immediately input to the tail operator, but are buffered before it. We call this buffer a *pre-tail buffer* of the pipeline. Pre-tail buffers enable tuples to be replicated and fed to the tail operator at the granularity of package. Only tuples generated from committed packages can be fed to the tail operator.

When a processing node knows that it has buffered all the tuples generated from a package and are to come to the node, it sends a copy of the tuples to its logically neighboring node. The receiver again buffers the tuples before the tail operator which leads to the backup copy of the sender node's result partition. We call this buffer the backup copy of the sender node's pre-tail buffer.

When all the processing nodes have successfully replicated tuples made from a certain package, those tuples become fault-safe. Thus, operator output replication allows each processing node to send an ACK for the package to the storage node when its copying to the neighbor node has been completed.

Each processing node periodically asks the storage node which of the tuples in their pre-tail buffers can be forwarded to the tail operator. This query is performed by sending a set of package identifiers to which the buffered tuples correspond. The storage node responds by sending back the set of committed packages. On receiving the answer, the processing node proceeds to deliver the ready tuples (corresponding to committed packages) in its master pre-tail buffer to the tail operator. Similarly, the neighbor node selects tuples in the backup copy of the pre-tail buffer and deliver them to the backup tail operator.

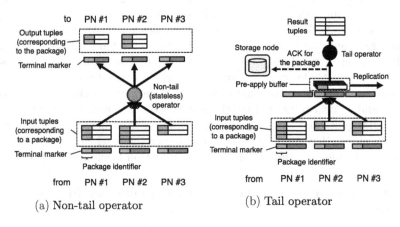

Fig. 4. PhoeniQ tags in-flight tuples so that it can identify the origin package of every tuple and detect the end of tuples generated from every package.

3.3 Tuple Tagging for Package-Level Commit

Package-level checkpointing requires additional information to be attached to in-flight tuples. For a processing node to decide when to send an ACK for a package, the following two conditions must be satisfied.

1. Each processing node can identify the origin package of every in-flight tuple.
2. Each processing node can identify whether or not all the tuples in a package have undergone the processing necessary to become fault-safe.

The first condition is satisfied by tagging each tuple in the pipeline. When a processing node fetches a package of tuples from the storage, the tuples get tagged with the identifier of the package. When these tuples undergo non-tail (stateless) operators, the result tuples are tagged with the same identifier as the input.

The second condition is satisfied by introducing marker tuples denoting that no following tuples correspond to a certain package (called *terminal marker* for a package), as shown in Fig. 4(a)(b). When a processing node has fetched a package of tuples, it appends to the fetched tuples a terminal marker tagged with the package identifier. When an operator redistributes (i.e., shuffles) output, terminal markers passed to it get broadcast.

If the pipeline has multiple shuffles, tuples made from a package can arrive from all the processing nodes. In this case, an operator knows that it has received all the tuples derived from the package only when it has received a terminal marker for the package from all the processing nodes. If the operator is non-tail (stateless) and redistributes output (Fig. 4(a)), it merges the N markers and then broadcasts it (where N is the number of processing nodes). If the operator is the tail operator (Fig. 4(b)), it knows it has gathered all the tuples corresponding to a package when it has seen N terminal markers for the package. It then can (replicate the tuples if operator output replication is enabled, and) issue its ACK for the package.

3.4   Query Processing Recovery

When a processing node fails, the remaining processing nodes invalidate all the in-flight tuples in the executed pipeline whose origin packages are in the active state. The storage node proceeds to rewind all the tuples in the active state to the unprocessed state, so that they can be refetched and reprocessed after recovery[3]. When a spare processing node has joined the query processing to replace the failed node, the pipeline can be restarted immediately if operator output replication is not enabled. Otherwise, the new processing node proceeds to receive the execution states from the two logically neighboring nodes before restarting the pipeline. In Fig. 1(b), for example, when processing node #2 fails, the new node receives the master result of processing node #1 and the backup result of processing node #3.

4   Evaluation

To demonstrate the feasibility and evaluate the effectiveness of our approach, we conducted a series of experiments with our prototype implementation. The experiments consist of two parts: an evaluation of the reduction of failure-recovery time and an evaluation of runtime overhead. In Sect. 4.1, we will describe the experimental setup and the benchmark query for the evaluations, and explain the execution plan. Then, Sect. 4.2 presents the experimental results.

---

[3] As long as all the non-tail operators are stateless as we have assumed, the reprocessing causes only marginal overhead compared to the entire pipeline processing.

4.1 Experimental Setup and Workload

We built our experimental system on public cloud services provided by Amazon Web Services. We used EC2 instances as the processing nodes and the storage node, whose specifications are shown in Table 1. For the storage, we used the instance store of the storage node instance (a SSD drive connected via NVMe).

Table 1. Experimental setup. The prototype system consists of up to sixteen processing nodes (PNs) and one storage node (SN).

	Processing node	Storage node
Instance	t2.medium	i3.xlarge
CPU	2 vCPUs	4 vCPUs
Memory	4 GiB	30.5 GiB
Storage	8 GB	950 GB
	(General purpose, SSD)	(Instance store, NVMe SSD)
OS	Amazon Linux 2	Amazon Linux 2

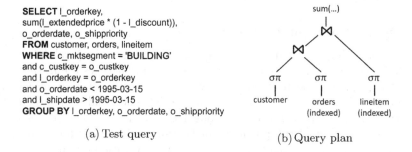

```
SELECT l_orderkey,
sum(l_extendedprice * (1 - l_discount)),
o_orderdate, o_shippriority
FROM customer, orders, lineitem
WHERE c_mktsegment = 'BUILDING'
and c_custkey = o_custkey
and l_orderkey = o_orderkey
and o_orderdate < 1995-03-15
and l_shipdate > 1995-03-15
GROUP BY l_orderkey, o_orderdate, o_shippriority
```

(a) Test query

(b) Query plan

Fig. 5. The test query and its query plan.

Our implementation was configured to run with or without PhoeniQ enabled. When enabled, our system ran with package-level checkpointing and operator output replication introduced in the previous section. In each experiment, we compared the results obtained with our approach enabled and disabled.

We prepared three relations and two indexes in the shared storage. The relations were `customer`, `orders` and `lineitem` from the TPC-H Benchmark [3] with scale factor 100. `orders` had an index file created on its primary key field `o_custkey`, and `lineitem` on `l_orderkey`. The relations were stored as arrays of C structures, and the indexes were Berkeley DB [1] B+ trees (version 18.1.32).

In the experiments, we ran a benchmark query shown in Fig. 5(a) to this dataset. The query involved selections and a joining of the three relations, followed by an aggregation, as illustrated in Fig. 5(b).

The entire query was processed with a single pipeline. The joining of the three relations was performed by scanning `customer` and taking advantage of the indexes. Each processing node first fetched `customer` tuples and applied the selection, and queried the storage node for joining `orders` tuples by join attribute values. The storage node looked up `order` index and provided joining `orders` tuples. Similarly, each processing node demanded joining `lineitem` tuples, and the storage node answered the query by reading `lineitem` index. Joined tuples underwent a hash-based shuffle before input to the aggregation.

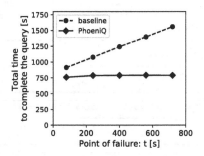

Fig. 6. PhoeniQ incurs almost zero penalties for the failure regardless of the query processing progress, whereas the conventional case incurs much longer execution time if the failure happens at later points in time.

When PhoeniQ was enabled, the computation and the result of the aggregation were replicated among two nodes. The system tracked the progress of the pipeline by managing the states of `customer` tuples. The processing nodes fetched 4096 `customer` tuples in each request, creating 4096-tupled packages.

The storage node ran worker threads, each of which was in charge of processing for each processing node. Each processing node ran two threads when PhoeniQ was disabled: one for the selections and the joining, and one for the aggregation. With PhoeniQ, each processing node ran one extra thread for checkpointing and replicating pre-tail buffers.

4.2 Experimental Results

We performed a scenario where a single node failed during a query execution. The benchmark query was run with 16 processing nodes. After t seconds, the program on one of the processing node instances was terminated. Three seconds after the failure, the failed program was restarted and joined the system. Depending on whether PhoeniQ was enabled or not, the system handled the failure differently. When PhoeniQ was disabled, all the nodes terminated their program, waited for the spare to join, and restarted the query from the beginning. When PhoeniQ was enabled, the system performed the recovery procedure and resumed the query. In either case, after the failure handling, the query was completed without failure.

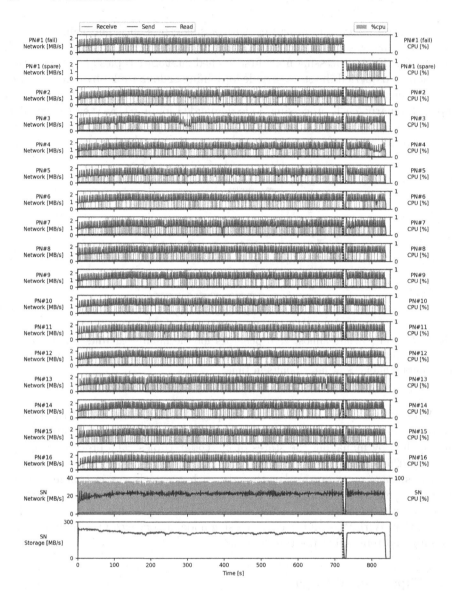

Fig. 7. PhoeniQ dynamically continues query execution even in the face of node failure. PN#1 (fail) terminates the execution at 720 s, but PN#1 (spare) immediately recovers the execution.

Figure 6 shows the total execution time with varying points of failure t (seconds). Without our approach, more time was spent to get the result when the failure took place later. In contrast, no noticeable penalty was present with our approach. At $t = 720$ s (when about 90% of customer relation has been scanned), our approach almost halved (-45%) the total execution time.

Figure 7 shows a recovery behavior at $t = 720$ s. In the recovery procedure, the spare node joined the cluster, and received around 80 K tuples in total from the two logically neighboring nodes in around 1.4 s. The spikes in the network throughput of the two nodes after the recovery were caused by the restoration of the result onto the new node. It could also be seen from the network and storage throughput that the system regained processing speed shortly after the recovery. During the entire execution, the CPUs of the storage node were almost fully utilized, whereas those of the processing nodes were underutilized because of the I/O bound characteristics of the workload. It can therefore be inferred that the additional CPU cost added by operator output replication did not affect the execution time in this workload.

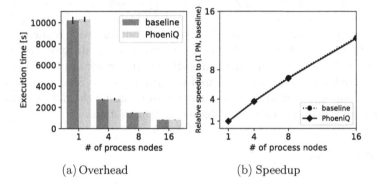

(a) Overhead   (b) Speedup

Fig. 8. PhoeniQ enables query continuation with negligible execution overhead and does not disturb scale-out.

Fig. 8(a) shows execution times without failure with varying numbers of processing nodes. Our approach incurred at most 1.6% execution overhead (at 4 processing nodes). This result implies that the overhead introduced by package-level checkpointing, as well as operator output replication, was quite marginal. Figure 8(b) shows that PhoeniQ did not disturb the scale-out.

These experiments demonstrated that PhoeniQ can continue the query processing in the face of node failures with significantly smaller cost than the conventional approach.

5   Related Work

This section outlines the previous work on query restarting techniques.

The previous work for centralized systems mostly aims to restart interrupted queries in favor of those of higher priority [8,10].

For distributed systems, a variety of methods have been proposed to support query fault tolerance. Early MapReduce [12] frameworks write out the output of every process stage to storage. While this allows the query to restart from the latest persisted state, additional I/O cost is not negligible, as demonstrated in [22]. Fault-tolerant query on systems where input data is dynamically provided (as known as stream-based systems [4,7,9]) have been relatively well studied. For example, [24] replicates every computation to backup nodes. In [19,21], master nodes take periodical checkpoints into spare nodes.

For parallel database systems that run queries on static data, several methods [18,25] aim to reduce reprocessing. However, [25] does not consider aggregation operators, and [18] allows a fair amount of recomputation of aggregation. OTPM [17] by B. Han et al. is close to our approach in that it curtails reprocessing of aggregation. In OTPM, operators track the progress of their upstream operator by monitoring the IDs of incoming tuples. The system requires additional nodes to store intermediate results. They have shown promising results from simulation-based evaluation, but a working implementation is not shown. One of the major differences between this approach and ours is that PhoeniQ does not track the progress of every operator. Furthermore, our approach replicates results in a way that does not require additional nodes. Lastly, our approach assumes shared-storage systems, while theirs and all the other work mentioned proposed for distributed settings assumes shared-nothing systems. Shared-storage approach is advantageous in that it does not require sending the data partitions of failed nodes to spare nodes as in a shared nothing system.

6   Conclusion

In this paper, we have proposed a method for parallel database systems to restore execution states on a spare node and to resume query processing. This is achieved by package-level checkpointing and operator output replication. We have implemented a prototype system and performed an experiment with up to 16 processing nodes in a cloud environment. The result shows that our approach successfully reduces restarting temporal penalty on failures with negligible overhead under I/O bound workload. Future work includes conducting experiments with an increased variety of queries. Hash join workloads, for example, are an interesting target. They involve multiple pipelines (separate pipelines for hash build and probe), and a single pipeline can involve multiple shuffles. Moreover, the performance overhead of operator output replication needs careful investigation, because hash joins are generally more CPU-heavy than index joins.

References

1. Oracle Berkeley DB. https://www.oracle.com/database/berkeley-db/db.html
2. The Internet of Things: Data from Embedded Systems Will Account for 10% of the Digital Universe by 2020. https://www.emc.com/leadership/digital-universe/2014iview/internet-of-things.htm
3. The TPC-H benchmark. http://www.tpc.org/tpch/
4. Abadi, D.J., et al.: The design of the borealis stream processing engine. In: Proceedings CIDR, pp. 277–289 (2005)

5. Boral, H., et al.: Prototyping bubba, a highly parallel database system. IEEE Trans. Knowl. Data Eng. **2**(1), 4–24 (1990)
6. Borthakur, D.: Petabyte scale databases and storage systems at facebook. In: Proceedings SIGMOD, pp. 1267–1268 (2013)
7. Carney, D., et al.: Monitoring streams - a new class of data management applications. In: Proceedings VLDB, pp. 215–226 (2002)
8. Chandramouli, B., Bond, C.N., Babu, S., Yang, J.: Query suspend and resume. In: Proceedings SIGMOD, pp. 557–568 (2007)
9. Chandrasekaran, S., et al.: Telegraphcq: continuous dataflow processing for an uncertain world. In: Proceedings CIDR (2003)
10. Chaudhuri, S., Kaushik, R., Ramamurthy, R., Pol, A.: Stop-and-restart style execution for long running decision support queries. In: Proceedings VLDB, pp. 735–745 (2007)
11. Daniel Weeks: Netflix: Integrating Spark at petabyte scale. https://conferences. oreilly.com/strata/big-data-conference-ny-2015/public/schedule/detail/43373
12. Dean, J., Ghemawat, S.: Mapreduce: simplified data processing on large clusters. Commun. ACM **51**(1), 107–113 (2008)
13. DeWitt, D.J., Gray, J.: Parallel database systems: the future of high performance database systems. Commun. ACM **35**(6), 85–98 (1992)
14. DeWitt, D.J., Madden, S., Stonebraker, M.: How to build a high-performance data warehouse how to build a high-performance data warehouse. http://db.csail.mit. edu/madden/high_perf.pdf
15. Ghandeharizadeh, S., DeWitt, D.J.: Hybrid-range partitioning strategy: a new declustering strategy for multiprocessor database machines. In: Proceedings VLDB, pp. 481–492 (1990)
16. Goda, K., Tamura, T., Oguchi, M., Kitsuregawa, M.: Run-time load balancing system on san-connected PC cluster for dynamic injection of CPU and disk resource - a case study of data mining application. Proc. DEXA. **2453**, 182–192 (2002)
17. Han, B., Omiecinski, E., Mark, L., Liu, L.: OTPM: failure handling in data-intensive analytical processing. In: Proceedings CollaborateCom, pp. 35–44. IEEE (2011)
18. Hauglid, J.O., Nørvåg, K.: Proqid: partial restarts of queries in distributed databases. In: Proceedings CIKM, pp. 1251–1260. ACM (2008)
19. Hwang, J., Xing, Y., Çetintemel, U., Zdonik, S.B.: A cooperative, self-configuring high-availability solution for stream processing. In: Proceedings ICDE, pp. 176–185 (2007)
20. Jeff Barr: Migration Complete - Amazon's Consumer Business Just Turned off its Final Oracle Database. https://aws.amazon.com/blogs/aws/migration-complete-amazons-consumer-business-just-turned-off-its-final-oracle-database/
21. Kwon, Y., Balazinska, M., Greenberg, A.G.: Fault-tolerant stream processing using a distributed, replicated file system. Proc. VLDB **1**(1), 574–585 (2008)
22. Pavlo, A., et al.: A comparison of approaches to large-scale data analysis. In: Proceedings SIGMOD, pp. 165–178 (2009)
23. Reza, S.: Uber's Big Data Platform: 100+ Petabytes with Minute Latency. https:// eng.uber.com/uber-big-data-platform/
24. Shah, M.A., Hellerstein, J.M., Brewer, E.: Highly available, fault-tolerant, parallel dataflows. In: Proceedings SIGMOD, pp. 827–838. ACM (2004)
25. Smith, J.E.T., Watson, P.: A rollback-recovery protocol for wide area pipelined data flow computations (2004)
26. Stonebraker, M.: The case for shared nothing. IEEE Database Eng. Bull. **9**, 4–9 (1985)

Consistency, Integrity, Quality of Data

A DSL for Automated Data Quality Monitoring

Felix Heine[1]([✉]), Carsten Kleiner[1], and Thomas Oelsner[2]

[1] University of Applied Sciences and Arts, Hanover, Germany
{felix.heine,carsten.kleiner}@hs-hannover.de
[2] mVISE AG, Hamburg, Germany
thomas.oelsner@mvise.de

Abstract. Data is getting more and more ubiquitous while its importance rises. The quality and outcome of business decisions is directly related to the accuracy of data used in predictions. Thus, a high data quality in database systems being used for business decisions is of high importance. Otherwise bad consequences in the form of commercial loss or even legal implications loom.

In this paper we focus on automating advanced data quality monitoring, and especially the aspect of expressing and evaluating rules for good data quality. We present a domain specific language (DSL) called *RADAR* for data quality rules, that fulfills our main requirements: reusability of check logic, separation of concerns for different user groups, support for heterogeneous data sources as well as advanced data quality rules such as time series rules. Also, it provides the option to automatically suggest potential rules based on historic data analysis. Furthermore, we show initial optimization approaches for the execution of rules on large data sets and evaluate our language based on these optimizations.

All in all the language presents a novel approach for a flexible and powerful management of data quality in practical applications while meeting the needs of actual data quality managers in being pragmatic and efficient.

Keywords: Data quality · Domain specific language · Data quality monitoring · Rule based data quality · Data heterogeneity

1 Introduction

Data is the core of many modern businesses. Data is typically stored in different distributed databases and business decisions rely on an integrated view on this

The project IQM4HD has been funded by the German Federal Ministry of Education and Research under grant no. 01IS15053A. We would also like to thank our partners SHS Viveon/mVise for implementing the prototype and CTS Eventim for providing important requirements and reviewing practical applicability of the prototype and concepts.

S. Hartmann et al. (Eds.): DEXA 2020, LNCS 12391, pp. 89–105, 2020.
https://doi.org/10.1007/978-3-030-59003-1_6

data. In order to take right decisions it is important that data used is of high quality. While there are many different definitions of data quality in the literature (e.g. [14] defines 16 dimensions), we focus, in line with practical requirements from our project partners, on the quality dimensions of accuracy, completeness, consistency and integrity. The DSL presented in this paper is applicable both to classical data warehouses as well as to heterogeneous, distributed data sources including non-relational data. An important part of data quality monitoring is to check internal consistency of a database and its conformance with external datasets. These checks can be rather simple checks like NOT NULL, structural checks like referential integrity checks, or complex statistical checks like conformance to a distribution or time series model. Our DSL integrates all these kinds of checks.

In order to detect quality problems reliably, it is important to check the data regularly. An extreme solution would be to run checks as database constraints, effectively avoiding the insertion of wrong data. However, this approach has multiple drawbacks: 1) The check logic slows down all processes that modify the data. 2) The check logic must be implemented inside the database and thus depends on the capabilities of the database. Not every check can be implemented on every system, especially more complex checks like statistical checks. 3) These kinds of checks often rely on proprietary languages like stored procedure languages. 4) Implementing these checks in a modular way, in order to apply the same logic to different databases, is often impossible. 5) Some checks, esp. statistical checks, do not always indicate clear errors. They rather result in warnings indicating an unusual data distribution that should be checked. Thus always rejecting non-conformant data is not possible. 6) Checks that span multiple databases (e.g. a relational database and a document database) are impossible with this approach.

Thus, our approach is to run the checks from an external engine that has its own language to encode the checks, called **rules** in our system. For this, we designed the $RADAR$[1] DSL. It has been developed in the context of the project $IQM4HD$[2], which not only targets the execution engine for $RADAR$, but also an approach to create and maintain the set of rules with low effort. This approach includes a profiling module that can analyze existing data to make suggestions for rules and a feedback system where the data steward can classify detected data quality problems and warnings (e.g. as real quality problems, exceptions, or new patterns). The feedback system can then use these classifications to suggest new or adapted rules. Note that automated data correction is not in the scope of our work.

The focus and contribution of this paper is the DSL $RADAR$ which we designed and implemented for our data quality engine, together with optimization approaches to execute the rules efficiently also on large data sets. The profiling module and the feedback system are not discussed in this paper due to space constraints.

---

[1] Rule Language for Automated Data Quality Assessment and Reporting.
[2] http://iqm4hd.wp.hs-hannover.de/english.html.

The remaining paper is structured as follows. First, the main concepts and design decisions of the DSL along with illustrative examples are presented in Sect. 2. Approaches to efficient execution code are described in the following section. Then we show the achievements and effectiveness of first optimization approaches in Sect. 4. After that, Sect. 5 covers related work and compares it with the specific contributions of this paper. Finally, the paper closes with a conclusion and ideas for future work.

2 Data Quality Rule Language RADAR

This chapter introduces the proposed language *RADAR* to specify data quality rules. For presenting details about the language we at first look at what information is needed to compose data quality rules, and secondly which types of users are present in our system. Information needed for quality rules includes (i) *what data should be checked*, (ii) *what quality aspect should the data be checked for* and lastly (iii) *what should be done with the result*. Regarding user types, we suggest a separation between the technical aspects of data retrieval and the subject-specific knowledge needed for writing coherent rules. The technical user has technical knowledge concerning the databases while the data quality manager has business knowledge about the data's meaning. Whereas the technical user just prepares the data for the data quality manager, the latter will - with his business knowledge - compose data quality rules or adjust automatically generated rules. The three aforementioned aspects are reflected in the quality rules in *RADAR* by three different components, namely Sources, Checks and Actions. An overview of the system architecture is given in Fig. 1. Note that a developer is only needed in case the Checks already provided with the system

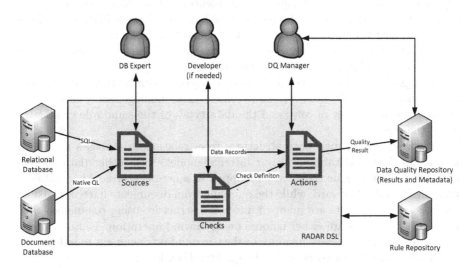

Fig. 1. Overview of DQ Monitoring with *RADAR* .

need to be extended. Details about these components will be explained in subsequent sections focusing on illustrative examples rather than a complete formal description.

It is important to note though, that for implementation we used a formal specification of *RADAR* in terms of a EBNF grammar definition that has been used to generate a parser for the quality rules language using `antlr`. Details on this have been omitted here due to space constraints, but can be found in [13] for the interested reader. We try to present both simple rules that are easy to understand and more complex ones to document the potential of the language.

2.1   Internal Data Model

The basic data type in our DSL is a **record**, which is similar to a relational tuple, however, with one important distinction. Each record carries so-called *roles* instead of attribute names. Each attribute can have multiple roles, and each role can be assigned to multiple attributes. As an example, the following illustrates a customer record:

Group identifier	Id identifier	Firstname name	Lastname name	Birthdate	Telephone
A	12345	Jon	Doe	5/7/1980	111-222-33333

In this case, the first attribute has the roles "group" and "identifier". However, the role "identifier" is assigned to the first two columns. When we access this role in our DSL, we effectively access the two-attribute record ('A', 12345). This is a very flexible and convenient way to write code that can handle either attributes or compound data using a single identifier. The order of the attributes is relevant. The attributes of the record have basic data types, like strings, integers, doubles or booleans, and there is a NULL value. There is no explicit data type definition, each value carries its own type information. A non-existing role in a record is implicitly treated as value NULL.

The next higher level structure is a **list** of records, similar to a relation. The records in a single list, however, do not need to be homogeneous, each record can have a different set of roles and the data types of the same role could vary across the records.

This way, we can map both structures from relational databases as well as document-oriented databases to our internal model. Even other data sources are supported as long as they can be mapped to a list of records. The relational mapping is straightforward, while the mapping from documents flattens the documents. Sub-documents are mapped into a flat structure using role names with dot notation. Arrays are either ignored or an unwind operation[3] is used to map each array element to a new instance of the surrounding document together with this element, according to needs of the quality check logic.

---

[3] https://docs.mongodb.com/manual/reference/operator/aggregation/unwind.

```
 1  SOURCE ageHistogram TYPE LIST CONST ROLES (frm, too, perc):
 2      [0, 18, .05], [18, 25, .20], [25, 35, .15], [35, 50, .25],
 3      [50, 65, .19], [65, 80, .10], [80, 95, .05], [95, NULL, .01]
 4  END
 5
 6  SOURCE Customer TYPE LIST QUERY ROLES (id: IDENTIFIER):
 7  DATABASE mondrian NATIVE
 8      SELECT id, name, date_part('year', age(dob)) AS age, email
 9      FROM customer
10  END
```

Listing 1. DSL code for a reference histogram and a relational Customer Source.

Since we only need to map results of queries to our internal data model for use in quality checks and not the entire database schemata, we do not have to consider more generic schema matching models here.

2.2 Sources

Sources are the interface to data from databases or other reference values a domain expert may work with when composing data quality rules. There are two main types of Sources. On one hand, there is the domain mapping layer, which maps database structures to Sources, like the **customer** Source (lines 6–10 in Listing 1), referring to a customer table in a relational database system. On the other hand, there are constant values, patterns or complex values that contain the parameters for statistical models. A simple example for this kind of Source could be an email pattern used to check customers' emails for invalid entries. A statistical parameter set could be a histogram containing a reference distribution to check the age distribution of customers against, as shown in lines 1–4 in Listing 1.

The domain mapping layer has been designed to create an interface between the technical and the domain-specific side of data quality rules. The goal is to provide an easy way for domain experts to work with the underlying data without having to know details about where and how the original data is stored. Since non-constant Sources are specified in the database's own language, it is an assignment for technical users to prepare Sources for domain experts. The customer example in lines 6–10 in Listing 1 shows a generic SQL statement selecting the **id**, **name**, **dob** (date of birth), converted to an age, and **email** from the **customer** relation. Additional meta information is added by declaring the field *id* to be an identifier. Having the source layer as an intermediate layer between rules and the source database allows a very flexible way to define data access for any system using the system's native query language and to map structures to the domain layer accessed in the Checks of the DSL.

Beside the simple selection of e.g. relations and fields, it is also possible to use the full power of the data source's query language. This may range from simple pre-filtering of data for use in quality rules up to assembling data in a way that is best suited for the future quality checks, by e.g.. using joins in relational sources. Data filtering may be beneficial for either legal reasons (i.e. the DQ manager is only allowed to access data for his country) or practical reasons, e.g.. quality rules that are varying over countries due to different formats. Data assembly may be beneficial to create a quality specific view on the source data in an efficient way in the source system. However, a Source should be defined as simple as possible and should not contain calculations related to data quality checks but be constrained to retrieving data for a future quality check only.

2.3 Checks

Checks are the functional skeletons of data quality rules, they can be compared to functions/methods from general purpose programming languages. A Check is basically a method for data quality control. They operate on data in the internal data model (cf. Sect. 2.1) and are either already implemented in the system core such as the examples presented below or can be implemented by a developer using the constructs of the Check part of *RADAR* . Note that while the simple Check examples below at first sight seem similar to database constraints, there are two important differences: for one, the Checks can be used on arbitrary Sources, including data sources that do not provide constraints such as many NoSQL systems. Secondly, while constraints will typically be enforced on database operation level (which is not always desriable, esp. in DWH scenarios) our Checks provide a "softer" view on the current state of a quality aspect without direct influence on operations. Checks always include RETURN statements, which yield detected quality problems. Thus the overall result is a list of potential quality problems that will later be evaluated by the Action.

Lines 1–3 in Listing 2 shows one of the most basic Checks, a check for NULL values. Lines 5–9 in Listing 2 compare two lists and check whether they have the same amount of entries. Note that the lists may originate from completely different data sources as assigned in the Actions below. Lines 11–20 in Listing 2 show a check for distribution matches. The dist parameter is a reference distribution, that defines bins for the values and a probability p for each bin. The age histogram in Listing 1 is an example. In the check code, first the number of data items is counted. Then, each value val is translated to a bucket name (bin) using the build-in function val2bin according to the buckets defined in the reference distribution dist. In the next step, a histogram is constructed and stored in the local LIST-type variable hist. Thereafter the score for each bucket is computed by $(O_i - E_i)^2/(O_i - E_i)$. In this case, the observed count O_i is the bucket count from the previous statement, and the expected count E_i is calculated using the size of the relation times the expected fraction from the reference distribution. The bucket scores are finally aggregated into a χ^2 score. This part shows the SQL-like syntax of *RADAR* for defining the Checks which works nicely with the lists of records internal data model produced by the Sources.

Since Checks are just archetypes of data quality rules, they provide parameters to utilize them for many different concrete cases. There are two kinds of

```
1   CHECK NotNullCheck ON value:
2       RETURN value IS NULL;
3   END
4
5   CHECK QuantityCheck ON LIST list1 WITH LIST list2:
6     cnt1 := SELECT COUNT(*) FROM list1;
7     cnt2 := SELECT COUNT(*) FROM list2;
8     RETURN cnt1 != cnt2;
9   END
10
11  CHECK DistChk ON LIST data(val) WITH dist(frm, to, bin, perc):
12    cnt := SELECT COUNT(*) FROM data;
13    bindata := SELECT val2bin(val, dist) ROLE val FROM data;
14    hist := SELECT bin, perc, COUNT (val) ROLE acnt
15           FROM bindata RIGHT JOIN dist ON val = bin
16           GROUP BY bin, perc;
17    scorebin := SELECT (acnt - cnt*perc) * (acnt - cnt*perc)
18                       / (cnt*perc) ROLE score FROM hist;
19    RETURN SELECT SUM(score) ROLE score FROM binscore;
20  END
```

Listing 2. DSL code for different Checks.

parameters for Checks, the first parameter provides the information, which data has to be checked for quality issues. The second type of parameters are control parameters, which are used to aid in checking. For example the `QuantityCheck` (lines 5–9 in Listing 2) expects two parameters, the first being a list of values, e.g. the content of a customer table from a Data Warehouse (DWH), whose quantity should be checked against the amount of values of the second parameter, another list of values (e.g. the content of the original customer table before inserted into the DWH). These parameters are divided via the keywords `ON` for "data-to-be-checked" and `WITH` for control parameters.

Thus the definition of the Checks is still generic in two ways. For one, the same Check can be applied to many different data sets. Also, the same Check can be used with different control parameters. The connection between Sources and Checks will be defined in the Actions explained below. Hence there is no reference to a specific age distribution in the `DistChk`. The exact same Check can be used for checking against an age reference distribution as well as a salary reference distribution depending on which parameter is passed to the Check as `WITH` parameter later. This significantly reduces the number of Checks required in a data quality monitoring system. Most basic and some complex analysis Checks are predefined in a rule catalog provided with the IQM4HD prototype. The goal is to provide a substantial foundation in order to cover most needs. Additional Checks can be implemented based on the provided language by developers as needed.

To cover all needs without creating a too complex language the expressiveness is three-tiered. First, there are boolean and mathematical expressions described directly by the language in combination with variable usage, IF statements and FOR loops as known from normal programming languages. Second, there is an SQL-like language to process data sets using typical operations like projection, selection, join and grouping as shown in lower part of Listing 2 already. Third, there are built-in methods to hide complexity, like a method to check multidimensional cube data.

Accessing Sources within the SQL-like statements in the Checks works straightforward for lists of homogeneous records as internal representation of list Sources originating, e.g. from relational or document databases. The name of a Source can be used in the FROM part to reference it. Individual columns within such Sources are identified by their attribute names. Attributes are assumed to be of non-complex data types. If a referenced attribute name is not present in any of the list elements (e.g. in the case of heterogeneous documents), this element is excluded from consideration in this Check.

For attributes of complex data types (e.g. sub-documents from a document database) the classical dot notation may be used to access key-value pairs within such an attribute. As documents may be nested hierarchically, path expressions separated by dots may be used to navigate into the hierarchy. In most cases the value specification using dot notation will rather be used in the Action part, because the Checks should be generic and operate on roles that are assigned in the Action part.

In order to present potential data quality issues to the DQ manager later, the RETURN statement is used. For proper storage and evaluation of the detected quality problems, a unified format to identify problems is necessary. Thus we classify quality problems according to the unit of data that is related to the problem: 1) Problems that are related to individual records like NULL values; 2) Problems that are related to values that relate to multiple records, like uniqueness problems; 3) Problems that are related to a whole data source, like skewed distributions. All returned values are bundled and will later be evaluated by the result part of the Action invoking this Check.

2.4   Actions

Actions are responsible for two things. Firstly, they connect Sources and Checks resulting in data quality rules. Secondly, they deal with the result and therefore conclude data quality rules. Actions were designed with natural language in mind, so that a rule composed in the DSL reads as a self-explaining sentence. Following are a few examples of Actions which cover different kinds of Checks.

Lines 1–3 in Listing 3 show an Action testing the customer names for NULL values. The NotNullCheck is designed for single values, because there is no LIST keyword in front of the parameter in the Check. Thus the action includes the EACH keyword in order to call the check for each item of the data source individually. Lines 5–8 in Listing 3 are an example of a check of the result of an Extract-

```
1  ACTION NotNullCheckCustomerName:
2      EXECUTE NotNullCheck ON EACH Customer(name) RESULT IN ERROR
3  END
4
5  ACTION DWHQuantityCheckCustomer:
6      EXECUTE QuantityCheck ON DWH_Customer WITH Customer
7      RESULT IN ERROR
8  END
9
10 ACTION AgeDistributionCheckCustomer:
11     EXECUTE DistributionCheck ON Customer(age) WITH ageHistogram
12     RESULT IN WARNING ABOVE 14.07 AND ERROR ABOVE 18.48
13 END
14
15 ACTION NullCheckCustomerStreetName:
16   EXECUTE NotNullCheck
17   ON EACH CustomerAddress(address.street.name: value)
18   RESULT IN ERROR
19 END
```

Listing 3. DSL code example Actions.

Transform-Load (ETL) process, it checks whether the amount of tuples in the DWH table fits the original table.

A more complex example is in lines 10–13 in Listing 3. It uses the customers dates of birth and a reference age distribution to compare the current customer ages with a saved reference distribution. The result is a score computed by a $\mathcal{X}^2$ test, which is evaluated by the RESULT IN part of the Action. The chosen threshold values 14.07 and 18.48 will issue a warning when the 95% value of the $\mathcal{X}^2$ test is not met and an error if even the 99% value of the test is not achieved. These concrete values are typically computed by the profiling component of the IQM4HD system.

The notation to access parts of a source definition follows the same logic as described for the Checks previously. Key-value pairs in nested documents within a document can be accessed by using the Source name together with a path expression. Lines 15–19 in Listing 3 show a NOT NULL Check on the name key inside the street sub-document of the address sub-document of the Source CustomerAddress.

3   Implementation / Optimization

An important goal of our execution engine implementation is flexibility with respect to location of execution of check logic. In principle, checks can be executed in the quality engine itself or moved to the source databases by rewriting the source query. On the one hand, the execution engine should be able to execute all logic on its own, in cases where the target database does not have enough capabilities, either because of resource constraints or because of a lack of features. On the other hand, if possible, logic should be moved to the database for more efficient execution due to co-location with the data.

To reach this goal, we distinguish two cases. First, we have checks for individual records, that are called in a loop for each item of the source database. In this case, we extract what we call *safe conditions* from the Check, i.e. conditions that need to be fulfilled for the Check to find quality problems. These conditions are then translated to filter conditions that are added to the native database query in the source database. This step needs individual code for each native query language and might fail, if the target dialect does not support the relevant conditions. However, if it succeeds, we avoid fetc.hing large numbers of records that will not produce output for the quality check anyway. As a very simple example, consider the `NotNullCheckCustomerName` in Listing 3. From the referenced check, we extract the condition `value IS NULL`. The role `value` is then back-substituted to `name` and the original query is modified to

---

```
SELECT * FROM (... original query ...) WHERE name IS NULL
```

For checks that work on `LIST`s, we follow a similar strategy. The first component is deferred execution. This means, that list expressions are evaluated only symbolically, i.e. variables are substituted with their current values. However, the list expression itself is not evaluated until it is actually needed (i.e. because it is part of a return statement or part of a condition that determines control flow). This leads to larger expressions that are build up part by part.

Once execution is requested, the whole expression is optimized quite similarly to normal database optimizers. Our optimizer tries to push as much logic towards the data sources, where the logic is added to the original database query (similar to the above simple modification for `NotNullCheckCustomerName`). However, as soon as different source databases are involved in an evaluation, no further pushing of operations to the Sources is possible and the quality engine has to take over evaluation from this point on. For example, a join of two sources from different databases has to be processed internally in the data quality engine. While the execution optimization already provides promising results (cf. Sect. 4), this part still remains an important issue for future improvements (cf. Sect. 6).

4   Evaluation

4.1   Functional Evaluation

In this section, we review the main features of our proposed DSL *RADAR* and discuss how they meet important DQ monitoring requirements.

Extensibility and Reusability. We think that an extensible rule logic is essential, so that users can specify new rule types in a flexible and reusable way. To achieve reusability, the logic behind the user-defined rules must be decoupled from the actual data source, so that the same logic be applied to multiple data sources. The separation between Sources, Checks and Actions fulfills this require-ment. The check logic is independent of the data source, and of the type of the data source.

Heterogeneous Data Sources. Data is stored using various technologies and different data models. A DQ system must be able to access and compare data from such heterogenous sources. This requires flexibility w.r.t. technological aspects and aspects of the data model. We resolve this by translating the data into the common internal format, so that all further components can be agnostic of the particular type of data source. A single Check can e.g. use data from a relational and a document database to test consistency constraints between the two databases.

Optimizability. As data sources might be huge, execution speed is an important issue, even though the checks are typically scheduled and executed in batch mode. For this, it is beneficial to make use of the features of the underlying DBMS like query optimization and available indices. However, as the available features differ, our optimizer component can decide which parts of the logic are executed using native DBMS features or within our execution engine.

User Groups. In DQ teams, multiple persons with different expertise are involved. First, there are experts for the data sources to be monitored, that understand the data models of these systems. Second, we have domain experts that understand the business logic and can define which data is valid from a business point of view. Third, there are programmers that understand how to code the business logic programmatically. We need the respective parts of the language to be accessible by the corresponding groups. Again, the separation in Sources, Checks and Actions supports this requirement. Sources are defined by database experts that know the underlying database models; Checks are either pre-defined by our system or can be extended by technical experts, and Actions are defined by domain experts.

Advanced Quality Rules. Apart from rather simple data quality checks such as non-null or range checks the rule language should also provide the option to specify advanced data quality checks. With the DistributionCheck one example of an advanced rule is given in this paper. Other examples not shown here include outlier identification in time series data and a check on cube data analyzing relevant aggregations of the data.

4.2   Performance Evaluation

In this section, we look at the efficiency of our execution engine and especially at the effectiveness of our optimization approaches as performance will be a critical feature in large scale use cases. The goal is not an exhausitive evaluation of all types of rules, but rather to show the potential of our optimization strategies. For this, we use three different scenarios. First, we look at a simple NOT NULL check (NN), which is an example of a single record check. In this case, the optimizer modifies the source statement so that only critical records (i.e. NULL values) are

Table 1. Evaluation results (averages over 5 runs)

	Errors	0%			1%			10%		
	Size	1%	10%	100%	1%	10%	100%	1%	10%	100%
NN	Std.	0.25	1.57	16.45	0.22	1.35	18.15	0.21	1.36	17.22
	Optim	0.06	0.05	0.33	0.04	0.06	0.48	0.05	0.23	2.85
	SQL	0.01	0.04	0.53	0.01	0.04	0.53	0.01	0.12	1.85
RI	Std.	0.66	5.44	106.38	0.56	5.62	142.90	0.67	11.98	458.45
	Optim	0.15	0.79	16.38	0.12	0.84	14.11	0.11	1.28	16.74
	SQL	0.11	0.76	11.24	0.10	0.87	10.75	0.12	1.37	11.01
Dist	Std.	0.82	6.40	74.01						
	Optim	1.07	7.80	73.04						
	SQL	1.23	7.97	70.21						

selected. Second, we look at a referential integrity check (RI) which checks which records in one list do not reference valid records in the other list. This contains more complex list expressions that are translated to SQL by the optimizer in case of a relational data source. As a final, more complex, example of this list expression to SQL translation, we use a distribution check (Dist) checking to which degree records in a source comply with a reference distribution.

For evaluation, we use a person (approx. 5,000,000 records) and a plays relation (approx. 8,400,000 records) from a movie database[4]. For each scenario, we look at different sizes of the data to check (1%, 10% and 100% of the original data). As the runtime might be influenced by the number of errors in the first two scenarios, we also use different numbers of errors (no errors, 1% errors, 10% errors).[5] The distribution check calculates an overall score for the target relation taking each record into account, thus here we have no difference in effort related to error ratio. As benchmark, we also list runtimes of SQL statements running directly on the database that perform the same check. This is only possible as long as all data originates from a single database and also sacrifices all other benefits of *RADAR* as explained in Sect. 4.1.

The results are summarized in Table 1. For the NN check, we see that the optimized version performs much better than the standard version. The standard version suffers from a large number of calls to the check (for each person record) and is almost independent of error ratio. The optimized version is a significant improvement that is the better the smaller the error ratio (since fewer data has to be transferred to the engine). Also, the scaling behavior looks good as the runtime is roughly proportional to the data size. Runtimes for the optimized version are within the same order of magnitude as the raw SQL statement (that

---

[4] ftp://ftp.fu-berlin.de/pub/misc/movies/database/.
[5] The errors were introduced by updating values to NULL or to non-existing FK values for a random subset of the records.

does not include any DSL processing). As the optimized version can only filter correct records, the runtime becomes larger with larger error ratio.

For the RI check in the standard version the scaling is not as good, both in terms of data size and error ratio. We assume that this is mainly due to our join implementation in the engine, which is currently not very efficient and subject to future improvements. However, the optimizer works quite well for this scenario again, reaching runtimes in the same order of magnitude as the raw SQL benchmark and similar scaling behaviour. Also, as expected, the runtime is almost agnostic of the error ratio in this case.

Finally, the Dist scenario shows a different picture. Here, the scaling works well, however the optimized version does not yield any improvement. A comparison with the raw SQL statement shows, that this statement already consumes nearly all of the runtime so that no improvement is possible. A closer look revealed that this is due to the stored procedure we used to translate the `val2bin` method to SQL. A hand-crafted SQL version that avoids this procedure can be much faster, so we plan to modify our SQL generation to output this result in the future.

Overall, we can conclude that the optimizer works very well in general and establishes runtimes close to raw SQL statements that implement the same logic. However, as we cannot always count on it in cases when the underlying database has limited query capabilities (NoSQL) or when data from multiple sources is combined, we will also continue to improve the DSL interpreter. However, the efficiency improvements here are more tied to improvements in small details (e.g. join implementation, implementation of the expression evaluation) and not to general architectural issues. Thus the engine to execute *RADAR* can be considered sufficiently efficient.

5 Related Work

Older papers, including our own predecessor project Data Checking Engine (DCE) [11], that address partial aspects discussed in this paper (e.g. heterogeneous data integration) are typically not applicable for non-RDBMS data sources and thus cannot satisfy an important requirement of our solution. However, many solutions and tools for DQ measurement and monitoring exist that also address data not stored in RDBMS. For a recent tool survey, e.g. see [6]. We now describe examples of individual tools and then summarize the key differences.

In Apache Griffin[6] (based on Spark and Hive), there is a DSL. However, it only contains the parameters to predefined rule types. E.g., for the type "accuracy", the rule essentially contains a logical expression that provides the link between two tables that are to be compared. In general, the parameters are used to instantiate predefined SQL patterns according to the rule type. For more specific rules, direct SQL can be inserted.

---

[6] http://griffin.apache.org.

Our main components Sources, Checks, and Actions are structurally similar to MobyDQ[7], however our Checks and Actions can be much more complex than indicators and alerts there and, in addition, data sources can be more diverse in *RADAR*.

Endler et al. developed an architecture for continuous data quality monitoring in medical centers [8]. They handle data quality monitoring with a rule-based approach. The most basic shown rules are simple boolean operations directly defined for the underlying database.

In [16], an approach with the goal to automate data quality checks in large data sets is presented. A wide range of basic types of checks is present, there are methods to automatically suggest potential quality rules, and integration of heterogeneous data sources is possible. They use a sophisticated optimization procedure to efficiently process the checks based on Spark and address differential quality checking in detail.

The work of Ehrlinger [7] focuses on automated and frequent execution of the DQ checks. Similarly, [5] also focuses on a highly efficient evaluation of rather simple rules (and with a more specific target, namely improvement of machine learning algorithms). In principle, these approaches complement each other as these efficient execution engines could be applied in principle to our simple rules part as well. However, the actual implementation technologies differ.

In summary, the main difference of our approach to these works is our DSL *RADAR* . It allows to extend the core functionality of the system with user-defined rule types. The logic of the rules is independent of the concrete data sources and independent of the target execution environment (i.e. Hive or a RDBMS or MongoDB). The action part of the DSL allows to instantiate rules in combination with specific sources. With its well-defined internal data model, the DSL also independent of the data model of the underlying DBMSes. Each rule can even execute on heterogeneous environments, combining data from different systems for cross-checking. The engine can execute the logic either by itself, thereby retrieving the source data without any transformations, or it can decide to execute parts of the logic using the features of the underlying DBMS. To the best of our knowledge, these are unqiue features, making our approach extremely flexible with respect to rule types and heterogeneous execution environments, while still allowing less technical users to define quality measurements. Furthermore, our DSL allows for advanced rule types like distribution checks (cf. [12]).

From the operational and query side, our work touches multiple lines of work in the database research that we are going to mention here. First of all, our *RADAR* language allows to formulate queries integrated into an imperative language. This is similar on the one hand to approaches like Oracle's PL/SQL [9] or other stored procedure languages, and on the other hand to LINQ-approaches (language integrated queries), see [4]. Yet another approach is to integrate SQL with functional programming [2]. However, our combination is specific to the domain of data quality and offers specific features.

---

[7] https://github.com/ubisoftinc/mobydq.

Furthermore, *RADAR* can act as a mediator [10] allowing to query and integrate multiple sources using a single piece of code. Thus the query optimization issues are quite similar to those in mediator-based systems. However, we don't define a global schema but allow the checks to directly access the individual sources, which is more appropriate for data quality checks. Similarly, a lot of work has been published in federated query processing which can complement our system in the future.

RADAR can access both relational and NoSQL databases from the same code, which again is a generic problem in heterogeneous environments. Here, both query languages (i.e. UnQL, [3]) and systems have been developed. There are multiple mediator approaches, e.g. [1,15,18] that allow to query relational and NoSQL-data from a unified SQL-like language. They are to some degree comparable, however, they lack the parametrizing features vital for our approach.

Parametrizing SQL in the way we do it – i.e. parameterizing also schema elements like attributes and relations and with the option to e.g. use a compound foreign key as actual parameter for a single attribute parameter – is, as far as we know, a new feature. From our point of view it is vital to achieve the flexible modularization and reuse-capabilities we are targeting. The only comparable approach uses the Maude system [17] to rewrite parametrized SQL statements. However, we integrated the parametrization capabilities into our core language.

6   Conclusion and Future Work

In this paper we have presented the DSL *RADAR* to specify data quality rules. The DSL has been developed based on a fixed set of requirements that have been set up together with company partners as potential users. We have shown the benefits of separating quality rules into Sources, Checks and Actions for different user groups and varying technical expertise. We have also illustrated how simple (NOT NULL check) as well as more complex (distribution check) quality rules are expressed in this language. More advanced quality rules (multidimensional checks) have already been discussed in a previous paper [12]. Our prototype is able to execute these quality checks on heterogeneous types of databases providing the source data. We have also shown how our engine optimizes rule execution by pushing parts of the execution logic towards the sources whenever possible. The implementation of the DSL allows for easy extensibility in case additional functionality is required. Other complex quality rules can also be specified and simple as well as complex quality rules can also be generated automatically based on analysis of existing data (profiling). Both has not been explained in detail here due to space constraints.

Within the project *RADAR* rules to check quality of real-world data provided by our project partner CTS Eventim, a large European online ticket seller, have been successfully used. In particular, rules have been applied to web tracking data to detect potential data quality issues, e.g. caused by a malfunctioning detection of user agent types. However, the need for further execution optimization on such large datasets has also been disclosed. In summary, *RADAR* together with the prototypical implementation provides an important step forward in efficiently managing data quality in data warehouses.

Several issues remain to be resolved in the future. While the applicability to different types of data sources has been shown in principal by using relational as well as document databases in the prototype, an extension to more types of data sources needs to be implemented. As stated before, execution optimization remains a major task in future work. From the practical experiments with the CTS Eventim data we can conclude that further optimization for relational data sources is beneficial. Separating functionality between DQ system and native source has to be addressed for non-relational data sources as well. Nevertheless, the DSL itself will remain unchanged even for different types of sources. Based on the formalized `antlr` grammar managing Sources, Checks and Actions could be simplified for the DQ manager with advanced editing capabilities such as syntax highlighting and automated code completion. Finally, future extensions of the DSL capabilities, particularly in the body of Checks, might be necessary to provide functionality that has not been necessary so far.

References

1. Atzeni, P., Bugiotti, F., Rossi, L.: SOS (Save Our Systems): a uniform programming interface for non-relational systems. In: Proceedings of the 15th International Conference on Extending Database Technology, pp. 582–585. ACM, New York (2012)
2. Binnig, C., Rehrmann, R., Faerber, F., Riewe, R.: FunSQL: it is time to make SQL functional. In: Proceedings of the 2012 Joint EDBT/ICDT Workshops, pp. 41–46. ACM, New York (2012)
3. Buneman, P., Fernandez, M., Suciu, D.: UnQL: a query language and algebra for semistructured data based on structural recursion. VLDB J. **9**(1), 76–110 (2000)
4. Cheney, J., Lindley, S., Wadler, P.: A practical theory of language-integrated query. In: Proceedings of the 18th ACM SIGPLAN International Conference on Functional Programming, ICFP 2013, pp. 403–416. ACM, New York (2013)
5. Ehrlinger, L., Haunschmid, V., Palazzini, D., Lettner, C.: A DaQL to monitor data quality in machine learning applications. In: Hartmann, S., Küng, J., Chakravarthy, S., Anderst-Kotsis, G., Tjoa, A.M., Khalil, I. (eds.) DEXA 2019. LNCS, vol. 11706, pp. 227–237. Springer, Cham (2019). https://doi.org/10.1007/978-3-030-27615-7_17
6. Ehrlinger, L., Rusz, E., Wöß, W.: A survey of data quality measurement and monitoring tools. arXiv preprint 1907.08138 (2019)
7. Ehrlinger, L., Wöß, W.: Automated data quality monitoring. In: MIT International Conference on Information Quality, vol. 22, pp. 19-1-19-8 (2017)
8. Endler, G., Schwab, P.K., Wahl, A.M., Tenschert, J., Lenz, R.: An architecture for continuous data quality monitoring in medical centers. In: MEDINFO (2015)
9. Feuerstein, S., Pribyl, B.: Oracle PL/SQL Programming. O'Reilly Media, Inc., Sebastopol (2005)
10. Garcia-Molina, H., et al.: The TSIMMIS approach to mediation: data models and languages. J. Intell. Inf. Syst. **8**(2), 117–132 (1997)
11. Heine, F., Kleiner, C., Koschel, A., Westermayer, J.: The Data Checking Engine: Complex Rules for Data Quality Monitoring. Int. J. Adv. Software **7**(12), 171–181 (2014)

12. Heine, F., Kleiner, C., Oelsner, T.: Automated detection and monitoring of advanced data quality rules. In: Hartmann, S., Küng, J., Chakravarthy, S., Anderst-Kotsis, G., Tjoa, A.M., Khalil, I. (eds.) DEXA 2019. LNCS, vol. 11706, pp. 238–247. Springer, Cham (2019). https://doi.org/10.1007/978-3-030-27615-7_18

13. Oelsner, T., Heine, F., Kleiner, C.: IQM4HD Concepts. Technical report, University of Applied Sciences and Arts Hannover, Germany (2018). http://iqm4hd.wp.hs-hannover.de/ConceptsIQM4HD_v09.pdf

14. Pipino, L.L., Lee, Y.W., Wang, R.Y.: Data quality assessment. Commun. ACM **45**(4), 211–218 (2002). https://doi.org/10.1145/505248.506010

15. Rith, J., Lehmayr, P.S., Meyer-Wegener, K.: Speaking in tongues: SQL Access to NoSQL systems. In: Proceedings of the 29th Annual ACM Symposium on Applied Computing, SAC 2014, pp. 855–857. ACM, New York (2014)

16. Schelter, S., Lange, D., Schmidt, P., Celikel, M., Biessmann, F., Grafberger, A.: Automating large-scale data quality verification. Proc. VLDB Endowment **11**(12), 1781–1794 (2018)

17. Sobieski, S., Zieliński, B.: Using maude rewriting system to modularize and extend SQL. In: Proceedings of the 28th Annual ACM Symposium on Applied Computing, SAC 2013, pp. 853–858. ACM, New York (2013)

18. Zhang, C., Xu, J.: A unified SQL middleware for NoSQL databases. In: Proceedings of the 2018 International Conference on Big Data and Computing, ICBDC 2018, pp. 14–19. ACM, New York (2018)

Fast One-to-Many Reliability Estimation for Uncertain Graphs

Junya Yanagisawa[1(✉)] and Hiroaki Shiokawa[2]

[1] Graduate School of Science and Technology, University of Tsukuba,
Tsukuba, Japan
y.junya@kde.cs.tsukuba.ac.jp
[2] Center for Computational Sciences, University of Tsukuba, Tsukuba, Japan
shiokawa@cs.tsukuba.ac.jp

Abstract. Uncertain graphs commonly represent noisy and unreliable real-world datasets in various applications. One fundamental primitive on uncertain graphs is reliability, which evaluates the connection robustness between two specific nodes. Although reliability is helpful to analyze uncertain graphs, it is computationally expensive because (1) the reliability estimation is #P-hard problem and (2) many applications require reliability computations among all possible pairs of nodes. To overcome the aforementioned problems, we present a novel algorithm called *Sharing RCSS+*, which efficiently computes the reliability among all possible pairs of nodes. Our extensive experiments on both real-world and synthetic uncertain graphs clarified that Sharing RCSS+ achieves a more efficient estimation than the state-of-the-art methods.

Keywords: Uncertain graphs · Reliability · Sampling algorithm

1 Introduction

Recently, *uncertainty* has been recognized as an essential concept to model real-world data as graphs since those data generally involve measurement noises, prediction errors, etc. For instance, protein-protein interactions (PPIs) are usually modeled as a graph, where nodes and edges represent proteins and interactions between proteins, respectively [12]. To infer the interactions, electrical simulations are necessary, but they are usually error prone. Thus, a PPI graph is inherently uncertain. Similarly, various applications of uncertainty in graphs arise in the fields of mobile ad-hoc networks, text mining, etc. [13]

Graphs accompanied with uncertainty are modeled by *uncertain graphs* $\mathcal{G} = (V, E, p)$, where V is a set of nodes, E is a set of edges, and p is a probability function $p : E \rightarrow (0, 1]$. Graph $\mathcal{G}$ is regarded as a probability space whose outcomes (called as *possible worlds*) are graphs $G = (V, E')$, where edge $e \in E$ is included in E' with probability $p(e)$ and is independent of the other edges. Because many real-world applications often involve uncertainty, and it is

© Springer Nature Switzerland AG 2020
S. Hartmann et al. (Eds.): DEXA 2020, LNCS 12391, pp. 106–121, 2020.
https://doi.org/10.1007/978-3-030-59003-1_7

more preferable to model data obtained from such applications as the uncertain graphs. Thus, various applications call for the development of fundamental primitives for mining uncertain graphs [3,14].

This work introduces a novel strategy for computing *reliability* [6] on uncertain graphs. Reliability is a fundamental primitives, which measures the probability that two given nodes are reachable [1]. This primitive can be regarded as a similarity between two nodes since it measures a connection robustness. Thus, reliability has been used to mine uncertain graphs such as uncertain graph clustering [3], and top-k similarity searches [14]. However, the complexity of the exact reliability computation is #P-hard [2] because all possible worlds G must be materialized from $\mathcal{G}$.

1.1   Existing Works and Challenges

To reduce the computational costs, sampling-based methods (*e.g.,* Monte-Carlo (MC) sampling [4], Lazy sampling [8], and Prob Tree [9]) have been proposed. These methods first sample K possible worlds $G_1, G_2, \ldots, G_K$ from $\mathcal{G}$ according to the independent edge probabilities. Given two nodes s and t in V, they estimate the reliability between s and t by the fraction of s-t reachable possible worlds among K samples. By letting $n = |V|$ and $m = |E|$, these methods require $O(K(n+m))$ time [6] to estimate the reliability between a pair of nodes.

The computational cost to analyze large uncertain graphs is hefty if such methods are used to mine uncertain graphs (*e.g.,* graph clustering and top-k search) due to two main reasons:

1. The sampling methods must sample many possible worlds to achieve a reasonable estimation accuracy [5,10] because the estimated reliability has a large variance. Furthermore, large uncertain graphs impose high computational costs to check the s-t reachability in each possible world. Consequently, it is time-consuming to estimate the reliability between a pair of nodes.
2. Sampling methods estimate the reliability only for a specific pair of nodes (*one-to-one reliability*), even though many applications require that the reliabilities from one node to all of the others (*one-to-many reliabilities*) are computed. For example, the clustering [3] and the top-k search [14] need to compute the one-to-many reliabilities for all nodes in the graph. To compute such reliabilities, the sampling methods have to perform multiple one-to-one reliability estimations, each of which requires $\Omega(nK(n + m))$ time.

To overcome the above problems, Zhu *et al.* recently proposed BFS Sharing (BFSS) [14], which integrates a *bit-wise offline sampling* with the breadth-first search. Although BFSS enables one-to-many estimations, a large number of possible worlds must be sampled to ensure the estimation accuracy since BFSS estimates the reliabilities based on the MC sampling technique. That is, BFSS is also time-consuming on large uncertain graphs. Therefore, realizing an efficient one-to-many reliability estimation remains a challenge.

1.2   Our Contributions

This work focuses on the problem of speeding up the one-to-many reliability estimations on uncertain graphs. We present a novel reliability estimation algorithm, *Sharing RCSS+*, which is designed to efficiently estimate the reliabilities while maintaining a high estimation accuracy. The basic idea underlying Sharing RCSS+ is to achieve a higher estimation accuracy while sampling fewer possible worlds compared to other methods. Existing methods (*e.g.,* MC sampling and BFSS) require sampling of many possible worlds to guarantee the accuracy because they repeatedly sample duplicate or unpromising possible worlds, which do not increase the estimation accuracy. However, such samples do not increase the estimation accuracy. To overcome this issue, Sharing RCSS+ employs recursive cut-set sampling (Sect. 3.2) before estimating one-to-many reliabilities. Our algorithm partitions a given uncertain graph into small subspaces to avoid sampling unpromising possible worlds. By discarding duplicate and unpromising samplings, Sharing RCSS+ tries to achieve fast and accurate one-to-many reliability estimations with a small number of possible worlds. As a result, our proposed method achieves the following attractive characteristics:

- **Efficient:** Sharing RCSS+ achieves a faster one-to-many reliability estimation than the state-of-the-art methods (Sect. 4.3).
- **Accurate:** Sharing RCSS+ has a smaller variance in the estimated reliabilities compared to the state-of-the-art methods (Theorem 1). Consequently, our proposed method achieves an accurate estimation with a smaller number of samples than the others (Sect. 4.2).
- **Scalable:** As the size of the uncertain graph increases, the time required to execute Sharing RCSS+ increases almost linearly (Sect. 4.3), and it is more scalable than the other estimation algorithms.

Our experiments show that our proposed method provides up to a 200 times faster estimation than MC sampling without sacrificing the estimation accuracy. Additionally, Sharing RCSS+ achieved 71 times faster estimations than those of the state-of-the-art algorithm [14]. Although the reliability estimation effectively enhances the application quality, it has been difficult to apply existing methods to real-world uncertain graphs. However, Sharing RCSS+ should improve the effectiveness of a wider range of applications because it is scalable and appropriate for a one-to-many reliability estimation.

Organization: The rest of this paper is organized as follows. Section 2 introduces basic notations and the backgrounds. Section 3 presents our proposed method, Sharing RCSS+, which achieves a fast reliability estimation on uncertain graphs. Then the experimental results are reported to verify the effectiveness of our approach in Sect. 4. Finally, Sect. 5 concludes this work.

2   Preliminary

2.1   Basic Notations

We here introduce the basic notation used in this paper. Table 1 summarizes the main symbols and their definitions. Let $\mathcal{G} = (V, E, p)$ be an uncertain graph,

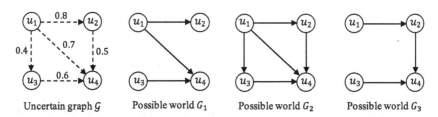

Fig. 1. Example of an uncertain graph.

Table 1. Definition of main symbols.

Symbol	Definition
$\mathcal{G}$	Uncertain graph
V	Set of nodes
E	Set of directed edges
p	Probability function such that $p : E \rightarrow (0, 1]$
n	Number of nodes in V, $i.e.$ $n = |V|$
m	Number of edges in E, $i.e.$ $m = |E|$
G_i	Possible world of $\mathcal{G}$, $i.e.$ $G_i \sqsubseteq \mathcal{G}$
E_i	Set of edges included in G_i
$Pr(G_i)$	Probability that $\mathcal{G}$ generates a possible world G_i
$\mathcal{G}_i$	Probability subspace in $\mathcal{G}$ (Definition 2)
$Pr(\mathcal{G}_i)$	Generating probability of subspace $\mathcal{G}_i$ (Definition 3)
$C(\mathcal{G}_i)$	Set of cut-set edges in $\mathcal{G}_i$ (Definition 4)
K	Number of sampled possible worlds
θ	Sampling threshold, $\theta \in \mathbb{N}$
r	Upper bound size of $|C(\mathcal{G}_i)|$
$R_{\mathcal{G}}(s, t)$	Exact reliability between nodes s and t in $\mathcal{G}$
$\hat{R}_G(s, t)$	Estimated reliability between nodes s and t in $\mathcal{G}$
$Var(\hat{R}_{\mathcal{G}}(s, t))$	Variance of the estimated reliability $\hat{R}_{\mathcal{G}}(s, t)$

where V and E are sets of nodes and directed edges, respectively. For convenience, we denote $n = |V|$ and $m = |E|$. p is a probability function $p : E \rightarrow (0, 1]$. That is, each edge $e \in E$ has a probability $p(e) \in (0, 1]$. An uncertain graph $\mathcal{G}$ can be viewed as a probability space whose outcomes are subgraphs of $\mathcal{G}$ where any edge $e \in E$ occurs with probability $p(e)$, which is independent of the other edges. These subgraphs are called *possible worlds*. A possible world of $\mathcal{G}$ is defined as $G_i = (V, E_i)$, where E_i is a set of edges occurring in G_i such that $E_i \subseteq E$. For convenience, we denote $G_i \sqsubseteq \mathcal{G}$ if and only if G_i is a possible world of $\mathcal{G}$.

From a given uncertain graph $\mathcal{G}$, up to 2^m possible worlds can be generated. For each possible world G_i, we define the generating probability of $\mathcal{G}_i$ as follows:

$$Pr(G_i) = \prod_{e \in E_i} p(e) \prod_{e \in E \setminus E_i} (1 - p(e)). \tag{1}$$

Note that $\sum_{G_i \sqsubseteq \mathcal{G}} Pr(G_i) = 1$ clearly holds.

Figure 1 shows an example uncertain graph $\mathcal{G}$, and its possible worlds G_1, G_2, and G_3. The dashed arrows are edges in $\mathcal{G}$ that are associated with a probability. By contrast, the solid arrow represents a materialized edge in possible worlds. As shown in Eq. (1), $\mathcal{G}$ generates the possible world G_1 with the probability $Pr(G_1) = 0.8 \times 0.7 \times 0.6 \times (1 - 0.4) \times (1 - 0.5) = 0.1008$. Similarly, G_2 and G_3 are generated from $\mathcal{G}$ by $Pr(G_2) = 0.0672$ and $Pr(G_3) = 0.0432$, respectively. Note that the possible worlds are not limited to G_1, G_2, and G_3. In fact, $\mathcal{G}$ has $2^5 = 32$ possible worlds.

We then define *reliability* on uncertain graphs. Given two nodes s and t in V, the reliability is the probability that node t is reachable from node s on an uncertain graph. Let $I_{G_i}(s, t)$ be an indicator function that returns 1 if node t is reachable from node s on a possible world G_i. Otherwise, $I_{G_i}(s, t) = 0$. The reliability $R_{\mathcal{G}}(s, t)$ is defined as follows:

$$R_{\mathcal{G}}(s, t) = \sum_{G_i \sqsubseteq \mathcal{G}} I_{G_i}(s, t) \cdot Pr(G_i). \tag{2}$$

That is, $R_{\mathcal{G}}(s, t)$ is a sum of the generating probability of G_i with a path from node s to node t among all possible worlds. The complexity of the exact reliability detection is #P-hard since all possible worlds need to be enumerated from $\mathcal{G}$.

Finally, we introduce the problem definition that we tackle in this work.

Definition 1 (One-to-many reliability). *Given uncertain graph $\mathcal{G}$ and node $s \in V$, one-to-many reliability is a problem to efficiently compute reliabilities $R_{\mathcal{G}}(s, t)$ for all nodes $t \in V$.*

As shown in Definition 1, the reliability $R_{\mathcal{G}}(s, t)$ needs to be computed multiple times to answer the one-to-many reliability problem.

2.2   Existing Sampling Algorithms

The reliability estimation is #P-hard problem. This leads to approximation techniques based on possible world sampling such as MC sampling [4] and BFSS [14]. Here, we briefly review these algorithms.

Monte-Carlo (MC) Sampling [4]: MC sampling is the most standard approximation algorithm to estimate the reliability between two nodes s and t. By following the probability of each edge, MC sampling randomly selects K possible worlds, $G_1, G_2, \ldots, G_K$, from a given uncertain graph $\mathcal{G}$. Estimated reliability $\hat{R}_{\mathcal{G}}(s, t)$ is computed from these possible worlds as follows:

$$\hat{R}_{\mathcal{G}}(s, t) = \frac{1}{K} \sum_{i=1}^{K} I_{G_i}(s, t). \tag{3}$$

The above equation indicates that MC sampling approximates the reliability by the fraction of s-t reachable possible worlds among all samples. If $R_{\mathcal{G}}(s,t)$ represents the exact reliability, Eq. (3) follows a Bernoulli trial with the probability of success $\hat{R}_{\mathcal{G}}(s,t)$. Therefore, in MC sampling, the estimator $\hat{R}_{\mathcal{G}}(s,t)$ has a variance given by

$$Var(\hat{R}_{\mathcal{G}}(s,t)) = \tfrac{1}{K^2} Var\left(\sum_{i=1}^{K} I_{G_i}(s,t)\right) = \tfrac{R_{\mathcal{G}}(s,t)\cdot(1-R_{\mathcal{G}}(s,t))}{K}. \tag{4}$$

The variance decreases as the size of K (*i.e.*, a number of sampled possible worlds) increases. However, Eq. (4) also implies that MC sampling needs to sample a large number of possible worlds to reduce the variance because the standard deviation of $\hat{R}_{\mathcal{G}}(s,t)$ decreases as the size of $\sqrt{K}$ increases. That is, MC sampling requires $4K$ samples to reduce the standard deviation to half of that obtained from K.

To solve the one-to-many reliability (Definition 1), we need to compute $\hat{R}_{\mathcal{G}}(s,t)$ for all nodes in V, which entails totally $\Omega(nK(n+m))$ time. Hence, MC sampling requires a large computation time for the one-to-many reliability.

BFS Sharing [14]: Zhu *et al.* recently proposed an efficient one-to-many reliability estimation method called BFSS [14] by extending MC sampling. By using bit-wise operations, BFSS estimates multiple reliabilities from a single breadth-first search (BFS) traversal.

Given a source node $s \in V$, BFSS performs the following offline sampling before the reliability estimation. First, BFSS randomly samples K possible worlds from $\mathcal{G}$. BFSS then assigns a K-bit vector $\mathcal{B}_e$ for each edge $e \in E$, whose i-th bit $\mathcal{B}_e[i]$ is 1 if e is included in a possible world G_i (*i,e.*, $e \in E_i$). Otherwise $\mathcal{B}_e[i] = 0$. Similarly, each node $v \in V$ also has a K-bit vector $\mathcal{B}_v$ to maintain reliability of node v from node s. The i-th bit $\mathcal{B}_v[i]$ is 1 only if node v is reachable from node s in the possible world G_i. Otherwise, $\mathcal{B}_v[i] = 0$. That is, for all $i \in K$, BFSS initially sets to $\mathcal{B}_s[i] = 1$, and $\mathcal{B}_v[i] = 0$ for $v \neq s$.

After that, BFSS estimates the one-to-many reliabilities by performing a BFS traversal from the source node s. If the BFS traversal reaches node v from node u, BFSS updates $\mathcal{B}_v$ by a bit-wise AND operation (*i.e.*, $\mathcal{B}_v = \mathcal{B}_u \text{AND} \mathcal{B}_e$, where $e = (u,v)$). BFSS continues the above BFS traversal until all bit vectors of nodes are updated. Finally, BFSS estimates the one-to-many reliabilities. By letting $\mathbb{1}(\mathcal{B}_v)$ represents the number of 1's in a bit vector $\mathcal{B}_v$, BFSS computes the reliability from node s to node v by $\hat{R}_{\mathcal{G}}(s,v) = \mathbb{1}(\mathcal{B}_v)/K$. Since every node $v \in \mathcal{G}$ maintains $\mathcal{B}_v$, the one-to-many reliabilities can be estimated from node s by computing $R(s,v) = \mathbb{1}(\mathcal{B}_v)/K$ for all nodes in $\mathcal{G}$.

However, as described in Sect. 1, BFSS requires to sample a large number of possible worlds to ensure estimation accuracy. This is because, BFSS still randomly samples possible worlds from $\mathcal{G}$, which is the same way as MC sampling. Consequently, BFSS also yields the same variance as MC sampling.

3   Proposed Method: Sharing RCSS+

The one-to-many reliability estimation requires a long computation time on large uncertain graphs. To address this issue, we present an efficient estimation algorithm called *Sharing RCSS+*.

3.1   Overview

To efficiently compute one-to-many reliabilities, we extend BFSS so that we can estimate the reliabilities from a smaller number of sampled possible worlds than BFSS. The proposed method has two main components:

(Step 1) Possible World Sampling: Our proposed method, Sharing RCSS+, performs offline sampling to construct a bit vector for each edge in $\mathcal{G}$. Given uncertain graph $\mathcal{G}$, number of samples K, and node $s \in V$, this step samples at most K possible worlds from $\mathcal{G}$. The goal of this step is to find K possible worlds that achieve a smaller variance than those of MC sampling and BFSS. To sample effective possible worlds, we present *a recursive cut-set sampling* in Sect. 3.2. Because the uncertain graph $\mathcal{G}$ can be regarded as a probability space, our algorithm recursively partitions the probability space into subspaces, and subsequently samples possible worlds from each subspace. In this manner, Sharing RCSS+ avoids sampling unpromising possible worlds.

(Step 2) Reliability Estimation: Sharing RCSS+ estimates the one-to-many reliabilities from node s to all of the other nodes in the uncertain graph $\mathcal{G}$. To detect the reliabilities, our algorithm uses the BFS-based estimation similar to BFSS shown in Sect. 2.2 on the K possible worlds obtained in (Step 1). Herein details of this step are omitted due to space limitations.

3.2   Recursive Cut-Set Sampling (RCSS)

We propose a recursive cut-set sampling (RCSS) that reduces the number of sampled possible worlds necessary to guarantee the estimation accuracy. The main idea underlying RCSS is that it discards duplicated and unpromising possible worlds containing many non-reachable nodes. To achieve this strategy, RCSS employs two key techniques: (1) *cut-set-based partitioning* and (2) *cut-set bounding*. By the cut-set-based partitioning, RCSS divides an uncertain graph into non-overlapping probability subspaces so that each one has many reachable nodes. Then RCSS samples a proportional number of possible worlds to the generating probability of each subspace. To further improve the estimation efficiency, RCSS selects several effective subspaces by the cut-set bounding. Here, we first define the probability subspace and then provide the detailed descriptions of the two key techniques.

Probability Subspace: First, we define the probability subspace as follows:

Definition 2 (Probability subspace). *Given uncertain graph $\mathcal{G} = (V, E, p)$, we suppose that sets of sampled edges E_i^1 and non-sampled edges E_i^0 exist such that $E_i^1 \cup E_i^0 \subset E$ and $E_i^1 \cap E_i^0 = \emptyset$. We define the probability subspace of $\mathcal{G}$ as $\mathcal{G}_i = (V, E_i^1, E_i^0, E_i^*, p)$, where $E_i^* = E \backslash \{E_i^1 \cup E_i^0\}$.*

By determining whether several edges are sampled or non-sampled, subspaces are generated from $\mathcal{G}$. For example, Fig. 2 shows three probability subspaces, $\mathcal{G}_1$, $\mathcal{G}_2$, and $\mathcal{G}_3$ generated from an uncertain graph $\mathcal{G}$. The solid arrows are sampled edges (*i.e.*, $E_1^1 = \{(u_1, u_2)\}$, $E_2^1 = \{(u_1, u_3)\}$, and $E_3^1 = \{(u_1, u_4)\}$). If the edges are non-sampled, then they are not shown in the subspaces (*i.e.*, $E_1^1 = \emptyset$, $E_2^0 = \{(u_1, u2)\}$, and $E_3^0 = \{(u_1, u_2), (u_1, u_3)\}$).

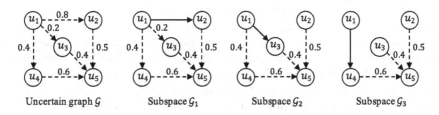

Fig. 2. Example of probability subspaces.

Definition 3 (Generating probability of $\mathcal{G}_i$). *Let $Pr(\mathcal{G}_i)$ be the generating probability of a subspace $\mathcal{G}_i$. Then probability $Pr(\mathcal{G}_i)$ is given as*

$$Pr(\mathcal{G}_i) = \prod_{e \in E_i^1} p(e) \prod_{e \in E_i^0} (1 - p(e)). \qquad (5)$$

The generating probability of each subspace is given by the probability of sampled and non-sampled edges. For instance, the subspace $\mathcal{G}_1$ can be generated by sampling edge (u_1, u_2) (Fig. 2). Thus, the generating probability of $\mathcal{G}_1$ is $Pr(\mathcal{G}_1) = 0.8$. Similarly, $\mathcal{G}$ outputs $\mathcal{G}_2$ by non-sampling (u_1, u_2) and sampling (u_1, u_3), resulting in $Pr(\mathcal{G}_2) = (1 - 0.8) \times 0.2 = 0.04$.

Cut-Set-Based Partitioning: From a given uncertain graph $\mathcal{G}$, Sharing RCSS+ recursively partitions $\mathcal{G}$ into non-overlapping subspaces by fixing the states of several edges in E. To effectively generate the subspaces, Sharing RCSS+ determines the state of *cut-set edges* for each subspace. A cut-set is a set of edges that splits a given graph $\mathcal{G}$ into two non-connected subgraphs by removing all edges in the cut-set. Formally, we define the cut-set edges as follows:

Definition 4 (Cut-set edges). *Given $\mathcal{G}_i = (V, E_i^1, E_i^0, E_i^*, p)$, a set of cut-set edges is denoted as $C(\mathcal{G}_i) = \{(u, v) \in E_i^* | u \in V_i^1, v \in V_i^0\}$, where V^1 is a set of nodes that can be reached from the source node s via edges in E_i^1, and V_i^0 denotes the rest of the nodes, i.e., $V_i^0 = V \backslash V_i^1$.*

For example, suppose we have subspace $\mathcal{G}_2$ shown in Fig. 2. If node u_1 is a source node, the cut-set edges should be $C(\mathcal{G}_2) = \{(u_1, u_4), (u_3, u_5)\}$ since $V_2^1 = \{u_1, u_3\}$ and $V_2^0 = \{u_2, u_4, u_5\}$. Similarly, if $\mathcal{G} = \mathcal{G}_o$ is given, $C(\mathcal{G}_0) = \{(u_1, u_2), (u_1, u_3), (u_1, u_4)\}$ is cut-set edges.

Once we receive probability subspace $\mathcal{G}_i$, a set of cut-set edges $C(\mathcal{G}_i)$ is obtained by Definition 4. Then, $|C(\mathcal{G}_i)| + 1$ edge sampling patterns are generated so that the patterns do not yield overlapping subspaces. Suppose $C(\mathcal{G}_i) = \{e_1, e_2, e_3, \ldots, e_{|C(\mathcal{G}_i)|}\}$, Table 2 shows the sampling patterns generating non-overlapping subspaces. In Table 2, 1 and 0 denote that $e_i \in C(\mathcal{G}_i)$ is and is not sampled, respectively. If e_j is $*$, $e_j \in E_i^*$ in a subspace $\mathcal{G}_i$.

Table 2. Sampling patterns.

Subspace	e_1	e_2	e_3	$\cdots$	$e_{|C(\mathcal{G}_i)|}$
$\mathcal{G}_0$	0	0	0	$\cdots$	0
$\mathcal{G}_1$	1	$*$	$*$	$\cdots$	$*$
$\mathcal{G}_2$	0	1	$*$	$\cdots$	$*$
$\mathcal{G}_3$	0	0	1	$\cdots$	$*$
$\vdots$				$\ddots$	
$\mathcal{G}_{|C(\mathcal{G}_i)|}$	0	0	0	$\cdots$	1

In Table 2, $\mathcal{G}_0$ can be regarded as an unpromising subspace because most nodes in $\mathcal{G}_0$, except for V_0^1, are not reachable due to the lack of cut-set edges. Hence, for such unpromising subspace $\mathcal{G}_0$, (1) only $\langle t, Pr(\mathcal{G}_0) \rangle$ is kept for each node $t \in V_0^1$ to balance the estimation later, and (2) $\mathcal{G}_0$ is removed from RCSS regardless of its generating probability. That is, $|C(\mathcal{G}_i)|$ subspaces are finally obtained from $\mathcal{G}_i$.

After generating $|C(\mathcal{G}_i)|$ subspaces (i.e., $\mathcal{G}_1, \ldots, \mathcal{G}_{|C(\mathcal{G}_i)|}$), we compute their generating probabilities by Definition 3. Let $\theta \in \mathbb{N}$ be a user-specified threshold and K be the maximum sample size. We perform the following steps for each subspace $\mathcal{G}_j \in C(\mathcal{G}_i)$:

1. If $Pr(\mathcal{G}_j) \cdot K < \theta$, we sample $\lceil Pr(\mathcal{G}_j) \cdot K \rceil$ possible worlds from $\mathcal{G}_j$ by MC sampling.
2. Otherwise, $\mathcal{G}_j$ is recursively partitioned in the same way.

The above partitioning is repeated until all non-partitioned subspaces are sampled. Consequently, we obtain at most K possible worlds.

Cut-Set Bounding: Sharing RCSS+ partitions a subspace $\mathcal{G}_i$ into at most $|C(\mathcal{G}_i)|$ subspaces. However, this strategy fails to improve the accuracy if $|C(\mathcal{G}_i)|$ is too large because a large $|C(\mathcal{G}_i)|$ generates many subspaces with a small generating probability. That is, we need to sample so many possible worlds with quite low probability without discarding unpromising possible worlds. To avoid this issue, we introduce an upper bound size of $|C(\mathcal{G}_i)|$ as r. Once we have $|C(\mathcal{G}_i)| > r$, we first sort all edges in $C(\mathcal{G}_i)$ by their probability in ascending order. Then, we select smaller r edges from $C(\mathcal{G}_i)$ by removing edges whose probability is larger than r-th smallest one.

3.3 Algorithm

Algorithm 1 shows the pseudo-code of Sharing RCSS+, which consists of the possible world sampling (lines 1–19) and the reliability estimation (lines 20–22).

Algorithm 1. Sharing RCSS+

Require: Graph $\mathcal{G}$, source node $s \in V$, parameters K, r, and θ.
Ensure: One-to-many reliabilities $\hat{R}_{\mathcal{G}}(s, t)$ for all $t \in V$.
 1: Queue $Q \leftarrow \{\mathcal{G}\}$, and Queue $P \leftarrow \emptyset$;
 2: **while** $Q \neq \emptyset$ **do**
 3:     Dequeue $\mathcal{G}_i$ from Q, and Set $S \leftarrow \emptyset$;
 4:     Obtain $C(\mathcal{G}_i)$ by Definition 4;
 5:     **if** $|C(\mathcal{G}_i)| > r$ **then**
 6:         Sort $C(\mathcal{G}_i)$;
 7:         Select r edges, $e_1, e_2, \ldots, e_r$, from $C(\mathcal{G}_i)$ by the cut-set bounding;
 8:         $C(\mathcal{G}_i) \leftarrow \{e_1, e_2, \ldots, e_r\}$;
 9:         Generate subspaces $\mathcal{G}_0, \mathcal{G}_1, \ldots, \mathcal{G}_r$ by Table 2;
10:         $S \leftarrow \{\mathcal{G}_0, \mathcal{G}_1, \ldots, \mathcal{G}_r\}$;
11:     **else**
12:         Generate subspaces $\mathcal{G}_0, \mathcal{G}_1, \ldots, \mathcal{G}_{|C(\mathcal{G}_i)|}$ by Table 2;
13:         Enqueue $\langle t, Pr(\mathcal{G}_0) \rangle$ to P for each reachable node $t \in \mathcal{G}_0$;
14:         $S \leftarrow \{\mathcal{G}_1, \ldots, \mathcal{G}_r\}$;
15:     **for each** $\mathcal{G}_j$ in S **do**
16:         **if** $Pr(\mathcal{G}_j) \cdot K < \theta$ **then**
17:             Sample $\lceil Pr(\mathcal{G}_j) \cdot K \rceil$ possible worlds from $\mathcal{G}_j$ by MC sampling;
18:         **else**
19:             Enqueue $\mathcal{G}_j$ into Q;
20: Estimate $\hat{R}_{\mathcal{G}}(s, t)$ for all $t \in V$ by BFSS (Section 2.2);
21: **for each** $\langle t, Pr(\mathcal{G}_0) \rangle \in P$ **do**
22:     $\hat{R}_{\mathcal{G}}(s, t) \leftarrow \hat{R}_{\mathcal{G}}(s, t) + Pr(\mathcal{G}_0)$;

In the possible world sampling, Sharing RCSS+ starts the cut-set based partitioning shown in Sect. 3.2. First, it obtains cut-set edges $C(\mathcal{G}_i)$ by Definition 4 (line 4). If $|C(\mathcal{G}_i)| > r$, our algorithm selects r edges from $C(\mathcal{G}_i)$ by the cut-set bounding (lines 5–8), and then generates the subspaces (lines 9–10). Otherwise, it first generates the probability subspaces, $\mathcal{G}_0, \mathcal{G}_1, \ldots, \mathcal{G}_r$ from Table 2 (line 12). Then, our algorithm removes unpromising subspace $\mathcal{G}_0$ after saving $\langle t, Pr(\mathcal{G}_0) \rangle$ to balance the reliability estimation later (lines 12–13). For each subspace $\mathcal{G}_j$, Sharing RCSS+ samples $\lceil Pr(\mathcal{G}_j) \cdot K \rceil$ possible worlds from $\mathcal{G}_i$ by MC sampling if $Pr(\mathcal{G}_j) \cdot K < \theta$ (lines 16–17); otherwise, it recursively partitions $\mathcal{G}_j$ into subspaces (lines 18–19). Finally, Sharing RCSS+ obtains at most K possible worlds after the termination.

After that, our algorithm starts the reliability estimation (lines 20–22). Sharing RCSS+ initially estimates the reliabilities by using BFSS (line 20). Then, our proposal balances the estimated reliabilities (lines 21–22) by using $\lceil Pr(\mathcal{G}_j) \cdot K \rceil$ obtained by unpromising subspace $\mathcal{G}_0$ (line 13). Finally, Sharing RCSS+ outputs the estimated reliabilities.

3.4 Theoretical Analysis

Finally, we assessed the variance of our reliability estimation approach.

Theorem 1. *If uncertain graph $\mathcal{G}$ is divided into r subspaces, $\mathcal{G}_1, \mathcal{G}_2, \ldots, \mathcal{G}_r$, the estimator $\hat{R}_{\mathcal{G}}(s,t)$ of Sharing RCSS+ has the following variance:*

$$Var(\hat{R}_{\mathcal{G}(s,t)}) = \sum_{i=1}^{r} Pr(\mathcal{G}_i) \frac{R_{\mathcal{G}_i}(s,t)(1-R_{\mathcal{G}_i}(s,t))}{K}, \tag{6}$$

where $R_{\mathcal{G}_i}(s,t)$ is the exact reliability between nodes s and t on $\mathcal{G}_i$.

Proof. For each subspace $\mathcal{G}_i$, Sharing RCSS+ samples $Pr(\mathcal{G}_i) \cdot K$ possible worlds from $\mathcal{G}_i$ by MC sampling. That is, by following Eq. (4), Sharing RCSS+ has the estimator $\hat{R}_{\mathcal{G}}(s,t)$ is given as follows:

$$\hat{R}_{\mathcal{G}}(s,t) = \sum_{i=1}^{r} Pr(\mathcal{G}_i) \cdot \hat{R}_{\mathcal{G}_i}(s,t), \tag{7}$$

where $\hat{R}_{\mathcal{G}_i}(s,t)$ is the estimated reliability between nodes s and t on a subspace $\mathcal{G}_i$. Hence, estimator $\hat{R}_{\mathcal{G}}(s,t)$ has the following variance:

$$Var(\hat{R}_{\mathcal{G}}(s,t)) = \sum_{i=1}^{r} Pr(\mathcal{G}_i) \frac{R_{\mathcal{G}_i}(s,t)(1-R_{\mathcal{G}_i}(s,t))}{K}, \tag{8}$$

which completes the proof. $\square$

By comparing Theorem 1 with Eq. (4), our proposed algorithm has a smaller variance than MC sampling and BFSS. Therefore, Sharing RCSS+ can achieve accurate estimations with fewer samples.

4   Experimental Analysis

We conducted extensive experiments to evaluate the effectiveness of Sharing RCSS+. Our experiments demonstrate that:

- **Sampling size:** Sharing RCSS+ requires fewer samples to estimate accurate reliabilities compared to the state-of-the-art methods (Sect. 4.2).
- **Efficiency and scalability:** Sharing RCSS+ performs faster estimations than the state-of-the-art algorithms (Sect. 4.3). Also, Sharing RCSS+ has a nearly linear scalability for graph sizes.
- **Effectiveness:** The key techniques of Sharing RCSS+, cut-set-based partitioning and the cut-set bounding, improve the estimation time on real-world graphs (Sect. 4.4).

4.1   Experimental Setup

We compared Sharing RCSS+ with the baseline algorithm MC sampling [4] and the state-of-the-art method BFSS [14]. All algorithms were implemented in C++ and compiled with GNU gcc 8.2.0 using -O3 option. All experiments were conducted on a server with an Intel Xeon CPU (3.50GHz) and 128 GiB RAM. Unless otherwise stated, we used default parameters of $r = 50$ and $\theta = 5$.

Reproducibility: Other researchers can confirm the reproducibility as we plan to share our codes publicly after the acceptance of this paper.

Datasets: We evaluated the algorithms on four real-world uncertain graphs, which were published by Ke *et al.* [6] and Sasaki *et al.* [11]. Table 3 summarizes the statistics of real-world datasets. [1] In addition, we also used synthetic graphs generated by LFR-benchmark [7], which is considered as the *de facto* standard model for generating graphs. The settings will be detailed later.

Table 3. Statistics of real-world datasets.

Dataset	$|V|$	$|E|$	Average degree	Mean probability (±SD)	Data source
LastFM	6,899	23,696	3.43	0.29 (±0.25)	[6]
NetHEPT	15,233	62,774	4.12	0.04 (±0.04)	[6]
Tokyo	26,370	64,596	2.45	0.39 (±0.15)	[11]
NYC	180,188	416,880	2.31	0.29 (±0.13)	[11]

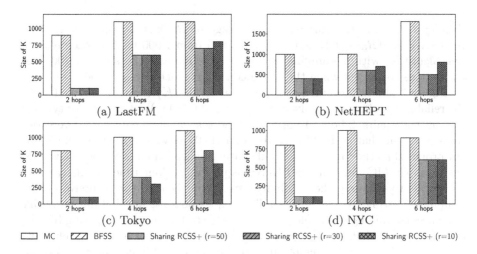

Fig. 3. Number of sampled possible worlds until convergence.

4.2 Sampling Sizes to Estimate Converged Reliabilities

We first assessed the number of sampled possible worlds to obtain converged reliabilities on the real-world graphs. To define when the estimated reliabilities are converged, we introduced the dispersion index ρ_K, which is used in [6]. Given 100 distinct pairs of nodes (*i.e.*, $\langle s_1, t_1 \rangle$, $\langle s_2, t_2 \rangle$, ..., $\langle s_{100}, t_{100} \rangle$), the dispersion index is computed as $\rho_K = \sum_{i=1}^{100} \frac{Var(\hat{R}_{\mathcal{G}}(s_i, t_i))}{Avg(\hat{R}_{\mathcal{G}}(s_i, t_i))}$, where $Var(\hat{R}_{\mathcal{G}}(s_i, t_i))$ and

---

[1] In Table 3, "SD" means the standard deviation of the edge probabilities.

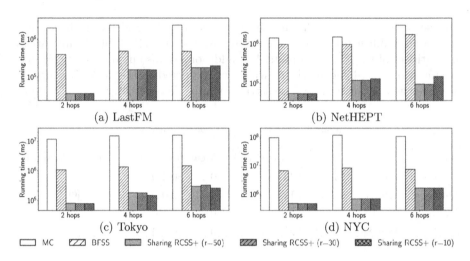

Fig. 4. Reliability estimation time.

$Avg(\hat{R}_{\mathcal{G}}(s_i, t_i))$ are the variance and average of $\hat{R}_G(s_i, t_i)$, respectively. Here, we derived $Var(\hat{R}_{\mathcal{G}}(s_i, t_i))$ and $Avg(\hat{R}_{\mathcal{G}}(s_i, t_i))$ from 1,000 trials of each estimation algorithm with a sampling size K. K was increased in steps of 100. The estimations converged when the dispersion index reached $\rho_K < 0.001$.

Figure 3 shows the number of possible worlds (K) that are sampled when the reliability estimations converged (*i.e.*, $\rho_K < 0.001$). We tested three types of distinct node pairs: 2-hops, 4-hops, and 6-hops away node pairs. Also, we varied the size of r in Sharing RCSS+ as 10, 30, and 50. As we can see from the figure, our proposed method requires a significantly smaller numbers of possible worlds. For example, MC sampling and BFSS require up to four times as many possible worlds to converge the reliability estimation. Due to the recursive cut-set sampling, Sharing RCSS+ successfully samples possible worlds from promising subspaces containing reachable paths from s_i to t_i. Furthermore, Sharing RCSS+ avoids sampling that overlaps possible worlds, which causes inefficient reliability estimations. On the other hand, MC sampling and BFSS must generate possible worlds from probability space $\mathcal{G}$, resulting in overlapping possible worlds. Consequently, they require many samples to preserve low variances ρ_K. In summary, our proposed method requires fewer number of samples to estimate accurate reliabilities compared to those necessary for the state-of-the-art methods.

4.3   Efficiency and Scalability

Efficiency: Next, we assessed the efficiency of each algorithm to estimate the reliabilities. Similar to Sect. 4.2, we measured the running time to estimate the reliabilities of 100 distinct node pairs. We tested on the three types of node pairs: 2-hops, 4-hops, and 6-hops away node pairs, and varied the size of r in Sharing RCSS+ as 10, 30, and 50. In addition, we set the sampling sizes K for each experimental setting to the same sizes as shown in Fig. 3.

Figure 4 shows the running times to estimate the reliabilities on the real-world graphs. Our proposed method outperforms all the other examined algorithms. Of particular interest, Sharing RCSS+ achieves up to a 200 times faster estimation than MC sampling. Our algorithm is up to 71 times faster than the state-of-the-art method BFSS. Additionally, Sharing RCSS+ effectively reduces the sampling sizes compared to those of MC sampling and BFSS. As proven in Theorem 1, our proposed algorithm has a better variance than those of MC sampling and BFSS. Hence, Sharing RCSS+ achieves faster one-to-many reliability estimations than the other competitive algorithms.

Scalability: We also assessed the scalability of Sharing RCSS+ by using the synthetic graphs generated by LFR-benchmark [7]. We generated four synthetic datasets by varying the number of nodes from 10^3 to 10^6 with an average degree 5 and a maximum degree 20. For each directed edge from node u, we assigned $\log(d_u + 1)/\log(d_{\max} + 2)$ as the edge probability by following [11], where d_u and $d_{\max}$ are the degree of u and the maximum degree in each graph, respectively.

Figure 5 shows the running time to the estimate reliabilities of 100 distinct 4-hops away node pairs. We used the same sampling sizes K for each algorithm as those in Fig. 3. In this evaluation, we also tested the scalability for 2-hops away and 6-hops away node pairs, but the results are omitted because they are similar to those for 4-hops away node pairs. As we can observe from Fig. 5, Sharing RCSS+ is still faster than the others even when the graph sizes increases. Furthermore, our proposed algorithm shows a nearly linear scalability in terms of the number of nodes, demonstrating that Sharing RCSS+ shows a better scalability than the state-of-the-art methods.

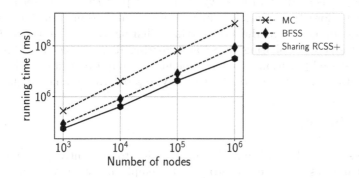

Fig. 5. Scalability by varying node sizes.

4.4 Effectiveness of Key Techniques

Finally, we examined the effectiveness of the key techniques of Sharing RCSS+. Sharing RCSS+ employed (1) the cut-set-based partitioning and (2) the cut-set bounding to effectively sample possible worlds. We compared the estimation

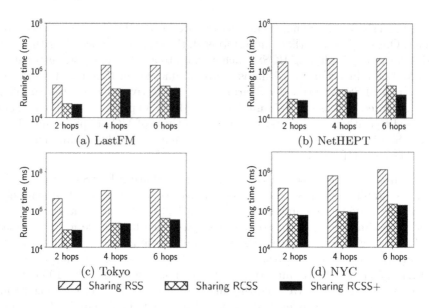

Fig. 6. Effectiveness of key techniques.

time of our proposed algorithm with two variants of Sharing RCSS+: *Sharing RSS* and *Sharing RCSS*. Sharing RSS lacks the cut-set-based partitioning and the cut-set bounding. It randomly samples r edges without considering the cut-set edges shown in Definition 4 and then partitions a subspace in the same way. Sharing RCSS is a minor variant, which does not perform the cut-set bounding.

Figure 6 shows the estimation time when each algorithm obtained converged reliabilities for each experimental setting. Sharing RCSS+ is up to 60 and 15 times faster than Sharing RCSS and Sharing RSS, respectively. Moreover, the cut-set-based partitioning achieves significant improvements in all datasets, demonstrating that our approach effectively improve the estimation efficiency.

5 Conclusion

We propose a novel algorithm, Sharing RCSS+, to efficiently compute one-to-many reliabilities on uncertain graphs. By employing recursive cut-set-based partitioning, Sharing RCSS+ reduces the number of sampled possible worlds to ensure the estimation accuracy (Theorem 1 and Sect. 4.2). Our proposal estimates the reliabilities within a shorter running time than the state-of-the-art methods (Sect. 4.3).

Acknowledgement. This work was supported by JSPS KAKENHI Early-Career Scientists Grant Number JP18K18057, and JST ACT-I.

References

1. Aggarwal, K.K., Misra, K.B., Gupta, J.S.: Reliability evaluation a comparative study of different techniques. Microelectron. Reliab. **14**(1), 49–56 (1975)
2. Ball, M.O.: Computational complexity of network reliability analysis: an overview. IEEE Trans. Reliab. **35**(3), 230–239 (1986)
3. Ceccarello, M., Fantozzi, C., Pietracaprina, A., Pucci, G., Vandin, F.: Clustering uncertain graphs. PVLDB **11**(4), 472–484 (2017)
4. Fishman, G.S.: A comparison of four monte carlo methods for estimating the probability of s-t connectedness. IEEE Trans. Reliab. **35**(2), 145–155 (1986)
5. Jin, R., Liu, L., Ding, B., Wang, H.: Distance-constraint reachability computation in uncertain graphs. PVLDB **4**(9), 551–562 (2011)
6. Ke, X., Khan, A., Quan, L.L.H.: An in-depth comparison of S-t reliability algorithms over uncertain graphs. PVLDB **12**(8), 864–876 (2019)
7. Lancichinetti, A., Fortunato, S., Radicchi, F.: Benchmark graphs for testing community detection algorithms. Phys. Rev. E **78**, 046110 (2008)
8. Li, Y., Fan, J., Zhang, D., Tan, K.: Discovering your selling points: personalized social influential tags exploration. Proc. SIGMOD **2017**, 619–634 (2017)
9. Maniu, S., Cheng, R., Senellart, P.: An indexing framework for queries on probabilistic graphs. ACM Trans. Database Syst. **42**(2), 1–34 (2017)
10. Potamias, M., Bonchi, F., Gionis, A., Kollios, G.: K-nearest neighbors in uncertain graphs. PVLDB **3**(1–2), 997–1008 (2010)
11. Sasaki, Y., Fujiwara, Y., Onizuka, M.: Efficient network reliability computation in uncertain graphs. Proc. EDBT **2019**, 337–348 (2019)
12. Shiokawa, H., Fujiwara, Y., Onizuka, M.: SCAN++: efficient algorithm for finding clusters, hubs and outliers on large-scale graphs. PVLDB **8**(11), 1178–1189 (2015)
13. Shiokawa, H., Onizuka, M.: Scalable graph clustering and its applications. Encyclopedia of Social Network Analysis and Mining, pp. 2290–2299 (2018)
14. Zhu, R., Zou, Z., Li, J.: Top-k reliability search on uncertain graphs. Proc. ICDM **2015**, 659–668 (2015)

Quality Matters: Understanding the Impact of Incomplete Data on Visualization Recommendation

Rischan Mafrur[1](✉) [iD], Mohamed A. Sharaf[2] [iD], and Guido Zuccon[1] [iD]

[1] The University of Queensland, Brisbane, Australia
{r.mafrur,g.zuccon}@uq.edu.au
[2] United Arab Emirates University, Al Ain, UAE
msharaf@uaeu.ac.ae

Abstract. Incomplete data is a crucial challenge to data exploration, analytics, and visualization recommendation. Incomplete data would distort the analysis and reduce the benefits of any data-driven approach leading to poor and misleading recommendations. Several data imputation methods have been introduced to handle the incomplete data challenge. However, it is well-known that those methods cannot fully solve the incomplete data problem, but they are rather a mitigating solution that allows for improving the quality of the results provided by the different analytics operating on incomplete data. Hence, in the absence of a robust and accurate solution for the incomplete data problem, it is important to study the impact of incomplete data on different visual analytics, and how those visual analytics are affected by the incomplete data problem. In this paper, we conduct a study to observe the interplay between incomplete data and recommended visual analytics, under a combination of different conditions including: (1) the distribution of incomplete data, (2) the adopted data imputation methods, (3) the types of insights revealed by recommended visualizations, and (4) the quality measures used for assessing the goodness of recommendations.

Keywords: Incomplete data · Visualization recommendation · Data exploration

1 Introduction

To support effective data exploration, there has been a growing interest in developing solutions that can automatically recommend data visualizations that reveal important data-driven insights. Several visual analytic tools have been introduced such as Tableau [9], Spotfire [8], Power BI [7]. The aim of those tools is to provide aesthetically high-quality visualizations that reveal interesting insights. Without any prior knowledge of the explored data, it is a challenging task for the analyst to manually select the combinations of attributes and measures that lead to interesting visualizations. Clearly, manually looking for insights in each visualization is a labor-intensive and time-consuming process. Such challenge motivated research efforts that focused on automatic recommendation of visualizations based on some metrics that capture the utility of recommended visualizations (e.g., [15,17–19,22,23,28,35,36]). However, all of those approaches

© Springer Nature Switzerland AG 2020
S. Hartmann et al. (Eds.): DEXA 2020, LNCS 12391, pp. 122–138, 2020.
https://doi.org/10.1007/978-3-030-59003-1_8

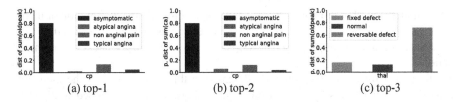

Fig. 1. Top-k recommended visualizations obtained from complete heart disease dataset, $k = 3$

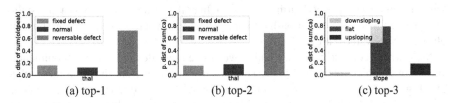

Fig. 2. Top-k recommended visualizations obtained from incomplete heart disease dataset (20% missing values), $k = 3$, NaN values are ignored

operate under the assumption that the analyzed data is clean and overlook the data quality problems that might impair the recommendation process.

Data quality is a crucial challenge to data exploration and analytics. Poor data quality would distort the analysis and reduce the benefits of any data-driven approach. That is, *garbage in, garbage out (GIGO)* phenomenon. In real world settings, most datasets exhibit data quality problems, such as incomplete data, which in turn leads to incorrect analytical results (e.g., [22,26]). This is true for *descriptive analytics*, in which incomplete data leads to incorrect results for aggregate and statistical queries [39]. It is also equally true for *predictive analytics*, where reduced accuracy in classification and prediction are common side effects of working with incomplete data (e.g., [10,16]). Moreover, in the general context of recommendation systems, incomplete data has been shown to result in inaccurate rankings, which has the expected effect of producing poor and misleading recommendations [31].

Several *data imputation* methods have been introduced to handle the incomplete data challenge (e.g., [13,24,27]). However, it is well-known that those methods cannot fully solve the incomplete data problem, but they are rather a mitigating solution that allows for improving the quality of the results provided by the different analytics operating on incomplete data [10]. Hence, in the absence of a robust and accurate solution for the incomplete data problem, it remains especially important to study the impact of incomplete data on different visual analytics, and how those visual analytics are affected by the incomplete data problem. This has been the focus of several research studies, including assessing the impact of incomplete data on analytics that rely on aggregate and statistical queries [39], predictions and classifications (e.g., [10,16]), or recommendation [31].

To the best of our knowledge, this work is the first to explore the impact of incomplete data on the quality of recommended visualizations. In particular, our focus in this work is to study the interplay between incomplete data and recommended visual analytics, under a combination of different conditions including: the distribution of incomplete data, the adopted data imputation methods, the types of insights revealed by those visualizations, and the quality measures used for assessing the goodness of recommendations.

To further illustrate the problems addressed in this work, consider the motivating example shown in Figs. 1 and 2. Both figures show the recommended top-k visual insights from a heart disease dataset [4] under two different settings: (1) complete data (Fig. 1), versus (2) incomplete data, with 20% missing values (Figure 2). The detail about the figures is explained further in Fig. 3a. In both settings, the top-k visual insights are generated using the deviation-based approach [36], where $k = 3$, and any missing cells (i.e., NaN values) are ignored.

Meanwhile, comparing Figs. 1 and 2, we notice the following: 1) the recommendations from complete data (Fig. 1) are significantly different from those on incomplete data (Fig. 2); 2) the two sets of recommendations have only one visualization in common (i.e., visualization based on *sum oldpeak* vs. *thal*[1]); and 3) that one common visualization was ranked top-3 based on the complete data, whereas it is ranked top-1 based on the incomplete data!

Based on the example above, a user who is analyzing an incomplete data with 20% missing values, would obtain a top-k recommended visualizations that are significantly different from those obtained from a complete dataset, and in turn gaining "false" insights from the data. Since incomplete data is a prevailing problem that can only be slightly mitigated by data imputation methods, it becomes essential to evaluate and quantify its impact on the insights gained from visual data analytics approaches. That is precisely the goal of this work, in which our main contributions are summarized as follows:

1. We study the different types of visual insights that are generally sought by data analysts in their data exploration workflows (Sect. 2).
2. We present three quality measures to quantify the impact of incomplete data on the quality of visualization recommendation (Sect. 3).
3. We conduct an extensive experimental evaluation on real datasets and present the impact of incomplete data on recommended visualizations with different data cleaning methods and different type of visual insights (Sect. 4).

2   Recommending Visual Insight

To recommend visual insight, we consider a multi-dimensional database D, which consists of a set of dimensional attributes $\mathbb{A}$ and a set of measure attributes $\mathbb{M}$. Also, let $\mathbb{F}$ be a set of possible aggregate functions over measure attributes. Hence, specifying different combinations of dimension and measure attributes along with various aggregate functions, generates a set of possible visualizations $\mathbb{V}$ over D. For instance, a possible visualization V_i is specified by a tuple $< A_i, M_i, F_i >$, where $A_i \in \mathbb{A}$, $M_i \in \mathbb{M}$, and $F_i \in \mathbb{F}$, and it can be formally defined as: Vi : VISUALIZE bar (SELECT A, F(M) FROM D WHERE T GROUP BY A). Where VISUALIZE specifies the

---

[1] thal: Thallium heart scan (normal, fixed defect, reversible defect).

visualization type (i.e., bar chart), SELECT extracts the selected columns which can be dimensional attributes $A \in \mathbb{A}$ or measures $M \in \mathbb{M}$, T is the query predicate (e.g., disease = 'Yes'), and GROUP BY is used in collaboration with the SELECT statement to arrange identical data into groups.

Figure 1 shows the top-k recommended visual insights obtained from the complete heart disease dataset where $k = 3$. Figure 1a is obtained from V_i : VISUALIZE bar (SELECT cp, SUM(oldpeak) FROM HeartDiseaseDB WHERE disease='Y' GROUP BY cp). However, obtaining this visualization V_i is only possible if the analyst knows exactly the parameters, which specify some aggregate visualizations that lead to those valuable visual insights (e.g., dimensional attributes, measures, aggregate functions, grouping attributes, etc.). Hence, it is time-consuming to iteratively create and refine visualizations to search for the ones that are useful and interesting.

Motivated by the need for efficient data analysis and exploration, several solutions for recommending visualizations have recently emerged (e.g., [14,15,18,28,29, 35,36]). In such solutions, a large number of possible data visualizations $\mathbb{V}$ are generated and ranked according to some metrics that capture the *utility* of recommended visualizations. Towards this, the utility of each visualization V_i in $\mathbb{V}$ is calculated according to the type of insight, which is described next.

In this work, we study three types of visual insights: The first type is the *aggregate-based insight* which has been shown to be effective in recommending visualizations based on some metrics that capture the utility of a recommended visualizations (e.g., [15,35,36]). The second type is the *correlation-based insight*. This insight type is generally sought by data analysts looking for the attribute pairs with the highest correlations [14]. The third type is the *distribution-based insight* (e.g., *skewness* and *kurtosis*) (e.g., [14,32]). In general, data analysts utilize distribution-based insight in order to find the dimensions that deviate from the normal distribution. Hence, by considering those insight types, we study insights based on single dimension (i.e., distribution-based insight), pairs of measures (i.e., correlation-based insight) and combination of dimensional attributes and aggregate functions of measures (i.e., aggregate-based insight). An example of those three types of visual insights can be seen in Fig. 3. Given three types of the insights above, our problem definition as follows:

Definition 1. Recommending top-k visual insights: *Given a dataset D, insight type Y, the goal is to recommend a set top-k visual insight $S \subseteq \mathbb{V}$, where $|S| = k$, and $\mathbb{V}$ is the set of all possible generated visualizations from D, such that the overall utility $U(S)$ based on Y is maximized.*

Meanwhile, the utility of each visualization V_i is computed based on the type of insight shown by recommended visualizations, which are explained next.

2.1 Aggregate-Based Insight

In this paper, we address two types of aggregate-based insight: *outstanding* and *similarity* (e.g., [34,36]). *Outstanding-based insight* recommends the most outstanding visualizations based on *deviation-based* approach (e.g., [17,29,36]). The deviation-based approach is able to provide analysts with interesting visualizations that high-

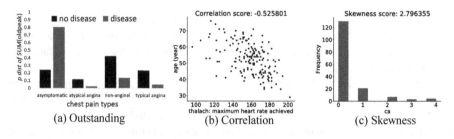

Fig. 3. Top-1 recommended visualization of (a) *aggregate/outstanding*, (b) *correlation* and (c) *skewness-based insight* from heart disease dataset where NaN values are ignored

light some of the particular trends of the analyzed datasets. The deviation-based app-roach compares an aggregate visualization generated from the selected subset dataset D_Q (i.e., target visualization $V_i(D_Q)$) to the same visualization if generated from a reference dataset D_R (i.e., reference visualization $V_i(D_R)$). To calculate the outstanding/deviation score, each target visualization $V_i(D_Q)$ is normalized into a *probability distribution* $P[V_i(D_Q)]$ and similarly, each reference visualization into $P[V_i(D_R)]$. In particular, consider an aggregate visualization $V_i =< A, M, F >$. The result of that visualization can be represented as the set of tuples: $< (a_1, g_1), (a_j, g_j), ..., (a_t, g_t) >$, where t is the number of distinct values (i.e., groups) in attribute A, a_j is the j-th group in attribute A, and g_j is the aggregated value $F(M)$ for the group a_j. Hence, V_i is normalized by the sum of aggregate values $G = \sum_{j=1}^{t} g_j$, resulting in the probability distribution $P[V_i] =< \frac{g_1}{G}, \frac{g_2}{G}, ..., \frac{g_t}{G} >$. Finally, the utility score of V_i is measured in terms of the distance between $P[V_i(D_Q)]$ and $P[V_i(D_R)]$, and is simply defined as: $U(V_i) = dist(P[V_i(D_Q)], P[V_i(D_R)])$

Figure 3a shows the top-1 recommended visualization of outstanding-based insight which is generated by [36] from heart disease dataset. The figure shows that an aggregate visualization based on *sum oldpeak* (i.e., pressure of the ST segment, where ST segment is an isoelectric section of the ECG) vs. *chest pain types* exhibits a large deviation between the target visualization (disease) and reference visualization (no-disease). That is, patients with a heart disease often suffer more from asymptomatic chest pains, in comparison to those without heart disease.

Meanwhile, *similarity-based insight* is the opposite to the outstanding-based insight. This insight type recommends the closest visualizations compared to the reference dataset [34].

2.2 Correlation-Based Insight

In the context of data exploration, data analysts generally derive insights from the data by iteratively computing and visualizing correlations looking for the attribute pairs with the highest correlations [14], either high positive or negative correlated [32]. Hence, the correlation-based insight recommends visualizations with the high correlated pair of measures. A visualization of correlation-based insight V_i is specified by a tuple $< B, C >$, where B and $C \subseteq$ M. The result of that visualization can be represented as the set of tuples: $< (b_1, c_1), (b_2, c_2), ..., (b_n, g_n) >$. Finally, the utility score of V_i is measured in terms of correlation coefficient of a tuple $< B, C >$. We use the Pearson correlation coefficient, which is formally defined as: $U(V_i) = \frac{\sum_{i=1}^{n}(b_i - \bar{b})(c_i - \bar{c})}{\sqrt{\sum_{i=1}^{n}(b_i - \bar{b})^2}\sqrt{\sum_{i=1}^{n}(c_i - \bar{c})^2}}$. Figure 3b shows top-1 recommended visualization V_i of correlation-based insight which is generated from the heart disease dataset, where V_i : VISUALIZE scatter (SELECT thalach, age FROM HeartDiseaseDB WHERE disease='Y'). The figure shows the high negative correlation of two measures (thalach: maximum heart rate achieved vs. age) where the correlation score is -0.53.

2.3 Distribution-Based Insight

Many classical statistical tests depend on normality assumptions [3]. Significant skewness and kurtosis clearly indicate that the data is not normaly distributed. Skewness is a measure of the lack of symmetry, while kurtosis is a measure of whether the data is heavy-tailed or light-tailed relative to a normal distribution. Generally, data analysts utilize values of skewness and kurtosis in order to find the attributes and measures that deviate from the normal distribution [12].

The distribution-based insight recommends the dimensional attributes or measures that most deviate from the normal distribution (e.g., [12,14]). A visualization of distribution-based insight V_i is specified by a tuple $< E,$ COUNT(E)$>$. The utility score for V_i is measured in terms of the third standardized moment μ_3 of V_i for the skewness-based insight and the fourth standardized moment μ_4 of V_i for the kurtosis-based insight. Hence, $U(V_i)$ for the skewness-based insight is $\frac{\mu_3}{\sigma^3}$, where $\mu_3 = \frac{\sum_{i=1}^{n}(e_i - \bar{e})^3}{n}$ and $\sigma = \frac{\sum_{i=1}^{n}(e_i - \bar{e})^2}{n}$. Meanwhile, $U(V_i)$ for the kurtosis-based insight is $\frac{\mu_4}{\sigma^4}$, where $\mu_4 = \frac{\sum_{i=1}^{n}(e_i - \bar{e})^4}{n}$ and $\sigma = \frac{\sum_{i=1}^{n}(e_i - \bar{e})^2}{n}$. In all cases, μ is the mean, σ is the standard deviation. Figure 3a shows the top-1 recommended visualization V_i of the skewness-based insight, where V_i : VISUALIZE bar (SELECT ca, COUNT(ca) FROM HeartDiseaseDB WHERE disease='Y' GROUP BY ca). The figure shows ca is the dimension with the highest skewness score: $+2.8$, where ca is the number of major vessels colored by flourosopy.

3 Incomplete Data and Visualization Recommendation Quality

In this section, we first discuss the incomplete data problems (Sect. 3.1). Then, we introduce the quality measures used for assessing the quality of recommendations. (Sect. 3.2).

3.1 The Incomplete Data Problem

Data quality is a crucial challenge to data exploration and analytics. Poor quality data would distort the analysis and reduce the benefits of any data-driven approach causing profound economic impact. Research has shown that the average cost of poor data on a business is 30% or more of its revenue [1]. The New York Times has also reported that analysts spend 50%–80% of their time preparing dirty data before it can be used for data analytics [6]. Common examples of data quality challenges include multiple representations as a result of merging data from a variety of sources, incomplete data, anomalies, invalid, extreme, erroneous or duplicate values (e.g., [22,26]).

In this paper, we focus on the incomplete data challenge. Incomplete data is common problem for data analytics (e.g., [10,16,26]). In *descriptive analytics*, incomplete data can lead to misleading conclusions such as wrong results for aggregate queries [39]. Meanwhile, in *predictive analytics*, incomplete data can introduce bias into a prediction or classification models (e.g., [10,16]). Moreover, in the context of *recommendation systems*, incomplete data has been shown to result in inaccurate rankings, which has the expected effect of producing misleading recommendations [31].

Several data cleaning techniques have been introduced to overcome incomplete data issues include substituting missing data values by mean, median, or the most frequent value (e.g., [24,27]), or using k-Nearest Neighbor [11], or association rules [38]. However, it is well-known that those imputation methods cannot fully solve the incomplete data problem. For instance, recent studies such as [10,20] compared the performance of several imputation methods (e.g., median, linear regression) and showed the reduction of prediction and classification accuracy using those imputation methods.

Instead of proposing a new imputation method, this work investigates the impact of incomplete data on the quality of recommended visualizations. To the best of our knowledge, there is no prior work that focuses on that area. Existing work (e.g., [25,33]) used sampling techniques to generate data visualizations and inspect the quality of the visualizations. However, our problem differs from those studies. Those studies focus on the quality of visualization while our work focuses on the quality of recommended visualizations. Another work is Profiler [22], which visualizes the data quality problems. This study also differs from ours. Profiler recommends visualizations that reveal data quality problems while our work recommends visualizations that reveal insights.

Recent data quality studies investigated the impact of incomplete data in predictive analytics (e.g., [10,20]). Those studies compared the performance of various imputation methods on different supervised classifiers and explored the impact of incomplete data on the quality of classification and prediction models. Our problem differs from those studies in two ways. First, those studies focus on the impact of incomplete data in predictive analytics while our work is studying the impact of incomplete data in descriptive analytics. Second, the context of those studies are on general classification and prediction problems while our context is on visualization recommendation.

Toward investigating the impact of incomplete data on the quality of visualization recommendation, we introduce three measures for assessing the quality of recommendations, which explained next.

3.2 Quality of Recommended Visualizations

Recall from definition 1 that the goal of visualization recommendation is to recommend a set of top-k visualizations that reveal insights, in particular, as formulated in the definition 1, given a multi-dimensional dataset D, the set of top-k visualizations S is

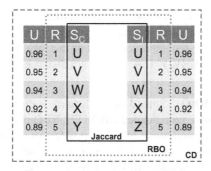

Fig. 4. A set of visualizations generated from complete data $S_C = (U, V, W, X, Y)$ and visualizations generated from incomplete data $S_I = (U, V, W, X, Z)$, R is ranking and U is utility score. (Color figure online)

recommended. Let us consider D_I is the incomplete version of D. To facilitate the discussion, let us assume S_C is the set of top-k visualizations from the complete data, and it is equally to S. Moreover, S_I is the set of top-k visualizations from the complete data D. In order to understand the interplay between incomplete data and recommended visualizations, the top-k set obtained from an incomplete data S_I is compared to the top-k set obtained from the complete data S_C.

In this work, we utilize various metrics to assess the quality of the recommended visualizations in S_I compared to S_C. First, we utilize the *Jaccard distance* [30], which compares the composition of two sets as in Fig. 4. The score of Jaccard distance is calculated by the number of visualizations in common, divided by the total number of visualizations. Accordingly, when applied to the set comparison, two sets with the same composition will have the same similarity score. However, in our work, the order of visualizations in the top-k set is essential. For instance, the top-1 visualization is more important than the top-10 visualization. Hence, we utilize the second metric, *Rank Biased Overlap (RBO)* [37], to consider the visualization ranking when assessing the quality of recommendations. As shown in Fig. 4, RBO considers the composition of the two sets and their ranking, and it can be seen within the blue dotted line.

Finally, we have two metrics to evaluates our recommended visualizations. However, both metrics only compare the composition of the sets without considering the utility score of each visualization inside the set. Thus, we utilize the third metric called *Cumulative Distance (CD)* [21]. This metric captures both the utility score of each visualization $U(V_i)$ and visualization ranking. Figure 4 within the red dashed line illustrates the scope of the CD metric. The detail of those three metrics is explained next.

Jaccard Distance. Jaccard distance [30] is defined as the magnitude of the intersection divided by the magnitude of the union of the two sets, which is formally defined as: $Jaccard(S_I, S_C) = \frac{|S_I \cap S_C|}{|S_I \cup S_C|}$. This distance is bounded by 1. The value is between 0 for no similarity and 1 for identical sets. According to Figure 4, consider $S_C = (U, V, W, X, Y)$ and $S_I = (U, V, W, X, Z)$, Jaccard distance score of S_I to S_C is $\frac{4}{6} = 0.66$. The score is obtained from the number of intersection (i.e., four visualizations in common U, V, W, X) divided by the union (i.e., six visualizations in total U, V, W, X, Y, Z). This computation is based on the composition of both sets, the visualization ranking inside the set is not counted. For instance, if the visualizations in S_I is reversed (i.e., $S_I = (Z, X, W, V, U)$), the Jaccard distance score is still same 0.66.

Rank Biased Overlap (RBO). Since Jaccard distance is discounting the visualization order, we utilize the second metric called RBO [37]. RBO is a popular metric in Information Retrieval, which commonly used for the problem of comparing two ranked lists. RBO is compatible with item order and also compatible with the dis-jointness problem (i.e., an item is present only in one ranked list). In this work, we adopt RBO to quantify the quality of recommended visualizations in S_I compared to S_C.

To calculate RBO score, RBO determines the fraction of content overlapping at different depths. Consider at each depth d, the intersection of sets S_I and S_C to depth d is: $I_{S_I, S_C, d} = S_{I:d} \cap S_{C:d}$. The size of this intersection is the overlap of sets S_I and S_C to depth d, $X_{S_I, S_C, d} = |I_{S_I, S_C, d}|$ and the proportion of S_I and S_C that are overlapped at depth d is their agreement, $A'_{S_I, S_C, d} = \frac{X_{S_I, S_C, d}}{d}$. Hence, the RBO score of S_I and S_C is defined as: $RBO(S_I, S_C, p) = (1 - p) \sum_{d=1}^{\infty} p^{d-1} * A'_{S_I, S_C, d}$. Similar to Jaccard, RBO has a range between 0 and 1, where 0 means disjoint, and 1 means identical. The parameter p models the user's persistence which is the probability of the user continuing to the next visualization. In particular, the smaller p, e.g., $p = 0$, only the top-ranked visualization is considered, and the RBO score is either zero or one. Meanwhile, if $p = 1$, the evaluation becomes arbitrarily deep due to the probability of deciding to stop is 0. The suggested p value is 0.95 or 0.97 [37]. In this work, we used $p = 0.95$, it means that the first 20 ranks have 86% of the weight of the evaluation.

Consider the example in Fig. 4, using RBO the effectiveness score of S_I in comparison to S_C is 0.84 due to the both sets S_I, and S_C have only one different visualization on the tail. The Y is the last visualization in S_C, while the Z is the last visualization in S_I. However, if both sets have different on the head (i.e., top-1 visualization), the RBO score is 0.7. This example shows the visualization ranking is counted in RBO.

Cumulative Distance (CD). We utilize Cumulative Distance as our third metric. We adopt CD from DCG (Discounted Cumulative Gain) [21]. Similar to RBO, the DCG metric is generally used in Information Retrieval. This metric is a popular method for measuring the quality of search results. It assumes that highly relevant results are more valuable than marginally relevant results, and the top result is more important than the tail. The DCG works by combining the degree of relevance and the rank of the search results in a coherent way. Meanwhile, the DGC score is unbounded. Hence, we can use the *normalized DCG (nDCG)*. The nDCG is defined as the actual DCG performance for a search query divided by the ideal DCG performance. To the best of our knowledge, this work is the first to use CD (i.e., mapped from nDCG) in the context of visualization recommendation. In our work, the degree of relevance of the visualization V_i is the utility score $U(V_i)$, where the utility score $U(V_i)$ is calculated according to the type of insight as explained in Sect. 2. The CD score of S_I to S_C is defined as the DCG of S_I

divided by DCG of S_C : CD $= \frac{\sum_{i=1,i\in S_I}^{n} \frac{1}{\log_2(i+1)} *U_i}{\sum_{i=1,i\in S}^{n} \frac{1}{\log_2(i+1)} *U_i}$, where U_i is utility score of each visualization from the complete dataset D.

Accordingly, Jaccard and RBO score from the example in Fig. 4 are 0.66 and 0.84. Those scores indicate that both sets have quite a lot of differences. However, when we look at the CD score, it provides a different perspective. The score of the CD is 0.99. It is close to 1 (i.e., almost identical). That is because the utility score of Y and Z is precisely same $(U(Y) = U(Z) = 0.89)$, which means the degree of importance of both visualizations (Y and Z) is the same.

4 Experimental Evaluation

In this section, we first discuss our experimental testbed (Sect. 4.1). Then, we present and discuss our experimental evaluation. (Sect. 4.2).

4.1 Experimental Testbed

Data cleaning methods: In this work, we utilize and compare various well-known data cleaning methods, which are summarized as follows:

1. *Ignore cell*: The top-k visual insights are generated directly from the incomplete dataset by ignoring missing cells. In this approach, the process of handling incomplete data is on the cell level (e.g., [10,20]).
2. *Eliminate row* : The process of handling incomplete data is on the row or tuple level. Particularly, a row that contains a missing cell is dropped. If the amount of missing cells is large, it may end up eliminating a huge amount of data [27].
3. *Impute cell*: In this approach, we utilize two common imputation techniques:
   (a) *Median and most frequent imputation*: This approach works by calculating the median of the non-missing values in a column and then replacing the missing values with the median within each column if the missing values are numerical data. Meanwhile, if the missing values are categorical data (strings or numerical representations), the missing values are imputed with the most frequent values within each column (e.g., [10,20]).
   (b) *KNN imputation*: This approach imputes the missing data by finding the k closest neighbors to the observation with missing data and then imputing them based on the non-missing values in the neighbors.

Datasets: We conduct our experiments over the following datasets: (1) The Cleveland heart disease dataset is comprised of 8 dimensional attributes, 6 measures, and 299 tuples [4]. (2) The New York Airbnb dataset is comprised of 4 dimensional attributes, 4 measures, and 30249 tuples [5]. (3) The Diabetes 130 US hospital dataset consists of 14 dimensional attributes, 13 measures and 100 thousand tuples [2]. We conduct our experiments over those three datasets, however, due to space limit, the Cleveland heart disease dataset is the default dataset for presenting the results in this paper.

Incomplete Data: We simulate missing data completely at random (MCAR) with different settings: (1) the distribution of missing values is on dimensional attributes $\mathbb{A}$, (2)

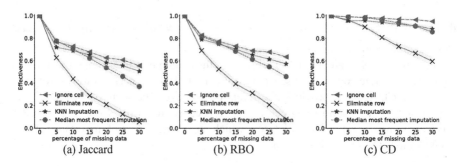

Fig. 5. Impact of data cleaning methods on effectiveness of outstanding-based insight using different data cleaning methods, $k = 10$

the distribution of missing values is on dimensional measures $\mathbb{M}$, and (3) the distribution of missing values is on the whole data $\mathbb{A} + \mathbb{M}$. Recall from definition 1, in this experiment, we create an incomplete version data D_I from D. Then compare the top-k set S_I, which generated from the incomplete data D_I to the top-k set S_C, which generated from complete data D. In order to avoid bias, 100 versions of D_I with different random missing seed are generated. Finally, we repeat the experiments with different settings including: the percentage of missing values (i.e., 0%– 90%), the number of k, the type of insights, the data cleaning methods, and the quality measures used for assessing the quality of recommendation.

Default Parameters: The default parameters used in our evaluation are $k = 10$, the percentage of missing data is 10%, the default of data cleaning method is *ignore cell*, the default dataset is Cleveland heart disease. The final result is the average from 100 versions of D_I and we present the results with confidence interval $CI = 0.95$.

Aggregate-Based Insight: In the case of aggregate-based insight, we use five aggregate functions (COUNT, AVG, SUM, MIN and MAX) where COUNT is only COUNT(*). We use different query predicates T to understand the impact of input queries on the quality of recommendation with different percentages of missing values. For example, we want to compare an aggregate visualization generated from the selected subset dataset *chest pain types = 'typical angina'* to the visualization if generated from a reference dataset *chest pain types != 'typical angina'*. In this work, to study the impact of query predicate T on the quality of recommendation, we use three different queries for heart disease dataset: 1) q_1: *cp = typical angina* vs *cp != typical angina*; 2) q_2: *sex = Female* vs *sex = Male*; 3) q_3: *exang = exercise induced angina* vs *exang != exercise induced angina*.

4.2   Experimental Evaluation

In this section, we discuss our experiment results under a combination of different settings including: (1) the adopted data imputation methods, (2) the distribution of incomplete data, (3) the types of insights revealed by those visualizations, and (4) the quality measures used for assessing the quality of recommended visualizations.

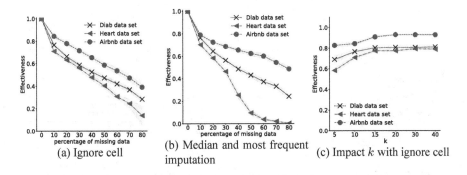

Fig. 6. Impact of Data Cleaning Methods on Effectiveness with Different Datasets - (a,b) *ignore cell vs. median and most frequent imputation,* (c) impact of k on effectiveness using *ignore cell* method

Impact of the Data Cleaning Methods on Effectiveness. In this experiment, we analyze the effectiveness of data cleaning methods under different percentage of missing values and the quality measures (Jaccard, RBO, and CD). We compare four common data cleaning methods (e.g., *ignore cell, eliminate row, median and most frequent imputation,* and *KNN imputation*). Since the *eliminate row* method is included, the maximum percentage of missing values for this experiment is 30%. Moreover, the missing values are distributed on the whole data and the results of this experiments are generated based on the outstanding-based insight. As shown in Fig. 5, the best data cleaning method is *ignore cell* and the worst is *eliminate row.* That is because that *eliminate row* leads to eliminate a huge amount of data. To the contrary, by ignoring missing cells without eliminating row, *ignore cell* outperforms other data cleaning methods. Meanwhile, in terms of imputation methods, *KNN imputation* has a better effectiveness than *Median and most frequent imputation* method. The result shows that the patterns are consistent for the three quality measures.

Impact of the Data Cleaning Methods on Different Datasets. In this experiment, we analyze the effectiveness of data cleaning methods under different datasets. We compare two data cleaning methods, which are *ignore cell* and *median and most frequent imputation* and the results of this experiments are generated based on the outstanding-based insight. The missing values are distributed on the whole data and maximum percentage of missing values for this experiment is 80%. As shown in Fig. 6, overall, the pattern from three datasets are similar. In particular, in terms of the impact of missing values (Figs. 6a and 6b), the effectiveness is decreasing when the number of missing values are increased. Moreover, in terms of the impact of k (Fig. 6c), the effectiveness is increasing when k is increased. Meanwhile, if we compare Figs. 6a and 6b, the effectiveness of *ignore cell* is better than *median and most frequent imputation,* especially for heart disease dataset. That is because the heart disease dataset has more dimensional attributes rather than measures. Imputing missing values on categorical data using *most frequent* method reduces the effectiveness. Further, the result of the Airbnb dataset is contrary to the result of the heart disease dataset. That is because the Airbnb dataset has more measures rather than dimensional attributes. The airbnb dataset consists of four dimen-

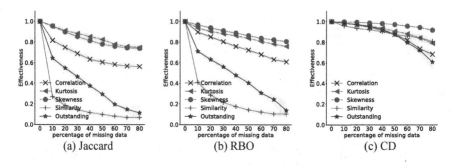

Fig. 7. Impact of incomplete data on effectiveness of different insight types, $k = 10$

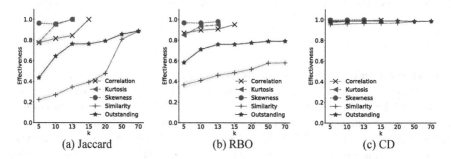

Fig. 8. Impact of k on effectiveness of different insight types, 10% missing values

sional attributes and four measures. However, since no incomplete data on predicate, the missing values are distributed on three dimensional attributes and four measures. Based on the results, we can conclude that *median and most frequent imputation* outperforms *ignore cell* if the data has more missing values on measures.

Impact of Incomplete Data on Effectiveness. Figure 7 shows the impact of incomplete data on effectiveness under different types of insights. The figure shows that if the percentage of missing data is higher then it reduces the quality of visualization recommendation. The most resilient insight type to incomplete data is distribution-based insight (i.e., skewness, kurtosis), then the correlation-based insight, and the less resilient is aggregate-based insight. The skewness-based insight and kurtosis-based insight are specified by a single attribute or measure. Hence, losing a certain percentage of data will not change much of the data in each dimension. Meanwhile, the correlation-based insight is based on a pair of measures. Hence, the correlation-based insight less tolerance to the incomplete compared to the distribution-based insight. The aggregate-based insight is the most complex insight type. It is specified by the combination of dimensional attributes and the aggregate function of measures. Hence, the aggregate-based insight is the most sensitive to incomplete data, especially the similarity-based insight.

Impact of k on Effectiveness. As shown in Fig. 8, the higher number of k results in the higher effectiveness due to the probability of the top-k set from the incomplete data having same content to the top-k set from the complete data is higher if the number of

k is larger. For instance, Jaccard score is equal to 1, if $k = |\mathbb{V}|$, where the number of k equal to the number of candidate visualizations, however, it is only applies to Jaccard not to RBO and CD

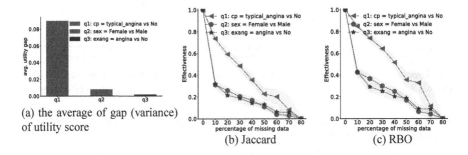

(a) the average of gap (variance) of utility score

(b) Jaccard

(c) RBO

Fig. 9. Impact of input queries on effectiveness of outstanding-based insight, $k = 10$

Impact of Input Queries on Effectiveness. Figure 9 shows the impact of predicate queries on the quality of visualization recommendation. Three different queries are used: 1) q_1: *cp = typical angina* vs *cp != typical angina*; 2) q_2: *sex = Female* vs *sex = Male*; 3) q_3: *exang = exercise induced angina* vs *exang != exercise induced angina*. Figure 9a shows that $q1$ is more resilient to the incomplete data compared to other input queries (Fig. 9b and 9c). The results show that if the input query generates top-k set that the variance among utility score of visualizations is very low, this low variance leads to more loss on effectiveness especially if the number of missing values is high.

Impact of Dimensional Attributes, Measures, and Attributes+Measures on Effectiveness. Do the incomplete data on dimensional attributes have more impact rather than on measures? If so, when data analyst has a dataset with missing values on both dimensional attributes and measures, then she should give more attention to dimensional attributes rather than measures. Based on the experiment results, missing values on attributes and measures have the same impact on the effectiveness. Figure 10 shows the impact of dimensional attributes, measures, and both on effectiveness with different percentage of missing values. The results are generated based on the heart disease dataset with the distribution of missing values on attributes and measures are equal. The results show that categorical and numerical data are equally important.

Impact of Recommendation Quality Metrics on Effectiveness Using Different Number of k and Different Missing Data Distributions. Figure 11 shows the impact of k on effectiveness if the incomplete data only on attributes, only on measures, and on both attributes and measures. As mentioned above, missing values on attributes and measures have the same impact on effectiveness (Figure 11a and 11b). The results also show how the performance of our three quality measures (i.e., Jaccard, RBO, and CD) under different number of k. Cumulative distance CD always performs above Jaccard and RBO. It is because of the default of percentage of missing values is quite small (10%). Meanwhile, there is an interesting pattern in Fig. 11c, the figure shows that if

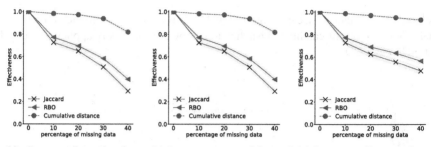

(a) Impact dimensional at- (b) Impact measures $\mathbb{M}$ on ef- (c) Impact attributes and mea-
tributes $\mathbb{A}$ on effectiveness   fectiveness   sures $\mathbb{A} + \mathbb{M}$ on effectiveness

Fig. 10. Impact of dimensional attributes, measures, and attributes + measures on effectiveness of outstanding-based insight, $k = 10$

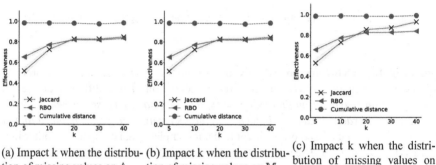

(a) Impact k when the distribu- (b) Impact k when the distribu- (c) Impact k when the distri-
tion of missing values on $\mathbb{A}$ tion of missing values on $\mathbb{M}$ bution of missing values on
                                                                          $\mathbb{A} + \mathbb{M}$

Fig. 11. Impact of k on effectiveness of outstanding-based insight 10% missing values

the number of k is small (e.g., $5, 10$), Jaccard performs under RBO, however, when k is large (e.g., > 20), Jaccard performs above RBO and there is a crossover between both of them. Hence, the higher number of k results in the higher effectiveness in terms of Jaccard but not RBO. Jaccard score is equal to 1, if $k = |\mathbb{V}|$ where the number of k equal to the number of candidate visualizations. To the contrary, RBO has a different pattern, RBO score can be equal to 1 if visualizations inside the two top-k sets are in the same order, which is hard to be achieved. Hence, by increasing the number of k does not necessarily result in increased effectiveness in terms of RBO.

5   Conclusions

In this work, we investigate the interplay between incomplete data and recommended visual analytics under a combination of different conditions. This study lays the foundation for further exploring appropriate ways to deal with incomplete data and minimize the impact of incomplete data on visualization recommendation. We believe that this work can provide valuable insights for data analysts rather than blindly believing a recommendation result over low-quality data.

Acknowledgments. Rischan Mafrur is sponsored by the Indonesia Endowment Fund for Education (Lembaga Pengelola Dana Pendidikan / LPDP)(201706220111044). Dr Mohamed A. Sharaf is supported by UAE University Grant (G00003352). Dr Guido Zuccon is the recipient of an Australian Research Council DECRA Research Fellowship (DE180101579) and a Google Faculty Award.

References

1. Bad data cost. https://www.entrepreneur.com/article/332238
2. Diabetes 130 us hospitals 1999–2008. https://www.kaggle.com/brandao/diabetes
3. e-handbook of statistical methods. http://www.itl.nist.gov/div898/handbook/
4. Heart disease data set. https://archive.ics.uci.edu/ml/datasets/heart+Disease
5. Inside airbnb. http://insideairbnb.com/new-york-city/
6. Janitor work is key hurdle to insights. https://nyti.ms/1mZywng
7. Power bi. https://powerbi.microsoft.com/en-us/
8. Spotfire. https://www.tibco.com/products/tibco-spotfire/
9. Tableau. https://public.tableau.com/s/
10. Barata, A.P., et al.: Imputation methods outperform missing-indicator for data missing completely at random. In: ICDM (2019)
11. Batista, G.E.A.P.A., et al.: A study of k-nn as an imputation method. In: HIS (2002)
12. Bono, R., et al.: Bias, precision, and accuracy of skewness and kurtosis estimators for frequently used continuous distributions. SYMMAM **12**(1), 19 (2020)
13. Cambronero, J., et al.: Query optimization for dynamic imputation. PVLDB **10**(11), 1310–1321 (2017)
14. Demiralp, Ç., et al.: Foresight: recommending visual insights. PVLDB **10**(12), 1937–1940 (2017)
15. Ding, R., et al.: Quickinsights: quick and automatic discovery of insights from multidimensional data. In: SIGMOD (2019)
16. Ehrlinger, L., Haunschmid, V., Palazzini, D., Lettner, C.: A DaQL to monitor data quality in machine learning applications. In: Hartmann, S., Küng, J., Chakravarthy, S., Anderst-Kotsis, G., Tjoa, A.M., Khalil, I. (eds.) DEXA 2019. LNCS, vol. 11706, pp. 227–237. Springer, Cham (2019). https://doi.org/10.1007/978-3-030-27615-7_17
17. Ehsan, H., et al.: Muve: efficient multi-objective view recommendation for visual data exploration. In: ICDE (2016)
18. Ehsan, H., et al.: Efficient recommendation of aggregate data visualizations. TKDE **30**(2), 263–277 (2018)
19. Ehsan H., Sharaf M.A., Demartini G. (2020) QuRVe: query refinement for view recommendation in visual data exploration. In: Darmont J., Novikov B., Wrembel R. (eds.) New Trends in Databases and Information Systems. ADBIS 2020. Communications in Computer and Information Science, vol 1259. Springer, Cham. https://doi.org/10.1007/978-3-030-54623-6_14
20. Garciarena, U., et al.: An extensive analysis of the interaction between missing data types, imputation methods, and supervised classifiers. Expert Syst. Appl. **89**, 52–65 (2017)
21. Järvelin, K., et al.: Cumulated gain-based evaluation of IR techniques. TOIS **20**(4), 422–446 (2002)

22. Kandel, S., et al.: Profiler: integrated statistical analysis and visualization for data quality assessment. In: AVI (2012)
23. Key, A., et al.: Vizdeck: dashboards for visual analytics. In: SIGMOD (2012)
24. Khatri, H., et al.: QPIAD: query processing over incomplete autonomous databases. In: ICDE (2007)
25. Kim, A., et al.: Rapid sampling for visualizations with ordering guarantees. PVLDB **8**(5), 521–532 (2015)
26. Kim, W.Y., et al.: A taxonomy of dirty data. KDD **7**(1), 81–99 (2003)
27. Little, R.J.A., et al.: Statistical Analysis with Missing Data. Wiley, Hoboken (1986)
28. Luo, Y., et al.: Deepeye: towards automatic data visualization. In: ICDE (2018)
29. Mafrur, R., et al.: Dive: Diversifying view recommendation for visual data exploration. In: CIKM (2018)
30. Manning, C.D., et al.: Introduction to Information Retrieval. Cambridge (2008)
31. Miao, X., et al.: SI2P: a restaurant recommendation system using preference queries over incomplete information. PVLDB **9**(13), 1509–1512 (2016)
32. Mirkin, B.: Divisive and separate cluster structures. Core Data Analysis: Summarization, Correlation, and Visualization. UTCS, pp. 405–475. Springer, Cham (2019). https://doi.org/10.1007/978-3-030-00271-8_5
33. Park, Y., et al.: Viz-aware sampling for very large databases. In: ICDE (2016)
34. Siddiqui, T., et al.: Effortless data exploration with zenvisage: an expressive and interactive visual analytics system. PVLDB **10**(4), 457–468 (2016)
35. Tang, B., et al.: Extracting top-k insights from multi-dimensional data. In: SIGMOD (2017)
36. Vartak, M., et al.: SEEDB: efficient data-driven visualization recommendations to support visual analytics. In: PVLDB (2015)
37. Webber, W., et al.: A similarity measure for indefinite rankings. TOIS **28**(4), 20–38 (2010)
38. Wu, C., et al.: Using association rules for completing missing data. In: HIS (2004)
39. Zhang, A., et al.: Interval estimation for aggregate queries on incomplete data. J. Comput. Sci. Technol. **34**(6), 1203–1216 (2019). https://doi.org/10.1007/s11390-019-1970-4

Data Analysis and Data Modeling

ModelDrivenGuide: An Approach for Implementing NoSQL Schemas

Jihane Mali[1], Faten Atigui[2], Ahmed Azough[1], and Nicolas Travers[2,3(✉)]

[1] Université Sidi Mohamed Ben Abdellah, Fès, Morocco
jihane.mali@usmba.ac.ma, ahmed.azough@gmail.com
[2] CEDRIC, Conservatoire National des Arts et Métiers (CNAM), Paris, France
{faten.atigui,nicolas.travers}@cnam.fr
[3] Léonard de Vinci Pôle Universitaire, Research Center, Paris La Défense, France
nicolas.travers@devinci.fr

Abstract. With data evolution in terms of volume, variety and velocity, Information Systems (IS) administrators need to find the best solution to store and manipulate data with respect to their requirements. So far, existing approaches provide rules to transform a source model to a target model, but none of them propose a method to lead the choice of the most suitable solution. ModelDrivenGuide suggests a model transformation approach that focuses on proposing the different relevant solutions to the case of study. It is based on a common meta-model for the 5 families (Relational & NoSQL) and a generation heuristic. Our approach is validated using the TPC-C benchmark.

Keywords: NoSQL · Information Systems · User guide · MDA · Model transformation · Schema denormalization

1 Introduction

For several decades, the storage and the exploitation of data has mainly relied on Relational Databases (RDB). With the advent of Big Data, the volume of data has exploded, the heterogeneity has increased, causing problems of transformation from traditional databases to new storage on the Cloud, whether in terms of storage management, data query, cost or performance. These new data management systems are called *NoSQL* systems since 2009. The *NoSQL* data models correspond to various families of data structures: key-value oriented (KVO), column oriented (CO), document oriented (DO), and graph oriented (GO).

With more than 225 different *NoSQL* solutions, it is difficult for a company's CIO to determine the most suitable solution for its functional needs. Indeed, transferring the database to a NoSQL solution is an extremely heavy and costly process. Inadequate choices can lead to problems of scalability, data consistency or pricing. Our present work aims to provide an answer by driving the choice of digital transformation solutions.

© Springer Nature Switzerland AG 2020
S. Hartmann et al. (Eds.): DEXA 2020, LNCS 12391, pp. 141–151, 2020.
https://doi.org/10.1007/978-3-030-59003-1_9

Existing solutions for the digital transformation of a relational database remain essentially limited in their approach. Indeed, proposed methods are based either on a transformation of the physical model causing problems of scaling, or a purely logical model providing transformations with respect to the business logic. However, they do not take into account the optimization problems inherent to the origin problem of this transformation (scaling up).

Our `ModelDrivenGuide` approach provides logical modeling suitable for models refinement in order to generate all types of optimized physical schemas. Based on rules, it provides a decision making process that integrates the use case to guide the implementation choice of the SQL and/or NoSQL solution(s). The main contributions of this paper are: (i) an approach that subtly combines conception with optimization in DB transformation process; (ii) a common meta-model that supports all families (relational and NoSQL) and schema transformation; (iii) a heuristic to reduce the search space of model generation.

The rest of this paper is organized as follows: the state of the art is presented in Sect. 2. Then our approach is developed in Sect. 3. We finally validate our approach on the TPC-C benchmark in Sect. 4 and conclude in Sect. 5.

2 Related Works

Most of the existing studies have proposed either (i) a comparative study between RDB and NoSQL DB and/or how to transform relational data into NoSQL data or (ii) how to transform a conceptual model into a specific NoSQL DB; (iii) while very few studies have proposed criteria for the choice of physical model and implantation platforms.

Transformation of a Relational DB into NoSQL DB. These approaches define a set of mapping rules that transform a relational schema into a NoSQL schema. We can cite [11] for CO families with *HBase*, [18] with several CO DB, or [16] for DO families with *MongoDB*. Other techniques are concerned with queries to define the target data model [8,10].

Transformation of a Conceptual Model into NoSQL DB. [7] proposes to transform an `Entity Relationship` (ER) model into a CO model, based on the definition of a CO schema using primary and foreign keys, and on transformation rules. Similarly, [2] suggests a query-driven approach for modeling `Cassandra` starting from an ER model. They define dedicated logical and physical models, as well as transformation rules between models. [5] proposes a conceptual transformation approach which converts an ER model into one of the 4 NoSQL families with an abstract formalization of the mapping rules.

Other studies adopt a model-driven architecture (MDA) to transform a class diagram into NoSQL DB. [1] transforms a class diagram into a NoSQL DB. The authors present a common logical model which describes the four families of NoSQL DB. Then, this logical model is transformed into physical models related to the four families. In [3], the authors propose `UMLtoNoSQL`, a MDA based framework to map conceptual models to several data storage solutions.

Their approach combines model mapping techniques with a set of rules to translate OCL constraints into various query languages. [12] and [4] systematically transform a class diagram into a CO DB with HBase and GO DB respectively.

NoSQL DB Choice Orientation. To guide the choice of the NoSQL DB, we find performance comparisons for dedicated needs [9,15,17], studies of applicability to specific scenarios or software quality choices [13]. A comparative classification model [6] proposes to link functional and non-functional requirements to the techniques of each NoSQL family. The result is presented as a decision tree that guides the choice of the NoSQL system based on sharding, volume of data and CAP properties.

To conclude on the state of the art, existing approaches that compared RDB and NoSQL DB are basically based on technical criteria, and do not cover all the factors that can impact performance such as the data model, the type and the frequency of queries, the optimization structures, and the case of study. Model transformation approaches provide rules that transform a source model into a target model. However, these approaches do not favor the flexibility of the schema offered by NoSQL DB as a whole, and do not allow to move from one model to another, or from one NoSQL family to another *via* transformations; splitting a source concept into several target concepts, or merging several source concepts into one target concept. Thus, the flexibility of models has not been sufficiently exploited to facilitate transformations and improve performances. Finally, the few existing works on NoSQL DB choice orientation are basically based on technical criteria and do not consider data model nor functional requirements.

3   ModelDrivenGuide: Implementing Conceptual Model

We propose an approach for the generation of logical models for each family (NoSQLs and relational). Our approach is based on transformation rules starting from the conceptual model, then going from one logical model to an other by refinement. We adopt a Model-Driven Architecture[1] that offers 3 types of models, namely 1) the *Computation Independent Model* (CIM) that describes the requirements; 2) the *Platform Independent Model* (PIM) which describes the components of the system independently of platforms; 3) and the *Platform Specific Model* (PSM) which describes the components of the system using a precise technical platform. MDA also recommends transforming models by formalizing the transformation rules in a language such as the *Query-View-Transformation* (QVT)[2], which is the standard proposed by the OMG for models transformation.

3.1   Overview of Our Approach

Our approach provides a modeling framework based on those multiple dimensions of choice as illustrated in Fig. 1.

---

[1] MDA: https://www.omg.org/mda/index.htm.
[2] https://www.omg.org/spec/QVT/About-QVT/.

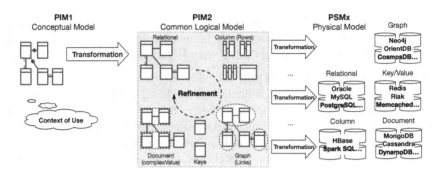

Fig. 1. ModelDrivenGuide: From conceptual to physical models

The **PIM1** integrates the functional needs of the IS, both in terms of data and queries. This traditional UML class diagram serves as a basis for modeling the user requirements.

The **PIM2** is the second level independent model, common to the five families of models (NoSQL & relational). It allows to carry out *refinements* by generating all the possible denormalized models using merge and split rules. To reduce the search space, a generation heuristic keeps only the effective solutions by simplification based on the use case.

The **PSMx** are obtained by the transformation of compatible PIM2 into the target data family (e.g. nesting for DO, rows for CO, edges for GO, etc.). Choices of sharding and indexing strategies are obtained using a generic cost model for the 5 families. All solutions can be proposed and sorted in relevance order.

This article focuses especially on the PIM2 5Families meta-model used to generate all possible data models, the transformation and refinement rules (between and inside meta-models) and also the heuristic that reduces the search space of PIM2 data-models by removing useless solutions. The choice of proper data-models determined by a global cost model is out of the scope of this paper.

3.2   The 5Families Meta-Model

The common meta-model seeks to produce different schemas compatible with the PSM constraints while remaining independent. The subtlety of the common PIM2 is to integrate the 5 families of data models. A major advantage of this meta-model is that if a denormalization (or normalization) solution proves to be well suited to a case of study using a family, it may also be the suited one to the four other families. Thanks to the *refinement* rules, it will be possible to merge or split concepts to adapt them to all the data models.

Figure 2 shows the PIM2 5Families meta-model integrating all the concepts used in the 5Families data models: the concepts contain rows (for column-oriented models), key-values with simple or complex values. Concepts can also be linked by edges to facilitate the integration of a graph database.

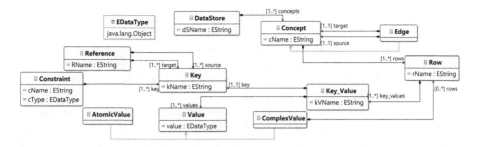

Fig. 2. PIM2 common meta-model for the 5 families of data models

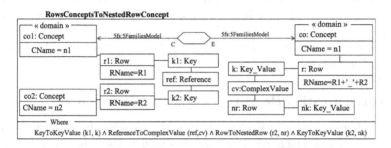

Fig. 3. PIM2 `5Families` Model Refinement - Concepts' Rows Merge QVT Rule

Complex Values are represented by the composition of rows. A loop occurs between rows and complex values, it shows that a value can be multiple, and is structurally identical to a concept. This modeling choice enables the application of refinement rules without iteration limits, allowing the production of new schemas. Meta-constraints are associated with the Complex Value to avoid the mutual composition of instances. Constraints and references are associated with keys, which can belong to distinct concepts. The idea is to be able to transform these constraints into new representations using refinement rules. For example, a *Reference* can be transformed into an *Edge* for graph-oriented models or into a *Foreign Key* in relational databases. UML classes from the PIM1 are transformed to the PIM2 using traditional mappings between concepts, keys and their values. Rules are expressed in `QVT` (for more details, see [1]).

3.3 Common Model Transformation and Refinement

The common meta-model helps to refine models iteratively while respecting the meta-model. The PIM2 refinement relies on three transformation rules:

- **Merge** *rows* between concepts to produce *complex values* for nesting (`CO`, `DO`) or to merge keys for values concatenation (`KVO`),
- **Split** *rows* to produce new columns in a same *concept* (`CO`),
- Transform *references* into the equivalent *edges* (`GO`).

At PIM2 level, models' refinement uses endogenous transformations where both the source model and the target model are conform to the same meta-model, here the 5Families meta-model. Here, we focus on the RowConceptsTo-NestedRowConcept *QVT* rule (Fig. 3) that merges rows belonging to two different concepts. Thus, two rows from two different concepts referring to each other (initially classes associated at the PIM1 level) can be merged into a single row with a complex value. Row *r2* of concept *co2* is linked by reference *ref* to concept *co1* through row *r1*. To merge rows, row *r2* is then nested into a complex value *cv* which corresponds to the transformation of the reference into a new key value *k*. The latter corresponds to the transformation of the reference key *k1*. This rule can be applied in both directions by switching *co1* and *co2* (generating a list of complex values in *k*). As detailed in the following, the generation of data models will rely on those rules by applying them recursively on various combinations of transformations. Each new schema produced by QVT rules can be transformed to target PSMx.

3.4   Transformation Heuristic to Generate Data Models

Models' refinement produces plenty of new data models in which we will find the target solution. The number of generated models and solutions is huge, since splits can be applied on each key and merges can be done in several ways. We need to apply a heuristic to reduce the search space and avoid cycles. It is based on the idea of avoiding to produce the same data models with two different paths. In fact, applying splits and merges in different orders will produce the same effects on the resulting data models. Moreover, one can reverse the effect of a path and produce a cycle in the production of data models. Thus, four main rules have to be adopted and can be summarized in the following:

- A *split* cannot be applied on a *Row* with *Complex Values*. Schemas generation should not go back after merges to avoid cycles.
- A *merge* cannot be applied between two *Rows* of the same *Concept*, which would cancel a split and generate a cycle.
- A *split* should not separate keys if queries from the use case express a link between them. This will avoid solutions which require instance reconstruction with costly joins.
- The *merge* of *Concepts* is led by the use case with queries combining *rows* of different concepts through *references*. The goal is to reduce the number of joins. It should be noted that the merge can be done in different ways.

4   Experimentations

To illustrate our approach, we used the TPC-C[3] benchmark giving a full use case mixing at the same time transactions, joins and aggregations. For the implementation of our approach, we generated different denormalized schemas that

---

[3] http://www.tpc.org/tpc_documents_current_versions/pdf/tpc-c_v5.11.0.pdf.

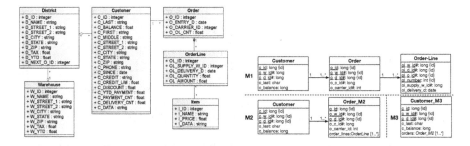

Fig. 4. PIM1 of TPC-C benchmark

Fig. 5. The Chosen PSM documents after several PIM2 fusions of TPC-C

we have integrated into a MongoDB database. The choice was led by the fact that MongoDB is one of the rare NoSQL solutions integrating ACID transactions in *sharding* required by the TPC benchmark. All experiments were run on Mon-goDB 4.2 on a cluster of 30 physical servers with a 3.60 GHz quad-core processor with 16 GB memory. For distribution, each shard corresponds to a distinct server. All showed processing times correspond to the average time given by 10 runs.

The TPC-C benchmark simulates the behavior of a logistic DB on user orders with transaction-oriented stock management (OLTP). To simplify, we will focus on the three main classes (*Customer*, *Order* and *OrderLine*) as illustrated in the PIM1 depicted in Fig. 4. TPC-C covers four transactions describing the activities of a management environment: **T1** - create an order and its delivery, **T2** - update by payments, **T3** - process 10 new orders and **T4** - supply warehouses' stock monitoring. When looking for a client, he is identified by his identifier (*c_id* in T2.1 and T4.1) or by his last name (*c_last* in T2.2 and T4.2).

Our experiments focus on the three most relevant document-oriented PSM models presented in Fig. 5. They correspond to three refinements of PIM2 for which were applied: the normalized model (**M1**), a merge of *Order* & *Order-Line* concepts (**M2** by nesting the list of *Order-Line*), and a second merge of *M2* & *Customer* concepts (**M3** with a list of *Order_M2*). Sharding keys are placed on *c_id* or *o_id*, and indexes on secondary keys *c_last* or nested keys.

4.1 Data Volume Variation

The increase in volume is tested on the M1 schema (equivalence). The data is distributed over two *shards* and the volume of data varies from 5 to 100 warehouses, representing 16 GB of data. Figure 6 shows the execution time of transactions' queries T1 to T4. The T3 query witnesses a different dimension due to its costly grouping operation thus it is shown separately.

We can notice that T1, T2.2 and T4.1 give weakly growing execution times *wrt.* the volume of data as a result of a good use of the sharding key which supports well the load of small updates. However, T2.2 and T4.2 transactions show a strong increase in execution time due to the use of secondary indexes on the *c_last* key. Their low selectivity does not support the increase of volume.

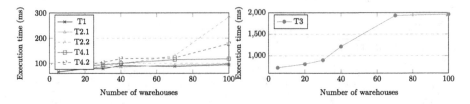

Fig. 6. Variation of data volume with M1

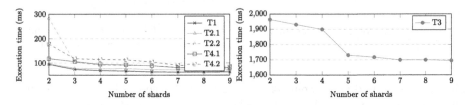

Fig. 7. Variation of the number of servers with M1 for 100 warehouses

The T3 must, on each iteration, process 10 new orders and the associated order lines. In fact, each order must be searched for, modified, and updated according to linked order lines. Thus, the volume of data has a huge impact on this heavy transaction requiring a strong monopolization of *chunks* and reducing the cache effect [14]. The threshold obtained after 70 warehouses corresponds to a null cache effect given the amount of data and the number of available *chunks*.

4.2   Number of Shards Variation

Here, we show the distribution effect by increasing the number of *shards* (nodes/servers). The M1 collection with 100 warehouses is distributed in order to define the number of servers necessary to reach convergence. Figure 7 shows the effect of parallelism until reaching stability starting from 7 *shards*. The gain is low for T1, T2.1 and T4.1 transactions (twice faster) since these are short transactions. For T2.2 and T4.2 based on *c_last*, a significant gain is obtained, since the query is carried out on smaller sets on each server (local indexes). In fact, time was divided by 4. The T3 transaction distributes updates for 10 orders and achieves a constant processing time per order (1,699 ms).

4.3   Data Models Performance

The last test case is to compare the 3 data models in order to see the impact of data modeling on the performance of the system. To do so, we used 100 warehouses on 9 *shards*. Each of the transactions was tested on the 3 data models M1, M2 and M3. Figure 8 shows the execution times for the M1 normalized model, the semi-denormalized M2 and the fully denormalized M3.

The first transaction T1 deals with order's insertion and associated order lines. In the M1 model each element is individually inserted. On the other hand,

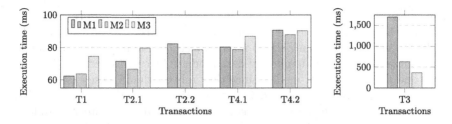

Fig. 8. Data models performance

the M2 model must gather the command lines to nest them as a block in the command, generating a delay. The M3 model generates a substantial additional cost (+20%). In fact, rewriting *chunks* force the placement of the modified document (more space), and thus adds processing time at concerned *shard*'s level.

For transaction T2.1 that consults orders and their order lines for a given customer, the M2 model is more efficient, since it merges them. The M1 model must perform the "join" during the process. However, the M3 model goes through all the orders to retrieve the last one, so it is more expensive. Transaction T2.2 searches for customers by last name. We note that the M3 model becomes more efficient because the additional cost of searching for orders is compensated by the search for the last name. This is not the case for the M1 and M2 models that must search for all of the homonyms before consulting the commands.

For T4.1 and T4.2 which update the payment of customers order, we observe a similar behavior with the additional cost of the updates of orders' "states".

Transaction T3 is very representative of the denormalization gain. Indeed, we can see the significant gain by merging concepts of −63% for M2 and −78% for M3. Indeed this transaction requires to join all concepts in order to process order lines' updates. Consequently, the normalized model M1 requires many "joins", while M2 and M3 merge order lines into orders and thus update a single document to process the transaction. M3 is slightly more efficient, as the delivery status is also notified to the customer (data linked to the document).

We conclude from above experiments, that : a) as expected, normalized standard modeling increases the number of treatments between collections, in particular on linked updates, b) complete denormalization of the schema increases the processing of a certain number of queries due to 1) generating insertions in lists, and 2) browsing lists for each consultation. However, queries requiring to link all the collections are optimal, c) the semi-denormalized data model brings a good compromise between the targeted search with and without "joins".

5   Conclusion and Future Work

In this paper, we have proposed ModelDrivenGuide, an IS modeling and implementation MDA approach that guides the user from conceptual model to platform technical choices. The cornerstone of our approach is the common logical meta-model that describes the 5 families of models. We have presented QVT refinement rules, particularly for merging concepts' rows. Those rules generate optimized schema and produce potential target models. This mix between data modeling and optimization rises to an approach that aims to find the efficient target model among the 5 families of data models. A heuristic reduces the search space and targets relevant models since they rely on the IS's use case.

In future work, we wish to define a generic cost model for PIM2, allowing to compare the solutions produced. This cost model will integrate different dimensions (storage, bandwidth, CPU/energy impact, etc.). We also work on the definition of the eligibility of a PIM2 in the target PSM. Indeed, the generated models must be compatible with a dedicated PSM model and it is necessary to define mapping rules (a key & a value for KVO, Concepts & Links for GO, etc.).

References

1. Abdelhedi, F., Ait Brahim, A., Atigui, F., Zurfluh, G.: MDA-based approach for NoSQL databases modelling. In: Bellatreche, L., Chakravarthy, S. (eds.) DaWaK 2017. LNCS, vol. 10440, pp. 88–102. Springer, Cham (2017). https://doi.org/10.1007/978-3-319-64283-3_7
2. Chebotko, A., Kashlev, A., Lu, S.: A big data modeling methodology for apache cassandra. In: 2015 IEEE ICBD, pp. 238–245. IEEE (2015)
3. Daniel, G., Gomez, A., Cabot, J.: UmltoNoSQL: mapping conceptual schemas to heterogeneous datastores. In: 2019 13th International Conference on Research Challenges in Information Science (RCIS), pp. 1–13 (2019)
4. Daniel, G., Sunyé, G., Cabot, J.: UMLtoGraphDB: mapping conceptual schemas to graph databases. In: Comyn-Wattiau, I., Tanaka, K., Song, I.-Y., Yamamoto, S., Saeki, M. (eds.) ER 2016. LNCS, vol. 9974, pp. 430–444. Springer, Cham (2016). https://doi.org/10.1007/978-3-319-46397-1_33
5. de Freitas, M.C., Souza, D.Y., Salgado, A.C.: Conceptual mappings to convert relational into NoSQL databases. In: ICEIS (1), pp. 174–181 (2016)
6. Gessert, F., Wingerath, W., Friedrich, S., Ritter, N.: NoSQL database systems: a survey and decision guidance. CSRD **32**(3–4), 353–365 (2017). https://doi.org/10.1007/s00450-016-0334-3
7. Hamouda, S., Zainol, Z.: Document-oriented data schema for relational database migration to NoSQL. In: Innovate-Data 2017, pp. 43–50. IEEE (2017)
8. Hanine, M., Bendarag, A., Boutkhoum, O.: Data migration methodology from relational to NoSQL databases. Int. J. IJECE **9**(12), 2369–2373 (2016)
9. Jatana, N., Puri, S., Ahuja, M., Kathuria, I., Gosain, D.: A survey and comparison of relational and non-relational database. Int. J. IJERT **1**(6), 1–5 (2012)
10. Lee, C.H., Zheng, Y.L.: SQL-to-NoSQL schema denormalization and migration: a study on content management systems. In: SMC 2015, pp. 2022–2026 (2015)
11. Li, C.: Transforming relational database into HBase: a case study. In: ICSESS 2010, pp. 683–687. IEEE (2010)
12. Li, Y., Gu, P., Zhang, C.: Transforming UML class diagrams into HBase based on meta-model. In: ISEEE 2014. vol. 2, pp. 720–724. IEEE (2014)
13. Lourenço, J.R., Cabral, B., Carreiro, P., Vieira, M., Bernardino, J.: Choosing the right NoSQL database for the job: a quality attribute evaluation. J. Big Data **2**(1), 1–26 (2015). https://doi.org/10.1186/s40537-015-0025-0
14. du Mouza, C., Travers, N.: Relevant filtering in a distributed content-based publish/subscribe system. NoSQL Models Trends Challenges **1**, 203–244 (2018)

15. Raut, A.: NoSQL database and its comparison with RDBMS. Int. J. Comput. Intell. Res. **13**(7), 1645–1651 (2017)
16. Rocha, L., Vale, F., Cirilo, E., Barbosa, D., Mourão, F.: A framework for migrating relational datasets to NoSQL. Procedia Comput. Sci. **51**, 2593–2602 (2015)
17. Tauro, C.J., Aravindh, S., Shreeharsha, A.: Comparative study of the new generation, agile, scalable, high performance NOSQL databases. IJCA **48**(20), 1–4 (2012)
18. Vajk, T., Fehér, P., Fekete, K., Charaf, H.: Denormalizing data into schema-free databases. In: CogInfoCom 2013, pp. 747–752. IEEE (2013)

Automatic Schema Generation for Document-Oriented Systems

Paola Gómez[1]([✉]), Rubby Casallas[2]([✉]), and Claudia Roncancio[1]([✉])

[1] Univ. Grenoble Alpes, CNRS, Grenoble INP, Grenoble, France
{paola.gomez-barreto,claudia.roncancio}@univ-grenoble-alpes.fr
[2] TICSw, Universidad de los Andes, Bogotá, Colombia
rcasalla@uniandes.edu.co

Abstract. Popular document-oriented systems store JSON-like data (e.g. MongoDB). Such data formats combine the flexibility of semi-structured models and traditional data structures like records and arrays. This allows numerous structuring possibilities even for simple data. The data structure choice is important as it impacts many aspects such as memory footprint, data access performances and programming complexity. Our work aims at helping users in selecting data structuring from a set of automatically generated alternatives. These alternatives can be analyzed considering complexity metrics, query requirements and best practices using such "schemaless" databases. Our approach for "schema" generation has been inspired from Software Product Lines strategies based on feature models. From a UML class diagram that represents user's data, we generate automatically a feature model that implicitly contains the structure alternatives with their variations and common points. This feature model satisfies document-oriented constraints so as user constraints reflecting good practices or particular needs. It leads to a set of data structuring alternatives to be considered by the user for his operational choices.

Keywords: NoSQL · Document-oriented systems · Variability · Feature models

1 Introduction

Flexibility of semi-structured models supported by document-oriented NoSQL systems opens the door to many possibilities for data representation. The choice of data structuring has a significant impact on the database and the applications. These concerns aspects such as performance, readability, usability, maintainability and evolution of code and system [5]. The data structuring alternatives can be numerous and the choice is not obvious in many cases. Our proposal aims at helping developers to carry out design and analysis phases despite the many possible

Grenoble INP—Institute of Engineering Univ. Grenoble Alpes, LIG, 38000 Grenoble, France.

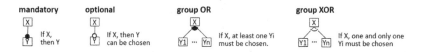

Fig. 1. Links between the features of a model of features

alternatives of a system without database schema (*Schemaless*). We propose the SCORUS approach [4] where the developer models the data using UML. This model is processed to automatically generate a set of *document-oriented structuring alternatives* presented as a set of AJSchema. It is a simple and readable format, which without playing the role of DB schema, highlights the data types and facilitates reasoning about the structure. SCORUS automatically evaluates metrics for each AJSchema. These metrics are objective indicators of the characteristics of the structures. An analysis phase, completes the approach based on metrics and user preferences as presented in [6,7]. The ScorusTool prototype implements this approach.

This article focuses on the automatic generation of structuring alternatives. The approach has been inspired from Software Product Lines strategies based on feature models [9]. These models represent the variability between different alternatives of a product. We consider document-structures as products and generate several alternatives to represent the UML class diagram given by the user. The UML is automatically processed to create a feature model reflecting the document-oriented structuring possibilities. The feature model includes constraints imposed by the document-oriented model definition and known good practices to discard invalid or irrelevant structures[1]. Each possibility is derived into an AJSchema to facilitate further analysis.

Section 2 presents the principles of product lines and feature models whereas Sect. 3 details how we use it to obtain document-oriented structuring alternatives. In Sect. 4 we introduce the AMISS algorithm to create the feature model. Section 5 details user restrictions allowing to customize the set of alternatives. Section 6 introduces the ScorusTool prototype. Related work is described in Sect. 7. Our conclusions and perspectives are presented in Sect. 8.

2   Background: Feature Models and Variability

It is essential to Software product lines (SPLs) to state common characteristics and variations among the products to build. To solve this, most strategies are based on feature modelling [8]. SPL define steps for feature modeling, product configuration and product derivation. We describe those steps briefly below.

Feature Modeling. A feature model, fm, consists of a *set of features* $\mathscr{F}$, links connected them $\mathscr{L}$, a root rc and a *set of constraints* $\mathscr{D}$ that force or prohibit

---

[1] Specific user preferences can also be added as constraints.

Fig. 2. a) Class model for agencies and business b) Basic case: two classes and one association

two or more features to be in the same product. Constraints follow the form $X \implies Y$.

$$fm = (\mathscr{F}, \mathscr{L}, rc, \mathscr{D})$$

Features are organized in a tree structure forming groups. Figure 1 shows the links used to designate the variations between features: *Mandatory, Optional, OR* and *XOR* groups.

Configuration Process. Once the feature model fm is created, the configuration step consists in selecting the set of valid combination of features $\|fm\|$. Each combination or configuration C consists of a subset containing the names of the chosen features. A valid configuration conforms to the constraints $\mathscr{D}$ specified in the feature model fm.

$$C(fm) = \{f | f \in \mathscr{F}(fm)\}$$

Derivation Process. Once configurations identified, each one can be used to produce an "artifact" with the selected features. For instance, an application or a hardware.

3   Generating Document-Oriented Structuring Alternatives

We adopt the product line approach because of the flexibility and the large number of structure possibilities offered by a document-oriented NoSQL system as MongoDB. In our work, a *product* is a possible "schema". This section presents how we build a feature model to derive document-structuring alternatives for a UML diagram.

3.1   Data Types and Data Modeling: Example and Notation

Document-oriented systems manage data as collections of documents. A document is a set of `attribute:value` pairs. The value type can be atomic or complex. A complex type is either an array of values of any type or a *nested document*. An attribute value can be the identifier of a document. This allows *referencing* one or more documents. This type system provides a lot of flexibility in creating complex structures.

Let's consider Fig. 2. The class `Agency` represents an agency handling business, represented by the class `Business` with the role `bLines`. A business is managed by a single agency (role `ag`). There are several ways to model such

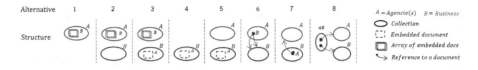

Fig. 3. Semi-structuring alternatives for the UML model of Fig. 2

data in a document-oriented base. Each class can lead (but is not mandatory) to a collection whose documents contain the class attributes. For instance, collection **Agencies** would have documents containing the attributes of class **Agency**. Relationship r_1 can be materialized either by nesting or by referencing information from the related classes. This allows several ways to structure collections. Figure 3 shows some representations. Alternative 2 shows nesting and possible duplication of business data. Some alternatives should be discarded because they do not show completeness with respect to the model. For example, collection **Agencies** does not reference or nested the related documents. The space of alternatives can be reduced according appropriate criteria such as *structuring constraints* following modeling guidelines, good practices or data-access priorities.

In the following we will use UML class models (Fig. 2b) noted as $mU = (E, R)$ where:

- $E = \{e_1, ..., e_n\}$ is the set of classes
- $R = \{r_1, ..., r_n\}$ is the set of relationships. We limit to binary relationships without attributes.
- $R(e_i) = \{r_1, ..., r_n\}$ is the set of relationships of the class e_i
- $E(r_n) = \{e_i, e_j\}$ are the classes connected by the relationship r_n. $e_i \neq e_j$
- $card(r_n, e_i)$ and $rol(r_n, e_i)$ are the cardinality and the role of r_n regarding to e_i.
- $A(e_i) = \{a_1, ..., a_n\}$ are the attributes of the class e_i. $a_i = \{a_i.name : a_i.type\}$
- $te_i = \{A(e_i), a_{id}\}$ is the type of e_i. It contains the class attributes and a default identifier[2].

3.2 Modeling Document-Oriented Structuring Variations with Feature Models

We use a feature model to represent structuring differences and to control the number of possibilities. These differences are guided by document-oriented structuring guidelines such as nesting, referencing, depth, and duplication. The valid configurations are treated as variants of our product, the schema. Products can be used for structure analysis.

We create the feature model, noted fm_s, from a UML model. The structuring alternatives express different collection definitions considering the following main modeling choices:

---

[2] This type is introduced to facilitate the explanation of our proposals.

1. A class e_i can lead to the creation of a collection of documents of type te_i.
2. A relationship r between e_i and e_j, can be materialized by nesting or referencing of e_i in e_j or vice versa.
3. A relationship r can be materialized by a separate collection called *collection-link*.

In the following, we incrementally present the process to create the feature model.

Modeling a Class
To model the variability of alternatives for structuring a collection for a class e_i, we propose the feature model fm_{col}, depicted by the gray area of Fig. 4a. Its features allow to treat the specific information of the class e_i and a relationship r:

- The "root" feature, named $cole_i$, defines the collection for the class e_i.
- The type of documents in this collection (the first level) is defined as a mandatory child feature, named te_i. This type includes the attributes of e_i and the identifier a_{id}.

Modeling Class Relationships introduces the extension of the type te_i by adding attributes to the documents at level 0. It uses the optional feature $te_i\_l_0\_ext$, child of the feature te_i. This extension introduce the materialization of the association r regarding documents e_j by an XOR group depending of feature r_role_j. Two alternatives are considered:

1. Feature $te_i\,EMB^{ce_j}\,te_j$ indicates the *nesting* in te_i of one or more documents of type te_j. ce_j is the cardinality. "many" leads to the nesting of an array of documents whereas cardinality 1 corresponds to the nesting of a single document.
2. Feature $te_i\,REF^{ce_j}\,te_j$ indicates the *referencing* form. Type te_i is extended with an attribute referencing one or several documents of type te_j according cardinality ce_j. One reference or an array of references. References use the identifier a_{id}.

Modeling Two Classes and a Relationship
Based on modeling features of a collection, fm_{col}, we model the alternatives of two classes and a relationship. We propose the feature model fm_{asso} (see Fig. 4) as follows:

- The root feature, called *Schema*, has two child features, one mandatory, named *colByClass*, and another optional, named *colByAssoc*.
- *colByClass* defines an OR group, the collections that may exist according to the UML classes. A child feature corresponds to an fm_{col} per class.
- *colByAssoc* defines a *collection-link* per relationship by the feature $colr_{e_i e_j}$. Their documents reference those of type te_i and te_j to represent the relationship.

Section 4 presents our algorithm *AMISS* that is a generalization of the strategy presented above to create incrementally the feature model for a complete class diagram.

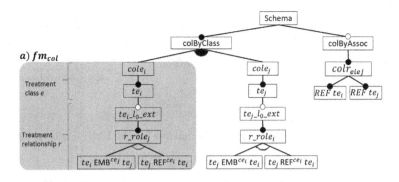

Fig. 4. Feature model fm_{asso} : two classes and a relationship

3.3   Valid Structures Definition

To reduce the combinatorial of configurations in the created feature model, we define additional constraints, based on characteristics related to the document-oriented model and good practices about data structuring:

Isolation Constraints $\mathscr{D}^I$, prevent the isolation of a collection. If a collection of type te_i, at the first level, does not represent the relationship r_n, then r_n has to be represented elsewhere. From a *collection-link* or from a nested level belonging to another collection:

$$\mathscr{D}^I \;=\; \{\, te_i \wedge \neg te_i\_lo\_ext \implies cole_i e_j \vee r\_role_j \;,\; te_j \wedge \neg te_j\_lo\_ext \implies cole_j e_i \vee r\_role_i \,\}$$

Existence Constraints $\mathscr{D}^E$ guarantee the existence of a referenced collection. If a type te_i is referenced, then a collection of the same type must exist.

$$\mathscr{D}^E \;=\; \{\, te_i\ REF^{ce_j}\ te_j \implies cole_j \;,\; te_j\ REF^{ce_j}\ te_i \implies cole_i \,\}$$

Nested Loop Constraints $\mathscr{D}^{BI}$ prevent collections from being referenced together. If te_i refers the type te_j, then te_j can not nest te_i.

$$\mathscr{D}^{BI} \;=\; \{\, te_i\ REF^{ce_j}\ te_j \implies \neg te_j\ EMB^{ce_i}\ te_i \;,\; te_j\ REF^{ce_i}\ te_i \implies \neg te_i\ EMB^{ce_j}\ te_j \,\}$$

Looped References Constraints $\mathscr{D}^{RB}$ prevent collections that nest documents of type te_i being referenced by another collection of the same type. If te_i refers the type te_j, then te_j can not refer te_i.

$$\mathscr{D}^{BR} \;=\; \{\, te_i\ REF^{ce_j}\ te_j \implies \neg te_j\ REF^{ce_i}\ te_i \,\}$$

Relationship *Links* Constraints $\mathscr{D}^L$ guarantee that a *collection-link* concerning a relationship r_n refers only collections whose type is not extended by the second type defined by r_n. If the collection-link $cole_i e_j$ exists, there is a collection of type te_i not extended by te_j and a collection of type te_j not extended by te_i.

$$\mathscr{D}^L \;=\; \{\, cole_i e_j \implies (cole_i \wedge \neg r\_role_j) \wedge (cole_j \wedge \neg r\_role_i) \,\}$$

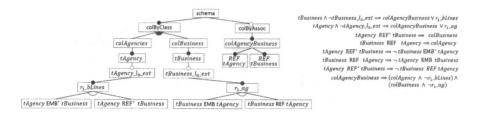

Fig. 5. Model fm_{asso} representing alternatives for the UML model of Fig. 2

The constraints of feature model fm_{asso} include constraints of isolation, existence, nested loops, looped references and links: $\mathscr{D}(fm_{asso}) = \{\mathscr{D}^I, \mathscr{D}^E, \mathscr{D}^{BI}, \mathscr{D}^{BR}, \mathscr{D}^L\}$.

3.4 Configuration and Derivation of Alternatives

The feature model fm_{asso} with constraints $\mathscr{D}(fm_{asso})$ provides 8 configurations[3] for two classes and one relationship, $\|fm_{asso}\| = \{C_1, C_2, C_3, C_4, C_5, C_6, C_7, C_8\}$.

$C_1 = \{Schema, colByClass, \mathbf{cole_i}, te_i, te_i\_lo\_ext, r\_role_j, te_iEMB^{ce_j}te_j\}$
$C_2 = \{Schema, colByClass, \mathbf{cole_i}, te_i, te_i\_lo\_ext, r\_role_j, te_iEMB^{ce_j}te_j, \mathbf{cole_j}, te_j\}$
$C_3 = \{Schema, colByClass, \mathbf{cole_i}, te_i, te_i\_lo\_ext, r\_role_j, te_iEMB^{ce_j}te_j,$
       $\mathbf{cole_j}, te_j, te_j\_lo\_ext, r\_role_i, te_jEMB^{ce_i}te_i\}$
$C_4 = \{Schema, colByClass, \mathbf{cole_j}, te_j, te_j\_lo\_ext, r\_role_i, te_jEMB^{ce_i}te_i\}$
$C_5 = \{Schema, colByClass, \mathbf{cole_i}, te_i, \mathbf{cole_j}, te_j, te_j\_lo\_ext, r\_role_i, te_jEMB^{ce_i}te_i\}$
$C_6 = \{Schema, colByClass, \mathbf{cole_i}, te_i, te_i\_lo\_ext, r\_role_j, te_iREF^{ce_j}te_j, \mathbf{cole_j}, te_j\}$
$C_7 = \{Schema, colByClass, \mathbf{cole_i}, te_i, \mathbf{cole_j}, te_j, te_j\_lo\_ext, r\_role_i, te_jREF^{ce_i}te_i\}$
$C_8 = \{Schema, colByClass, \mathbf{cole_i}, te_i, \mathbf{cole_j}, te_j, colByAssoc, \mathbf{colr_{e_ie_j}}\}$

Configurations C_1 to C_5 correspond to structures with nested documents, whereas C_6 to C_8 use document referencing. For example, configuration C_2 contains the features $cole_i$ and $te_iEMB^{ce_j}te_j$ indicating the existence of a collection $cole_i$ with nested documents of type te_j. C_2 also contains the feature $cole_j$ without features EMB or REF related to it. That implies the existence of a collection $cole_j$ without extension its type te_j. These configurations provided by the feature model of Fig. 5 correspond to the alternatives illustrated by Fig. 3.

Structure alternatives are derived from the feature model as AJSchemes. Figure 6 shows those produced by the feature model in Fig. 5. The AJSchema format facilitates the analysis of the structures and allows to automatically evaluate structural metrics [7]. Metrics quantify the complexity of the structure to help developers deciding about data structures to use.

---

[3] We used the *SAT solver* S.P.L.O.T. to calculate the number of valid configurations http://www.splot-research.org/ .

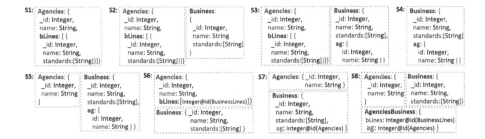

Fig. 6. Products of fm_{asso} of Fig. 5 in the form of AJSchema

4 *AMISS*: Algorithm for Modeling Semi-structured Schemes

We generalize our approach so as to create a feature model fm_s for all classes and relationships of a given class diagram. We propose the *A*lgorithm for *M*odel*I*ng *S*emi-Structured *S*chemes, *AMISS* here after. The creation of the feature model starts from

Algorithm 1 - AMISS modeling semi-structured schemes
Input: UML model m, classes with binary associations ER
Output: Feature model fm_s
1: $fm_s \leftarrow null$
2: $e_i \leftarrow random(ER)$
3: $fm_s \leftarrow treatAssociations(fm_s, e_i)$
4: $fm_s \leftarrow treatIsolatedTypes(fm_s, E - ER)$
5: **function** TREATASSOCIATIONS(fm_s, e_i) : fm_s
6:    **foreach** $r_n \in R(e_i)$ **do**
7:       $e_j \leftarrow getTarget(r_n, e_i)$
8:       $fm_{asso} \leftarrow generateFMassociation(e_i, e_j, r_n)$
9:       **if** fm_s is null **then** $fm_s \leftarrow fm_{asso}$
10:      **else** $fm_s \leftarrow$ FUSIONASSOCIATION($fm_s, fm_{asso}, r_n, e_i, e_j$)
11:      **end if**
12:      $fm_s \leftarrow$ TREATASSOCIATIONS(fm_s, e_j)
13:    **end foreach**
14:    **return** fm_s
15: **end function**

one class having relationships. One relationship of the class is treated and gives rise to a feature model fm_{asso} (line 8). The other classes and relationships are then treated. Each association leads to a feature model fm_{asso} which is then integrated into the complete model fm_s (line 10). The feature models are "merged".

FusionAssociation, see Algorithm 2, modifies fm_s to integrate the modeling (represented in fm_{asso})

of a new association r_n with its classes e_i and e_j. Figure 7 illustrates the main actions (numbered with stars) that enrich fm_s. A stands for `Agency`, B for `Business` and C for `CoWorker`.

1. Integrate the partial model r_n into collection $cole_i$. For this, add the branch r_role_j of the fm_{asso} to the features of fm_s that nest documents of type te_i (see Lines 1–9).

Algorithm 2 - FusionAssociation $(fm_s, fm_{asso}, r_n, e_i, e_j) : fm_s$

                                                        ▷ Complete types te_i at level 0 or nested

1: $fm_{branch} \leftarrow$ extractExtension(fm_{asso}, e_i, r_n) ▷ Extract branch r_role_j and constraints
2: $fm_s \leftarrow$ insertBranchCommonClass(fm_s, fm_{branch}, e_i)
3: **foreach** $f \in \mathscr{F}(fm_s) \wedge f = \bigotimes emb^{ce_i} te_i$ **do**
4:     **if** f *is leaf* **then**
5:         $fm_s \leftarrow$ addFeatureExtension$(fm_s, f, e_i, cse(e_i))$
6:     **end if**
7:     $fm_{branchV} \leftarrow$ createBranchVersion$(fm_{branch}, e_i, e_j, csr(r_n, e_j))$
8:     $fm_s \leftarrow$ embedBranch$(fm_s, fm_{branchV}, f.child)$
9: **end foreach**
10: $fm_{col} \leftarrow$ extractFMcol(fm_r, e_j, r_n) ▷ Add new collection $cole_j$
11: $fm_s \leftarrow$ insertFMcolNewClass(fm_s, fm_{col}, e_j)

                                          ▷ Complete type te_i nested with $cole_j$
12: $fm_s \leftarrow$ addFeatureExtension$(fm_s, te_j emb^{ce_i} te_i, e_i, cse(e_j))$
13: **foreach** fm_{branch} *filsOf* $(te_i\_lo\_ext) \in fm_s \wedge fm_{branch}.root \neq r\_role_j$ **do**
14:     $fm_{branchV} \leftarrow$ createBranchVersion$(fm_{branch}, e_i, e_j, csr(r_n, e_i))$
15:     $fm_s \leftarrow$ embedBranch$(fm_s, fm_{branchV}, e_j, cse_t)$
16: **end foreach**
17: $fm_s \leftarrow$ insertBranchRelation$(fm_s, fm_{asso}, r_n, e_i, e_j)$ ▷ Add *collection-link* for r_n

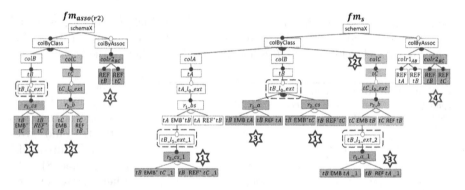

Fig. 7. Fusion of fm_{asso} (left) to obtain the global model fm_s (right) integrating fm_{asso}

2. Add the model of the new collection created for e_j. The model fm_{col} corresponding to the branch $cole_j$ of fm_{asso} is added to the group OR of feature $ColByClass$ of fm_s (see Lines 10–11).

3. Complete the type te_i nested in the new branch $cole_j$ of fm_s. For this, add to it the already existing model of relationships other than r_n. These relationships are in type te_i of $cole_i$ in fm_s (see Lines 12–16).

4. Add the *collection-link* for r_n. This is represented by the feature $colasso_{e_i e_j}$ of fm_{asso} and must be inserted in the tree of the feature $colByAssoc$ of fm_s (Line 17).

The resulting fm_s model represents a set of valid configurations corresponding to structuring alternatives. The feature tree can yield to a large number of valid configurations. The following Section discusses how to include user knowledge to reduce and to customize the set of alternatives.

5   Customizing Structuring Alternatives

The alternatives produced by the feature model created by AMISS can be filtered and customized to capture user preferences. For example, good practices issued by the software developer community (e.g. *Avoid nesting more than N documents*) or application specific preferences. Globally accepted constraints can be predefined as packages to be included automatically. We help constraint definition by providing an interactive module. Users simply choose the constraint type and value and the system automatically adds it to the feature model without effort for the user. Feature model fm_s (Fig. 7) leads to 119 valid structuring alternatives if no user constraint is used.

Avoiding References From a Collection: To discard alternatives considering references from `Agencies`, a user can choose the option "avoid references starting from a certain level of a collection". Then, she selects collection `Agencies` (*colA* in fm_s) and level 0. The following constraint is automatically created and added to the model:

$$coltA \implies \neg tA\_REFn\_tB \wedge tB\_REFn\_tC\_1$$

This constraint discards all the structuring alternatives where collection `Agencies` refers to documents of type `Business` or `CoWorker`. After that, 97 structuring alternatives still available.

Forcing Collections Existence: If a user wants to preserve the existence of collections corresponding to embedded types into collection `Agencies` (*colA*) until level 2, she selects the appropriate option and gives the collection name and level. The following constraint will be automatically created: $colA \implies coltB \wedge coltC$

Users may also introduce preferences according to a given application context. The interactive module, helps an easy declaration of constraints without conflict with the *mandatory structuring constraints* enforced by AMISS. Due to space limitation no further explanation are given in this paper.

6   *ScorusTool* Prototype

ScorusTool implements the schema generator and a module to evaluate structural metrics. Given a user provided UML class diagram, the tool creates a feature model and translates configurations of the fm_s model into AJSchemes (as in Fig. 6) and a tree structure, AJTree. We use the *FAMILIAR* language [2] to manage the feature models and to implement the AMISS algorithm. An interactive interface helps users to easily declare constraints and related values.

7   Related Work

The use of feature models to generate document-oriented data structuring alternatives has proven to be relevant. This approach is, to the best of our knowledge, original for the design of databases. It allows to explore and to manage several structuring alternatives and offers interesting possibilities for taking into account a variety of constraints. Related work on NoSQL data modeling use other approaches. The motivation for most of these works was the almost ad-hoc choice of a data structure favorable to the performance for a given set of queries. [13] presents an algorithm for the systematic creation of a structure for an entity-relationship model. The structure is "denormalized" and relationships are represented by nesting documents. This implies a kind of pre-calculation of natural joins. The solution usually leads to data redundancy.

Works such as [10–12] focus on alternative data representation in Cassandra's "big table" model. The objective is storage and performance optimization. [10] proposes the creation of copies of the data with different structures according to pre-calculated queries.

The approaches presented in [1,3] target heterogeneous systems. They are based on a meta-model representing the characteristics of NoSQL data models. For the same information, they propose structuring alternatives according to the target system. [1] uses a model driven approach. They transform a UML model into a generic logic model containing the common concepts of Cassandra, Neo4J and MongoDB. A transformation phase creates structuring alternatives for each NoSQL system. Specific rules are used for transforming a relationship for each system. Five solutions are proposed for MongoDB but the process for several relationships is not explained.

8   Conclusion and Future Work

The flexibility of data structuring offered by document-oriented NoSQL systems, such as MongoDB, allows many modeling options. Taking into account the stakes and the difficulty of the choice, we offer to the user a solution to generate automatically alternatives and to constraint them according to user criteria. Alternatives are created as a feature model for a UML class diagram. The feature model captures (1) the flexibility and restrictions of the document-oriented type systems and (2) the context needs, given as constraints. Constraints allow better targeting the set of alternatives. The set of constraints can evolve to incorporate new criteria appropriate to specific contexts. Our research perspectives focus on large scale experiments on the whole Scorus approach and in handling irregular structures.

Acknowledgments. We thank G. Vega, J. Chavarriaga, JP. Giraudin and anonymous reviewers for their valuable feedback.

References

1. Abdelhedi, F., Brahim, A.A., Atigui, F., Zurfluh, G.: MDA-based approach for NoSQL databases modelling. In: International Conference on Big Data Analytics and Knowledge Discovery (2017)

2. Acher, M., Collet, P., Lahire, P., France, R.B.: FAMILIAR: a domain-specific language for large scale management of feature models. Sci. Comput. Program. J. **78**(6), 657–681 (2013)
3. Atzeni, P., Bugiotti, F., Cabibbo, L., Torlone, R.: Data modeling in the NoSQL world. Comput. Standards Interfaces J. **67**, 103149 (2016)
4. Gomez, P.: Analyse et évaluation de structures orientées document. Ph.d. thesis, Thèse. Université Grenoble Alpes (2018)
5. Gómez, P., Casallas, R., Roncancio, C.: Data schema does matter, even in NoSQL systems! In: 10th International Conference on Research Challenges in Information Science (RCIS). IEEE (2016)
6. Gómez, P., Roncancio, C., Casallas, R.: Métriques structurelles pour l'analyse de bases orientées documents. In: Actes du XXXVIème Congrès INFORSID (2018)
7. Gómez, P., Roncancio, C., Casallas, R.: Towards quality analysis for document oriented bases. In: Trujillo, J.C., et al. (eds.) ER 2018. LNCS, vol. 11157, pp. 200–216. Springer, Cham (2018). https://doi.org/10.1007/978-3-030-00847-5_16
8. Kang, K.C., Cohen, S.G., Hess, J.A., Novak, W.E., Peterson, A.S.: Feature-oriented domain analysis (FODA) feasibility study. Technical report, Carnegie-Mellon Univ Pittsburgh (1990)
9. Kang, K.C., Lee, J., Donohoe, P.: Feature-oriented product line engineering. IEEE Softw. Mag. **19**(4), 58–65 (2002)
10. Lombardo, S., Nitto, E.D., Ardagna, D.: Issues in handling complex data structures with NoSQL databases. In: 14th International Symposium on Symbolic and Numeric Algorithms for Scientific Computing (SYNASC) (2012)
11. Mior, M.J., Salem, K., Aboulnaga, A., Liu, R.: Nose: schema design for NoSQL applications. IEEE Trans. Knowl. Data Eng. **29**(10), 2275–2289 (2017)
12. Vajk, T., Feher, P., Fekete, K., Charaf, H.: Denormalizing data into schema-free databases. In: 4th International Conference on Cognitive Infocommunications (CogInfoCom). IEEE (2013)
13. Zhao, G., Lin, Q., Li, L., Li, Z.: Schema conversion model of SQL database to NoSQL. In: International Conference on P2P, Parallel, Grid, Cloud and Internet Computing (3PGCIC) (2014)

Schedule for Detour-Type Demand Bus with Regular Usage Data

Yuuri Sakai[1]([✉]), Yasuhito Asano[2], and Masatoshi Yoshikawa[1]

[1] Graduate School of Informatics, Kyoto University, Yoshida-honmachi, Sakyo-ku, Kyoto 606-8501, Japan
yuuri.sakai@db.soc.i.kyoto-u.ac.jp
[2] INIAD, Toyo University, Akabanedai 1-7-11, Kita-ku, Tokyo 115-0053, Japan

Abstract. We propose a scheduling method for a "detour-type demand bus" as a public transportation service suitable for densely populated areas with a "star-type demand distribution" that has a demand accumulation point. To address regular usage, which is the key to public transportation on-demand buses, we consider construction of a schedule for each vehicle from trip data. We designate the schedule for each vehicle constructed from regular usage data as the "basic schedule". However, one-time reservation usage, which is a feature of a demand bus, is handled by adding a transit point to the basic schedule. The schedule for each vehicle in the stage of responding to this reservation usage is designated as the "operation schedule". We conducted simulation experiments using data from the Katsura area, a residential suburb of Kyoto, Japan. The proposed method has good results in terms of stable transportation of regular usage, but it is inferior to conventional demand buses in terms of flexible response to reservation usage.

Keywords: Demand responsive transport · Dial-a-ride problem · Detour-type demand bus · Public transportation

1 Introduction

Development of public transportation networks has not kept pace with rapid population concentrations in many areas. Efficient intra-regional transportation systems functioning as feeder networks will be established in such densely populated areas.

These areas have a certain level of regular trip demand for commuting. It causes a characteristic tendency of the distribution of demand, a so-called star-type demand distribution, for which the demand for moving in and out of a demand accumulation point, such as a station from within the area occupies a certain percentage of traffic during a specific time period. Noda et al. [5] demonstrated by simulation that a "demand bus" is effective for such a star-type demand distribution of a certain scale. A "demand bus" accepts reservations from a user, such as a departure and arrival point, and a departure and

© Springer Nature Switzerland AG 2020
S. Hartmann et al. (Eds.): DEXA 2020, LNCS 12391, pp. 164–176, 2020.
https://doi.org/10.1007/978-3-030-59003-1_11

arrival time, in the form of a reservation. Then the service constructs an operation schedule that is able to satisfy as many reservations as possible. However, demand buses have an important shortcoming: it is difficult to accommodate regular usage stably because of the characteristics of dynamic schedules. To resolve this difficulty, we construct schedules for a detour-type demand bus using regular usage data, which represent trip demand information collected from users who make the same trip every day.

To confirm the effectiveness of our proposed method, we conducted a simulation experiment using data from the Katsura in Kyoto, Japan, which has a star-type demand distribution. We compared the proposed method, non-detour-type demand bus, and Kyoto City Bus, which is an existing bus service in the area, and discussed their benefits and shortcomings. In terms of the acceptance rate of regular usage data, the proposed method outperforms both a non-detour-type demand bus and the Kyoto City Bus. However, it is slightly inferior to a non-detour-type demand bus in terms of the acceptance rate of the entire demand including one-time reservation usage. Whereas the existence of the basic schedule contributes to the stable response to regular usage, it also causes constraints for the operation schedule. Therefore, it becomes an obstacle to flexible response to the reservation usage. As described herein, construct the operation schedule without modification is done using the Insertion Heuristics reported by Solomon [8]. Some room for improvement remains.

2   Basic and Operation Schedules

We construct a detour-type demand bus schedule according to the following two stages. In the first stage, a basic schedule is constructed for each vehicle using regular usage data. This schedule is constructed before beginning the service using it. In the second stage, during application period of the basic schedule, the operation schedule is constructed by dynamically modifying the basic schedule according to reservation data. We adopt this design to handle regular usage data representing long-term demand and reservation data representing one-time demand.

Basic Schedule. For simplicity, assuming that the vehicles stop only at intersections, the basic schedule can be represented as a sequence of intersections and transit times assigned to each vehicle. The route that is run between these intersections is free as long as the fixed transit time is known.

We construct the basic schedule according to the usage of regular users to meet the demands of regular users, who are crucially important for public transportation. By collecting "regular usage data" indicating the regular user trips every day in advance, one can ascertain when and where the regular users embark on the bus and disembark from it. Of course, because of the limitations of the speed and the number of vehicles, not all regular usage data are satisfied in the basic schedule. It is important to construct an algorithm generating a basic schedule that satisfies as many regular usage data as possible.

Operation Schedule. Whereas the basic schedule responds to regular usage data as long-term demands, the operation schedule responds to reservation data as one-time demands. Because the reservation data are input in real-time by reservations from users during operating hours, the operation schedule must be dynamically changed accordingly.

The representation format of the operation schedule we propose is the same as that of the basic schedule, a sequence of intersections and their transit times given to each vehicle. The operation schedule is changed by inserting the source and destination intersections and times of the reservation data into the transit sequence defined by the basic schedule and the reservation data inserted earlier. Insertion should be done by selecting the target vehicle so that as much reservation data as possible can be satisfied, and as the basic schedule, so that not all reservation data can be satisfied because of resource limitations.

3   Algorithms for Generating Schedules

Each vehicle has one basic schedule and one operation schedule. A basic schedule set is constructed by allocating regular usage data to an optimal vehicle. An operation schedule set is constructed by allocating reservation data to an optimal vehicle.

3.1   Preliminaries

Map Data. We employ a representation of the road network in that area as a weighted undirected graph as map data. The weight represents the road length. In addition, an intersection that is regarded as the depot position and an intersection that is regarded as the demand accumulation point are determined.

Regular Usage Data. We assume that each regular user uses the same source and destination intersections simultaneously every day for commuting. Regular usage data indicate usage information of the corresponding regular user, i.e., source and destination intersections, and the source and destination time.

Reservation Data. Reservation data include usage information of the corresponding reservation user. Their representations are the same as those of regular usage data.

Vehicle Data. Each vehicle is assigned a basic schedule and an operation schedule to follow. The number of vehicles is determined in advance.

3.2   Basic Schedule Construction Algorithm

The basic schedule set is constructed by allocating regular usage data to appropriate vehicles. Although allocation procedure details differ among methods, the basic strategy of repeating the allocation recursively between the regular usage dataset and the vehicle set is the same. Regular usage data that do not meet the allocation constraints explained later will not be accepted. As a result, a sequence of transit intersections and times is assigned to each vehicle as a basic schedule.

Notation. Let $G = (V, E)$ denote the map data, for which V is the intersection set and E is the road set. The distance between intersections $v_a \in V$ and $v_b \in V$ by the shortest path is represented as $|v_a, v_b|$. In addition, set depot $v_{base} \in V$ and the demand accumulation point $v_{accum} \in V$.

Each regular usage data c_k of user k consists of $(v_k^s, t_k^s, v_k^d, t_k^d)$, where v_k^s and v_k^d represent the source and destination intersections, and t_k^s and t_k^d respectively represent the source and destination time. Consequently, the regular usage dataset C consists of $\{c_1, \ldots, c_{|C|}\}$.

For these analyses, we assume that $vehicle_i$ denotes vehicle i. All vehicles have a uniform speed limit $Speed$. The basic schedule for $vehicle_i$ is b_i. The basic schedule b_i has a sequence of intersections V_i^b through which $vehicle_i$ must pass, and a corresponding sequence of transit times T_i^b. Consequently, the vehicle set $Vehicle$ consists of $\{vehicle_1, \ldots, vehicle_{|Vehicle|}\}$. The basic schedule set B consists of $\{b_1, \ldots, b_{|Vehicle|}\}$.

Constraints in Allocation. Vehicle $vehicle_i$ must pass on time at all intersections specified by the accepted regular usage data because the vehicle must observe the speed limit $Speed$. Furthermore, the operation start time t_{start} and end time t_{end} are set for vehicle $vehicle_i$. If all constraints can be satisfied, then the regular usage data c_k can be allocated to vehicle $vehicle_i$.

Minimum Cost Priority Allocation Method. The Minimum Cost Priority Allocation method finds which vehicle to use for regular usage data based on the allocation cost, i.e., the increased mileage of the vehicle attributable to the allocation. The initial value of the mileage of each vehicle is calculated based on the assumption that the vehicle runs back and forth along the shortest route between the depot v_{base} and the demand accumulation point v_{accum}. This temporary calculation formula is used for setting an appropriate initial value. For vehicles with at least one set of regular usage data allocated, the vehicle mileage $vehicle_i$ is calculated assuming that the vehicle runs on the shortest route while passing through the sequence of transit intersections V_i^b in order according to the basic schedule b_i.

In this method, regular usage data and a vehicle allocation pair are selected such that the total mileage of vehicles is reduced using the allocation cost. By making the total mileage as small as possible, a basic schedule with more time to spare is constructed. More reservation data can be accepted at the stage of constructing the operation schedule.

The outline of the method is to repeat the allocation using the combination of the regular usage and the vehicle, i.e., the allocation pattern with the smallest allocation cost. For regular usage data not yet allocated to any vehicle $C_{unallocated}$ and vehicle set $Vehicle$, the allocation cost in all allocation patterns is calculated. Then the process of allocation according to the pattern that minimizes the cost is repeated. If the allocation cannot be made with the minimum cost pattern, then the allocation process ends because the same applies to any allocation pattern.

Algorithm 1. Minimum Cost Priority Allocation Method

1: $C_{unallocated} = C$
2: **while** $|C_{unallocated}| > 0$ **do**
3:     **for all** $vehicle_i \in Vehicle$ **do**
4:         **for all** $c_k \in c_{unallocated}$ **do**
5:             Calculate the allocation cost when c_k is allocated to $vehicle_i$.
6:         **end for**
7:     **end for**
8:     Let $(vehicle_{min}, c_{min})$ be a pair achieving the minimum allocation cost where $vehicle_{min} \in Vehicle, c_{min} \in C_{unallocated}$.
9:     **if** c_{min} can be allocated to $vehicle_{min}$ **then**
10:        Allocate c_{min} to $vehicle_{min}$
11:        Remove c_{min} from $C_{unallocated}$
12:    **else**
13:        break
14:    **end if**
15: **end while**

Transport Density Priority Allocation Method. The Transport Density Priority Allocation method is not aimed at optimizing only the allocation cost, i.e., the mileage; it also specifically examines the transport density of each vehicle. The transport density of $vehicle_i$ is the sum of the boarding distance of the regular usage data accepted by $vehicle_i$ divided by the total mileage of $vehicle_i$. Transport density increases when vehicles transport users densely, such as when many users board the same vehicle simultaneously. This method is devised to make the transport density of each vehicle uniform. As a result, each vehicle with various transit intersections and times has the same margin for accepting new reservations in the operation schedule construction; each is expected to contribute to the reservation data acceptance rate.

To make the transport density uniform, regular usage data are allocated to vehicles with the lowest transport density in this method. The allocation pattern by which the allocation cost is calculated for each recursion is "the vehicle with the lowest transport density $vehicle_{low}$ and each unallocated regular usage data

$c_k \in C_{unallocated}$". From the calculated $|C_{unallocated}|$ allocation costs, select regular usage data c_{min} that minimize the cost; if c_{min} can be allocated to $vehicle_{low}$, then allocate c_{min} to $vehicle_{low}$ and recalculate the vehicle transportation density of $vehicle_{low}$. If allocation is not possible, then $vehicle_{low}$ cannot accept any unallocated regular usage data. It will be ignored in subsequent recursion. This process is repeated there are no unallocated regular usage data or no vehicle can accept further regular usage data.

Algorithm 2. Transport Density Priority Allocation Method

1: $C_{unallocated} = C$
2: $Vehicle_{allocatable} = Vehicle$
3: **while** $|C_{unallocated}| > 0$ and $|Vehicle_{allocatable}| > 0$ **do**
4:     $vehicle_{low} = vehicle_{low} \in Vehicle_{allocatable}$ with the lowest transport density
5:     **for all** $c_k \in C_{unallocated}$ **do**
6:         Calculate the allocation cost when c_k is allocated to $vehicle_{low}$.
7:     **end for**
8:     $c_{min} = c \in C_{unallocated}$ with the lowest allocation cost to $vehicle_{low}$
9:     **if** c_{min} can be allocated to $vehicle_{low}$ **then**
10:         Allocate c_{min} to $vehicle_{low}$
11:         Remove c_{min} from $C_{unallocated}$
12:         Recalculate the transport density of $vehicle_{low}$
13:     **else**
14:         Remove $vehicle_{low}$ from $Vehicle_{allocatable}$.
15:     **end if**
16: **end while**

3.3   Operation Schedule Construction Algorithm

We adopt insertion heuristics reported by Solomon [8] for construction of the operation schedule. Insertion heuristics allocates reservation data to vehicles in the order of receipt. The initial operation schedule of each vehicle is equal to its basic schedule. When reservation data are input, their allocation cost is calculated for all vehicles. It will be allocated to the vehicle with the lowest cost. If no vehicle can meet the constraints, then the reservation data will be rejected. The reservation will not be established. The constraints and definition of allocation cost are almost identical, as in basic schedule construction.

The reason for adopting insertion heuristics is that allocation can be performed quickly. In constructing the operation schedule, one can process reservation data input in real-time without delay. Therefore, insertion heuristics has been adopted for studies of demand buses with real-time processing of reservations, such as the model of the demand bus for feeder networks in Okinawa by Uehara et al. [11] and the detour-type demand bus design by Uesugi et al. [12].

For reservation users, source and destination intersections might be different from their desired intersections represented in the reservation. These intersections must each be reachable within walking distance from the indicated intersections. To select an optimal vehicle and intersections, the allocation cost is calculated for all allocation patterns of intersection candidates and vehicles. Then the one that minimizes the cost is adopted. For simplicity, deviation of departure and arrival times that is attributable to the time for walking is ignored.

4   Experiment and Discussion

We conducted a simulation experiment using data from the Katsura area of Kyoto, Japan.

Fig. 1. Graph created from Katsura area map.

4.1   Used Data

Map Data. We created a graph as map data from latitude $34.9455° - 34.9944°$ and longitude $135.656° - 135.7206°$ in OpenStreetMap [6]. We deleted roads that are unsuitable for bus vehicles, such as residential roads. Additionally, some vertices and edges were deleted so that the graph is connected and so that no vertices of degree 1 or fewer exist. The number of vertices is 1072; the number of edges is 1540. The depot v_{base} is an intersection near Rakusai Bus Terminal: the largest operation base in this area. The demand accumulation point v_{accum} is the intersection near Hankyu Railway Katsura Station, which is the main station in this area.

Reservation Data. We estimated a reservation dataset for one day from the query log of the transfer search "Aruku machi Kyoto" (Jul. 1–Dec. 31, 2016) service and usage data of bus stops by the Kyoto Municipal Transportation Bureau.

Regular Usage Data. From the generated reservation data, we extracted regular usage data where the source or destination intersection is close to Katsura Station: the demand accumulation point. Considering the number of regular users of Katsura Station, we set the number of extractions as 5000 with the dataset at 200 m. With the dataset at 500 m, the extracted data were 4473.

Probe Data of Kyoto City Bus. Based on the GPS location data of Kyoto City Buses, we created a running result for Kyoto City Buses on July 1, 2016 in the Katsura area.

4.2   Settings

Methods to be Compared. Basic schedule construction algorithms of two types are proposed, along with the option of not using the basic schedule, i.e., not using the demand bus as a detour-type. We compared the schedules constructed using these three methods and the Kyoto City Bus schedule.

Parameter Settings. The demand bus speed of *Speed* is set to 19.8 km/h, referring to the scheduled speed of West No. 1, the main line of the Kyoto City Bus service in the Katsura area. Service hours of demand buses extend from 4:00 a.m. to 1:00 a.m. the following day. Assuming that vehicles of sufficient size can be prepared, the vehicle capacity is not considered for both demand buses and the Kyoto City Bus. The number of vehicles for demand buses is set at 75.

As discussed in Sect. 4.1, parameters of two types exist in data generation for the reservation and regular usage datasets. The reservation dataset selected from 200 m around Kyoto City Bus stop was influenced heavily by bus stops in the original search log. The distribution was biased. However, the dataset

Table 1. Minimum cost priority allocation method

Reservation dataset	200 m		500 m	
Walking distance limit	200 m	500 m	200 m	500 m
Acceptance rate of reservation data	56.6%	66.1%	51.8%	64.5%
Acceptance rate of regular usage data	99.8%	99.8%	99.8%	99.8%
Total mileage	17,524 km	16,730 km	17,874 km	16,849 km

Table 2. Transport density priority allocation method

Reservation dataset	200 m		500 m	
Walking distance limit	200 m	500 m	200 m	500 m
Acceptance rate of reservation data	57.0%	66.4%	52.2%	65.0%
Acceptance rate of regular usage data	93.8%	93.8%	93.4%	93.4%
Total mileage	17617 km	16945 km	17968 km	17049 km

at around 500 m is closer to an even distribution by expanding the selection of intersections. We prepared these two datasets to consider both the "dataset assuming the route network of Kyoto City Bus" and the "dataset including the potential trip demand". In general, many "potential usages" are likely to be those with a source or destination intersection that is not part of the roadway through which the buses run. These users gave up using Kyoto City Buses because of the distance to the bus stops.

According to the dataset, the walking distance limit is set as 200 m, which is a realistic walking distance for using the bus, and at 500 m, which allows most users to walk to the bus stop even with the reservation dataset at around 500 m.

4.3　Discussion

Effects of Reservation Dataset Settings. First, we consider the relation between the dataset used and the acceptance rate of the proposed detour-type demand bus. Tables 1 and 2 show that, in experiments using either the Minimum Cost Priority Allocation method or the Transport Density Priority Allocation method for constructing the basic schedule, the difference in the acceptance rate of reservation data between the two datasets is 4.8%, at most, if the walking distance limits are the same. Although the reservation dataset at 200 m is regarded

Table 3. Demand bus without basic schedule

Reservation dataset	200 m		500 m	
Walking distance limit	200 m	500 m	200 m	500 m
Acceptance rate of reservation data	66.0%	81.8%	61.2%	77.9%
Acceptance rate of regular usage data	72.6%	87.7%	66.5%	83.6%
Total mileage	18,912 km	15,658 km	19,431 km	16,500 km

Table 4. Kyoto City bus

Reservation dataset	200 m		500 m	
Walking distance limit	200 m	500 m	200 m	500 m
For users near bus route				
No. reservations	12,276	14,895	4829	13,661
No. regular uses	4255	4915	2420	4210
Acceptance rate of reservation data	48.0%	59.0%	39.1%	54.5%
Acceptance rate of regular usage data	50.6%	60.3%	43.6%	56.7%
For all users				
Acceptance rate of reservation data	38.8%	57.9%	12.4%	49.1%
Acceptance rate of regular usage data	43.0%	59.3%	23.6%	53.3%
Total mileage	16116 km			

as easier for riders because of the biased distribution of reservation data, the fact that the acceptance rate does not change much between the dataset at 200 m and the dataset at 500 m indicates that the proposed demand bus can respond flexibly to widely diverse user demands.

Second, we consider "potential reservation usage and usage" as described in Sect. 4.2. Results of the experimental pattern of "datasets at 500 m, walking distance limit: 200 m" show the number of data near the bus route as greatly reduced. To ensure convenience in terms of "shortening the walking distance before and after boarding and alighting the bus", one must increase the variety of bus stops. However, considering the fact that Kyoto City Bus has increased the number of routes in the Katsura area to ensure a variety of bus stops, it is difficult to establish a route network that covers more bus stops than the current one.

Effects of Basic Schedule. Because the basic schedule exists, the proposed detour-type demand bus guarantees that the regular usage accepted in the basic schedule is also accepted in the operation schedule. However, the operation schedule flexibility is reduced by constraints imposed by accepting regular usage. This constraint is also evident in the experimentally obtained results when comparing Tables 1 and 2, where basic schedules are used, and Table 3, where basic schedules are not used. Regarding acceptance rates of reservation data, absence of a basic schedule gives a higher rate than the presence of one. For the acceptance rate of regular usage data, the presence a basic schedule gives a higher rate.

The existence of the basic schedule markedly improves the acceptance rate of regular usage, leading to the purchase of commuter passes. Therefore, the basic schedule can contribute to the improvement of profitability in this respect. Comparison of the detour-type demand bus with the Transport Density Priority Allocation method (Table 2) and the demand bus without a basic schedule (Table 3) in the results with the pattern "dataset at 500 m, walking distance limit: 200 m" shows that the difference in acceptance rate of reservation data is 9.0% (1369 cases), whereas that of regular usage data is 26.9% (1204 cases). These represent the number of regular users secured by adopting the basic schedule and the number of reservations that are no longer accepted. In terms of profitability, we believe that the existence of the basic schedule is working positively because an earlier study showed that the availability of service for commuting is an important factor for evaluating demand buses from users [10].

However, the acceptance rate of reservation data is related directly to the great advantage of the demand bus. We adopt the insertion heuristics used in earlier studies [5,12] for the operation schedule construction algorithm. We also set the allocation cost as a simple increment of vehicle mileage. In this respect, much room exists for improvement. We would like to address this important issue in a future study.

Comparison of Basic Schedule Construction Algorithms. We consider basic schedule construction algorithms of two types: The Minimum Cost Priority Allocation method and the Transport Density Priority Allocation method.

Comparison of Table 1 and 2 shows that the Minimum Cost Priority Allocation method is favorable in terms of the acceptance rate of regular usage data. Although the computational complexity is greater than that of the Transport Density Priority Allocation method, the computational time is not a major hindrance to the construction of the basic schedule if the computation is completed within a realistic execution time. However, a slightly higher acceptance rate of reservation data is achieved using the Transport Density Priority Allocation method. Comparison of the acceptance rate of regular usage data and that of reservation data indicates that the superior method is the Minimum Cost Priority Allocation method, which achieves stable acceptance of regular usage. As described in this paper, we specifically examined only those vehicles with the lowest transport density during allocation with the Transport Density Priority Allocation method, but attempting allocation by checking more vehicles might have been better.

Comparison Between the Proposed Detour-Type Demand Bus and Kyoto City Bus. In terms of the acceptance rates of both regular usage data and reservation data, the proposed method outperformed the Kyoto City Bus. These have almost equal total mileage. Moreover, the acceptance rates are comparable in terms of profitability. As a comparative experiment, it would be preferable to use a optimized fixed-route bus that is also optimized, for example, with route planning particularly addressing many-to-one commuting trips such as the star-type demand distribution described in a study by Chien et al. [1], or a method using a genetic algorithm described by Sadrsadat et al. [7]. These comparisons are left as subjects for future work.

5   Related Work

Demand bus studies have been conducted for many years, yielding many examples of demonstrations and services. Yamato et al. [13] conducted a demonstration experiment in a sparsely populated area of Kashiwa City, Chiba, Japan, and administered a questionnaire survey of participants. Among the demand bus features, participants particularly appreciated that many potential source and destination points were provided. A demonstration experiment reported by Tsubouchi et al. [10] in Unzen, Nagasaki, Japan, revealed characteristics of demand buses that users value. They differ depending on which mode of transportation a person usually uses. These results suggest the importance of clearly assuming a mode of usage when designing a public transport service. Hiawata et al. [3] and Enoch et al. [2] respectively surveyed demand bus services provided to the public in Japan and the United Kingdom. The services presented in both papers are mainly operations in rural areas, operating with few resources, and often provided as an alternative to traditional bus service.

However, Noda et al. [5] demonstrated the effectiveness of demand bus service in densely populated areas through simulation experiments. They compared demand buses with fixed route buses under various conditions and found that demand buses are particularly good to address situations in which high demand is generated and in which its distribution tendency resembles a star-type demand distribution. As an earlier study of detour-type demand buses, Uesugi et al. [12] designed a demand bus that operates between two fixed points in parallel. They constructed a schedule by appropriately selecting allocation of destinations of reservations from multiple vehicles. As a study with an idea resembling ours, Uehara et al. [11] proposed a feeder network with a demand bus to transport commuters in coordination with a railway in the suburban areas of Naha, the central city of Okinawa in Japan.

In constructing the operation schedule, we use Insertion Heuristics using Solomon [8]. Insertion heuristics is designed for the Vehicle Routing and Scheduling Problem with time window constraints (VRSPTW), which can rapidly find a near-optimal solution.

6 Conclusion

We proposed a scheduling method for a detour-type demand bus with two stages of schedule construction. We aimed for stable acceptance of regular usage such as commuting by collecting regular usage data in advance and by using the data to construct basic schedules, which worked well and which outperformed the non-detour-type demand bus without a basic schedule in this respect. We proposed two methods for constructing a basic schedule set: Minimum Cost Priority Allocation method and Transport Density Priority Allocation method. The former can achieve both stable acceptance of regular usage and flexible handling of reservation usage to some degree, but the latter can be improved in the method design. However, we used Solomon's insertion heuristics [8] to construct the operation schedule. For handling reservation usage, the proposed detour-type demand bus performed worse than the non-detour-type demand bus. Improved insertion heuristics might alleviate this shortcoming. However, compared to a traditional bus service, Kyoto City Bus, the proposed method demonstrated a higher capacity for both regular and reservation usage. Future studies will be undertaken to improve our algorithm using new techniques such as an algorithm to address various vehicle routing problems reported by Subramanian et al. [9] and the algorithm particularly addressing demand uncertainty reported by Moghaddam et al [4]. In addition, real-world public roads present obstacles such as congestion and waiting at traffic lights. Therefore, the simulation should reflect these difficulties.

References

1. Chien, S.I.J., Dimitrijevic, B., Spasovic, L.: Optimization of bus route planning in urban commuter networks. J. Public Transportation **6**, 4 (2003)
2. Enoch, M., Potter, S., Parkhurst, G., Smith, M.: Intermode: innovations in demand responsive transport (2004)
3. Hiekata, K.: On-demand transportation: state of the art and future perspectives. Syst. Control Inf. **61**(12), 500–505 (2017)
4. Moghaddam, B.F., Ruiz, R., Sadjadi, S.J.: Vehicle routing problem with uncertain demands: an advanced particle swarm algorithm. Comput. Ind. Eng. **62**(1), 306–317 (2012)
5. Noda, I., Shinoda, K., Ohta, M., Nakashima, H.: Evaluation of usability of dial-a-ride system using simulation. IPSJ J. **49**(1), 242–252 (2008)
6. OpenStreetMap foundation: Openstreetmap (2020). https://www.openstreetmap.org/
7. Sadrsadat, H., Poorzahedi, H., Haghani, A., Sharifi, E.: Bus network design using genetic algorithm, vol. 1, pp. 210–225 (2012)
8. Solomon, M.M.: Algorithms for the vehicle routing and scheduling problems with time window constraints. Oper. Res. **35**(2), 254–265 (1987)
9. Subramanian, A., Uchoa, E., Ochi, L.S.: A hybrid algorithm for a class of vehicle routing problems. Comput. Oper. Res. **40**(10), 2519–2531 (2013)
10. Tsubouchi, K., Yamato, H., Hiekata, K.: On the advantage of the on-demand bus system in less populated area. Robot. Soc. Jpn. **27**(1), 115–121 (2009)
11. Uehara, K., Akamine, Y., Toma, N., Nerome, M., Endo, S.: A proposal of a cooperative transport system with demand responsive transits and mass transits. IPSJ J. **56**(1), 46–56 (2015)
12. Uesugi, K., Watanabe, T., Mukai, N.: K-means hou wo mochiita yorimichi gata demand bus no heiretsu unkou shuhou (in Japanese). In: DEWS2008 (2008)
13. Yamato, H., Tsubouchi, K., Hiekata, K., Honda, K., Sugimoto, C.: Evaluation of on-demand bus services applicability to older persons. JRM **20**, 810–817 (2008)

Algebra for Complex Analysis of Data

Jakub Peschel$^{(\boxtimes)}$, Michal Batko, and Pavel Zezula

Masaryk University, Brno, Czech Republic
{jpeschel,batko,pzezula}@mail.muni.cz

Abstract. In data science, the process of development focuses on the improvement of methods for individual data analytical tasks. However, their combination is not properly researched. We believe that this situation is caused by a missing framework, that would focus solely on data analytical tasks, instead of complicated transformation between individual methods. In this paper, a new analytical algebra is defined. This algebra is based on a flat structure of transaction file and operations over it. As a part of the paper, definitions of several data analytical tasks are proposed. Algebra is recursive and extendable. As an example of usability of the algebra, one complex analytical task created by a combination of analytical operators is described.

Keywords: Data analysis · Analytical algebra · Similarity · Pattern mining

1 Introduction

The technology advanced in such a way that a lot of data from many resources is produced daily. Users digitise enormous amounts of information in the form of images, messages, posts, trajectories, timestamps, geotags, systems produce logs about their behaviour, internet of things provides data about interactions of smart devices, etc. In the last decades, dynamicity of the data forms made it harder and harder to analyse data compared to the past. Database systems are no longer fully sufficient to capture whole information. The more complex forms of storage such as graph databases and other NoSQL databases emerged to handle new types of information.

Hand in hand with this change in storage a need for more complex analysis also emerged. Such analysis should not only analyse standard statistical properties of the data, which were sufficient in the past, but also focus on analysing structures of this complex data and should take into account also context of the data. Techniques like pattern mining or analysis based on similarity are often involved in proper data analysis.

Although there is a demand for more complex analyses, it is not easily deliverable. Most of the libraries focus on one specific area and provide a large amount of optimised methods solving one specific task. To create ensemble of methods

This research has been supported by the GACR project No. GA19-02033S.

S. Hartmann et al. (Eds.): DEXA 2020, LNCS 12391, pp. 177–187, 2020.
https://doi.org/10.1007/978-3-030-59003-1_12

provided by different libraries, nontrivial data transformations are necessary. In some cases, transformations are not even realisable.

Recently, tools like python's Jupyter notebooks [10] or complex multipurpose environments like R [21] or Matlab [12] have been used for the creation of such complex analytical processes. However, the development of pipelines takes a huge amount of focus on data type manipulation.

To tackle the before-mentioned problems, we propose a new analytical algebra for the description of complex analytical functions created by a combination of basic analytical tasks. This algebra is task-oriented to alleviate the need to focus on specific analytical algorithms. Another important aspect is the extensibility of this algebra, so it can be supplemented with new operations as needed.

The main contributions of this paper are the following:

- We introduce a new unifying structure for data analysis, transaction file.
- We define an extensible and recursive algebra based on the proposed structure.
- Several data analytical tasks are defined in the context of this new algebra.

2 Previous Work

Because data science is a broad area, we focus on its sub-part consisting of pattern mining and similarity-based methods. However the same approach is applicable to other areas of data analysis. For the purpose of this paper, we refer to tasks such as clustering, outlier detection and similarity searching as similarity-based analysis.

The main focus of researchers in these areas is invested in the development of optimised approaches for each of these tasks. In the case of pattern mining tasks, this can be seen in tasks such as frequent item-set mining algorithms [2,3,9,24], sequence mining algorithms [16,20,23], graph mining algorithms [14,22], and many others [1]. An extensive amount of algorithms for several clustering tasks can be seen in [7]. In the case of similarity searching, research is mostly divided into development of the indexing structures [5,15] or hashing techniques [11,13].

These techniques are often grouped in libraries such as pattern mining algorithms in SPMF [6] or clustering techniques in ELKI [19] or in WEKA [8]. Such libraries focus very often on one specific task and do not expect a combination of included methods. Because of that, input and output format differs to such extent, that complicated transformations must be developed by an expert with coding knowledge. In some cases, this transformation requires multiple accesses into underlying storage.

As mentioned in the introduction, this problem is often solved by the usage of multipurpose environment or language. An example of such an environment can be Jupyter notebook that allows creating complete pipelines written in python language with the usage of open-source libraries providing methods and transformation written by the programmer. Although this approach is a commonly accepted way to provide complex analytical pipelines with presentable results, it requires nontrivial time spent on the development of transformations and selection of best suited algorithms, which can be in many cases selected automatically.

Each task of the data analysis has an expected input and output format. So if the whole pipeline is defined and the conditions on data are implied, then transformations of such data can be done without further need to define them in multipurpose language.

3 Analytical Algebra

We propose a new data model, which shares similarities with relational model. The data model consists of unifying structure called transaction file and analytical algebra with potential extension to query language for ease of use. Although this restricts the number of useful tools available at the moment, the benefit of this approach leads to a more optimised system of analysis focused on the analysis itself.

3.1 Transaction File

The core of our proposal is a universal data-structure, transaction file, that has been introduced in [17]. It is a unifying structure for most of the data analytical tasks, as shown later in Sect. 3. Transaction file is a set of transactions:

$$T : \{t_1, ..., t_n\} | n \in \mathbb{N}$$

Transaction t is a pair of a list of data items $data$ and set of attributes.

$$t : (data, \{Attr_1, Attr_m\}) | n, m \in \mathbb{N}$$

Structure $data : (i_1, ..., i_n)$ is a list of variable length containing data items. The form of a list was used, because most of the other structures often used in data analysis can be derived, such as set, multi-set, vector, matrix, etc.

Examples of possible data stored in $data$ are DeCAF descriptors for image analysis, sets of neighbouring e-mail accounts, DNA sequences, etc.

Data item, denoted as i, is a distinguishable application dependent object. The structure of data item can be compared by specialised distance functions, but for most of the provided methods, the method for comparison of identity is sufficient for most of the analytical techniques.

Attributes $Attr$ are key:value pairs. Attributes serve as intermediate storage of information, which is not used for analytical operations itself, but is rather used for filtering or as a storage of an analytical result.

3.2 Properties of Analytical Algebra

This analytical algebra consists of operators working over a transaction file. The functions are constrained by this requirement. Definition of the operator is:

$$f : ([T, ...], params) - > T$$

where *params* are additional parameters of a specific operator.

An important feature of this analytical algebra is extensibility. Any analytical function can be added into this analytical algebra without changing its behaviour.

Due to this strict definition of input and output, recursivity of operators is possible. The result of one operator can be used as an input to others. This property is used to create a complex operation by a combination of simpler ones.

At the moment, this algebra contains two types of operators, analytical operators and supporting operators. Due to the high modularity of the algebra, more groups of operators can emerge.

3.3 Analytical Operators

Analytical operators are analytical functions that define the semantics of the whole analytical pipeline.

Many of the analytical operators have individual transaction or pattern as an input parameter. There is a possibility to pass the whole collection of such parameters as an input for simplicity. The operators are then evaluated for each of such item individually, and a relevant query object is stored in attribute *"query"* in each resulting transaction.

The proposed operator supports the high modularity of this algebra. Due to no restrictions on the process, a lot of analytical operations can be defined and added. We provide an example of several such operators:

Range Search. One of the traditional tasks in data analysis is a similarity search task. The goal of the search task is to obtain the relevant objects in database $\mathcal{D}$. The similarity to the query is measured to define the relevance of the object. The two most used query types are range query and k-nearest neighbours query.

The goal of the range query is to obtain all objects with distance to the query object less than a user-defined threshold. To measure distance, distance function *dist* must be provided. At the moment we expect usage of function that meets the metric postulates: identity, symmetry, non-negativity, triangle inequality.

For purposes of our algebra, database $\mathcal{D}$ corresponds to transaction file T. Definition of a respective analytical operator is:

$$RS : (T, q, dist, r) \rightarrow \{(t_i, \{\text{"distance"} : dist(q, t)\}) | t_i \in T : dist(q, t_i) \leq r\};$$
$$r \in \mathbb{R}_0^+$$

where q is an query object in the form of transaction.

Distance function *dist* is expected to follow:

$$(T, T)-> \mathbb{R}_0^+$$

As mentioned before, most of the methods also expect, that distance function is a metric function. An example of such distance functions can be Jaccard coefficient or Edit distance. If additional conditions on *data* of transaction files, as the same length or addition of zeroes to the same length, are set, distance functions like L^2 distance can be used.

Example: To show the usability of this operator, a small example follows. Let assume e-mail communication as input data. Data in transactions will consist of communication partners of each e-mail account. Information about e-mail account can be stored in the attribute for simple access.

The goal is to find all the users whose composition of communication partners has 80% overlap with the pre-selected group of e-mail accounts. Query will look as:

$$RS(T, t_q, dist_{Jaccard}, 0.2)$$

Result of such analysis is a set of e-mails where 80% of items in *data* are from the originally pre-selected set.

K-NN Search. Another task from the similarity search area is K-NN search. The goal of this task is to find k objects for provided query object in database $\mathcal{D}$. To search for objects, similarity measure must be provided as in range search.

In this algebra, the role of the database $\mathcal{D}$ is supplemented by the transaction file T. Definition of respective function is:

$$KNNS : (T, q, dist, k) \rightarrow \{(t_i, \{\text{"distance"} : dist(q, t_i.data)\}) | t_i \in R; |R| = k;$$
$$\forall t_j \in T - R : dist(q, t_i.data) \leq dist(q, t_j.data)\}$$

Query object q and amount of search neighbours k must be provided by user beforehand. Distance function *dist* has same restrictions as in range search.

Example: To show this operator, we assume data in form of objects having coordinates stored. The goal is to find the 10 closest objects in our vicinity. Each object is transformed into the transaction with longitude and latitude stored in *data* of the transaction. Our location is stored in the data in query object q. To measure the distance between objects, euclidian distance will be used. The query will be:

$$KNNS(T, q, dist_{euclidian}, 10)$$

As a result of this query, transaction file with ten transactions will be returned. Transactions contain distance in the form of an attribute of the respective transaction.

Frequent Item-Set Mining. The second group of tasks we are focusing on in this paper is pattern mining. In pattern mining, the main goal is to obtain hidden information about patterns occurring in database $\mathcal{D}$. Patterns are a collection of data item occurring in database records that occurs frequently. Frequency is defined by user-defined threshold θ. The threshold can be either absolute or relative to the size of the database.

There are several types of pattern mining tasks, first of them frequent item-set mining. The goal of this mining is to obtain subsets of original sets stored in the database.

Database for the frequent item-set mining contains sets of individual items. To cope with this expected format, *data* in transactions are viewed as sets. Duplicity and order are ignored. The definition of frequent item-set mining follows:

$$IM : (T, \theta) \rightarrow \{(p, \{\text{``support''} : s_p\}) \,|\, p \subseteq t.data; t \in T; s_p \geq \theta\}$$

where p is a set of items and s_p is support computed over the original transaction database.

Example: To show the usability of this definition, we assume data in the form of e-mails. Each e-mail has in header a set of receivers. The goal of the analysis is to uncover if there is a set of e-mails, that are very often part of the correspondence. To establish what is very often, we define the threshold of one per cent to be frequent enough.

Transactions in the transaction file T will contain a set of e-mail addresses in the *data*. Ordering is not relevant in this case, because the IM operator ignores it. The query will be in the form of:

$$IM(T, 0.1)$$

The result of such an operator is a set of transactions, each containing a group of e-mail addresses with respective frequency stored in attribute "support". Support of each such transaction is bigger than the selected threshold, more than one per cent in this case.

Sequence Mining. Similarly to frequent item-set mining, sequence mining is the task of pattern mining, where frequent patterns are searched. In the case of sequence mining, the goal is to obtain frequent subsequences in sequence database $\mathcal{D}$. Frequency is defined by a user-selected threshold of θ.

For the purpose of analytical algebra, database $\mathcal{D}$ is provided by the transaction file. Sequences are stored in *data* of individual transactions. The task of sequence mining is formalised as:

$$SM : (T, \theta) \rightarrow \{(p, \{\text{``support''} : s_p\}) | p \subseteq t.data; t \in T; s_p \geq \theta\}$$

where s_p is support computed over original transaction database.

Example: In genomics, one of the tasks is to discover sequence motifs in DNA. These are subsequences frequently occurring within nucleic acids. For this task, a database of DNA sequences is assumed. Individual transactions correspond to respective proteins. Sequences of amino-acids are stored in transaction's *data*. In this case, the threshold is expected to be high at around 0.9%. The query will be in the form of:

$$SM(T, 0.9)$$

The result of such query is transaction file with transactions corresponding to possible motifs. Motifs are expected to be not only frequently occurring, but also of biological significance, so post-evaluation of individual sequences by an expert is needed.

Association Rule Mining. Another task of pattern mining is association rule mining. The goal of this task is to obtain association rules. This task is often connected with other specific mining task and is considered as post-process.

Association rule describes the probability that the presence of one pattern leads to the occurrence of the other. The probability is computed based on the measures such as confidence, lift, all-confidence and many others. This measure is a function:

$$conf : (p_q, p_{sub}, T) \rightarrow \mathbb{R} | p_{sub} \subseteq p_q$$

Association rule mining in context of this analytical algebra is understood as searching for all sub-patterns of a user-selected pattern in transaction file T that has their evaluation higher than user-defined threshold θ. Association rule mining is defined in this analytical algebra as:

$$ARM : (T, p_q, conf, \theta) \rightarrow \{(p_i, \{\text{"}confidence\text{"} : conf(p_q, p_i, T)\}) |$$
$$\exists t_j : t_j \in T, p_i \subseteq t_j.data\}$$

Example: As a part of the analysis, there often may be interest in deriving rules for the purpose of optimisation. As an example, the network of communication partners from a range search can be used. The goal is to analyse if all the members of top management are informed together. As a query, there is a prepared set of e-mail addresses of top management. The threshold will be set, if in at least 99% of time the subset was a part of the whole group. Confidence is used as a measure. The query looks like:

$$ARM(T, addresses, conf, 0.99)$$

As a result, we get groups of e-mail addresses that were most of the time (in 99%) a part of the e-mail conversation addressing all the selected e-mails.

Similarity Join. The last analytical operator defined in this paper is similarity join. The goal of the join operation is to assign objects from one set to objects from another. In similarity join, the assignment is done based on the similarity of individual items. In this paper, we assume that similarity join assigns for each item the most similar one. Other variants of the task can be defined in a similar way.

$$SJ : (T_q, T, dist, k) \rightarrow \{(t_i, \{\text{``}distance\text{''} : d, \text{``}query\text{''} : t_q\}) |$$
$$\forall t_q \in T_q : t_i \in KNNS(T, t_q, dist, k)\}$$

Example: In online games, the creation of multiple accounts is often prohibited. For this purpose, the administrators of the game are analysing the behaviour of the player. One possibility of uncovering duplicit accounts is to analyse similar behaviour of different accounts such as interaction with the same accounts. Similarity join can be used for discovery of such accounts. The interactions are often recorded for every account. These can be transformed into a list and a transaction for each account can be created. Then the similarity of each item is compared with the other set, and the selected amount of similar ones is captured. The query for such a task looks like:

$$SJ(T, T, dist_{Jaccard}, 2)$$

For every item from first set the result of this query contains two most similar items from the second set. Because the query is using the same sets, the first transaction for each pair will consist of two identical items, because their distance is zero.

3.4   Supporting Operators

During analysis, it is necessary to filter and modify data in-between individual tasks. Although we try to minimise the amount of transformation needed, in some cases, it is more feasible to filter or transform data in-between instead of storing all the intermediate results and creating relevant transaction files from them. For this reason, analytical algebra contains the second type of operators called supporting operators. They share the same structure of universal operator defined in analytical algebra.

In this stage, we propose two supporting operators:

Filtration: This operator serves for definition of a new selection of transactions in the transaction file. Filtration is similar in behaviour to a selection in relational algebra. Operation is defined as:

$$\sigma : (T, \text{``condition expression''}) -> T$$

where transactions are filtered based on condition expression. In condition expression, standard application dependent operations can be used. For the selection of relevant attributes, dot notation is used.

Transformation Second operator serves for modification of existing attributes of transactions. This does not allow modification of analysed data, which should stay without modification.

$$\rho : (T, \text{``modification expression''}) -> T$$

3.5   Example of Complex Analytical Function

The main purpose of analytical algebra is to allow the creation of complex analytical functions based on the application of multiple analytical operators. As an example, we provide the task of community mining in this paper.

Community Mining. The problem of community mining consists of detection of potentially overlapping groups where items are related to one another. The non-exclusivity of the groups creates a difference compared to the classical problem of data science, clustering.

Community: There has not been a commonly accepted definition of community, but there are two major groups of definitions. [4,18] The first group of definitions is based on the density of connection between the nodes. Loosely defined, a group of nodes is marked as a community if the number of in-between connections is significantly higher than the number of outgoing connections. The second group of definitions is based on a combination of motifs, i.e. structures occurring more frequently than in a random graph. In this example, we incline more to the first group of definitions. The community is a group of densely interconnected nodes.

The data for community mining are expected to be in graph form. They are expected to have connections between them. For the proposed data model, we use an adjacency list as a suitable representation of the underlying graph. This representation can be transformed into a transaction file, where each transaction corresponds to one node in the original network. For the purpose of community mining, it is necessary to add node itself into the list of neighbours and store information about the source of the node.

Such a prepared transaction file is used as input into frequent item-set mining. The fully interconnected sub-graphs will occur in at least each transaction corresponding to the node in the group. This will result in the discovery of approximation of such highly interconnected groups. These interconnected groups are then used as cores of groups and nodes are assigned by range query.

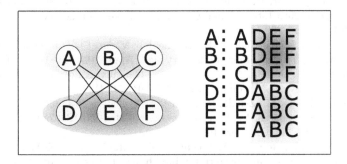

Fig. 1. Example of graph resulting in discovery of non-existing communities. (Left: Bipartite graph, Right: simplification of transaction file)

There is a possibility of obtaining groups of nodes, that are not interconnected at all. The example can be seen in Fig. 1. This situation can be then corrected by application of filtration on the result of the range query. Each node is then checked if it is part of the community. The combined query can be constructed, and it is as follow:

$$\sigma(RS(T, IM(T, 15), dist_{Jaccard}, 0.8)), (t)- > t.query.contains(t.source))$$

The result of this query is a transaction file containing transactions that have an overlap of communication with frequent item-sets of size at least 15, and they have an overlap bigger than the selected threshold of 80%.

4 Summary

In this paper, we presented the basis of new analytical algebra for the description of complex analytical operations. This algebra is based on a transaction file structure which is a part of the analytical system ADAMiSS. This algebra is modular and allows easy addition of possible operators both supporting, for better transition of data, and analytical, for semantic analysis of the data. Due to a suitable representation of data in the form of transaction file, this algebra allows a recursive application, thus easier creation of complex analytical operation.

This algebra is going to be integrated into the system ADAMiSS and similarly to the database system and relational algebra; it is going to provide simple means of creation of complex analytical functions. As a next step, there is expected to be provided a simple query language similar to SQL to allow an easier description of expected pipelines for complex data analysis.

References

1. Aggarwal, C.C., Han, J. (eds.): Frequent Pattern Mining. Springer, Cham (2014). https://doi.org/10.1007/978-3-319-07821-2
2. Agrawal, R., Srikant, R., et al.: Fast algorithms for mining association rules. In: Proceedings of 20th International Conference on Very Large Data Bases, VLDB. vol. 1215, pp. 487–499 (1994)
3. Aryabarzan, N., Minaei-Bidgoli, B., Teshnehlab, M.: negFIN: an efficient algorithm for fast mining frequent itemsets. Expert Syst. Appl. **105**, 129–143 (2018)
4. Chakraborty, T., Dalmia, A., Mukherjee, A., Ganguly, N.: Metrics for community analysis: a survey. ACM Comput. Surv. (CSUR) **50**(4), 1–37 (2017)
5. Ciaccia, P., Patella, M., Zezula, P.: M-tree: An E cient access method for similarity search in metric spaces. In: Proceedings of the 23rd VLDB Conference, Athens, Greece, pp. 426–435. Citeseer (1997)
6. Fournier-Viger, P., Gomariz, A., Gueniche, T., Soltani, A., Wu, C.W., Tseng, V.S.: SPMF: a java open-source pattern mining library. J. Mach. Learn. Res. **15**, 3569–3573 (2014). http://jmlr.org/papers/v15/fournierviger14a.html

7. Gupta, M.K., Chandra, P.: A comparative study of clustering algorithms. In: 2019 6th International Conference on Computing for Sustainable Global Development (INDIACom), pp. 801–805. IEEE (2019)

8. Hall, M., Frank, E., Holmes, G., Pfahringer, B., Reutemann, P., Witten, I.H.: The weka data mining software: an update. ACM SIGKDD Explor. Newslett. **11**(1), 10–18 (2009)

9. Han, J., Pei, J., Yin, Y.: Mining frequent patterns without candidate generation. ACM SIGMOD Rec. **29**, 1–12 (2000)

10. Kluyver, T., et al.: Jupyter notebooks-a publishing format for reproducible computational workflows. In: ELPUB, pp. 87–90 (2016)

11. Leskovec, J., Rajaraman, A., Ullman, J.D.: Mining of Massive Data sets. Cambridge university press, New York (2020)

12. MATLAB: version 7.10.0 (R2010a). The MathWorks Inc., Natick, Massachusetts (2010)

13. Mitzenmacher, M., Pagh, R., Pham, N.: Efficient estimation for high similarities using odd sketches. In: Proceedings of the 23rd International Conference on World Wide web, pp. 109–118 (2014)

14. Nijssen, S., Kok, J.N.: A quickstart in frequent structure mining can make a difference. In: Proceedings of the Tenth ACM SIGKDD International Conference on Knowledge Discovery and Data Mining, pp. 647–652. ACM (2004)

15. Novak, D., Batko, M., Zezula, P.: Metric index: an efficient and scalable solution for precise and approximate similarity search. Inf. Systems **36**(4), 721–733 (2011)

16. Pei, J., et al.: Prefixspan: mining sequential patterns efficiently by prefix-projected pattern growth. In: Proceedings of 17th International Conference on Data Engineering, pp. 215–224. IEEE (2001)

17. Peschel, J., Zezula, P.: ADAMiSS: advanced data analysis, mining and search, system. In: Amato, G., Gennaro, C., Oria, V., Radovanović, M. (eds.) SISAP 2019. LNCS, vol. 11807, pp. 351–355. Springer, Cham (2019). https://doi.org/10.1007/978-3-030-32047-8_31

18. Plantié, M., Crampes, M.: Survey on social community detection. In: Ramzan, N., van Zwol, R., Lee, J.S., Clúver, K., Hua, X.S. (eds.) Social Media Retrieval. CCN, pp. 65–85. Springer, Lodon (2013). https://doi.org/10.1007/978-1-4471-4555-4_4

19. Schubert, E., Zimek, A.: ELKI: a large open-source library for data analysis - ELKI release 0.7.5 "heidelberg". CoRR abs/1902.03616 (2019). http://arxiv.org/abs/1902.03616

20. Srikant, R., Agrawal, R.: Mining sequential patterns: Generalizations and performance improvements. In: Apers, P., Bouzeghoub, M., Gardarin, G. (eds.) EDBT 1996. LNCS, vol. 1057, pp. 1–17. Springer, Heidelberg (1996). https://doi.org/10.1007/BFb0014140

21. Team, R.C., et al.: R: a language and environment for statistical computing (2013)

22. Yan, X., Han, J.: gSpan: graph-based substructure pattern mining. In: 2002 IEEE International Conference on Data Mining, 2002. Proceedings, pp. 721–724. IEEE (2002)

23. Zaki, M.J.: Spade: an efficient algorithm for mining frequent sequences. Mach. Learn. **42**(1–2), 31–60 (2001)

24. Zaki, M.J.: Scalable algorithms for association mining. IEEE Trans. Knowl. Data Eng. **12**(3), 372–390 (2000)

FD-VAE: A Feature Driven VAE Architecture for Flexible Synthetic Data Generation

Gianluigi Greco⬤, Antonella Guzzo$^{(\boxtimes)}$⬤, and Giuseppe Nardiello⬤

University of Calabria, Rende, Italy
{gianluigi.greco,antonella.guzzo}@unical.it, giuseppe.nardiello@yahoo.it

Abstract. Variational autoencoders (VAEs) are artificial neural networks used to learn effective data encodings in an unsupervised manner. Each input x provided to a VAE is indeed mapped to an internal representation, say z, in a low-dimensional space, called the *latent space*, from which an approximate version $\tilde{x}$ of x can be eventually reconstructed via a decoding phase. VAEs are very popular generative models because, by randomly sampling points from the latent space, we can generate novel and unseen data that still reflect the characteristics of the dataset used for the training. In many application domains, however, generating random instances is not enough. Rather, we would like mechanisms that can generate instances enjoying some high-level features that are desired by the users. To accomplish this goal, a novel VAE architecture—named Feature Driven VAE—is presented. Internally, it uses Gaussian Mixture Models to structure the latent space into meaningful partitions, and it allows us to generate data with any desired combination of features, even when that specific combination has been never seen in the training examples. The architecture is orthogonal to the underlying application domain. However, to show its practical effectiveness, a specialization to the case of image generation has been presented and implemented. Results of experimental activity conducted on top of it are eventually discussed.

Keywords: Data generation · Generative models · Deep learning

1 Introduction

Synthetic data generation is a challenging task aiming at producing artificial data, either from scratch or by using advanced data manipulation techniques that synthesize novel and possibly unseen examples from real ones. In fact, in real world applications, it might happen that the possibility of implementing some advanced reasoning mechanism, based on Artificial Intelligence techniques, is limited by the lack of sufficient training data, for instance due to privacy constraints or legal restrictions, or just because the underlying application has been not yet deployed. In these cases, it is of utmost importance to use synthetic data,

ⓒ Springer Nature Switzerland AG 2020
S. Hartmann et al. (Eds.): DEXA 2020, LNCS 12391, pp. 188–197, 2020.
https://doi.org/10.1007/978-3-030-59003-1_13

generated in compliance with the properties or conditions that are envisaged to emerge in practice or with a few examples that have been collected to this end.

Generative models are increasingly used in the literature as infrastructures to generate complex data having properties that are as close as possible to those characterizing the samples in some given dataset at hand. In particular, *variational autoencoders (VAEs)* [9,14] are well-known generative models that learn the dimensional dependencies among data and produce new samples that are similar to those in the original dataset, but not exactly the same. A trained variational autoencoder can generate new samples by taking a random vector in its latent space. This is fundamental because those models that do not generate a continuous latent space cannot be easily used for sampling novel data. Moreover, unlike other generative models, which either require strong assumptions about the structure of the data or rely on computationally expensive inference procedures, VAEs only make weak assumptions on the data, and the training procedure is fast via back-propagation.

In many application domains, however, generating a random instance is not enough. The latent space of classical VAEs, indeed, has no special structure and it does not reflect in a clear and interpretable manner the high-level features that might be desired by the user. Consider, for instance, the task of generating some images of faces of persons. Typical high-level features are related to whether the person wear glasses, to the color of her/his eyes, to the kind of hair he/she has, and so on. An ideal generative mechanism should be capable of taking as an additional input the specific combination of such high-level feature desired by the user and produce a synthetic phase accordingly. With current VAEs, however, this is not possible. Our only hope is to have an image with that specific combination of features, so that we can sample a point close to the point into which such image is mapped in the latent space.

To overcome this drawback, we move forward the current research on VAEs by proposing a neural model that can force the latent space to some pre-fixed form, so that the random sampling can be guided on the basis of the high-level features that are desired. Our model, named *Feature Driven* VAE (FD-VAE), allow users to select the features that are the most desired and produce novel data according to them, even in the cases where the specific combination of features has been not registered at all in the given training examples. Technically, this is achieved by using a VAE architecture where each possible value of each feature is dealt with by a specific encoder, and where the resulting latent space is modeled by using Gaussian Mixture Models (GMM). Indeed, trained GMMs can be used to probabilistically produce new samples from the latent space with high precision according to the features that are desired by the user.

Organization. The overall FD-VAE architecture is illustrated in detail in Sect. 3, after having introduced works that are relevant in the context of generative models (cf. Sect. 2). Note that the architecture is entirely orthogonal to the underlying application domain. For the sake of concreteness, however, we present in Sect. 4 a specialization to the problem of image generation, and we report results of experimental activity conducted in this specific setting. A few final comments and remarks are eventually illustrated in Sect. 5.

2   Related Work

Generative adversarial networks (GANs) are generative models used to generate synthetic data [2]. They achieve very good performances in increasing the quality of images. However, the latent space associated with GANs is discrete, while we are interested in approaches that allows us to have a continuous latent space to get samples characterized by different features and to navigate new samples similar to training data. *Variational autoencoders (VAE)* [9,14] are other popular generative models that facilitate the generation of novel examples and that are capable to deal with complex data distributions. Differently from GAN, VAEs use Bayesian probabilistic loss functions to enforce a continuous latent space and, in fact, they found numerous applications in different real world domains. VAEs found several application in data generation [6,7,10], including for instance the Graph Variational Autoencoder proposed in [10] for generating and optimizing chemical molecules, and the Action Point Process VAE [12] that generates action sequences and to model categorical and temporal variability.

Our solution approach will use VAEs by coupling them with *Gaussian Mixture Models (GMMs)* [1]. These models are probabilistic models that are widely used in pattern recognition, data mining, image analysis, machine learning and in many problems involving clustering and classification methods [17–19]. A VAE-based architecture based on GMMs has been earlier proposed in [15], where the authors actually use GMMS to preprocess the data in order to improve the controllability and interpretability in text generation. That work extends the results on an earlier model presented in [4,8], whose goal is to prevent the mode-collapse problem in language generation, where the multiple Gaussian priors tend to concentrate during training and eventually degenerate into a single Gaussian. Clearly enough, the usage of GMMs in these works is entirely different from ours, where indeed GMMs are not used in pre-processing phases but constitute an core part of the architectural infrastructure.

3   Overview of the Approach

Our approach to produce synthetic data in a way that complies with some high-level user-desired features is based on three different ingredients.

First, we use a novel kind of *variational autoencoder architecture (VAE)* that distinguishes itself from earlier approaches in the literature because of its ability of partitioning the latent space on the basis of the possible values taken by the features. After that the latent space is learned on the basis of some available training data, novel data can be generated by sampling from this space and by providing the sampled point to the decoder block (of the autoencoder). However, sampling must be carefully performed, since our goal is that the generation of the novel data must be guided by the high-level features that are desired by the user. This motivates the introduction of a second ingredient in our method, namely the use of some probabilistic models—in fact, *Gaussian mixture models*—whose parameters are learned on the basis of the characteristics of the latent

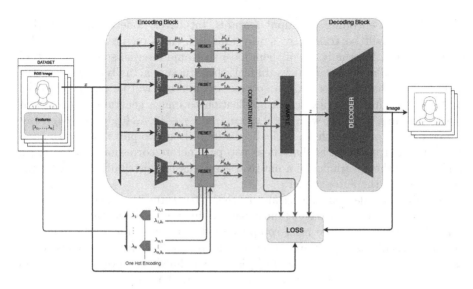

Fig. 1. FD-VAE architecture.

space, for each possible value taken by the features over the training data. By suitably combining the resulting probabilistic models, we can eventually conceive a sophisticated sampler that, in combination with the decoder, will act as our user-driven synthetic data generator.

The overall architecture is detailed in the rest of the section. Hereinafter, let D be the training dataset and let us assume that inputs are characterized by certain high-level features $F_1, ..., F_n$. For instance, over a dataset of (images of) faces, F_1 might be the feature indicating whether the person wear some glasses while F_2 might be the feature indicating the characteristics of his/her hair. Each feature F_i, with $i \in \{1, ..., n\}$, takes values from some discrete domain, say Λ_i. For instance, in the above exemplification, Λ_1 might be just the Boolean domain, while Λ_2 might be the domain $\{long, short, ...\}$.

3.1 Feature Driven VAE

The first building block of our architecture is a novel kind of variational autoencoder, which we call *Feature Driven VAE (FD-VAE)* and which is schematically described in Fig. 1. Hereinafter, let $x \in D$ be a training input in our dataset and assume that its features are $\lambda_1 \in \Lambda_1, ..., \lambda_n \in \Lambda_n$.

Each feature λ_i, with $i \in \{1, ..., n\}$, is firstly pre-processed with a one-hot encoding. That is, we define $\lambda_{i,1}, ..., \lambda_{i,k_i}$ Boolean values with $k_i = |\Lambda_i|$ such that $\lambda_{i,j} = 1$ (resp, $\lambda_{i,j} = 0$) if λ_i corresponds (resp., does not correspond) to the j-th value in Λ_i according to some given, but prefixed arbitrary order. Note that in total we have $K = k_1 + ... + k_n$ Boolean values.

The input point $x \in D$ is passed in parallel to K encoders, indexed from $ENC_{1,1}$ to ENC_{n,k_n}, which are precisely one-to-one associated with such

Boolean values encoding the input features. Intuitively, one might think that each encoder will be trained in a way that it focuses on precisely one value of the features. Internally, all of them have the same internal structure (but, of course, with different parameters). In fact, the structure actually depends on the kind of data that is processed—a specific architecture will be discussed in the next section while overviewing an application to image generation. The result of the encoding $ENC_{i,j}$, with $i \in \{1, ..., n\}$ and $j \in \{1, ..., k_i\}$, is a *compressed* representation of x, which we typically distinguish in two parts named $\mu_{i,j}$ and $\sigma_{i,j}$. The interpretation is that these two parameters correspond to the mean value and to the covariance of a multivariate Gaussian distribution defining the *latent* space of the autoencoder.

Since all encoders are identically in their structure, it is natural to expect that at some point their output must be combined with the features of x. This is precisely the goal of the *reset* components illustrated in Fig. 1. Basically, the output of $ENC_{i,j}$ is passed to a corresponding *reset* component, which also takes as input $\lambda_{i,j}$. The output $\mu'_{i,j}$ and $\sigma'_{i,j}$ is easily defined: we have $\mu'_{i,j} = \mu_{i,j}$ and $\sigma'_{i,j} = \sigma_{i,j}$ in the case where $\lambda_{i,j} = 1$; otherwise, we have $\mu'_{i,j} = \sigma'_{i,j} = 0$.

After these phase all values $\mu'_{i,j}$ and $\sigma'_{i,j}$ are respectively concatenated, then resulting in just to vectors μ' and σ'. At this point, the architecture of our FD-VAE resemble that of a standard VAE. In particular, we sample over the latent space from the Gaussian distribution with parameters μ' and σ', thereby getting a novel point z in the latent space. Eventually, z is passed to a single decoder whose output is meant to precisely correspond with the input data x— the architecture of the decoder is typical symmetric to that of the encoder, and depends on the specific data being processed.

The whole training phase of the FD-VAE is guided by the standard *loss function* for variational autoencoders aiming at minimizing the difference between the original point x and its reconstructed version.

3.2   Gaussian Mixture Models

After that the FD-VAE has been trained, it can be used in principle to generate novel data by exploring the latent space. However, our goal is more ambitious as we do not want to generate data at random, but we would like to guide the generation in a way that data is produced that complies with some desired high-level features. For instance, after that we learn the parameters of the FD-VAE on some images of faces, we might want to generate a novel face of a person wearing glasses having long hair. In particular, this is challenging, since we would like to achieve this goal even if the training data does not contain no example with such specific combination of values for the features—which is likely the case in presence of many input features.

To address this challenge, a further ingredient is needed to be put in place. Essentially, we need to learn how the latent space reflects the high-level features, which we accomplish by using the architecture depicted in Fig. 2.

The figure shows that we consider the encoding block of our FD-VAE, after that it has been trained (so, its parameters are frozen in this phase). Again,

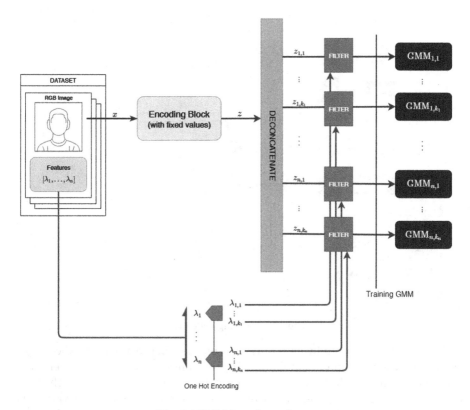

Fig. 2. GMM learning scheme.

let $x \in D$ be an input data and let $\lambda_{i,j}$ the one-hot encoding. We already know that, on input x, the encoding block produces a vector z in the latent space. This vector is now split in its coordinates, say $z_{1,1}, ..., z_{n,k_n}$, over the various values for the features, and each entry $z_{i,j}$ is filtered based on the actual value of $\lambda_{i,j}$. In particular, the *filter* component depicted in the figure blocks $z_{i,j}$ if, and only if, $\lambda_{i,j} = 0$. Whenever $z_{i,j}$ passes the filter, it eventually becomes the input for a module called $GMM_{i,j}$ whose role is to learn a Gaussian mixture model [5] that best reflects the distribution of its inputs.

3.3 Sampling Component

After that the various Gaussian mixture models $GMM_{i,j}$ have been learned, we have all ingredients in place to generate random data guided by the high-level features desired by the user. The last architectural element of our method which accomplish this task is shown in Fig. 3.

The input in this case is not longer some training data (x), but rather the features desired by the user, say $f_1, ..., f_n$. As usual, these feature are encoded then leading to the values $f_{1,1}, ..., f_{n,k_n}$. Now, the crucial observation is that, for

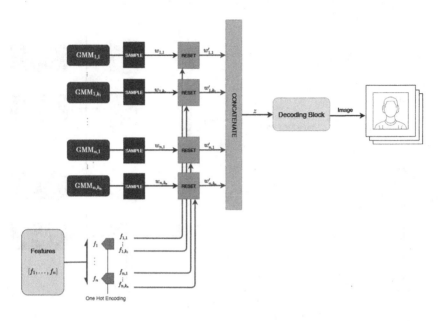

Fig. 3. Sampling scheme.

each possible value taken by each of the feature (leading to a pair of indices i, j), we use the corresponding module $GMM_{i,j}$ by sampling at random an element from the distribution it has learned from the data. Let $w_{i,j}$ be the value that has been sampled. This value is passed to a *reset* component that is activated via the value $f_{i,j}$—precisely as we have seen in the discussion of the FD-VAE architecture. The resulting values $w'_{1,1}, ..., w'_{n,k_n}$ are concatenated, thereby leading to a value z in the latent space. The intuition is now that z has been generated in a way that is fully coherent with the specific properties of the latent space when considering data having the features desired by the user.

Eventually, z is passed to the decoding block of the FD-VAE and we produce as output the desired novel data. By just repeating the sampling process, we can get an entirely novel dataset with the required characteristics.

4   Specialization and Experiments

The architecture we have discussed in Sect. 3 is orthogonal to the underlying application domain. In particular, depending on the specific data to hand, the encoders $ENC_{i,j}$ and the decoder have to be properly specialized. For the sake of concreteness and to show the practical effectiveness of our approach, in this section we discuss a simple specialization to the setting of image generation.

In fact, we point out that our goal is not to propose a highly-optimized solution for image generation, but rather to show a significant exemplification of the various ingredients discussed so far. In more details, in the context of image generation, each encoder has been implemented as a simple stack consisting of a flatten module, plus two convolutions layers, plus a final dense layer.

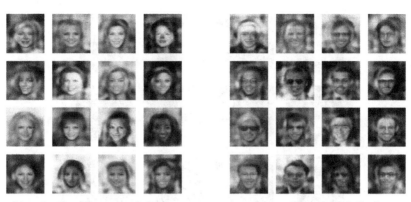

Female, No Glasses, and Smile.   Male, Glasses, and Smile

Fig. 4. Examples of generated images.

We implemented this kind of architecture by using the tensorflow library [3], and exploiting *numpy* [13] for the backend scientific calculations and *CUDA* for GPU parallelization. To evaluate the effectiveness of the implementation, we have considered the well-known Large-scale CelebFaces Attributes Dataset (CelebA) dataset, a benchmark of images of celebrity faces [11], where each image is annotated as a set of attributes/features. CelebA contains more than 200.000 celebrity images, each with 40 attribute annotations. The images in this dataset cover large pose variations and background clutters. CelebA has large diversities, large quantities, and rich annotations, including 10.177 number of identities, 202.599 number of face images, 5 landmark locations, and 40 binary attributes annotations per image. The dataset has been employed in a large number of research works, and it has been used to provide solutions in a number of computer vision tasks, including face attribute recognition, face detection, landmark (or facial part) localization, and face editing and synthesis. Moreover, correlations between attributes and their predictive quality of for user's personality traits has been investigated in [16].

Our experiments have been conducted at the varying of the number of features being considering, up to 10. Basically, we generated a large number of images whose features have been a-priori defined by the user and, for each of them, we actually checked (by human inspection) that the desired features actually occur in it. As a matter of facts, this happened in every single experiment we have conducted, hence confirming the validity of our approach.

As an example, in Fig. 4a we report some images that our system generated by considering the features *Female, No Glasses*, and *Smile*. Another example, for a different set of features, is shown in Fig. 4b. In all cases, it clearly emerges that the system is capable to produce the faces with the given features. Showing that this is the case is precisely the main goal of our experimentation. In fact, the specific quality of the images can be then further improved by considering

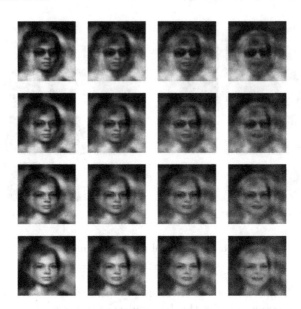

Fig. 5. Transformations to a test case: vertical axis toggles *Glasses*, while horizontal adds *Smile*.

deeper (and more involved) encoders and decoders and by considering specific heuristics for image generation—but this is an issue that is out of the scope of the paper and we leave it as a subject of further research.

Another kind of experimentation has been conducted by taking some initial (randomly) generated imaged and by transforming by adding/removing certain features. In this case, we start by applying the sampling component illustrated in Fig. 3 by considering some initial desired set of features. Let $w_{i,j}$ be the values sampled and corresponding to that features. Then, to transform the value of a feature f_i, we just re-apply the architecture in the figure, without sampling novel points, but just updating the values provided as input to the *reset* components. As an example, in Fig. 5 we show the results of toggling *Glasses* and add *Smile* to a given image.

5   Conclusions and Future Work

We have introduced a novel VAE architecture supporting the capability to control data generation according to desired sets of features. The features are high-level ones, and our architecture is capable to use them to structure the latent space in a way that data can be created to comply even to specific combination of features that have no occurrences in the training examples. An interesting avenue of further research is to look for other possible applications of our architecture, in particular to time series data where the various features can control the specific patterns of occurrence of the various event over the time

References

1. Bishop, C.M.: Pattern Recognition and Machine Learning. Information Science and Statistics. Springer, New York (2006)
2. Goodfellow, I.J., et al.: Generative adversarial nets. In: Proceedings of NeurIPS, pp. 2672–2680 (2014)
3. Abadi, M., et al. TensorFlow: large-scale machine learning on heterogeneous systems (2015). Software www.tensorflow.org
4. Dilokthanakul, N., et al.: Deep unsupervised clustering with Gaussian mixture variational autoencoders. CoRR, abs/1611.02648 (2016)
5. Gan, M.-T., Hanmandlu, M., Tan, A.H.: From a Gaussian mixture model to additive fuzzy system. IEEE T. Fuzzy Syst. **13**, 303–316 (2005)
6. Hou, X., Shen, L., Sun, K., Qiu, G.: Deep feature consistent variational autoencoder. In: Proceedings of WACV, pp. 1133–1141. IEEE Computer Society (2017)
7. Hou, X., Sun, K., Shen, L., Qiu, G.: Improving variational autoencoder with deep feature consistent and generative adversarial training. Neurocomputing **341**, 183–194 (2019)
8. Jiang, Z., Zheng, Y., Tan, H., Tang, B., Zhou, H.: Variational deep embedding: an unsupervised and generative approach to clustering. In: Proceedings of IJCAI, pp. 1965–1972 (2017)
9. Kingma, D.P., Welling, M.: Auto-encoding variational bayes. In: Bengio, Y., LeCun, Y. (eds.) Proceedings of ICLR (2014)
10. Liu, Q., Allamanis, M., Brockschmidt, M., Gaunt, A.L.: Constrained graph variational autoencoders for molecule design. In: Proceedings of NeurIPS, pp. 7806–7815 (2018)
11. Liu, Z., Luo, P., Wang, X., Tang, X.: Deep learning face attributes in the wild. In: Proceedings of ICCV (2015)
12. Mehrasa, N., Jyothi, A.A., Durand, T., He, J., Sigal, L., Mori, G.: A variational auto-encoder model for stochastic point processes. CoRR, abs/1904.03273 (2019)
13. Oliphant, T.: NumPy: A Guide to NumPy. Trelgol Publishing, New York (2006)
14. Rezende, D.J., Mohamed, S., Wierstra, D.: Stochastic backpropagation and approximate inference in deep generative models. In: Proceedings of ICML, pp. 1278–1286 (2014)
15. Shi, W., Zhou, H., Miao, N., Zhao, S., Li, L.: Fixing Gaussian mixture VAEs for interpretable text generation. CoRR, abs/1906.06719 (2019)
16. Torfason, R., Agustsson, E., Rothe, R., Timofte, R.: From face images and attributes to attributes (2016)
17. Viroli, C., McLachlan, G.J.: Deep Gaussian mixture models. Stat. Comput. **29**(1), 43–51 (2019)
18. Wang, Z., He, D., Li, B.: Clustering of copper flotation process based on the AP-GMM algorithm. IEEE Access **7**, 160650–160659 (2019)
19. Zhao, B., Zhong, Y., Ma, A., Zhang, L.: A spatial gaussian mixture model for optical remote sensing image clustering. IEEE J. Sel. Topics Appl. Earth Obser. Remote Sens. **9**(12), 5748–5759 (2016)

Data Mining

Bi-Level Associative Classifier Using Automatic Learning on Rules

Nitakshi Sood[1,2], Leepakshi Bindra[1], and Osmar Zaiane[1,2(✉)] (iD)

[1] University of Alberta, Edmonton, Canada
{nitakshi,leepaksh,zaiane}@ualberta.ca
[2] Alberta Machine Intelligence Institute, Edmonton , Canada

Abstract. The power of associative classifiers is to determine patterns from the data and perform classification based on the features that are most indicative of prediction. Although they have emerged as competitive classification systems, associative classifiers suffer from limitations such as cumbersome thresholds requiring prior knowledge which varies with the dataset. Furthermore, ranking discovered rules during inference rely on arbitrary heuristics using functions such as sum, average, minimum, or maximum of confidence of the rules. Therefore, in this study, we propose a two-stage classification model that implements automatic learning to discover rules and to select rules. In the first stage of learning, statistically significant classification association rules are derived through association rule mining. Further, in the second stage of learning, we employ a machine learning-based algorithm which automatically learns the weights of the rules for classification during inference. We use the p-value obtained from Fisher's exact test to determine the statistical significance of rules. The machine learning-based classifiers like Neural Network, SVM and rule-based classifiers like RIPPER help in classifying the rules automatically in the second stage of learning, instead of forcing the use of a specific heuristic for the same. The rules obtained from the first stage form meaningful features to be used in the second stage of learning. Our approach, BiLevCSS (**Bi-Lev**el **C**lassification using **S**tatistically **S**ignificant Rules) outperforms various state-of-the-art classifiers in terms of classification accuracy.

Keywords: Associative classification · Classification rules · Statistical significance

1 Introduction

Classification is the process of organizing and categorizing data into distinct classes. It involves various tasks like building a model based on the distribution of the data in consideration and further using this model for identification of the class label of new data. An associative classifier is a kind of supervised classification model that learns on association rules that attribute features with class

© Springer Nature Switzerland AG 2020
S. Hartmann et al. (Eds.): DEXA 2020, LNCS 12391, pp. 201–216, 2020.
https://doi.org/10.1007/978-3-030-59003-1_14

labels. The association rule mining identifies patterns in the data by extracting associations between items in a dataset. The class association rules (CARs) obtained from mining are represented in the form, $X \rightarrow Y$, where X and Y are the antecedent and consequent respectively. For an associative classifier, we choose the consequent to be the class label while the antecedent set includes the set of items that are highly indicative of their association with the class label based on association rules.

Most of the previously proposed associative classification algorithms like CMAR [16], CBA [17] and CPAR [19] have different rule discovery, rule pruning, rule prediction and evaluation methods. However, a predefined weighting scheme is required, for each of these methods in order to predict the class from the association rules. Heuristics like maximum/minimum of confidence, average of confidence or sum of confidence of the rules for the classes can be used to decide the predicted value for the new samples. However, the weighting scheme may differ for various applications when using associative classifiers. Deciding the heuristics to select rules to apply during inference and therefore to predict the class from the derived classification rules is a challenging task, and is typically fixed as part of the algorithm.

This form of classification offered by associative classifiers is easily understandable, flexible and does not assume independence between attributes, however, it requires prior knowledge to choose appropriate support and confidence threshold values for rule mining. Moreover, they contain a large number of noisy rules which are redundant, uninteresting and lead to longer classification time. Various pruning techniques have been designed to deal with this limitation, for instance, removing the low ranked specialized rules, removing conflicting rules or using database coverage based pruning strategy. A two level classification method was initially proposed by Antonie and Zaiane in [4], where the first stage used Apriori-based approach [1] to generates associative rule classification model which is followed by a stage of machine learning classifiers to learn the weights for classification in the second stage. We extended their work and compare the performance of SVM [10], Neural Networks [6] and RIPPER [9] in the second stage of learning. Although, this automatic approach of learning to use the rules is expected to give better classification results, it suffers with certain limitations. Firstly, the setting up of an optimal support and confidence threshold values to mine the rules in the first stage is a cumbersome task. Secondly, the rules generated using the former approach may contain noisy, non statistically significant rules and may not cover all the important features in the selected rules.

Therefore, in order to address the above given limitations, we propose BiLevCSS (**Bi-Lev**el **C**lassification using **S**tatistically **S**ignificant Rules), which uses statistically significant rules generated from a first stage, to form features that are made full use of, for classification in the second stage of learning. We follow the approach proposed by Li and Zaiane in [15] for generation of statistically significant CARs. We also use Fisher's exact test to obtain the p-value which is used to determine the statistical significance of the association rules. We further extract features from these significant association rules and then train the supervised learning classifiers like Neural Network, SVM and RIPPER on them. Finally, the trained model from the second stage is used to find the class label for a new data point.

Traditional association rules mining methods mostly prune the infrequent items on the basis of frequency of the itemset and thereafter calculate the strength of the rule in the form of its confidence values. This also ignores the statistically significant rules. Although most of the associative classifiers deal with this limitation by setting up small minimum threshold values, however, this leads to the generation of a huge number of insignificant rules. Therefore, in our proposed model, we use the instance-centric pruning strategy as used in SigDirect [15] to find globally optimal CAR (Class Association Rules) for each instance in the training dataset without compromising the classification accuracy.

Furthermore, we use Neural networks [6] and Support Vector Machines [10] in our approach as they are strong machine learning classifiers, that have proved their worth in various applications. With the aim to build an efficient classification strategy, we train them using meaningful features obtained from the first stage of learning. However, many real time applications specifically in healthcare and medicine require explainable models in order to interpret the results post classification. In our proposed strategy, although the statistically significant rules and derived features obtained in the first stage form an explainable model, Neural Network and SVM used in the second stage for classification might make the results un-explainable for such applications Therefore, in order to make our approach interpretable, we explored the applicability of a rule-based classifier like RIPPER in the second stage for classification of derived features. Ripper [9] is a rule-based classifier which was found to produce a minimal set of explainable classification rules when given meaningful features in the second stage of our proposed approach, without compromising on the classification accuracy.

Therefore, in our study we propose a novel bi-level classification model, which uses the association rule mining to produce statistically significant rules. Further these rules are used to form more meaningful and non redundant features to be given as input in the second stage of learning comprised of a second classifier. The proposed algorithm helps in automatic learning of non noisy, statistically significant rules and further, it leads to a higher classification accuracy. The main contributions of this work are:

- We propose BiLevCSS, (**Bi-Lev**el **C**lassification using **S**tatistically **S**ignificant Rules), which is an effective two stage learning model. In the first stage of learning, we build an associative rules classifier (ARC) model based on statistically significant rules, followed by a supervised learning classifier in the second stage of learning for classification.
- We evaluate the performance of Neural Networks and SVM against rule-based classifier RIPPER to compare their accuracy and suitability for different datasets when used in the second phase of BiLevCSS.
- We evaluate the proposed algorithm BiLevCSS on 10 UCI datasets and with other commonly used classifiers on the basis of classification accuracy. The

results show that our classifier gives better classification accuracy than various state-of-the-art classifiers.

The rest of the paper is organized as follows: Sect. 2 gives a literature review about some previously proposed associative classifiers; Sect. 3 explains the methodologies we have adapted in our algorithm; Sect. 4 shows the evaluation results of our proposed classifier on UCI Datasets; and lastly, Sect. 5 gives the conclusion and directions about our future work.

2   Related Work

The idea of associative classifiers was first presented by Liu et al. [17], while the concept of using association rules as CARs was proposed earlier by Bayardo Jr. [5]. Liu et al. proposed CBA, an approach to perform classification using the class association rules in [17]. The proposed work used Apriori based rule generation algorithm, involving the cumbersome process of tuning support and confidence values. Furthermore, CBA applies the paradigm of "database coverage" for rule pruning and uses highest ranked matching rules as the heuristic for classification. This work paved the way for the associative classification. Li et al. proposed another associative classifier called CMAR in [16]. CMAR uses FP-growth [12] which is a frequent pattern mining based approach to produce a set of association rules. The authors also use a novel data structure called CR-tree to store the CARs. Furthermore, CMAR determines the class label based on the set of matching rules using weighted chi-square measure. Antonie and Zaiane propose an associative rule-based classifier by category for automatic text categorization called ARC-BC [2]. ARC-BC forms association rules grouped by the category for each set of documents. The average confidence value is calculated for each category and finally the class label of the group with highest confidence value is considered as the predicted category. The proposed algorithm works for both single and multi class.

Antonie and Zaiane further proposed the first associative classifier that uses both the positive and negative CARs in [3]. They use Pearson's coefficient as the interestingness measure to mine positively and negatively correlated CARs. They were able to prove that a much smaller set of positive and negative CARs was efficient enough to compete and outperform various other categorization systems. The classification is made by using an average confidence heuristic.

Coenen and Leng have reviewed three case satisfaction mechanisms namely, Best First Rule, Best K Rules and All Rules in [8] and various alternative rule ordering strategies. The authors have evaluated these case satisfactions as they have been commonly used in numerous Classification Association Rule Mining (CARM) algorithms to use the classifier thus formed, for the prediction task.

A two stage classification model called 2SCARC was proposed in [4], which automatically learns to use the rules for classification. Antonie and Zaiane used an Apriori based algorithm in the first stage to generate features from class association rules, which are given to the next stage for training a Neural Network to automatically learn the weights for classification. The main aim of this work was to overcome the cumbersome task of tuning support and confidence values for every dataset. Although, the results obtained are interesting, however they are not convincing as they tend to ignore the statistical significance of the rules. Noisy and meaningless rules produced in the first stage might mislead the classification in the second phase. This forms the baseline of our work as described in further sections.

Furthermore, Li et al. presented a novel associative classifier which is built upon both positive and negative association classification rules in [14]. The proposed classifier incorporates, rule generation where statistically significant positive and negative CARs are discovered and a rule pruning phase where irrelevant rules are pruned. Further, these rules are used for the prediction of the unlabeled data. They propose a very efficient rule pruning strategy so as to prune both negative and positive CARs simultaneously. Li et al. concluded that summing up the confidence values of all matching rules and accordingly making the class label prediction proved to be the best classification strategy. Li et al. have also presented an associative classifier called SigDirect [15] which produces statistically significant and meaningful rules for classification. The authors have obtained globally optimal association rules using a novel instance-centric rule pruning strategy instead of more prevalent pruning strategy like database coverage. Li et al. evaluate various heuristics for the classification and infer that SigDirect, with a specific heuristic, gives high classification accuracy using a minimum set of association rules.

3 Methodology

In this section, we introduce the details about the proposed Bi-Level classification technique. We initially describe the baseline technique of developing a two level classifier by using the Apriori algorithm for building the ARC model in the first level. However, this technique was found to suffer limitations with regard to selecting the optimum support and confidence threshold values for different datasets. Therefore, we extended our baseline to include the approach proposed by Li et al. [15]. In our proposed method we use statistically significant CARs to obtain rule features that are used in the second stage of learning.

3.1 Notations and Definitions

Definition 1. Dependency of a CAR [15]
If a transaction database T consists of a set of items $I = \{i_1, i_2, ..., i_m\}$ and a set of class labels $C = \{c_1, c_2, ..., c_L\}$, a transaction X in T consists of a set of items $A = \{a_1, a_2, ..., a_n\}$ and a particular class label c_k such that $A \subseteq I$ and $c_k \in C$. A CAR R in the form of $A \rightarrow c_k$ is called dependent if the antecedent part and the consequent class label of the CAR satisfy $P(A, c_k) \neq P(A)P(c_k)$, where $P(A)$ denotes the probability of occurrence of itemset A.

Definition 2. Fisher's exact test [14]
Consider a null hypothesis in which A and c_k are assumed to be independent

of each other. The dependency of the CAR A $\rightarrow c_k$ is said to be statistically significant at level α, if the probability p of obtaining an equal or stronger dependency in a dataset complying with a null hypothesis is not greater than α. The probability p, i.e., p-value, can be calculated by Fisher's exact test:

$$p_f(A \rightarrow c_k) = \sum_{i=0}^{min\{\sigma(A,\neg c_k)\sigma(\neg A,c_k)\}} \frac{\binom{\sigma(A)}{\sigma(A,c_k)+i}\binom{\sigma(\neg A)}{\sigma(\neg A,\neg c_k)+i}}{\binom{|T|}{\sigma(c_k)}} \qquad (1)$$

where $\sigma(X)$ denotes the support count of X. The significance level α is usually set to be 0.05.

Definition 3. Potentially Statistically Significant [15]
*The CAR A $\rightarrow c_k$ is defined as "Potentially Statistically Significant" (**PSS**), if it meets either of the following conditions:*

1. *$\sigma(A) \leq \sigma(c_k)$ holds, and the lower bound $\frac{\sigma(\neg A)!\sigma(c_k)!}{|T|!(\sigma(c_k)-\sigma(A))!}$ is smaller than or equal to α;*
2. *$\sigma(A) > \sigma(c_k)$ holds.*

where $A \subseteq I_{Remaining}$ and $c_k \in \{c_1, c_2, ..., c_L\}$
If a CAR is **PSS**, we need to calculate the exact p-value to see if it is indeed statistically significant.

3.2   Method 1

The aim of associative classification is to find knowledge from data in the form of association rules associating features and class labels. During inference one or a set of rules are selected and used to predict the class label. This selection is typically based on heuristics for ranking rules.

Using the proposed approach of two stage classification in [4], we have implemented the same technique for building a model which would learn to select and use the discovered rules automatically rather than relying on heuristics to select them. In brief, the first stage is to learn an associative classifier and the second stage is to extract features from the learned rules to learn a second predictor predicting which rule is best to use during inference. The initial training dataset is split into two parts, one used to derive rules with association rule mining and the second part to extract features for the second training level. These two sets are disjoint in order to avoid overfitting. On the TrainSet 1, the first stage of learning is performed. Here, our algorithm uses a constrained form of Apriori [1] to perform association rule mining to obtain a set of rules that have features on the left and class labels on the right side of the rule and that are above the minimum threshold values for support and confidence. This ARC Model is used to collect a set of features from the samples present in TrainSet 2. As proposed in [4], we have used two approaches namely, the class based and the rules based feature extraction, to get the set of features and class labels from the ARC model.

Class Based Features. For the class based feature extraction technique, we derive rules from TrainSet 1 and we match the features from our TrainSet 2 with the antecedents of the rules in the ARC Model. A rule is said to be *applicable* to a new instance of TrainSet2 if the antecedent of the rule is a subset of the features of the instance. Using the set of rules that apply to the instances in TrainSet 2, we count the number of rules that match for each class. Using this approach we derive a transformed feature set as shown in Table 1, where we state the average confidence and the count of all the matching rules for an example of three given class labels. This dataset of *class-based features* is given to the next level of learning in order to train a classification model that selects rules.

Table 1. Example for transformed set of features in Class based

Class1		Class2		Class3	
Avg Conf	#Rules	Avg Conf	#Rules	Avg Conf	#Rules
85	1	81.6	3	80	2

Table 2. Example for transformed set of features in Rule based

R1			R2			R3		
Conf	Sup	Match	Conf	Sup	Match	Conf	Sup	Match
80	10	0	90	10	1	85	15	1

Rule Based Features. For the rule based approach, we use the characteristics of the rules derived from TrainSet 1 to create a new feature space. For each instance in the dataset TrainSet 2, we check if each of the rules in the ARC model apply or not, that is we match the features from the sample with the antecedents of the rule. This feature is denoted by a boolean value 1 to represent a match, 0 for absent. Along with this, information of support and confidence is added as features in the new set. An example is shown in Table 2, where one row in the dataset is taken and a new feature is generated for 3 rules of the ARC Model.

The features derived using the ARC model are further given as a training input to the next level, consisting of the classifier, which learns how to use the rules in the prediction process. In the second level, machine learning based classifiers like Neural network (NN) and Support Vector Machine(SVM) or rule based classifier like RIPPER, are used to automatically learn on rules to determine the weighting scheme for classification and obtain the final model.

For testing, we use the ARC model to derive the set of features for the Test dataset. Further, these features are given to the trained model of Neural network,

SVM or RIPPER to classify the new samples. The ARC model and the trained model in the second level together predict the class for any new sample given for classification.

3.3   Method 2

In our second approach, we extend the bi-level classification technique by using statistically significant CARs. For this purpose, we derive positive and negative rules which are statistically significant [14]. Li et al. proposed to use Fisher's exact test to extract the statistical significance of rules. The proposed algorithm determines non-redundant association rules for classification which show statistical dependency between the antecedent items and the consequent labels by using the p-value.

We split our training dataset into two parts as illustrated in Algorithm 1. On the TrainSet 1, the first level of learning is performed. The association rule mining is done by building Apriori like tree to form the ARC model, which gives us the set of association classification rules. The rules described by this ARC model are statistically significant, giving us the p-value for each rule. The rules obtained in the first level are used to extract the transformed feature set from the TrainSet 2. We used rule-based approach as described in Method 1 to extract features for this classifier as well. This is because the rule-based features in Method 1 shows better results than class-based features, as will be discussed in Sect. 4.

As proposed in [15], initially, all the impossible items are removed. An item is termed as impossible to appear in a statistically significant CAR if it has support value below $\gamma|T|$, where $\gamma \leq 0.5$ and T is the transaction database. These items are removed and thereafter all the left over items ($I_{Remaining}$) are sorted in the ascending order of their support values. Further the tree is enumerated to generate class association rules and only those with one antecedent are listed. These rules are then checked for their PSS value (Definition 3). Rules that do not satisfy either of the PSS conditions are pruned and the other rules are checked for statistical significance. From PSS 1-itemset rules, PSS 2-itemset rules are generated considering the property that if a rule is PSS, then its parent rule will also be PSS, i.e. if CAR $A \rightarrow c_k$ is PSS, then any of its parent rule $B \rightarrow c_k$ is also PSS, where $B \subsetneq A$ and $|B| = |A| - 1$. The process repeats until no PSS rules are generated at a certain level. Also, if a rule is marked as minimal, the expansion from this rule is stopped because all of its children rules can not get a lower p-value.

The number of rules generated by the above approach may be large and might contain some unnecessary rules as well. In order to make the classification efficient and to obtain globally best rules from the training dataset, we use the proposed instance-centric rule pruning approach [15]. These pruned rules form the ARC model for this method.

We further apply the rule based approach to extract the features for the TrainSet 2 using this ARC model. An example for rule-based feature extraction for Method 2 is shown in Table 3 with just two rules. For each sample in

TrainSet 2, we take the boolean value representing whether the rule matches the sample or not. Along with this, we take the characteristics of the rule as features in the transformed feature set. These include support value of the rule, its confidence and the log of the p-value. The lower the p-value, the better the rule, and summing up the p-value is not a suitable heuristic for a set of rules. Hence, we take the log value of p-value in order to generalize the process for rule-based and class based feature extraction. The features are extracted for each row in the testing dataset using the ARC model and the learnt classification model predicts the class label for each data point.

Table 3. Example for transformed set of features for method 2

R1				R2			
Conf	Sup	ln(p-value)	Match	Conf	Sup	ln(p-value)	Match
80	10	−10.6	0	90	10	−5.1	1

We also evaluate the BiLevCSS with SigDirect associative classifier in the second level. However, SigDirect is found to have a limitation of not being able to work well with very high dimensional datasets. For some datasets when using BiLevCSS, the features extracted for the second phase are found to have a large dimensionality due to a sizeable number of generated rules. This greatly increases the runtime of the SigDirect algorithm when used in the second phase. Therefore, we do not report the results of SigDirect as a predictor in the second stage.

4   Experimental Results

We have evaluated our algorithm on 10 UCI datasets to compare the classification accuracy with other rule based and machine learning based algorithms that exist in the literature. We report the average of the results obtained for every dataset on the 10 fold cross validation in our experiments. We compare the performance with common machine learning techniques like SVM and Neural networks, rule-based classifiers like C4.5 and RIPPER and previously proposed associative classifiers like CBA, CMAR and CPAR. We also compare our baseline approaches 2SARC1 (NN) [4], 2SARC2 (NN) [4], 2SARC1 (SVM), 2SARC2 (SVM), 2SARC1 (RIPPER) and 2SARC2 (RIPPER) with these classifiers.

4.1   Classification Accuracy

We compare our proposed model BiLevCSS with the above stated contenders on the basis of classification accuracy. We evaluate the performance of BiLevCSS model with three different classifiers in the second level; Neural Network at the second stage (regarded as BiLevCSS (NN)), RIPPER in the second stage

---

Algorithm 1: Algorithm for BiLevCSS

Data: **Train Dataset**: Initial training dataset. **Test Dataset**: Initial testing
dataset. **TransformedTestSet**: Testing dataset for classification model.
TrainSet1: Training set used to build the ARC Model. **TrainSet2**:
Training set used to build features using the ARC model and train the
classication model.

Result: Predict class label of each instance in **TestSet**.

1 Use TrainSet1 to generate all statistically significant CARs $\mathbf{A} \rightarrow c_k$. ; ▷ Follow
   the Algorithm 1 and 2 in [15]
2 classLabelsSet ⟵ Unique set of class labels in dataset
3 ARC Model = { CARs A $\rightarrow c_k$ | $c_k \in$ classLabelSet}
4 **for** *each instance T in TrainSet2* **do**
5   NewFeature=[]
6   **for** *each rule R in ARC model* **do**
7   match(T, R) ; ▷ Determine if instance T matched the antecedent
            of rule R.
8   **if** *match(T, R)==True* **then**
9   NewFeature.append(Conf(R), Support(R), log(P-value(R)), 1);
10   **else**
11   NewFeature.append(Conf(R), Support(R), log(P-value(R)), 0);
12   **end**
13   **end**
14   TransformedTrainSet.append(NewFeature);
15 **end**
16 Train a supervised learning model using TransformedTrainSet dataset for
   classification.
17 Repeat steps 4 to 15, to extract features from Test Dataset using ARC model to
   build TransformedTestSet for second stage of learning.
18 Derive the accuracy of the classification model using the Test dataset.

(regarded as BiLevCSS (RIPPER)) and SVM in the second stage (regarded as
BiLevCSS (SVM)).

We follow the default parameter values for SVM [10], C4.5 [18], CBA [17],
CMAR [16], CPAR [19] as stated in the original papers. For RIPPER as a
standalone rule based classifier, we have used default parameters from Weka [13]
which are also stated to be the best by the authors in [9]. For vanilla Neural
Network, we use a single hidden layer with the number of nodes to be the average
of the number of input and output nodes and we also tune ReLU or sigmoid
activation functions with a learning rate of 0.1.

For our baseline Method 1, we perform experiments using Apriori [1] based
rule generation in the first level learning. Further, we test the accuracy of the
rule-based feature extraction approach to build the bi-level classifier with Neural
Network, SVM or RIPPER in the second stage. Similarly, we also measure the
accuracy, of the bi-level classifier, which uses class-based features. For Apriori,
we use a range of support values from 5% to 30% depending on the size of
the dataset. The threshold value for confidence is set around 50%. In Table 4,

Table 4. Comparison of classification accuracy using Rule-based and Class-based Features extraction in Method 1

Datasets	#cls	#rec	2SARC2 (NN)	2SARC2 (SVM)	2SARC2 (RIPPER)	2SARC1 (NN)	2SARC1 (SVM)	2SARC1 (RIPPER)
Iris	3	150	93.74	89.74	**94.28**	**94.11**	89.3	90.94
Glass	7	214	48.9	**52.2**	**69.17**	50	**52.2**	51.74
Heart	5	303	**63.5**	54.34	54.95	**62.34**	57.14	54.02
Hepati	2	155	**85**	**81.25**	79.97	70	75	80.48
Pima	2	768	66.45	65.2	**72.74**	64.39	67.53	**70.93**
Flare	9	1389	74.5	70.58	**84.35**	74.39	70.6	**83.96**
Anneal	6	989	77	82	**96.41**	79.5	78.04	**83.74**
Horse	2	368	67.6	63.3	**81.40**	**72.97**	70.96	63.75
Breast	2	699	89.7	93	93.14	93.75	**98.6**	**93.78**
Wine	3	178	**97.18**	77.97	85.15	**94.92**	72.02	53.84
Average			**76.35**	72.95	**81.15**	75.63	73.13	72.71

we report the accuracy obtained for the 10 UCI datasets using our baseline approach. Along with the classification accuracy values, the name of the dataset and the number of records have also been reported. As can be seen from Table 4, the overall accuracy does not follow a pattern and nothing conclusive could be derived from the results aforementioned. However, the results from Method 1 showed that, for most of the UCI datasets, the rule-based feature extraction approach is found to give altogether a better average accuracy over the class-based feature extraction approach.

Therefore, in the second method, we adapt the rule-based feature extraction approach to build the bi-level classification model with statistically significant rules. For the following experiments, we discretize the numerical attributes of the datasets as stated in [7]. All the results reported in this section have been performed on the same discretized dataset for fair comparison.

Moreover, as suggested by Li et al. in [15], we use the Fisher exact test to analyse the statistical significance of the class association rules. The threshold for p-value is set to be 0.05. The use of only statistically significant rules and the addition of p-value value along with support and confidence as a feature in the rule-based classification gives us much better results for Method 2 than the baseline Method 1. For the second layer of both the methods, we use Neural Network with single hidden layer, with 'ReLU' or 'sigmoid' as the activation functions and a learning rate of 0.1. We also tune the hyper parameter values of gamma, kernel and regularization parameters for the SVM classifier. We have performed 5 fold internal cross validation for SVM and NN to tune their respective hyper parameter values. For RIPPER at the second stage of learning, we use the default best parameters from Weka. It can be observed that, in Table 5, the BiLevCSS model gives the best overall classification accuracy for the considered

Table 5. Comparison of classification accuracy of BiLevCSS with other state-of-the-art classifiers

Datasets	BilevCSS (RIPPER)	BilevCSS (NN)	BilevCSS (SVM)	RIPPER	NN	SVM	C4.5	CBA	CMAR	CPAR
Iris	95.72	**100**	**98.66**	94	98.09	94.6	94	94.67	94	94.7
glass	69.27	**86.60**	59.52	68.69	70.14	68.6	71.47	73.9	70.1	**74.4**
Heart	56.51	**78.64**	52.84	53.97	56.72	55.4	**61.5**	57.8	56.2	53.8
Hepati	82.57	**84.95**	**88.41**	78.06	82.89	79.3	79.25	81.82	80.5	79.4
Pima	73.64	**81.24**	73.2	66.36	**75.95**	74	73.7	72.9	75.1	73.8
Flare	84.27	**96.1**	83.1	72.13	**84.61**	73.8	82.1	84.2	84.3	63.9
Anneal	**96.93**	96.96	96.25	95.8	93.96	85	89.87	**97.91**	97.3	98.4
Horse	83.34	**87.78**	77.27	84.23	81.321	72.5	**85.04**	82.36	82.6	84.2
Breast	93.05	94.26	92.80	95.42	**96.83**	95.7	94.71	96.28	**96.4**	96
Wine	89	**94.94**	84.20	91.57	91.66	**94.9**	71.7	49.6	92.7	88.2
Average	82.43	**90.14**	80.62	80.02	**83.21**	79.38	80.33	79.14	82.92	80.68

datasets. Our algorithm BiLevCSS(NN) outperforms all the other classification algorithms in the 10 UCI datasets with highest average accuracy.

We further perform a comparison between BiLevCSS with Neural Network at the second level against the vanilla Neural Network with 1 hidden layer, to validate the efficiency of the model. The results show that the proposed algorithm outperforms the vanilla NN. Similarly, BiLevCSS(SVM) was found to outperform vanilla SVM and BiLevCSS(RIPPER) outperformed the vanilla RIPPER algorithm. Figure 1 illustrates the comparison of results given by the best model BiLevCSS(NN) with vanilla Neural Network.

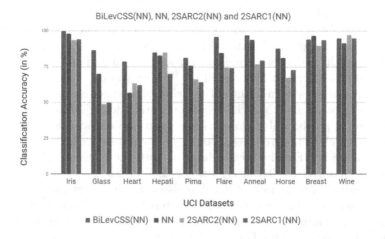

Fig. 1. Comparison of classification accuracy for BiLevCSS(NN) with vanilla Neural Network, 2SARC1(NN) and 2SARC2(NN).

The results shown in Table 5 highlight that the BiLevCSS model outperforms other rule based and associative classifiers on comparison. Next, we compared the three proposed strategies namely, BilevCSS (Ripper), BiLevCSS (NN) and BiLevCSS (SVM) with SigDirect. The results of this comparison are summarized graphically in Fig. 2. The graph shows that BiLevCSS (NN) performs better than the rest, which proves that, when meaningful, statistically significant and non-noisy rules are given to Neural Network, the classification accuracy of the classifier improves. The results obtained from BiLevCSS (Ripper) are motivating, however do not beat BiLevCSS (NN) in performance. Therefore, in the future we aim to evaluate more explanatory classification models in the second phase of learning, for a more explainable model since Neural Networks are more of a black box compared to Ripper.

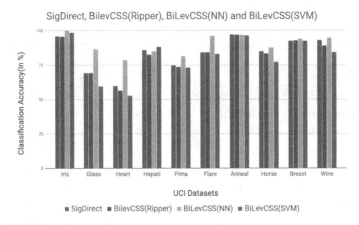

Fig. 2. Comparison of classification accuracy for BiLevCSS(NN) with BiLevCSS (RIPPER), BiLevCSS(SVM) and SigDirect.

4.2   Statistical Analysis

The accuracy values report that BiLevCSS performs better for most of the datasets. To confirm this statement, we perform statistical analysis as shown in Table 6. We follow Demsar's study [11] and use Friedman's test to compare the statistical significance of the results obtained from the comparison of all the algorithms on the basis of classification accuracy. Since the p-value obtained from this test was less than the critical value (alpha) which is equal to 0.05, it proves that the results are statistically significant and the algorithms are significantly different from one another.

Furthermore, to investigate the statistical significant of the proposed algorithm with other contenders pair-wise, we perform another non-parametric test called Wilcoxon signed ranked test [11]. In this test, for every pair of algorithm in consideration, the difference of their classification accuracy, D_i is calculated to analyse the ranks based on the absolute values of these differences, $|D_i|$. Further, positive ranks R_i^+ and negative ranks R_i^- are calculated based on the original values of D_i for two algorithms. Adding up all the values of R_i^+ and R_i^-, W_{stat} is calculated as $min(\sum R_i^+, \sum R_i^-)$ which gives us the critical value Z. For alpha value equal to 0.05, the corresponding Z-value is -1.96, therefore, the null hypothesis is rejected if the obtained critical value Z is less than -1.96.

Table 6 reports the p-values obtained by comparing the most accurate model, BiLevCSS(NN) against other classifiers using Wilcoxon test. We also compare the number of times the different algorithms win or lose against BiLevCSS(NN) and if there is a tie between them. The p-values obtained are less than 0.05 which show that BiLevCSS(NN) is statistically significantly better than all the contenders. The results show that the proposed BiLevCSS algorithm with Neural Network at the second stage of learning outperforms the rest of the algorithms by winning in at least 8 out of 10 instances.

Table 6. BiLevCSS(NN) compared to the rest of the algorithms on 10 UCI datasets

Classifiers	Wins	Losses	Ties	P-value
BiLevCSS(NN) vs BiLevCSS(SVM)	9	1	0	0.017
BiLevCSS(NN) vs RIPPER	9	1	0	0.007
BiLevCSS(NN) vs NN	9	1	0	0.013
BiLevCSS(NN) vs SVM	9	1	0	0.009
BiLevCSS(NN) vs 2SARC2(NN)	8	2	0	0.013
BiLevCSS(NN) vs 2SARC2(SVM)	10	0	0	0.005
BiLevCSS(NN) vs 2SARC2(RIPPER)	10	0	0	0.005
BiLevCSS(NN) vs 2SARC1(NN)	10	0	0	0.005
BiLevCSS(NN) vs 2SARC1(SVM)	9	1	0	0.007
BiLevCSS(NN) vs 2SARC1(RIPPER)	10	0	0	0.005
BiLevCSS(NN) vs BiLevCSS(RIPPER)	10	0	0	0.05
BiLevCSS(NN) vs C4.5	9	1	0	0.007
BiLevCSS(NN) vs CBA	8	2	0	0.013
BiLevCSS(NN) vs CPAR	8	2	0	0.013
BiLevCSS(NN) vs CMAR	8	2	0	0.013

5   Conclusion and Future Work

In this project, we have introduced a novel approach BiLevCSS, a two level classifier built on statistically significant dependent CARs. The proposed classification model consists of four steps of rule generation, rule pruning, transformed feature extraction for the next phase using the obtained rules and finally, the prediction

on the learned model using Neural Network in the second stage. Rule generation leads to the generation of all statistically significant rules which are further
used to train a second classification model to select appropriate rules. Since,
these rules might be noisy with some irrelevant information, they are pruned
using the instance-centric rule pruning strategy. Furthermore, the features are
extracted using rule based or class based techniques. Finally, the classification
is done by using the learned NN, SVM or RIPPER in the second level. The
idea of using statistically significant rules has made our algorithm more efficient by selecting only valuable CAR and providing new features for the second
stage. The experimental results are very encouraging. The proposed classifier
especially BiLevCSS(NN) is found to have achieved better prediction than other
state-of-the-art classification algorithms in terms of accuracy.

In the future, we aim to experiment our algorithm by incorporating more
features other than support, confidence, lift and p-value. We would also like to
evaluate the performance of our model with explainable associative classifiers
in the second stage of learning. We would also extend our work for multi-label
classification.

References

1. Agrawal, R., Srikant, R.: Fast algorithms for mining association rules. In: Proceedings of 20th International of Very Large Databases (VLDB), vol. 1215, pp. 487–499
   (1994)
2. Antonie, M., Zaiane, O.R.: Text document categorization by term association. In:
   proceedings of IEEE International Conference on Data Mining, pp. 19–26 (2002)
3. Antonie, M.L., Zaïane, O.R.: An associative classifier based on positive and negative rules. In: Proceedings of the 9th ACM SIGMOD Workshop on Research Issues
   in Data Mining and Knowledge Discovery, pp. 64–69 (2004)
4. Antonie, M.L., Zaiane, O.R., Holte, R.C.: Learning to use a learned model: a two-
   stage approach to classification. In: Sixth International Conference on Data Mining
   (ICDM), pp. 33–42. IEEE (2006)
5. Bayardo Jr., R.J.: Brute-force mining of high-confidence classification rules. KDD
   97, 123–126 (1997)
6. Beale, H.D., Demuth, H., Hagan, M.: Neural Network Design. PWS, Boston (1996)
7. Coenen, F.: The LUCS-KDD software library (2004). http://cgi.csc.liv.ac.uk/
   ~frans/KDD/Software/
8. Coenen, F., Leng, P.: An evaluation of approaches to classification rule selection.
   In: Fourth IEEE International Conference on Data Mining (ICDM 2004), pp. 359–
   362 (2004)
9. Cohen, W.: Fast effective rule induction. In: International Conference on Machine
   Learning, pp. 115–123. Elsevier (1995)
10. Cortes, C., Vapnik, V.: Support-vector networks. Mach. Learn. **20**(3), 273–297
    (1995)
11. Demšar, J.: Statistical comparisons of classifiers over multiple data sets. J. Mach.
    Learn. Res. **7**, 1–30 (2006)
12. Han, J., Pei, J. and Yin, Y.: Mining frequent patterns without candidate gener-
    ation. In: Proceedings of 2000 ACM SIGMOD International Conference on Man-
    agement of Data, vol. 29, pp. 1–12 (2000)

13. Holmes, G., Donkin, A., Witten, I.: Weka: a machine learning workbench. In: Proceedings of ANZIIS (1994)
14. Li, J., Zaiane, O.: Associative classification with statistically significant positive and negative rules. In: Proceedings of the 24th ACM International on conference on Information and Knowledge Management, pp. 633–642. ACM (2015)
15. Li, J., Zaiane, O.R.: Exploiting statistically significant dependent rules for associative classification. Intell. Data Anal. **21**(5), 1155–1172 (2017)
16. Li, W., Han, J. and Pei, J.: CMAR: accurate and efficient classification based on multiple class-association rules. In: IEEE International Conference on Data Mining, ICDM, pp. 369–376 (2001)
17. Liu, B., Hsu, W.Y.: Integrating classification and association rule mining. In: International Conference on Knowledge Discovery and Data Mining (1998)
18. Quinlan, J.R.: C4.5: programs for machine learning. Mach. Learn. **16**(3), 235–240 (1994)
19. Yin, X., Han, J.: CPAR: classification based on predictive association rules. In: SIAM International Conference on Data Mining, pp. 331–335 (2003)

RandomLink – Avoiding Linkage-Effects by Employing Random Effects for Clustering

Gert Sluiter[1], Benjamin Schelling[1,2,3(✉)], and Claudia Plant[1,4]

[1] Faculty of Computer Science, University of Vienna, Vienna, Austria
benjamin.schelling@univie.ac.at
[2] MCML, Munich, Germany
[3] Ludwig-Maximilians-Universität München, Munich, Germany
[4] ds:UniVie, Vienna, Austria

Abstract. We present here a new parameter-free clustering algorithm that does not impose any assumptions on the data. Based solely on the premise that close data points are more likely to be in the same cluster, it can autonomously create clusters. Neither the number of clusters nor their shape has to be known. The algorithm is similar to SingleLink in that it connects clusters depending on the distances between data points, but while SingleLink is deterministic, RandomLink makes use of random effects. They help RandomLink overcome the SingleLink-effect (or chain-effect) from which SingleLink suffers as it always connects the closest data points. RandomLink is likely to connect close data points but is not forced to, thus, it can sever chains between clusters. We explain in more detail how this negates the SingleLink-effect and how the use of random effects helps overcome the stiffness of parameters for different distance-based algorithms. We show that the algorithm principle is sound by testing it on different data sets and comparing it with standard clustering algorithms, focusing especially on hierarchical clustering methods.

1 Introduction

Most clustering algorithms are based on some kind of assumptions about the distribution or form of the data. K-Means [9], for example, is based on the assumption of Gaussian distributed clusters, with the variance in all directions being basically the same. EM [2], on the other hand, does not necessarily assume a uni-directional variance but is capable of finding lopsided, stretched Gaussian-distributed clusters. The assumption behind it though is a Gaussian distribution.

To overcome these restrictions, purely distance-based clustering techniques like SingleLink [12] and DBSCAN [3], which no longer make assumptions about the distribution of data, have been created. These techniques, however, often need at least one parameter to help them to estimate how dense the expected

---

G. Sluiter and B. Schelling are contributed equally to the paper and share first authorship.

S. Hartmann et al. (Eds.): DEXA 2020, LNCS 12391, pp. 217–232, 2020.
https://doi.org/10.1007/978-3-030-59003-1_15

clusters should be. Since the density of the data set might vary, this may cause that clusters with different densities can be cut out very poorly. DBSCAN is well known to have problems with varying densities [11]. SingleLink, on the other hand, suffers from the "SingleLink-Effect", where clusters are combined, if they have a "bridge" of data points connecting them. All these difficulties are caused by the strict focus on the given parameters, which does not always give the leeway needed.

The technique that we would like to present here, RandomLink, avoids such problems by using randomised effects to determine the clusters. It does not need parameter(s) or model assumptions to find the clusters. Its only premise is that the closer data points are, the more likely they belong into the same cluster. It starts with a fully connected data set (all data points are connected) and deletes all connections between data points. The order of deletion is remembered and inverted to connect data points until a certain number of clusters is found. This number of clusters would be a parameter which we want to avoid. Therefore, we also present a simple strategy to get rid of it. Connections are deleted depending on their length. A long connection is more likely to be deleted. Thus, every pair of data points can stay connected, but distant ones are less likely to. Direct connections between them are highly unlikely to consist for a long time, but they can nevertheless be connected via other data points. If the density between them is high, then there might be a path of data points linking them. If there is a path between them, then they are part of the same cluster, as clusters are defined as "connected components", i.e. all data points that are connected via paths.

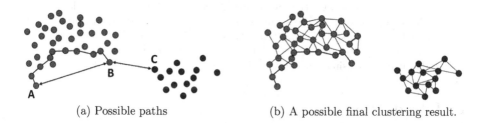

(a) Possible paths   (b) A possible final clustering result.

Fig. 1. The principle behind RandomLink.

See Fig. 1 for this. The direct connection between data points A and B is highly unlikely to exist for a long time, as it is rather long. Compared to it, the line between B and C has a relatively high chance to remain for longer. This would link two clusters, which clearly do not belong together, and split one cluster into two, which do. RandomLink removes – with a certain probability – longer links first, which means that while it does remove the link between A and B, it will eventually also remove the link between B and C. Since the links between clusters are – as a tendency – longer than many links inside a cluster, the connection between clusters will eventually be cut. Inside the cluster though, many links will consist for a long time and thus a path (like the one shown in the Fig. 1) between A and B may remain and link them indirectly. There is, of course, the chance that this specific path is also interrupted, but as there is a high number of possible paths between A and B, at least one will - most likely - remain.

Thus, not only the distance between data points is relevant for RandomLink, but also the placement and relative distances of the other data points are very important to decide if data points are considered part of the same cluster. While the direct connection will be – very often – erased, indirect paths will remain inside the cluster, that link different parts of the cluster with each other. One possible clustering result then might look like Fig. 1b. The direct connection between A and B is gone, but an indirect path remains. RandomLink potentially takes all connections between data points into account in every step. One might state that RandomLink considers the whole data set, not only the local environment.

The similarity to SingleLink is clear. SingleLink connects the closest data points, while RandomLink is only **likely** to connect them or might link them via other close data points. It has a broader approach because the length gives the probability of a link remaining, not the absolute 0/1-decision of SingleLink. With this RandomLink can e.g. alleviate the SingleLink-effect (or chain effect), from which SingleLink suffers and cluster a data set more naturally. The SingleLink-effect is caused by a sort of bridge between clusters, which will keep them connected, no matter how long the bridge. With RandomLink the connections of the bridge will be thinned out while clustering until it eventually is no longer connecting. The specifics, of course, depend on the length, density, etc of the bridge, but RandomLink can overcome, or at least lessen, the difficulty of the SingleLink-effect. RandomLink is as a whole not purely dependent on local density, as e.g. DBSCAN is, but evaluates on a broader spectrum, as stated before. This leads to a reduced dependency on the rigidity of fixed parameters compared to DBSCAN or SingleLink have. We will talk in more detail about this in Sect. 4, on how we intend to use random effects to our advantage.

1.1   Related Work

RandomLink computes the distances between all data points and determines an order by deleting them depending on their length combined with random numbers. After this order is fixed, it starts connecting the data points until the clusters are found. The closest related methods are clearly hierarchical clustering approaches like SingleLink. The differences in these methods are how they decide on the order of connecting the data points to clusters. SingleLink connects data points/clusters depending on the closest data points in the clusters, AverageLink [13] connects the clusters depending on the average distance between data points, CompleteLink [1] on the maximal distance between data points and Ward's Criterion [16] on the change in variance in the clusters. These methods all decide on the order of connecting data points deterministic while RandomLink uses random effects and connects the **likely** closest clusters. As a distance between clusters, we use the SingleLink approach of the minimal distance between data points in clusters. It could be easily adapted to other definitions of distance, as e.g. AverageLink uses.

One could consider Graph-clustering methods like Highly Connected Subgraphs (HCS) [5] or Chameleon [7] as related, as we plot various Graph-like Figures, but these are used to explain RandomLink. Graph-clustering methods are focused on clustering graphs, while RandomLink clusters numerical data. One can create a Graph out of numerical data and employ these methods, but other clustering approaches are more closely related. HCS divides graphs along a minimum cut [6] multiple times before it starts re-connecting them. Chameleon [7] creates a sparse kNN graph, partitions it into many sub-graphs and merges them using a minimum cut criterion.

Both these methods have parameters which are hard to tune. RandomLink needs no parameters at all. In the straightforward implementation, we would leave the number of clusters to be found to the user, but we created an approach for estimating the optimal stopping point in creating the final clusters (details in Sect. 2.2). This stopping criterion could be construed as an ensemble-approach (also called consensus clustering, see [8] for an introduction). Ensemble methods are a relatively new approach to data mining. They try to combine multiple clustering results into one and this single result should be better than any of the input clusterings. RandomLink uses k-Means to find the best stopping point, but it does not combine clustering results, so it is not exactly an ensemble-approach.

Spectral Clustering-approaches start with a similarity/distance matrix on which they eventually employ k-Means. RandomLink computes the distances and uses k-Means to decide the stopping point. There are some similarities and we take them as interesting comparison methods, in particular, the fundamental algorithm of Ng, Jordan and Weiss [10], which has become one of the classical clustering approaches by now. These standard methods should be used as comparison methods as they provide a baseline of what one can expect as a clustering result. Another standard approach is DBSCAN, one of the best-known examples of distance-based clustering methods, which we will discuss in more detail later.

1.2   Contributions

In this paper we propose the clustering algorithm RandomLink, which is extensively tested on various real world datasets. It performs especially well on data sets where other clustering approaches have difficulties to reach even a minimal level of clustering quality. Thus, it can be used as an approach for difficult data sets, where not much is known. The advantages of RandomLink are the following:

- RandomLink tends to find decent clusterings even when well-established algorithms fail.
- RandomLink is completely parameter-free and needs no input from the user. It does not need the number of clusters and stops completely automatic.
- RandomLink has no assumption about the shape of distribution of clusters. They can be of arbitrary shape and distributions.

– The source code of RandomLink is publicly available for everyone to corroborate our results.

2   The Algorithm

As stated, RandomLink determines the order of connecting clusters by first deleting them using their length as probability for being deleted. Thus, we start by computing the distances between all data points and store them in a distance matrix S. We can imagine the data as a fully connected data set, i.e. each pair of data points has an edge, i.e. link, e in between them with the length of this link defined by the Euclidean distance of the data points. The algorithm now deletes one link after another, with the length of an link determining the probability of that happening.

2.1   Deleting Links

All links are lined up and the total length of all links sum_S is computed. See Fig. 2 for a simple example. A random number in the interval $[0, sum_S]$ is drawn, here it is 3.1415, and the corresponding link e_2 is removed. It is simply the link into which' interval the number falls. Link e_2 is then removed from the concatenated links and the total length is now 18, i.e. $sum_S - l(e_2)$ with $l(e)$ being the length of link e. A new number is now drawn from the interval $[0, sum_S - l(e_2)]$ and the same step is repeated.

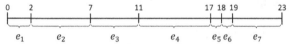
(a) Links listed with their length. The total length of all links is 23.

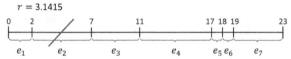
(b) We draw the random number r=3.1415 from the range $[0, 23]$. The corresponding link is e_2.

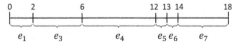
(c) We remove e_2. The total length of the links is now 18.

Fig. 2. How we determine the order of connecting clusters.

The probability of a specific link e_x being removed is thus

$$p(e_x) = \frac{l(e_x)}{\sum_e l(e)} \tag{1}$$

It is linearly dependent on the length of the e, i.e. longer links are more likely of being removed. This approach is commonly referred to as Roulette Wheel Selection [4].

RandomLink starts by creating a distance matrix $S_{n \times n}$ computed from a data set $D_{d \times n}$ using a distance function. We use the Euclidean distance, but any other distance or similarity measure could be used as well. Using Manhattan distance or Cosine and Gaussian similarity lead only to tiny differences. The results reported in Sect. 3 are almost uninfluenced by the used distance/similarity function. The values in S give the probability (if scaled as in Eq. (1)) for every link connecting two data points to be deleted. We calculate the sum of S and store it as sum_S. As explained earlier, we choose a random value r in the range $[0, sum_S]$. There is a link corresponding to r and this link is deleted from the data set and its value in S set to 0. Its index is stored in a stack and identifies the deleted link. This procedure is then continued until all links are deleted.

Algorithm 1. RandomLink

Require: Data D
    procedure RANDOMLINK(D)
        $S \leftarrow$ similarity matrix(D)
        $sum_S \leftarrow \sum(S)$, $score_{max} \leftarrow 0$
        while $sum_S > 0$ **do**
            delete random link e
            $sum_S \leftarrow sum_S$-length(e)
            set index of e 0 in S
            stack.push(index of e)
        end while
        $ds \leftarrow$ Disjoint Set.make_set(S)
        while ds.n_connected_components > 1 **do**
            ds.union(stack.pop())
            if (ds.n_connected_components changed) **then**
                $score \leftarrow$ stopping criterion(ds.connected_components, D)
            end if
            if ($score > max_{score}$) **then**
                $Labels \leftarrow ds$.connected_components, $score_{max} \leftarrow score$
            end if
        end while
        return $Labels$
    end procedure

The algorithm starts with a fully connected data set from which the links are deleted until every data point is a singleton. Thus, the order of the links between the data points is determined and the algorithm can start connecting them, i.e. the clusters will be created. The algorithm adds the first link as determined in the order of the links, reducing the number of connected components from n to $n - 1$. After a few links are added, we will have a situation like in Fig. 3d. Many clusters exist and the correct clusters are still separated. Adding a few more links

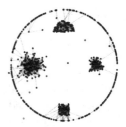
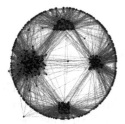

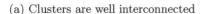

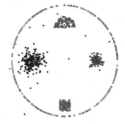

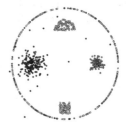

(a) Clusters are well interconnected   (b) Clusters are interconnected by few links

(c) More and more clusters appear.   (d) Few links are inserted

Fig. 3. Different amount of links and the connected components. The algorithm deletes more and more connections, until eventually all data points are singletons. The order of the connections is determined by the deletion of links and the clusters found by inserting the links in reverse order.

will lead to Fig. 3c, where the clusters start to make sense. In the beginning, most added links will be small until, eventually, longer links are added. Connections between the clusters become more and more likely. This is the situation depicted in Fig. 3b. The number of connected components decreases with those longer links being added and interconnecting clusters. The algorithm terminates as soon as the number of connected components $|cc|$ becomes one. All data points are connected and the clustering no longer changes.

Somewhere between the two extremes of $|cc| = n$ and $|cc| = 1$ is the point where we want to stop. If we were to look for exactly k clusters, we could simply stop as soon as we found k connected components, but we want to have a completely parameter-free clustering approach which determines the number of clusters automatically. It is possible to apply RandomLink with k as a parameter, but knowing the correct number of clusters is seldom trivial and we can not expect that the user always knows exactly for what he/she is looking for. We found a heuristic that helps us find the optimal stopping point.

We make here use of a union-find data structure which is also known as disjoint-set for an online connected components analysis and benefit from quasi constant time per operation [14]. Every time clusters are merged the number of connected components decreases by one and the root nodes for every data point in the disjoint-set data structure are used as cluster labels.

2.2 Stopping Criterion

In-between $|cc| = n$ and $|cc| = 1$ is the point where we want to stop. If $|cc| = n$ then every point is its own cluster. Adding links leads to the data points creating clusters. Adding even more links leads to the clusters becoming connected until eventually $|cc| = 1$. The typical progress of this can be seen in Fig. 4. The measure for clustering quality used there is Normalised Mutual Information (NMI) [15] which is widely used to evaluate clustering results. NMI scales between 1.0 (perfect clustering) and 0.0 (purely random cluster assignments). We see in Fig. 4 that there is a peak between the fully connected data set and fully isolated data points. The NMI is maximal there, i.e. it is the best result that can be reached with our approach and we find it the following way:

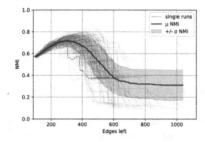

Fig. 4. Average NMI when deleting links for our running example.

Whenever the number of connected components $|cc|$ changes, a *score* is computed by our stopping criterion (Algorithm 2). This *score* changes for different values of $|cc|$. When the *score* has become maximal, max_{score}, the ideal state of the clustering has been reached and the connected components cc at that time will be returned as the clustering result. The score itself is computed by comparing our current clustering result with the result of k-Means by executing k-Means on the data D and using the number of connected components $|cc|$ as the number of clusters k. Thus, we compare the clusterings of RandomLink and k-Means with NMI and remember the comparison value. Finding the stopping point this way ensures that we need no further parameters (the k for k-Means is given by RandomLink) and although it is purely heuristic, it is easy to understand.

Algorithm 2. Stopping Criterion

Require: connected components cc, Data D
   procedure EVALUATE STOPPING CRITERION(cc,D)
      $Label \leftarrow$ k-Means $(D,k=|cc|)$
      return NMI $(Label,cc)$
   end procedure

Thus, we know when the algorithm should stop. It would be beneficial for RandomLink to know k, as we will see in Sect. 3.2, but this way we remove the last parameter at the cost of a relatively small loss in clustering quality. Hence, one can use RandomLink and set a number of clusters as e.g. SingleLink does it, or use it automatically in combination with this stopping criterion.

3 Experiments

3.1 Real World Data

We tested RandomLink with various real world data sets from the UCI Machine Learning Repository to evaluate its performance. The most important comparison method is clearly SingleLink, as it is closest to our approach. We also included AverageLink [13], CompleteLink [1] and Ward's method [16] as further representatives of hierarchical clustering methods. We also included the standard clustering methods, i.e. EM [2], as they give a realistic value of what can be expected from clustering for a data set. K-Means [9] is a standard approach and also employed to evaluate the optimal stopping point and therefore a necessary comparison method. DBSCAN [3] is one of the most prominent distance-based clustering methods which we will talk about more later on. We mentioned the similarities to Spectral Clustering-methods and have thus also included STSC [17] as a popular method and FUSE [18] as a more recent one. With Spectral Clustering we refer to the essential algorithm by Ng et al. [10], which was foundational for this type of clustering methods. Furthermore, the aforementioned Chameleon [7] is chosen to represent graph clustering methods.

The results are shown in Table 1. "RandomLink max" value stands for the best result which could have been reached with our approach, while "RandomLink" denotes the actual result we reach in combination with the stopping criterion. On the data sets, RandomLink always yields a good NMI value while the other algorithms partially completely fail and are sometimes clearly outperformed by RandomLink. RandomLink is the best choice on all of the 8 data sets and loses only once by a tiny deficit. The data sets have a wide range of dimensionality and number of clusters, which shows that RandomLink is not restricted in these regards. We want to especially emphasise the improvement over SingleLink. Including these random effects into it, massively improved the clustering results. The stopping criterion works most of the time as intended and returns a result close to the optimum. Furthermore, the clustering results are mostly stable. The small deviations in clustering quality show that the algorithm behaves predictably in a certain range.

Parameters. We tried to be as fair as possible to the comparison methods. The algorithms were given correct k if needed and using Euclidean distance. For DBSCAN we performed a grid search on the parameter range $\epsilon = [0.01 - -10.0]$ in 0.5 increments and $minPts = 2, 5, 8, 11, 14$ and report the best found NMI value. For Chameleon, the kNN graph was constructed with $k = 10$ and the

Table 1. Experimental results. All non-deterministic results have been repeated 100 times and the average is given. The correct value for k is always given for the comparing algorithms. Results given in NMI and best result bold marked.

Data set	Yeast	Fish	User know.	Crowdsourced.
# of dimensions	8	463	5	28
# of classes	10	7	4	2
RandomLink max	0.48	0.57	0.47	0.55
RandomLink	**0.45 ± 0.01**	**0.55 ± 0.01**	**0.46 ± 0.01**	**0.49 ± 0.02**
SingleLink	0.12	0.03	0.05	0.03
CompleteLink	0.23	0.19	0.29	0.34
AverageLink	0.11	0.13	0.32	0.40
Ward's method	0.27	0.35	0.28	0.43
k-Means	0.27	0.28	0.23	0.43
DBSCAN	0.12	0.39	0.11	0.00
EM	0.17	0.25	0.42	0.42
Chameleon	0.00	0.47	0.35	—
Spectral Clustering	0.28	0.39	0.23	0.43
STSC	0.06	0.10	0.04	0.12
FUSE	—	0.19	0.02	0.01
Data set	Glass Id.	Thyroid	Libras Move.	Arrhythmia
# of dimensions	9	13	90	278
# of classes	6	6	15	11
RandomLink max	0.50	0.54	0.68	0.61
RandomLink	**0.47 ± 0.02**	**0.52 ± 0.02**	**0.64 ± 0.02**	**0.56 ± 0.05**
SingleLink	0.07	0.51	0.12	0.35
CompleteLink	0.38	0.47	0.54	0.43
AverageLink	0.11	0.51	0.60	0.40
Ward's method	0.40	0.44	0.62	0.47
k-Means	0.43	0.46	0.59	0.44
DBSCAN	0.46	0.00	0.59	0.00
EM	0.34	0.44	0.59	0.43
Chameleon	0.00	0.44	0.00	0.54
Spectral Clustering	0.31	0.44	0.62	0.46
STSC	0.09	0.11	0.22	0.44
FUSE	0.28	—	0.18	0.31

default value of $\alpha = 2.0$ was used for the cluster merging, besides that, the authors stated that Chameleon is not very sensitive to the parametrization [7]. The similarity matrix for spectral clustering was created by using the Euclidean distance of the 10-nearest neighbours reassuring a connected graph. For Self

Tuning Spectral Clustering (STSC) [17] the default parameters were used. For SingleLink, FUSE, k-Means etc. the correct k is always given, as stated. RandomLink was executed 100 times for every data set listed in Table 1 and the mean NMI is reported as well as the standard deviation. We use NMI to evaluate clustering quality, as it is widely used and often considered the standard when evaluating clustering results.

Table 2. We compare the clustering result of the stopping criterion to knowing k in regard to clustering quality (NMI) and runtime. Best result in bold.

Dataset	Stopping criterion		Knowing the number of clusters	
	Mean NMI	Runtime	Mean NMI	Runtime
Yeast	0.45 ± 0.01	100%	$\mathbf{0.48 \pm 0.00}$	68.8%
Fish	0.55 ± 0.01	100%	$\mathbf{0.57 \pm 0.00}$	49.5%
User Know.	0.46 ± 0.01	100%	$\mathbf{0.48 \pm 0.00}$	53.0%
Crowdsourced.	0.49 ± 0.02	100%	$\mathbf{0.55 \pm 0.00}$	52.4%
Glass Id.	0.47 ± 0.02	100%	$\mathbf{0.53 \pm 0.00}$	50.5%
Thyroid	0.52 ± 0.02	100%	$\mathbf{0.58 \pm 0.04}$	37.6%
Libras Move	0.64 ± 0.02	100%	$\mathbf{0.69 \pm 0.00}$	34.8%
Arrhythmia	0.56 ± 0.05	100%	$\mathbf{0.65 \pm 0.02}$	24.1%

Sourcecode. Under the following links, one can find code and data sets:
https://github.com/53RT/RandomLink
https://dm.cs.univie.ac.at/research/downloads/
We do so as we feel that it is important that our claims can be validated and fellow researchers can build upon our results if they feel so inclined.

3.2 Adding the Number of Clusters as a Parameter

We pride ourselves on RandomLink being completely parameter-free, i.e. no density-parameter or number of clusters is needed. The question is whether this has deteriorated the clustering quality, and if so, by how much? Thus, we ran RandomLink to find exactly $|cc| = k$ clusters and compared it to the results of our stopping criterion. In Table 2 is the effect of supplying k described.

Two relevant effects can be observed here: 1) The runtime does clearly decrease if k is supplied as a parameter. Computing the stopping criterion is no longer necessary and one can stop the algorithm as soon as the correct number of clusters is reached, which means one has to perform less operations with the disjoint-set data structure used in the link insertion phase. 2) The difference in NMI is small. This means that our method either stops at the correct number of clusters or, if it stops at a different point, finds a stopping point that is comparable in regard to clustering quality. There is a tendency for the results to be better with given k, but this is not exactly surprising as the additional information makes things easier. Knowing when to stop, reduces the risk of generating poor clustering results.

A user has, therefore, the possibility to either let the algorithm find the number of clusters automatically or ask for a specific number of clusters, which entails a speed-up of a factor of roughly 2–3. Automatic setting of k is a major advantage in an unsupervised setting, as most of the time the data set is not very well understood and any decision a user has to make might be false. RandomLink takes this responsibility from the user, though, at the cost of runtime, but in similar clustering quality.

3.3 Runtime

We omit extensive experiments on runtime, as we are more interested if this approach is valid in principle, but we do calculate the estimations. For the algorithm RandomLink itself, we first need to compute the distances between all data points. This takes $\mathcal{O}(n^2)$ operations to do. Alternatively, we can also start with an adjacency matrix, and perform RandomLink on it, but this is not a massive overall improvement, as we still need to determine the order of links. There are n^2 many links. Selecting a specific one, as described in Sect. 2.1, entails a binary search, i.e. a runtime of $\mathcal{O}(\log_2(n^2))$. This link is now removed. Finding the next link entails again a binary search, but this time on $n^2 - 1$ many elements, thus the runtime for it is $\mathcal{O}(\log_2(n^2 - 1))$. Summing up over all binary searches from n^2 to 1 link(s) is $\sum_{i=0}^{n^2} \log_2(i)$. Using Stirling's approximation this can be estimated as

$$\mathcal{O}(n^2 \cdot \log_2(n)) \tag{2}$$

With this, the order of the links is determined and we start to create the clusters. The stopping criterion consists of executing k-Means, whenever the number of connected components changes. K-Means has an estimation of $\mathcal{O}(n)$ and has to be computed at most n times. Thus, it adds to the total actual runtime but does not add anything in regards to the $\mathcal{O}$-calculation.

The second phase of RandomLink - inserting the links in the reversed top-down order - can be solved efficiently using a disjoint-set data structure [14]. First the *make set* method initialises the data structure with creating a node for every item with the parent node pointing at itself. This takes $\mathcal{O}(n)$ time. The parent node is used for a recursive traversal to determine if two data objects are connected which is true if they have the same root. If path compression and union by size or rank is used in the data structure the complexity reduces to $\mathcal{O}(\alpha(n))$ for the *find* and *union* operations which is optimal and quasi constant. As we have $\approx n^2$ links to insert in a fully connected setting there will be at most $\mathcal{O}(n^2 \cdot \alpha(n))$ operations which is never reached in reality as the connected components reaches one with a fraction of inserted links. The number of connected components can be retrieved as a byproduct as it always decreases by one if clusters are united and the cluster labels can be easily extracted from the root node of every item.

Since $\alpha(n)$ is essentially constant, creating the clusters is $\mathcal{O}(n^2)$ and the dominating part of the estimation is computing the order of the links, i.e. $\mathcal{O}(n^2 \cdot \log_2(n))$. Equation (2) gives therefore the total of the runtime-estimation.

4   Using Random Effects

Every pair of data points entertains the possibility in RandomLink to not be connected with each other. This drastically lessens linkage-effects, which necessarily happens for SingleLink for data sets like the one depicted in Fig. 5. The SingleLink-effect, often referred to as chaining-effect or chain-effect, describes the tendency of SingleLink to create long chains of clusters, i.e. linking clusters through small bridges of data points which do not belong together. The example, shown in Fig. 5 is a prime, if somewhat extreme, example of this happening. The two clusters do have a bridge in between them and this bridge needs to be broken for them to be correctly clustered. SingleLink is not capable of this. Even if the bridge were far longer SingleLink would still connect the two clusters. RandomLink, on the other hand, will break this bridge and the further away the two clusters are, the faster this will happen.

Somewhat similar is the situation for DBSCAN. We see in Fig. 6 two Gaussian clusters with different density. DBSCAN is well known to have problems with this type of setting, where density varies. We see the difference in NMI and how much more capable RandomLink is in clustering this data set.

Distance-based techniques are most of the times deterministic and, therefore, forced to "obey" their parameters. Since these parameters are necessarily based on local density (that is either the closest neighbour (e.g. SingleLink) or the number of neighbours in a certain environment (e.g. DBSCAN)), the local density determines if data points are put into the same cluster. The difficulty now lies therein that only taking local density into account might lead to troubling clustering results. This is obvious for SingleLink with data sets like the one shown in Fig. 5, where the local density in the bridge between clusters is relatively high and SingleLink will, therefore, connect the clusters. This drawback is also present in DBSCAN, as it is not fit to handle clusters with different densities (see Fig. 6). Such a situation will lead to sub-par clustering results. RandomLink, on the other hand, can handle such situations due to its more "holistic" approach as it takes the whole data set into account. It splits the clusters in Fig. 5 apart, without falling into the same trap as SingleLink. It can also handle a situation like in Fig. 6, where DBSCAN (as well as SingleLink) would fare very badly. The idea for the future is to combine random effects with DBSCAN and SingleLink to overcome these difficulties these algorithms have with such data sets. The approach of RandomLink that employs randomised effects, helps overcome the restrictions of "fixed" parameters.

This is what we did with SingleLink: In the classical form, SingleLink first computes all distances and then continues linking the closest clusters until either

k connected components are created, with k as a given parameter, or a stopping criterion tells it to. RandomLink, on the other hand, computes all distances, deletes them and then continues linking the **likely** closest clusters until the stopping criterion tells it to stop. The similarities are obvious. One can consider RandomLink as an extension of SingleLink with the help of random effects, and a stopping criterion. SingleLink is essentially the expected result of RandomLink, but the random effects present in RandomLink help to overcome the chaining-effect and to break the bridge between clusters.

Figure 6 also suggests that the same approach is also possible for other deterministic distance-based methods like DBSCAN, i.e. that we can to combine DBSCAN with randomized effects to lessen the dependency on fixed parameters. The goal is to establish the inclusion of randomised effects into distance-based clustering algorithms as a general principle, which helps with overcoming certain restrictions, that these algorithms suffer as a consequence of their rigidity.

(a) Groundtruth   (b) RandomLink NMI: 0.86   (c) SingleLink NMI: 0.015

Fig. 5. SingleLink-Effect for two Gaussian cluster with a bridge in between. Random-Link separates the Link, while SingleLink cannot.

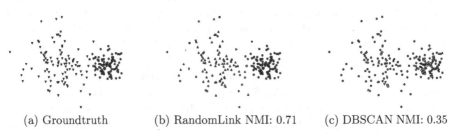

(a) Groundtruth   (b) RandomLink NMI: 0.71   (c) DBSCAN NMI: 0.35

Fig. 6. Gaussians with different density. RandomLink can separate them better than the carefully parametrized DBSCAN.

5   Outlook and Conclusion

RandomLink in combination with the stopping criterion is a completely parameter-free clustering approach that can handle a wide range of data sets. It assumes no specific distribution for a cluster, thus, can handle clusters of arbitrary shape. One might think that the use of k-Means to estimate the optimal stopping point, limits it to Gaussian clusters, but k-Means has no influence on how the clusters are constructed. The shape of the clusters found is determined solely by the order in which the links are added to the data set. Instead of k-Means, we also tried Spectral Clustering, which can handle non-convex clusters, and the results barely differ. Since it does increase runtime, we stuck with k-Means.

RandomLink does not have some of the drawbacks of other distance-based clustering approaches, as we have shown in comparisons with SingleLink and DBSCAN, the main representatives of this group. It can handle bridges between clusters and clusters with varying density. We outlined our idea about including randomised effects into other distance-based clustering algorithms, as we have done here with RandomLink for SingleLink. The idea to use random effects for clustering might not be the most obvious one, but we are convinced that we established here the usefulness of such an approach, especially when comparing the clustering results of SingleLink and RandomLink in Table 1. RandomLink can be taken as an extension of SingleLink with random effects (and a stopping criterion) and we are certain that this can also be done with other algorithms. Our main concern in this work has been to establish that the combination of clustering with random effects can prove useful, especially for overcoming the restrictions that these algorithms have. We are optimistic that this has been implied heavily by RandomLink for SingleLink and we are looking forward to combining this approach with other methods like DBSCAN.

References

1. Defays, D.: An efficient algorithm for a complete link method. Comput. J. **20**, 364–366 (1977)
2. Dempster, A.P., Laird, N.M., Rubin, D.B.: Maximum likelihood from incomplete data via the EM algorithm. J. R. Stat. Soc. **39**(1), 1–38 (1977)
3. Ester, M., Kriegel, H.P., Sander, J., Xu, X.: A density-based algorithm for discovering clusters in large spatial databases with noise. In: KDD (1996)
4. Goldberg, D.E.: Genetic Algorithms in Search, Optimization and Machine Learning. Addison-Wesley Longman Publishing Co., Inc., Boston (1989)
5. Hartuv, E., Shamir, R.: A clustering algorithm based on graph connectivity. Inf. Process. Lett. **76**(4–6), 175–181 (2000)
6. Karger, D.R.: Minimum cuts in near-linear time. J. ACM **47**(1), 46–76 (2000)
7. Karypis, G., Han, E.H.S., Kumar, V.: Chameleon: Hierarchical clustering using dynamic modeling. Computer **32**(8), 68–75 (1999)
8. Krawczyk, B., Minku, L.L., Gama, J., Stefanowski, J., Woźniak, M.: Ensemble learning for data stream analysis: A survey. Inf. Fusion **37**, 132–156 (2017)
9. MacQueen, J.B.: Some methods for classification and analysis of multivariate observations. In: Proceedings of the Fifth Berkeley Symposium on Mathematical Statistics and Probability, vol. 1, pp. 281–297. University of California Press (1967)
10. Ng, A.Y., Jordan, M.I., Weiss, Y.: On spectral clustering: Analysis and an algorithm. In: Dietterich, T.G., Becker, S., Ghahramani, Z. (eds.) Advances in Neural Information Processing Systems 14, pp. 849–856. MIT Press (2002)
11. Sang, Y., Yi, Z.: Motion determination using non-uniform sampling based density clustering. In: 2008 Fifth International Conference on Fuzzy Systems and Knowledge Discovery, vol. 4, pp. 81–85 (2008)

12. Sibson, R.: Slink: An optimally efficient algorithm for the single-link cluster method. Comput. J. **16**(1), 30–34 (1973)
13. Sokal, R.R., Michener, C.D.: A statistical method for evaluating systematic relationships. Univ. Kans. Sci. Bull. **38**, 1409–1438 (1958)
14. Tarjan, R.E.: A class of algorithms which require nonlinear time to maintain disjoint sets. J. Comput. Syst. Sci. **18**(2), 110–127 (1979)
15. Vinh, N.X., Epps, J., Bailey, J.: Information theoretic measures for clusterings comparison: Variants, properties, normalization and correction for chance. J. Mach. Learn. Res. **11**, 2837–2854 (2010)
16. Ward, J.H.: Hierarchical grouping to optimize an objective function. J. Am. Stat. Assoc. **58**(301), 236–244 (1963)
17. Yang, C., Zhang, X., Jiao, L., Wang, G.: Self-tuning semi-supervised spectral clustering. In: 2008 International Conference on Computational Intelligence and Security, pp. 1–5 (2008)
18. Ye, W., Goebl, S., Plant, C., Böhm, C.: Fuse: Full spectral clustering. In: Proceedings of the 22Nd ACM SIGKDD International Conference on Knowledge Discovery and Data Mining, KDD 2016, pp. 1985–1994. ACM, New York (2016)

Fast and Accurate Community Search Algorithm for Attributed Graphs

Shohei Matsugu[1]($\boxtimes$), Hiroaki Shiokawa[2], and Hiroyuki Kitagawa[2]

[1] Graduate School of SIE, University of Tsukuba, Tsukuba, Japan
matsugu@kde.cs.tsukuba.ac.jp
[2] Center for Computational Sciences, University of Tsukuba, Tsukuba, Japan
{shiokawa,kitagawa}@cs.tsukuba.ac.jp

Abstract. The community search algorithm is an essential graph data management tool to identify a community suited to a user-specified query node. Although the community search algorithms are useful in various applications, it is difficult for them to handle attributed graphs since (1) traditional algorithms ignore node attributes and (2) algorithms require strict topological constraints to find a community. In this paper, we define a novel class of the community search problem on attributed graphs called the flexible attributed truss community (F-ATC) problem. To overcome the aforementioned limitations, the F-ATC problem relaxes the topological constraints and evaluates node attributes. Since the F-ATC problem is NP-hard, we propose two greedy algorithms to solve it efficiently. Our extensive experiments on real-world graphs clarify that our approach achieves higher efficiency and accuracy than the state-of-the-art method.

Keywords: Graphs · Community search · Clustering.

1 Introduction

Given an attributed graph, how can we efficiently find the most suitable (or relevant) community to a user-specified query node among all possible communities? Recent advances in information and social sciences have shown that attributed graphs are becoming increasingly important as they represent complicated and schema-less data. For example, in the case of a friendship network (*e.g.*, Facebook), each user node has several attributes such as affiliation, residential area, and topics of interests.

To understand such complicated graphs, community search algorithms [2,4,17] play an important role in various applications. Once they receive a query node from a user, the community search algorithms explore a single community (cluster) that has dense inner-community connections with the largest relevance to the query node. Unlike traditional community detection algorithms (*e.g.*, modularity-based methods [1,13] and density-based methods [15,16]), community search algorithms can return a search result within a short computation time since

© Springer Nature Switzerland AG 2020
S. Hartmann et al. (Eds.): DEXA 2020, LNCS 12391, pp. 233–249, 2020.
https://doi.org/10.1007/978-3-030-59003-1_16

they do not need to compute the entire graph. Due to their efficiency, such algorithms have been applied to various applications, including social analysis and protein analysis.

Although community search algorithms are useful in various applications, they have a serious weakness when handling real-world attributed graphs. Traditional community search algorithms [2,4,17] cannot find accurate communities on attributed graphs. Many real-world graphs consist of relationships among various attributes [12,20]. However, traditional algorithms attempt to detect a dense subgraph such as k-core [17] or k-truss [2,4], which is the most relevant to the query node without measuring attribute similarities between the query and the community. That is why traditional algorithms fail to capture attributed-driven communities [3].

1.1   Existing Approaches and Challenges

To address the above issue, Huang *et al.* recently proposed LocATC [3], which is a novel community search algorithm for attributed graphs. Once parameters k and d are specified, LocATC searches a (k,d)-truss [3] that yields the largest attribute similarity for the query node. In this study, (k,d)-truss is a subgraph (1) whose nodes are at least d-hop reachable from the query node and (2) whose edges have at least $(k-2)$ triangles (*a.k.a.*, 3-cliques). Although LocATC successfully handles attributed graphs, its community search accuracy is limited for real-world graphs because LocATC searches (k,d)-trusses under the assumption that each community contains a sufficient number of triangles. However, this assumption is not suitable for real-world graphs since they have very diverse topological structures. For example, as Leskovec and Krevl reported in [6], the average fraction of triangles is only 5.68% in various real-world graphs. In addition, Shiokawa *et al.* reported that real-world graphs show a wide range of clustering coefficient values [15]; that is, real-world graphs may not contain many triangles. Consequently, LocATC fails to detect precise communities in various real-world graphs, which makes it difficult to efficiently find accurate communities in attributed graphs.

1.2   Our Approaches and Contributions

Our goal is to achieve fast and accurate community searches on large-scale attributed graphs. In this paper, we define a novel community search problem called the *flexible attributed truss community (F-ATC) problem* and present novel heuristic community search algorithms to efficiently solve it. To overcome the aforementioned limitations, the F-ATC problem finds a (k,d)-truss that maximizes the attribute similarity under all possible k settings, whereas LocATC explores communities only for a specific k value. Although such a relaxation increases the computational cost compared to LocATC, the F-ATC problem allows community search algorithms to explore diverse subgraphs regardless of the actual topological structures included in real-world graphs. To moderate the computational costs incurred by the F-ATC problem, herein we propose two heuristic community search algorithms based on the well-known beam-search algorithm [11]. Consequently, our proposed methods achieves the following attractive characteristics:

- **Accurate:** Our proposed method can identify more accurate communities than those obtained by the state-of-the-art method LocATC because parameter k of (k, d)-truss is relaxed (Sect. 4.2).
- **Fast:** Compared with the state-of-the-art method LocATC, our proposal achieves high-speed community searches on attributed graphs (Sect. 4.3). That is, our proposed method can find accurate communities without sacrificing the community search efficiency (Sect. 4.4).
- **Easy to deploy:** Our proposed method does not require parameter k, which determines the number of triangles included in each community (Algorithm 1). Therefore, our proposal provides a simple solution for diverse applications.

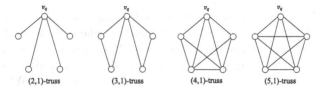

Fig. 1. Examples of $(k, 1)$-truss.

Table 1. Definition of main symbols.

Symbol	Definition
G	Connected attributed graph
V	Set of nodes in G
E	Set of edges in G
A	Set of attributes in G
$attr(u)$	Set of attributes attached on node $u \in V$
$V(H)$	Set of nodes in a subgraph H
$V_a(H)$	Set of nodes having an attribute $a \in A$ in a subgraph H
$E(H)$	Set of edges in a subgraph H
q	User-specified query
v_q	User-specified query node such that $v_q \in V$
A_q	User-specified query attributes such that $A_q \subseteq A$
$dist(v_q, H)$	Query distance between a query node v_q and a subgraph H (Definition 1)
$sup(e)$	Number of triangles (3-cliques) with an edge $e \in E$ in G
$\Delta_{k,d}$	Set of (k, d)-trusses in G (Definition 2)
$f(H, A_q)$	Attribute score function (Definition 3)
$N(H)$	Set of 1-hop neighbor nodes of a subgraph H (Definition 5)
$C(H)$	Set of candidate communities obtained from a subgraph H (Definition 6)

Our extensive experiments showed that our proposed algorithms run up to 50 times faster than LocATC without sacrificing the community search accuracy. For example, our algorithm can compute an attributed graph with 1.1 million nodes and 3 million edges in 0.1 seconds. Although previous community search algorithms have effectively enhanced application quality, they are difficult to apply to large-scale attributed graphs due to their accuracy and efficiency limitations. On the other hand, our propose method should improve the effectiveness of a wide range of applications and realize a fast and accurate approach appropriate to real-world graphs.

Organization: This paper is organized as follows. Section 2 introduces basic notations and definitions of this work. Section 3 defines the F-ATC problem, and presents two greedy algorithms; baseline algorithm and fast enumeration algorithm. Section 4 describes the experiments to verify the effectiveness of our approaches. Related works are briefly reviewed in Sect. 5. Finally, Sect. 6 concludes this paper.

2 Basic Notations and Definitions

Here, we formally define basic notations and definitions used in this paper. Let $G = (V, E, A)$ be a connected attributed graph, where V, E, and A are sets of nodes, edges, and attributes, respectively. Each node $u \in V$ has a set of attributes denoted by $attr(u) \subseteq A$. To simplify the representaions, each node is assumed to have one or two attributes (i.e., $1 \leq |attr(u)| \leq 2$). Without loss of generality, other types of attributed graphs can be handled even if each node has more than two attributes. For convenience, $V(H)$ and $E(H)$ are denoted as sets of nodes and edges included in subgraph H, respectively. Here, $a \in A$, $V_a(H)$ is a set of nodes with attribute a in subgraph H. Similarly, we define a user-specified query as $q = (v_q, A_q)$, where v_q is a query node included in $V(G)$, and A_q is a set of query attributes such that $A_q \subseteq A$. Table 1 summarizes symbols and their corresponding definitions used in this paper. The following basic definitions are necessary to discuss the new community search algorithms in the next section:

Definition 1 (Query distance)
Let $dist(u, v)$ be the shortest path distance between nodes u and v on graph G. Given subgraph $H \subseteq G$ and query node v_q, the query distance between query node v_q and subgraph H is defined as $dist(v_q, H) = \max_{v \in V(H)} dist(v_q, v)$.

Definition 2 ((k, d)-truss [3])
Let $sup(e)$ be the number of triangles containing the edge $e \in E$. Given query node v_q and parameters k and d, a set of (k, d)-trusses is defined as

$$\Delta_{k,d} = \{H \subseteq G | \forall e \in E(H), sup(e) \geq k - 2, dist(v_q, H) \leq d\}.$$

Definition 2 indicates that a (k, d)-truss is a subgraph such that (1) the nodes are d-hop reachable from the query node v_q and (2) each edge has more than $k - 2$ triangles, (i.e., $sup(e) \geq k - 2$ for each $e \in E(V)$). By controlling the values of k and d, we can determine the density and the size of (k, d)-trusses. Figure 1 shows examples of $(k, 1)$-trusses for various k settings. For instance, as shown in Fig. 1, all $(k, 1)$-trusses are 1-hop reachable from the query node v_q. If $k = 2$, each edge in $(2, 1)$-truss does not need to have any triangles. By contrast, in the case of $k = 5$, $(5, 1)$-truss should contain at least three triangles for each edge.

Finally, we introduce an attribute score function [3] that evaluates the attribute similarity between the query and a community.

Definition 3 (Attribute score function [3])
Given subgraph H and set of query attributes A_q, attribute score function $f(H, A_q)$ is defined as

$$f(H, A_q) = \sum_{a \in A_q} \frac{|V_a(H)|^2}{|V(H)|}.$$

Definition 3 implies that attribute score function $f(H, A_q)$ increases as the subgraph contains more attributes in query attributes A_q.

3   Proposed Method

Our goal is to efficiently find an accurate community in G that corresponds to the user-specified query. To achieve a highly accurate community search, we first present a novel class of the community search problem, called the *F-ATC problem*, in Sect. 3.1. In Sects. 3.2 and 3.3, we propose two heuristic search algorithms to efficiently solve the F-ATC problem.

3.1   The F-ATC Problem

LocATC imposes strict topological constraints such that each community should contain a sufficient number of triangles based on user-specified parameter k. However, this assumption is not suitable for real-world graphs, which generally have diverse topological structures and may not contain triangles [6]. Thus, we introduce a new class of the community search problem that relaxes the strict k setting in LocATC.

Definition 4 (the F-ATC problem)
Given graph $G = (V, E, A)$, query $q = (v_q, A_q)$, and parameter d, the F-ATC problem finds subgraph $H \in \bigcup_{k \geq 2} \Delta_{k,d}$ that yields the largest value of $f(H, A_q)$.

Unlike LocATC [3], the F-ATC problem does not require parameter k. It attempts to find (k, d)-truss maximizing the attribute score function under all possible k settings. For example, if $d = 1$, the F-ATC problem explores all $(k, 1)$-trusses shown in Fig. 1 and returns a single $(k, 1)$-truss that yields the largest score of the attribute score function.

By relaxing user-specified parameter k, the F-ATC problem can handle diverse typologies of real-world graphs. However, the F-ATC problem requires exhaustive subgraph searches to obtain a subgraph that maximizes the attribute score function. Let $\bar{k}$ be the maximum k setting for a given graph. We can reduce the F-ATC problem to the (k, d)-truss search problem in the polynomial time by performing LocATC [3] for $k = 2$ to $k = \bar{k}$. As discussed in [3], the (k, d)-truss search problem is NP-hard. Therefore, the F-ATC problem is also NP-hard. Below, we present two heuristic search algorithms to efficiently solve the F-ATC problem.

3.2   Baseline Algorithm

We refine [9] as our baseline algorithm, which is an algorithm to improve the accuracy of LocATC [3]. The baseline algorithm is based on the well-known beam search technique [11]. By letting β represent the beam width that controls a number of search results, a beam search explores graphs maintaining top-β search results under an objective function. Based on this search strategy, the baseline algorithm greedily explores top-β (k, d)-trusses by the attribute score function.

Before providing detailed descriptions of the baseline algorithm, we introduce the following definitions:

Definition 5 (1-hop neighbor nodes)
Given subgraph $H \subseteq G$, $N(H)$ is 1-hop neighbor nodes of H defined as

$$N(H) = \{v \in V(G) | (u, v) \in E(G) \text{ for } u \in V(H) \text{ and } v \notin V(H)\}.$$

Definition 6 (Candidate communities)
Given subgraph $H \subseteq G$ and beam width β, we denote a set of candidate communities as $C(H)$, which is defined as

$$C(H) = \{C_1(H), C_2(H), \dots, C_\beta(H)\} \subseteq 2^{|V(H) \cup N(H)|},$$

where $C_i(H)$ is a (k, d)-truss composed of nodes in $V(H) \cup N(H)$ such that $f(C_1(H), A_q) \geq f(C_2(H), A_q) \geq \cdots \geq f(C_\beta(H), A_q) \geq f(C_{\beta+j}(H), A_q)$ for all $j \in \mathbb{N}$.

Definition 6 indicates that (1) $C(H)$ expands subgraph H as $V(H) \cup N(H)$ and (2) $C(H)$ lists the top-β (k, d)-trusses from $V(H) \cup N(H)$ so that $C(H)$ maximizes the attribute score function among all possible (k, d)-trusses.

Algorithm: Based on the above definitions, we present the baseline algorithm to solve the F-ATC problem. Given graph $G = (V, E, A)$, query $q = (v_q, A_q)$, parameter d, and beam width β, we initially set a subgraph as $H = \{v_q\}$. Afterwards, the baseline algorithm performs the following three steps:

(**Step 1**) Obtain $N(H)$ from subgraph H by Definition 5.
(**Step 2**) Construct $C(H)$ from $V(H) \cup N(H)$ by Definition 6.
(**Step 3**) Select a (k, d)-truss $C_i(H)$ from $C(H)$, and set $H = C_i(H)$.

The baseline algorithm iterates the above steps until all the d-hop reachable nodes of v_q are computed. After the terminating, it returns a (k, d)-truss that yields the largest score of the attribute score function in $C(H)$.

By iteratively enumerating $C(H)$ in (Step 2), the baseline algorithm explores (k, d)-trusses for various k settings so that the trusses increase the attribute score function. However, (Step 2) requires $\Omega(2^{|V(H) \cup N(H)|}|E(N(H))|^{1.5})$ time to find top-β candidate communities from Definition 6. This is because (1) (k, d)-trusses need to be explored from all possible subgraph in $|V(H) \cup N(H)|$ and (2) (k, d)-truss detection requires $\Omega(|E(N(H))|^{1.5})$ time [7]. If a given graph is large, the size of $V(H) \cup N(H)$ clearly increases. To improve the efficiency, it is important to reduce the computational cost of (Step 2).

3.3   Fast Enumeration Algorithm

To solve the F-ATC problem on large attributed graphs, we present a fast enumeration algorithm to search the candidate communities in the baseline algorithm. Instead of enumerating all possible candidates, the fast enumeration algorithm directly lists the top-β candidates using attribute-aware candidate selection techniques.

For simplicity, we denote the query attributes as $A_q = \{a_1, a_2\}$ without loss of generality. To achieve fast top-β candidate enumeration, we have the following properties from Definition 3.

Lemma 1. *Given subgraph H and node $v \in N(H)$, $f(H \cup \{v\}, A_q) > f(H, A_q)$ holds, if $a_1, a_2 \in attr(v)$ holds.*

Proof. Since $a_1, a_2 \in attr(v)$, the following equation is derived from Definition 3,

$$
\begin{aligned}
f(H \cup \{v\}, A_q) - f(H, A_q) &= \sum_{a \in A_q} \left\{ \frac{(|V_a(H)| + 1)^2}{|V(H)| + 1} - \frac{|V_a(H)|^2}{|V(H)|} \right\} \\
&= \frac{\sum_{a \in A_q} \left\{ |V(H)|(|V_a(H)| + 1)^2 - (|V(H)| + 1)|V_a(H)|^2 \right\}}{|V(H)|(|V(H)| + 1)}.
\end{aligned}
$$

Clearly, $|V(H)| \geq |V_a(H)|$ for all attribute $a \in A_q$. Hence,

$$
|V(H)|(|V_a(H)| + 1)^2 - (|V(H)| + 1)|V_a(H)|^2 > 0.
$$

Therefore, we have $f(H \cup \{v\}, A_q) - f(H, A_q) > 0$, which completes the proof. $\qquad\square$

Lemma 2. *Given subgraph H and node $v \in N(H)$, $f(H \cup \{v\}, A_q) < f(H, A_q)$ holds, if $a_1, a_2 \notin attr(v)$ holds.*

Proof. Due to $a_1, a_2 \notin attr(v)$, $|V_a(H \cup \{v\})| = |V_a(H)|$ clearly holds for all attribute $a \in A_q$. That is, from Definition 3,

$$f(H \cup \{v\}, A_q) - f(H, A_q) = \sum_{a \in A_q} \left\{ \frac{|V_a(H \cup \{v\})|^2}{|V(H)| + 1} - \frac{|V_a(H)|^2}{|V(H)|} \right\}$$

$$= \sum_{a \in A_q} \left\{ \frac{|V_a(H)|^2}{|V(H)| + 1} - \frac{|V_a(H)|^2}{|V(H)|} \right\} < 0,$$

which completes the proof of Lemma 2.   □

For a given subgraph H and its 1-hop neighbor node v, Lemma 1 and Lemma 2 imply that (1) if node v has all query attributes in A_q, (k, d)-trusses composed of $H \cup \{v\}$ always increase the attribute score function, and (2) if node v has no query attributes in A_q, the (k, d)-trusses decreases the function. We also identify the following properties, which play essential roles in our fast enumeration algorithm.

Lemma 3. *Given subgraph H and node $v \in N(H)$ such that $a_1 \in attr(v)$ and $a_2 \notin attr(v)$, $f(H \cup \{v\}, A_q) > f(H, A_q)$ if and only if $|V(H)|(2|V_{a_1}(H)| + 1) > |V_{a_1}(H)|^2 + |V_{a_2}(H)|^2$ holds.*

Proof. We first prove the sufficient condition. From Definition 3,

$$f(H \cup \{v\}, A_q) - f(H, A_q) = \frac{2|V_{a_1}(H)| + 1}{|V_{a_1}(H)|^2 + |V_{a_2}(H)|^2} - \frac{1}{|V(H)|}.$$

Since we clearly have $f(H \cup \{v\}, A_q) - f(H, A_q) > 0$, we can derive $|V(H)|(2|V_{a_1}(H)| + 1) > |V_{a_1}(H)|^2 + |V_{a_2}(H)|^2$ from the above equation, which completes the proof.

Next, we prove the necessary condition. From $|V(H)|(2|V_{a_1}(H)| + 1) > |V_{a_1}(H)|^2 + |V_{a_2}(H)|^2$, the following condition is derived;

$$0 < \frac{2|V_{a_1}(H)| + 1}{|V_{a_1}(H)|^2 + |V_{a_2}(H)|^2} - \frac{1}{|V(H)|} = f(H \cup \{v\}, A_q) - f(H, A_q).$$

Thus, $f(H \cup \{v\}, A_q) > f(H, A_q)$ holds.   □

Lemma 3 leads the following corollary for given subgraph H and node $v \in N(H)$ such that $a_1 \in attr(v)$ and $a_2 \notin attr(v)$.

Corollary 1. $f(H \cup \{v\}, A_q) \leq f(H, A_q)$ *holds if and only if $|V(H)|(2|V_{a_1}(H)| + 1) \leq |V_{a_1}(H)|^2 + |V_{a_2}(H)|^2$ holds.*

Proof. We omit the proof of Corollary 1 due to space limitations.   □

From Lemma 3 and Corollary 1, several nodes in $N(H)$ can increase the attribute score function, even if the nodes have only a subset of A_q. Consequently, from Lemmas 1, 2, and 3, Theorem 1 can be theoretically derived.

Algorithm 1. Fast enumeration

Require: A subgraph H, a parameter β, and a parameter d
Ensure: A set of candidate communities $C(H)$
1: Obtain a subgraph H' from $V(H) \cup N(H)$ by Theorem 1;
2: **while** $|C(H)| < \beta$ **do**
3:　　$\bar{k} \leftarrow \max_{e \in E(H')} sup(e)$;
4:　　**for** $k = \bar{k}$ to 2 **do**
5:　　　　Add all (k, d)-trusses in H' into $C(H)$;
6:　　　　**if** $|C(H)| \geq \beta$ **then**
7:　　　　　　break;
8:　　　　**end if**
9:　　**end for**
10:　　Obtain a node $v \in V(H') \cap N(H)$ decreasing $f(H', A_q)$ by Corollary 1;
11:　　$H' \leftarrow H' \backslash \{v\}$;
12: **end while**

Theorem 1. *Given subgraph H and its 1-hop neighbor node set $N(H)$, $f(H, A_q)$ shows the largest score among all possible communities in $H \cup N(H)$ if the subgraph H is marged with all nodes satisfying Lemma 1 and 3 in $N(H)$.*

Proof. We can clearly prove Theorem 1 from Lemma 1, 2, and 3 and Corollary 1. Thus, we omit the details of the proof due to space limitations.　　　　□

Theorem 1 implies that a set of nodes that maximizes the attribute score function can be directly found from a given $H \cup N(H)$.

Algorithm: Based on the above properties, we design a fast enumeration algorithm for (Step 2) in Sect. 3.2. Algorithm 1 shows details of our algorithm. First, the fast enumeration algorithm obtains subgraph H' from $V(H) \cup N(H)$ so that $f(H', A_q)$ is maximized (line 1). As we proved in Theorem 1, such subgraph H' can be obtained by adding nodes in $N(H)$ into H if the nodes satisfy Lemma 1 or Lemma 3. Then the algorithm adds all (k, d)-trusses composed of H' into $C(H)$ (lines 4–9). Afterwards that the algorithm removes node $v \in V(H') \cap N(H)$ from H' so that removing node v decreases $f(H', A_q)$ based on Corollary 1 (lines 10–11). Finally, the algorithm terminates if $|C(H)|$ reaches β (lines 2 and 6–8).

Theoretical Analysis: Finally, we theoretically assess the time complexity of the fast enumeration algorithm.

Theorem 2. *The enumeration algorithm requires $\Omega(|V(N(H))| + \beta|E(N(H))|^{1.5})$ time to find the top-β candidates from $V(H) \cup N(H)$.*

Proof. As shown in Algorithm 1 (line 1), the algorithm obtains subgraph H' by Theorem 1 before starting the while loop. This procedure requires $\Omega(|V(N(H))|)$ time since all nodes in $V(H)$ must be checked using Theorem 1. Afterwards, that our algorithm explores (k, d)-trusses that yield large scores of the attribute

Table 2. Statistics of real-world datasets.

Name	$|V|$	$|E|$	$|A|$	Number of triangles	Fraction of triangles
Cornell	195	304	1,588	59	0.04
Texas	187	328	1,501	67	0.03
Amazon	335 K	926 K	157	667 K	0.08
YouTube	1.10 M	3.00 M	5,327	3.0 K	0.002

score function. In the worst case, the algorithm adds only a single (k, d)-truss to $C(H)$ in each while loop. That is, the while loop (lines 2–12) must be iterated $\Omega(\beta)$ time. In each while loop, the algorithm can find a (k, d)-truss, which incurs $\Omega(|N(H)|^{1.5})$ time [7], and it removes node v from H' in $O(1)$ time. Hence, the algorithm requires $\Omega(\beta|N(H)|)$ time. Therefore, Algorithm 1 incurs $\Omega(|V(N(H))| + \beta|E(N(H))|^{1.5})$ time.   □

Recall that the baseline algorithm requires $\Omega(2^{|V(H) \cup N(H)|}|E(N(H))|^{1.5})$ time for each (Step 2). By contrast, our enumeration algorithm consumes $\Omega(|V(N(H))| + \beta|E(N(H))|^{1.5})$ time which is clearly a smaller cost than the baseline. Thus, our enumeration algorithm can reduce the computational cost for the F-ATC problem.

4   Experimental Analysis

In this section, we experimentally discuss the effectiveness of our proposed algorithms. We designed our experiments to demonstrate that:

- **High accuracy:** Our proposed algorithms achieve higher community search accuracy than those of the state-of-the-art algorithm (LocATC) on real-world graphs.
- **High efficiency:** Although our fast enumeration algorithm outputs more accurate communities than LocATC, it outperforms LocATC and the baseline algorithm in terms of community search time on real-world graphs.

4.1   Experimental Setup

Methods: We compared our proposed algorithms (the baseline algorithm and the fast enumeration algorithm) with LocATC [3]. As we described in Sect. 1, LocATC is the state-of-the-art community search method for attributed graphs. Given a user-specified query and parameters k and d, LocATC finds a single (k, d)-truss that maximizes the attribute score function shown in Definition 3. In our experimental analysis, we used the k optimization technique recommended in the original paper [3].

All algorithms were implemented in C++ and compiled with gcc-8.2.0 using the -O3 option. All experiments were conducted on a server with an Intel Xeon CPU (3.50GHz) and 128 GiB RAM. Here we report the average results of 100 queries.

Datasets: We used four real-world graphs, which were published in a previous study [3] and the SNAP repository [6]. Table 2 shows their statistics. Since all datasets have ground-truth communities, they were used to evaluate the community search accuracy. Each node in Cornell and Texas has at most two node attributes. Because Amazon and YouTube do not provide node attributes to their nodes, we assigned synthetic attributes for each node by following the same method as the previous study [3]. Specifically, we assigned the synthetic attributes as follows:

- For each graph, we generated $|A| = 0.005|V|$ synthetic attributes.
- For each ground-truth community, we randomly selected three attributes in A, and assigned each one to 80% nodes in the community.
- To model noise attributes, we assigned randomly selected attributes to each node.

Queries: We generated 100 queries for each dataset by following the settings in the previous work [3]. Specifically, we randomly selected 100 query nodes from each graph. For each query node v_q, we set two attributes as query attributes A_q using the most frequent attributes in the ground-truth community, including node v_q.

4.2   Accuracy

To assess whether the F-ATC problem achieves higher accuracy than LocATC, we evaluated the community search accuracy on real-world graphs. We compared the community search results with the ground-truth through F1-measure [8]. Figure 2 shows the community search accuracy of each algorithm by varying the size of d from 2 to 5. We also varied β for our proposed algorithms since they require the beam width size β for the beam searches. Note that the results of our baseline algorithm are omitted from Fig. 2 because it did not return any results within one hour on Amazon or YouTube.

Figure 2 shows that our proposed algorithms outperform LocATC in terms of the F1-score if the beam width sizes are large. Moreover, our proposed algorithms show higher accuracies than LocATC, except for Texas, even if the β values are small. As we described in Sect. 1, LocATC assumes that real-world graphs contain a sufficient number of triangles although they can contain very diverse topological structures. Hence, LocATC fails to capture the ground-truth communities if those communities have a small number of triangles. By contrast, the F-ATC problem allows more diverse typologies along with the maximization of the attribute score function to be explored. Hence, our proposed method can achieve higher accuracy on a wider range of real-world graphs than LocATC. Specifically, our proposed method maintains higher accuracy than that of LocATC on YouTube even though the datasets have relatively smaller fractions of triangles than the others. These results imply that the F-ATC problem successfully captures diverse typologies in real-world graphs.

Figure 2 also indicates that the fast enumeration algorithm does not sacrifice the community search accuracy compared with the baseline algorithm. As theoretically discussed in Sect. 3.3, the fast enumeration algorithm can directly find a subgraph that maximizes the attribute score function (Theorem 1). Therefore, the fast enumeration algorithm does not degrade the community search accuracy compared with the baseline algorithm, which performs exhaustive searches.

4.3 Efficiency

We evaluated the community search time of each algorithm on four real-world datasets. Similar to the previous section, we varied beam width β of our proposed methods for each d setting. Figure 3 shows the community search time on the

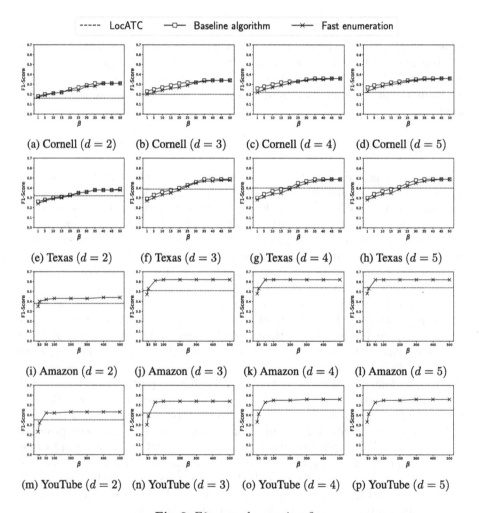

Fig. 2. F1-scores by varying β.

real-world datasets. The results for the baseline algorithm are omitted since it did not finish the community search on Amazon or YouTube within one hour.

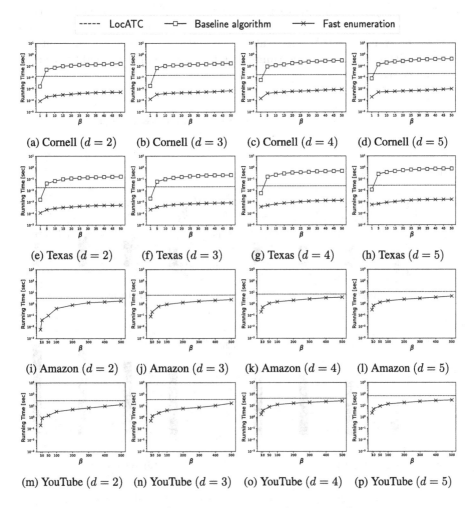

Fig. 3. Query processing time by varying β.

The fast enumeration algorithm outperforms LocATC and the baseline algorithm under all examined conditions (Fig. 3). Although the baseline algorithm is 10 times slower than the query processing time of LocATC, the fast enumeration algorithm successfully mitigates the expensive enumeration cost in the baseline algorithm. In our experimental results, the fast enumeration algorithm has an improved speed up to three orders of magnitude higher than the baseline algorithm. Furthermore, the fast enumeration algorithm has up to 50 times faster query processing time than the state-of-the-art method LocATC. By comparing the running time among different parameter settings, the fast enumeration algorithm gradually increases the running time as the sizes of β and d increase. This is because the F-ATC problem requires a large number of community candidates to be searched if those parameters are large. However, the community search accuracy reaches a plateau on the real-world graphs, even if β and d are small (Fig. 2). For instance, our proposed method shows an almost constant accuracy on Amazon and YouTube at $\beta = 50$ and $d = 3$. Hence, the fast enumeration algorithm can reduce the running time while keeping its highly accurate community search results.

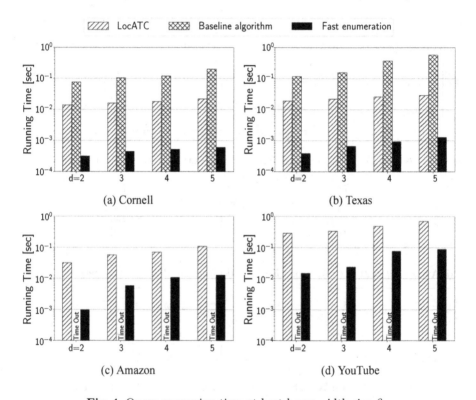

Fig. 4. Query processing time at best beam width size β.

4.4   Peak Performance Analysis

The fast enumeration algorithm shows a trade-off between the community search accuracy and processing time. Thus, we discuss the peak query processing performance. In this evaluation, we compared the running time of our proposed algorithms with the best β value, which returns the highest F1-score among all possible β settings. Figure 4 shows that our fast enumeration algorithm outperforms the community search time of LocATC for all settings. Specifically, our proposed method provides a community search that is up to 50 times faster than LocATC. Additionally, the fast enumeration algorithm outputs more accurate communities than LocATC in the case of the best β settings. That is, these results imply that our proposed method achieves higher peak performances than LocATC on real-world graphs.

5   Related Work

Community search algorithms are fundamental tools to analyze complex data structures obtained from various applications [5,17,18]. Unlike traditional community detection algorithms [10,14,19], community search algorithms do not compute the entire given graph. Consequently, they efficiently find a community for the user-specified query. Here, we briefly review some of the more successful community algorithms.

Traditionally, community search algorithms are considered as a problem to detect cohesive communities that contain user-specified query nodes on non-attributed graphs. For example, Sozio and Gionis [17] designed a community search problem to find k-core that includes query nodes. Similarly, Huang *et al.* proposed the k-truss search algorithm to reveal the most relevant communities against a given query. Because these algorithms assume that the community has dense and robust inner-community connections, they perform local search methods to retrieve dense subgraphs (*i.e.* k-core and k-truss). However, these methods are designed for non-attributed graphs. Hence, they are unsuited to extract attribute-driven communities.

To overcome the above issue, Huang *et al.* recently proposed another class of the community search problem, namely the ATC problem [3]. The ATC problem is designed to find the (k, d)-truss, which is shown in Definition 2 that yields the largest attribute similarity with the query. Since the ATC problem is NP-hard, Huang *et al.* proposed the LocATC algorithm. This is the state-of-the-art algorithm to solve such a problem within a short running time. By introducing the ATC problem, LocATC can efficiently extract communities while ensuring a high cohesiveness and a high attribute similarity. However, LocATC assumes that each community has a sufficient number of triangles. This is unrealistic because real-world graphs have very diverse topological structures [6]. In this paper, we experimentally confirm that the accuracy of LocATC reaches a plateau if a given graph is sparse and has a small fraction of triangles. By contrast, our proposed algorithms overcome these performance limitations by relaxing the topological constraints. Consequently, our proposed method achieves faster community searches and higher accuracies than LocATC.

6   Conclusion

Herein we propose a novel community search problem called the F-ATC problem for attributed graphs. By relaxing the topological constraints of the community search, the F-ATC problem can explore divergent community structures included in real-world graphs. Because the F-ATC problem is NP-hard, we also present two heuristic algorithms based on the beam search to solve the F-ATC problem efficiently. Our experiments on real-world graphs demonstrate the advantages of our proposed algorithms compared to the state-of-the-art method.

Acknowledgement. This work was supported by JSPS KAKENHI Early-Carrer Scientists Grant Number JP18K18057, and JST ACT-I.

References

1. Blondel, V., Guillaume, J., Lambiotte, R., Mech, E.: Fast unfolding of communities in large networks. J. Stat. Mech. Theory Exp. **2008**(10), P10008 (2008)
2. Huang, X., Cheng, H., Qin, L., Tian, W., Yu, J.X.: Querying k-truss community in large and dynamic graphs. In: Proceedings of SIGMOD 2014, pp. 1311–1322 (2014)
3. Huang, X., Lakshmanan, L.: Attribute-driven community search. PVLDB **10**(9), 949–960 (2017)
4. Huang, X., Lakshmanan, L., Yu, J.X., Cheng, H.: Approximate closest community search in networks. PVLDB **9**(4), 276–287 (2015)
5. King, A.D., Pržulj, N., Jurisica, I.: Protein complex prediction via cost-based clustering. Bioinformatics **20**(17), 3013–3020 (2004)
6. Leskovec, J., Krevl, A.: SNAP Datasets: Stanford large network dataset collection (2014). http://snap.stanford.edu/data
7. Li, Z., Lu, Y., Zhang, W.-P., Li, R.-H., Guo, J., Huang, X., Mao, R.: Discovering hierarchical subgraphs of k-core-truss. Data Sci. Eng. **3**(2), 136–149 (2018). https://doi.org/10.1007/s41019-018-0068-2
8. Manning, C.D., Raghavan, P., Schütze, H.: Introduction to Information Retrieval. Cambridge University Press, Cambridge (2008)
9. Matsugu, S., Shiokawa, H., Kitagawa, H.: Flexible community search algorithm on attributed graphs. In: Proceedings of the 21st International Conference on Information Integration and Web-based Applications & Services, iiWAS 2019, Munich, Germany, 2–4 Dec 2019, pp. 103–109. ACM (2019)
10. Onizuka, M., Fujimori, T., Shiokawa, H.: Graph partitioning for distributed graph processing. Data Sci. Eng. **2**(1), 94–105 (2017)
11. Reddy, D.R.: Speech understanding systems: A summary of results of the five-year research effort, Department of computer science. Technical report, Carnegie-Mellon University (1977)
12. Sato, T., Shiokawa, H., Yamaguchi, Y., Kitagawa, H.: Forank: Fast objectrank for large heterogeneous graphs. In: WWW 2018: Companion Proceedings of the The Web Conference 2018, pp. 103–104. International World Wide Web Conferences Steering Committee, Republic and Canton of Geneva, CHE (2018)
13. Shiokawa, H., Amagasa, T., Kitagawa, H.: Scaling fine-grained modularity clustering for massive graphs. In: Proceedings of the 28th International Joint Conference on Artificial Intelligence, IJCAI-19, pp. 4597–4604 (2019)
14. Shiokawa, H., Fujiwara, Y., Onizuka, M.: Fast algorithm for modularity-based graph clustering. In: Proceedings of the Twenty-Seventh AAAI Conference on Artificial Intelligence (AAAI 2013), pp. 1170–1176 (2013)
15. Shiokawa, H., Fujiwara, Y., Onizuka, M.: SCAN++: Efficient algorithm for finding clusters, hubs and outliers on large-scale graphs. PVLDB **8**(11), 1178–1189 (2015)

16. Shiokawa, H., Takahashi, T., Kitagawa, H.: ScaleSCAN: Scalable density-based graph clustering. In: Proceedings of the 29th International Conference on Database and Expert Systems Applications, pp. 18–34. DEXA (2018)
17. Sozio, M., Gionis, A.: The community-search problem and how to plan a successful cocktail party. In: Proceedings of KDD 2010, pp. 939–948 (2010)
18. Takahashi, T., Shiokawa, H., Kitagawa, H.: SCAN-XP: Parallel structural graph clustering algorithm on intel xeon phi coprocessors. In: Proceedings of the 2nd International Workshop on Network Data Analytics (NDA), pp. 6:1–6:7 (2017)
19. Zhang, X., Newman, M.E.J.: Multiway spectral community detection in networks. Phys. Rev. E **92**, 052808 (2015)
20. Zhou, Y., Cheng, H., Yu, J.X.: Graph clustering based on structural/attribute similarities. PVLDB **2**(1), 718–729 (2009)

A Differential Evolution-Based Approach for Community Detection in Multilayer Networks with Attributes

Clara Pizzuti$^{(\boxtimes)}$ and Annalisa Socievole

Institute for High Performance Computing and Networking (ICAR), National Research Council of Italy (CNR), via P. Bucci 8-9C, 87036 Rende, CS, Italy
{clara.pizzuti,annalisa.socievole}@icar.cnr.it

Abstract. A differential evolution based algorithm for detecting community structure in multilayer networks with node attributes is proposed. The method optimizes a fitness function that combines structural connectivity of each layer with node similarity to obtain multilayer communities with high link density and composed by nodes having similar attributes. Experiments on synthetic networks show that the method finds communities almost equal to the ground-truth ones. Moreover, we compared our approach with a clustering method using only the attribute information, and a method which clusters nodes using only the multilayer network structure, on four real-world multilayer networks enriched with attributes. The results point out that the exploitation of the information coming from both all the layers and the node features allows the identification of accurate network divisions.

Keywords: Community detection · Multilayer networks · Attributed networks

1 Introduction

Community detection in complex networks is one of the most studied research problems in the field of network science [17]. A lot of methods have been proposed since the appearance of the seminal paper of Girvan and Newman [18], presenting an approach to detect communities by identifying the inter-community edges to remove in order to isolate them. As research in this area regarding simple graphs, constituted by nodes and single links connecting them, has matured, the study of complicated structures, more similar to real-worlds networks, received a lot of attention in the last years. In fact, modeling a complex system with a network in which entities are interrelated by only one type of relationship, with no additional information regarding such entities, has been recognized a rather poor approximation of reality [2,4,20]. In complex networks, objects can be intertwined by multiple relationships, each representing a different kind of interaction, for instance working or friendship relationships in social networks. Moreover, often nodes are enriched with feature vectors representing information

© Springer Nature Switzerland AG 2020
S. Hartmann et al. (Eds.): DEXA 2020, LNCS 12391, pp. 250–265, 2020.
https://doi.org/10.1007/978-3-030-59003-1_17

content associated with such objets. Thus, the design of new methods, able to detect community structure by considering at the same time the information coming from all the layers and the node attributes, is a challenging research problem which deserves investigation.

In this paper, a method based on *Differential Evolution (DE)* [15], named *ML@NetDE*, to find a partition of attributed multilayer networks is proposed. The method employs a fitness function, introduced in [28], which optimizes, simultaneously for all the layers, the connectivity between nodes of the same community and, at the same time, the homogeneity of their features. Experiments on synthetic networks, for which the ground truth division is known, show that the method finds communities very similar to the ground-truth ones. Moreover, *ML@NetDE* has been compared with the *k-means* method, which groups nodes by considering only the node attributes, and the *Louvain* method, generalized to multilayer networks [24], which clusters networks by using only the network structure, on four real-world multilayer networks enriched with attributes. The results point out that the exploitation of the information coming from both all the layers and the node features allows the identification of network divisions with high edge density and node homogeneity.

The paper is organized as follows. The next section introduces the concept of multilayer network with attributes and defines the problem of community detection. Section 3 gives an overview of methods detecting communities in multilayer networks or attributed networks. Section 4 defines the fitness function optimized by the method. Section 5 describes the proposed approach. Section 6 reports the experimental evaluation and shows the results. Finally, Sect. 7 concludes the paper.

2   Preliminaries

In this section the definition of attributed multilayer network is presented and the problem of community detection in this kind of networks is introduced. The definition combines the concepts of multilayer/multiplex network and edge-attributed network given in [2,4].

An **attributed multilayer network** $\mathcal{G} = \{G_l\}_{l=1}^{L}$ is a family of L networks, called layers, representing different types of connections among the same set V of n nodes. Each layer is an attributed graph $G_l = (V, E_l, A, F)$, where $E_l = \{(i, j, l) : i, j \in V, 1 \leq l \leq L, i \neq j\}$ is the set of (eventually weighted) edges between the nodes of the same layer l, $A = \{\alpha_1, \alpha_2, \ldots, \alpha_\mathcal{A}\}$ is the set of numerical and categorical attributes (features), and $F = \{a_1, a_2, \ldots, a_\mathcal{A}\}$ is a set of functions. Each node of V is characterized by the same vector of features, whose values are determined by the functions $a_\alpha : V \to D_\alpha$, $1 \leq \alpha \leq \mathcal{A}$, with D_α the domain of attribute α.

The $n \times n$ adjacency matrix W^l of each layer G_l is such that the element $W_{ij}^l \geq 1$ if nodes i and j are connected in the l-th layer, i.e. $(i, j, l) \in E_l$, $W_{ij}^l = 0$ otherwise. If $W_{ij}^l = 1$ the network is unweighted.

The intuitive concept of community in a network, either monolayer or multi-layer, assumes that the interactions among the nodes of the same community are denser than those between nodes of different communities. However, as outlined in [17,20], there is no a generally accepted definition of what is a community. The most popular definition is based on the idea that a graph presents commu-nity structure if it differs from a random graph. To this end, a null model, i.e. a graph with structural features matching the original graph can be built, and then used as a term of comparison. The famous null model proposed by Newman and Girvan [26], which builds a random graph from the original one by rewiring edges at random, but assuring that the expected degree and the original degree of each node are the same, is at the base of the definition of *modularity*, the most popular criterion defining a community. The modularity of a graph is defined as follows:

$$Q = \frac{1}{2 \mid E \mid} \sum_{ij} \left(W_{ij} - \frac{k_i k_j}{2 \mid E \mid} \right) \delta_{ij} \tag{1}$$

where W is the adjacency matrix of the graph, k_i and k_j are the degrees of nodes i and j respectively, and $\delta_{ij} = 1$ if i and j are in the same commu-nity, zero otherwise. It is generally assumed that a good community structure has a high modularity value. An extension of the concept of good partition in multilayer networks has been proposed by Tang et al. [34,35]. Given a partition $C = \{C_1, \ldots, C_k\}$ of the set V of vertices, the modularity $Q_l(C), l = 1, \ldots, L$ on each of the layers is different. C is considered a good community structure if the modularity values on all the layers are high [34,35].

However, since each layer is a graph with attributes, another criterion to optimize to have a community structure of good quality is the intra-cluster node similarity.

Thus the problem we want to solve is the following.

Problem Definition: Find a division $C = \{C_1, \ldots, C_k\}$ of an attributed mul-tilayer network $\mathcal{G} = \{G_l\}_{l=1}^{L}$ such that:

- intra-cluster density is high and inter-cluster density is low for all the graphs $G_l \in \mathcal{G}$;
- for each layer, nodes belonging to the same community are similar, while nodes of different communities are dissimilar.

The similarity between nodes is computed by using the vector of feature values associated with each node.

3   Related Work

In the last years the interest in multilayer networks and attributed networks has sensibly increased and many approaches have been proposed to detect commu-nity structure [4,19,23]. However, the majority of the algorithms in the litera-ture consider either the multiple level aspect or the attribute information, while methods exploiting both different kinds of relationships and the rich contents of nodes are missing. In the following, we thus describe some of the most popular approaches for detecting communities in multilayer and/or attributed networks.

3.1 Community Detection in Multilayer Networks

A main strategy for mining multilayer networks consists in aggregating layers to obtain a single graph and then use classical methods for community detection [7,8,23,34]. However, the collapsed graph may not reflect the true community structure and important information may be lost. Another approach obtains communities for each layer by optimizing a score function and then finding a consensus partition [35]. In [30] a many-objective evolutionary algorithm which optimizes the modularity value on each layer and then chooses a solution from the Pareto front by applying the consensus clustering strategy, is proposed. Mucha et al. [24], generalized the concept of modularity to the multilayer case, thus allowing the use of algorithms which optimize modularity, like the *Louvain* method [1], for multilayer networks.

Boutemine and Bouguessa [5] proposed a method based on label propagation which identifies communities and relevant dimensions by using an objective function implementing the selection process of the dimensions during the optimization. Boden et al. [3] deal with multilayer graphs where edges are labelled, thus they propose a method to find clusters of densely connected nodes with similar edge labels.

3.2 Community Detection in Attributed Networks

Several methods for community detection in attributed networks have been proposed in recent years [4,9] and classified in different categories on the base of the adopted strategy. A common approach consists in reducing the network to a weighted graph by computing the similarity between nodes and then use a classical community detection method [10–12,25,33]. Another approach augments the graph with new nodes representing the attributes [36]. Elhadi and Agam [16] proposed an algorithm that, depending on the type of graph, either uses either the structure data and applies the *Louvain* method [1], or the attribute data, and then executes the *k-means* method. Li *et al.* [22] presented a multiobjective evolutionary algorithm for attributed networks which defines a new node similarity measure between nodes and optimizes the modularity function. In [31] a multiobjective genetic framework to uncover community structure in attributed networks, which evaluates different structural measures and different similarity measures between attributes, to obtain densely connected communities and homogeneous attributes, is proposed.

Neither of the described methods is able to find communities when the network has multiple layers and nodes are characterized by features. In the following, we introduce an objective function which takes into account both node information and multiple layers, whose optimization allows the detection of community structure on this kind of complex networks.

4   Unified Distance Measure

The *unified distance measure* is a measure introduced by Papadopoulos et al. [28] for *attributed multi-graphs*. In this kind of graph each couple of nodes can be connected by multiple relations and each node is characterized by more than one heterogeneous attribute, describing the properties of the object represented by that node. This definition of attributed multi-graph directly corresponds to the definition of attributed multilayer network given in Sect. 2 if we consider as many layers as the number of different links of the multi-graph, and on the l-th layer nodes are connected only through the l-th type of connection. The *unified distance measure* can then be applied also to multilayer networks. However, in a multilayer network the attribute values of nodes can be different for different layers, thus an attributed multilayer network is a more general concept of attributed multi-graph.

The *unified distance measure* is a combination of two terms: the *similar connectivity* of each couple of nodes, which considers the structural part of a network, and the *attribute distance*, which takes into account the information associated with nodes.

Given an attributed multilayer network $\mathcal{G} = \{G_l\}_{l=1}^{L}$ with L layers, each layer being an attributed graph $G_l = (V, E_l, A, F)$, the *similar connectivity* between two nodes i and j for the layer l, denoted $SC_l(i,j)$, measures how dissimilar they are on the l-th layer with respect to their shared neighbors. Low values mean that the two nodes i and j should be grouped together. It is defined as

$$SC_l(i,j) = \frac{1}{n} \sum_{k=1}^{n} [w_l(i,k) - w_l(j,k)]^2 \qquad (2)$$

where

$$w_l(i,j) = \begin{cases} w_l(i,j) \text{ if } & (i,j,l) \in E_l \\ 1 & \text{if } (i{=}j) \\ 0 & \text{otherwise} \end{cases} \qquad (3)$$

The *total similar connectivity* of two nodes is computed as:

$$SC(i,j) = \frac{1}{L} \sum_{l=1}^{L} SC_l(i,j) \qquad (4)$$

The *attribute distance* between two nodes i and j for the layer l, denoted $AD_l(i,j)$, measures how dissimilar they are with respect to their attribute values on the l-th layer. It is computed as:

$$AD_l(i,j) = \sum_{\alpha \in A} W_\alpha \cdot \delta_\alpha(i,j), \quad \sum_{\alpha \in A} W_\alpha = 1 \qquad (5)$$

where W_α is the weight of attribute α, and $\delta_\alpha(i,j)$ is the attribute distance between nodes i and j for attribute α. For numerical attributes scaled in the interval $[0, 1]$, $\delta_\alpha(i, j) = [a_\alpha(i) - a_\alpha(j)]^2$, while, for the categorical attributes

$$\delta_\alpha(i, j) = \begin{cases} 0 \text{ if } & a_\alpha(i) = a_\alpha(j) \\ 1 \text{ otherwise} \end{cases} \tag{6}$$

The *total attribute distance* of two nodes is computed as:

$$AD(i, j) = \frac{1}{L} \sum_{l=1}^{L} AD_l(i, j) \tag{7}$$

The *unified distance measure (udm)* combines the *attribute distance* (AD) and the *similar connectivity* (SC) between two nodes as follows:

$$d(i, j) = W_{attr} \cdot AD(i, j) + W_{links} \cdot SC(i, j) \tag{8}$$

where W_{attr} and W_{links} are weights representing the importance of attributes and edges, respectively.

Given a network division $\mathcal{C} = \{C_1, \ldots, C_k\}$, the *clustering unified distance measure cudm($\mathcal{C}$)* of the solution $\mathcal{C}$ is defined as

$$cudm(\mathcal{C}) = \frac{1}{k} \sum_{C \in \mathcal{C}} \sum_{\{i,j\} \in C \ i \neq j} d(i, j) \tag{9}$$

where k is the number of communities of the solution $\mathcal{C}$, i and j are nodes of a community $C \in \mathcal{C}$ and $d(i, j)$ is the *unified distance measure* between nodes i and j.

5   Method Description

In this section, the method for detecting communities in multilayer networks with attributes, based on DE, is described in detail.

Any general framework for solving a problem with differential evolution needs to choose the mutation and crossover operators, the representation of the problem and the fitness function to optimize.

ML@NetDE uses the *Indexed locus-based* representation, based on the *Relative Position Indexing* approach for combinatorial problems [27]. In this representation, a chromosome is a vector of size n, where n is the number of nodes. The value associated with node i, instead of being the identifier j of one of its neighbors, as in the locus-based representation [29], is the relative position of node j in the i-th row of the adjacency list of the graph.

Consider the two-layer network of Fig. 1(a) with the associated adjacency list. The locus-based and indexed locus-based representations of the network division in the two communities $\{1, 2, 3, 4\}, \{5, 6, 7\}$ are shown in Fig. 1(b). For instance, in the locus-based representation node 7 is connected with node 5, in the indexed one, 5 is substituted with 2 since 5 is the second neighbor of 7, as shown in the adjacency list.

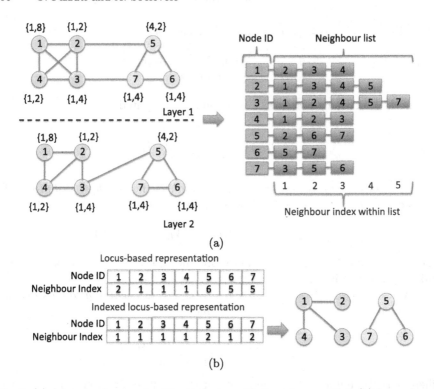

Fig. 1. (a) A two level network with attributes with the adjacency list. (b) Locus-based and Indexed locus-based representation of the network division $\{1, 2, 3, 4\}, \{5, 6, 7\}$.

Regarding the operators, *ML@NetDE* uses the *DE/current-to-rand/1* mutation operator and classical binomial crossover [14].

Let $P = \{x_1, \ldots, x_{nP}\}$ be a population of nP n-dimensional target vectors generated at random, where n is the number of nodes.

A target vector x_i^t, at a generic generation t, generates a *mutant* vector by applying the *DE/current-to-rand/1* mutation operator. This operator computes the mutant vector by using the current target vector x_i^t and three random parameter vectors, as follows:

$$v_i^t = x_i^t + F * (x_{r1}^t - x_i^t) + F * (x_{r2}^t - x_{r3}^t) \tag{10}$$

where F is a scaling factor. The binomial crossover operator generates the trial vector u_i^t from the mutant vector v_i^t as

$$u_{i,j}^t = \begin{cases} v_{i,j}^t & \text{if } rand_j \leq CR \text{ or } j = j_{rand} \\ x_{i,j}^t & \text{otherwise} \end{cases} \tag{11}$$

where $u_{i,j}^t$, $x_{i,j}^t$, and $v_{i,j}^t$ are the jth dimension of u_i^t, x_i^t, and v_i^t, respectively, $rand_j$ is a random number between 0 and 1, CR is the crossover control parameter, j_{rand} is a random number between 1 and n. The target vector x_i^{t+1} at

The *ML@NetDE Method*:
Input: The attributed multilayer network $\mathcal{G} = \{G_l\}_{l=1}^{L}$;
    nP: population size;
    T: maximum number of generations;
Output: $\mathcal{C} = (C_1, \ldots, C_k)$: network division in k communities.

```
1   nMax=5; //max number of generations the objective function is allowed to not improve
2   n₀=5; //number of trial vectors generated at each generation
3   Fmin = 0.01; //lower bound of the scaling factor;
4   Fmax = 0.1; //upper bound of the scaling factor
5   Initialize the population P with random individuals xi, 1 ≤ i ≤ nP,
6   evaluate each individual xi to obtain xi.Fit
7   Let t = 1 and xbest be the individual with the minimum fitness value xbest.Fit
8   while t ≤ T do
9   For i=1 to nP do
10    For j=1 to n₀ do
11      Select randomly from the population the individuals r₁ ≠ r₂ ≠ r₃ ≠ i
12      compute the mutant vector vⱼᵗ by applying the DE/current-to-rand/1 mutation operator
13      compute the trial vector uⱼᵗ = binomialCrossover(xiᵗ, vⱼᵗ);
14    end for
15    let ubestᵗ be the trial vector with the minimum fitness value among uⱼᵗ, j = 1, n₀
16    if (ubestᵗ.Fit < xiᵗ.Fit) then
17      xiᵗ⁺¹ = ubestᵗ;
18    else
19      xiᵗ⁺¹ = xiᵗ
20    end if
21    if ( xiᵗ⁺¹.Fit < xbest.Fit )
22      xbest = xiᵗ⁺¹
23    end for //end for each element of the population
24  if xbest does not improve for nMax generations then
25    Initialize the 50% of the population P with random individuals
26    Fmax = Fmax + 0.1
27  end if
28  t = t + 1;
29  end while
30  return the partition C = {C₁, ..., Ck} corresponding to the solution x* with the best fitness value
31  merge a community Ci with a neighboring community Cj if the number of connections
    from nodes of Ci to nodes of Cj is higher than the number of links among nodes inside Ci.
```

Fig. 2. The pseudo-code of the *ML@NetDE* algorithm.

the next generation is obtained by choosing the best, in terms of fitness value, between the mutant vector u_i^t and the target vector x_i^t:

$$x_i^{t+1} = \begin{cases} u_i^t \text{ if } cumd(u_i^t) \leq cumd(x_i^t) \\ x_i^t \text{ otherwise} \end{cases} \tag{12}$$

where *cumd* is the unified distance measure described in Sect. 4.

It is worth pointing out that, as already experimented in [32] for community detection in monolayer graphs, differential evolution may suffer of premature convergence, probably because of the number of discrete values which the positions of a mutant vector can have, i.e. at most the number of neighbors of the corresponding node. To dampen this problem, we adopt the same strategies proposed in [32]. First of all, the scaling factor is generated, different for each node, from a continuous uniform distribution with lower bound 0.01 and upper bound 0.1. Then, if the method does not improve the fitness function for a fixed number $nMax$ of generations, the upper bound is progressively incremented to allow more variation and the half of the population is reinitialized with random individuals. Moreover, instead of one, a number n_0 of trial vectors are generated and that having the best fitness value is chosen to compete with the target vector to survive at the $t + 1$ generation.

Table 1. mLFR-128 parameters setting.

Parameter	Value
Number of nodes	128
Node average degree	8
Node maximal degree	16
Mixing parameter μ	[0.1, 0.2, 0.3, 0.4, 0.5]
Attribute noise ν	[0.1, 0.2, 0.3, 0.4, 0.5]
Exponent for power law creating degree sequence	2
Exponent for power law creating community sizes	1
Maximal community size	32
Minimal community size	8
Number of layers	2
Number of attributes	2
Degree change chance	0.2
Exponent for power law of nodes through layers	2

The pseudocode of the algorithm is reported in Fig. 2. *ML@NetDE* needs as input parameters the population size nP and the maximum number of generations T. The population is initialized with random individuals $\boldsymbol{x}_i$ by assigning to each node the index of one of its neighbors and their fitness $\boldsymbol{x}_i.Fit$ is computed (steps 5–6). Then, for a maximum number T of generations and for each element of the population, a number n_0 of trial vectors are generated by applying the *DE/current-to-rand/1* mutation and binomial operators (steps 10–14). The best among these n_0 vectors is chosen to compete with the target vector to survive for the next generation (steps 16–20).

If the fitness value does not increase for $nMax$ generations, half of the population is substituted with new random individuals and the value of the upper bound of the scaling factor is incremented (steps 24–26). Finally, the returned solution is checked for merging (step 31).

6   Experimental Evaluation

In this section we evaluate the performance of *ML@NetDE* both on synthetic networks, for which the partitioning in communities is known, and on real-world networks where the ground truth is not known. For the first class of networks, since there is no a benchmark network generator for multiple layers and node attributes, we first generated synthetic multilayer networks by using the bench-mark generator[1] proposed by Bródka and Grecki [6], which is an extension of the LFR benchmark generator by Lancichinetti et al. [21], and then enriched the nodes of networks with attributes as follows. The attributes have been synthet-ically generated setting a certain level of attribute noise (i.e. a certain level of dissimilarity). For example, if a network has a value of attribute noise of 0.1, the 10% of the nodes will have attributes dissimilar with respect to those of their community. We generated numerical attributes by using a uniform distribution over an interval that we have initially set different for each community. Then, given an attribute noise level, we randomly chose the corresponding number of nodes on which varying the attributes accordingly. Specifically, for each node on which varying the attributes, we added to the initial attribute values an offset.

The method has been implemented in Matlab.The results of *ML@NetDE* are the average values obtained by running the method 10 times, crossover proba-bility 0.9, population size 100 and 50 generations. In the following subsections we first describe the datasets and the evaluation measures used, and then the results obtained.

6.1   Datasets

Synthetic Networks. The parameters used for generating the mLFR networks with attributes are shown in Table 1. Each layer consists of 128 nodes parti-tioned in different communities having variable sizes. Specifically, we generated networks with two layers by varying two parameters: the *mixing parameter* μ and the *degree change chance (DCC)*, as reported in Table 1. The mixing parameter μ is the fraction of links shared by a node with the nodes of its community. When $\mu < 0.5$ the number of neighbors of a node inside its group are more than the number of neighbors belonging to the other groups. The degree change chance controls how much different the network layers are in terms of node degree. The higher the DCC of a network, the more the nodes may have different degree values within different layers, and thus the more diverse the networks on differ-ent layers. Attributes have been generated by varying between 0.1 and 0.5 the noise level ν of attribute similarity between nodes of the same community. For instance, if $\nu = 0.1$ then the 10% of nodes have values different from the other nodes of the same community.

Real-World Networks. We considered the following four real-world attributed multilayer networks.[2] In Table 2 their main characteristics are summarized.

Krackhardt High-Tech Managers. These data were collected from the 21 managers of a United States high-tech company. The three layers have been obtained by asking each manager the three questions: "To whom do you go

---

[1] mLFR Java code is available at https://www.ii.pwr.edu.pl/~brodka/mlfr.php.
[2] See http://moreno.ss.uci.edu/data.html.

to for advice?", "Who is your friend?", and "To whom do you report?". The attribute information consists of managers age (in years), length of service or tenure (in years), level in the corporate hierarchy (coded as $1 = $ CEO, $2 = $ Vice President, $3 = $ manager) and department (coded 1,2,3,4).

Dutch College Freshmen. The data were collected among a group of university freshmen who were asked to rate their relationships at 7 time points. Each student is characterized by gender ($1 = $ F, $2 = $ M), education program (2-year, 3-year, 4-year), and smoking behavior ($1 = $ yes, $2 = $ no).

Lazega Law Firm. This dataset was generated from a network study of corporate law partnership, carried out in a Northeastern US corporate law firm. It includes measurements of networks among the 71 attorneys (partners and associates) of this firm, i.e. their strong-coworker network, advice network, friendship network. Each attorney is characterized by 7 attributes: status ($1 = $ partner; $2 = $ associate), gender ($1 = $ man; $2 = $ woman), office ($1 = $ Boston; $2 = $ Hartford; $3 = $ Providence), years with the firm, age, practice ($1 = $ litigation; $2 = $ corporate), law school (1: harvard, yale; 2: ucon; 3: other).

Table 2. Features of the real-world datasets.

Dataset	# nodes	# edges	# attributes	# layers
High-tech managers	21	312	4	3
Dutch college freshmen	32	880	3	7
Lazega law firm	71	2223	7	3
CKM physicians innovation	246	1551	13	3

Table 3. NMI values obtained by *ML@NetDE* for the the mLFR-128 networks.

	$\nu = 0.1$	$\nu = 0.2$	$\nu = 0.3$	$\nu = 0.4$	$\nu = 0.5$
$\mu = 0.1$	0.923 (0.064)	0.894 (0.035)	0.845 (0.041)	0.9 (0.005)	0.888 (0.035)
$\mu = 0.2$	0.858 (0.066)	0.907 (0.066)	0.902 (0.023)	0.845 (0.095)	0.911 (0.043)
$\mu = 0.3$	0.909 (0.031)	0.959 (0.024)	0.885 (0.101)	0.925 (0.06)	0.893 (0.064)
$\mu = 0.4$	0.876 (0.147)	0.812 (0.121)	0.876 (0.198)	0.865 (0.12)	0.822 (0.081)
$\mu = 0.5$	0.908 (0.067)	0.786 (0.236)	0.812 (0.129)	0.856 (0.127)	0.785 (0.138)

CKM Physicians Innovation. This dataset contains data collected by Coleman, Katz and Menzel on medical innovation, considering a set of physicians in four towns of Illinois (Peoria, Bloomington, Quincy and Galesburg). Data concerne with the impact of network ties on the physicians' adoption of a new drug, the tetracycline. Three social networks were generated, based on: (1) information or advice about questions of therapy, (2) physicians often asked for discussing therapy (3) friends most often seen. The resulting network has 246 nodes in total with 1551 connections, with 13 attributes: city of practice, recorded date of tetracycline adoption date, years in practice, meetings attended, journal subscriptions, free time activities, discussions, club memberships, friends, time in the community, patient load, physical proximity to other physicians and medical specialty.

6.2   Evaluation Measures

Normalized Mutual Information. For the synthetic networks the ground truth partitioning is known, thus, to evaluate the quality of the solutions we use the well known measure of *Normalized Mutual Information (NMI)* [13].

The normalized mutual information $NMI(A, B)$ of two divisions A and B of a network is defined as follows. Let C be the confusion matrix whose element C_{ij} is the number of nodes of community i of the partition A that are also in the community j of the partition B.

$$NMI(A, B) = \frac{-2 \sum_{i=1}^{c_A} \sum_{j=1}^{c_B} C_{ij} log(C_{ij} n / C_{i.} C_{.j})}{\sum_{i=1}^{c_A} C_{i.} log(C_{i.}/n) + \sum_{j=1}^{c_B} C_{.j} log(C_{.j}/n)} \qquad (13)$$

where c_A (c_B) is the number of groups in the partition A (B), $C_{i.}$ $(C_{.j})$ is the sum of the elements of C in row i (column j), and n is the number of nodes. If $A = B$, $NMI(A, B) = 1$. If A and B are completely different, $NMI(A, B) = 0$.

Regarding the real-world networks, the true division is not known, thus we use the two standard indexes of *density* and *entropy*, the former considers the network structure and measures the internal edge density of a partitioning, the latter the attribute homogeneity.

Density. It is defined as

$$D = \sum_{C \in \mathcal{C}} \frac{m_c}{m} \qquad (14)$$

where m_c is the number of edges of the community C and m is the total number of edges of the network.

Entropy. It is based on the information theory concept of entropy, and it measures the average Shannon information content of a set. A highly disordered set with different elements has a high entropy. Thus, the lower the entropy, the more homogeneous the attribute values. Entropy is defined as

$$E = -\sum_{C \in \mathcal{C}} \frac{n_c}{n} \sum_{a \in A} p_{ac} \log(p_{ac}) \qquad (15)$$

where p_{ac} is the percentage of nodes in community C with the attribute value a, n_c is the number of nodes on the community C and n is the number of vertices of the network.

6.3   Results

Table 3 shows the normalized mutual information results for the mLFR-128 networks with attributes for different values of the mixing parameter μ and the attribute noise ν. As outlined above, each value in the table corresponds to an NMI value averaged over 10 runs; between brackets we report the standard deviation. For this first experiment, we focused on understanding how *ML@NetDE* behaves as the structure of the communities of the multilayer network and the attribute homogeneity vary. By fixing a certain level of mixing parameter μ and varying the attribute noise ν, we observe that *ML@NetDE* is able to obtain high values of NMI in most cases around 0.9, even if the attributes are not so homogeneous within the communities. Hence, *ML@NetDE* independently from the attribute noise level, finds good partitions. Similarly, when fixing a certain level of attribute noise, we note that even if μ increases, i.e. the communities are not well separated, the algorithm returns high NMI values. Thus we can conclude that *ML@NetDE* is able to uncover good partitions, independently from the structure of the communities and the attribute noise.

The second experiment aimed to compare *ML@NetDE* with two methods on real-world networks for which the ground truth is not known. In this case, we measured the performance of all the methods in terms of density and entropy. The first method is the classical *k-means* clustering algorithm which considers only the attributes of the network. The second one is the extension of the *Louvain* method to multilayer networks[3] [24]. In Table 4, the results in terms of density are shown. Again, each value corresponds to an average over 10 runs. Here, we do not report the standard deviation since the values are very low. Since *ML@NetDE* and Louvain find 3, 3, 3 and 12 communities, over High-tech, Dutch college, Lazega and CKM, respectively, we set the k value for the *k-means* algorithm to these values. Overall, *ML@NetDE* outperforms the other two algorithms, thus showing that combining network structure and attributes, the algorithm is able to find more dense communities (i.e. better partitions). Instead, considering only the structure or the attributes such as in the case of *Louvain* and *k-means*, respectively, the resulting density values are lower. Table 5 shows the entropy results. For the High-tech network, the three algorithms obtain zero values. This means that all the algorithms find communities where the attributes

Table 4. Density values for the real-world datasets.

Dataset	ML@NetDE	k-means	Louvain
High-tech managers	0.531	0.449	0.531
Dutch college freshmen	0.476	0.351	0.442
Lazega law firm	0.642	0.438	0.628
CKM physicians innovation	0.941	0.127	0.899

---

[3] https://github.com/GenLouvain.

are homogeneous. For the other datasets, *ML@NetDE* outperforms *k-means* and *Louvain* resulting in the lowest entropy values.

Table 5. Entropy values for the real-world datasets.

Dataset	*ML@NetDE*	k-means	Louvain
High-tech managers	0	0	0
Dutch college freshmen	0	0	0.05
Lazega law firm	0.06	0.129	0.133
CKM physicians innovation	0.063	0.136	0.078

7 Conclusions

The paper proposed a new approach based on differential evolution for uncovering community structure in multilayer networks with node features. The experimentation of the method on synthetic networks showed that it is able to recover the true community structure with high accuracy. A comparison with the *k-means* method, which uses only the attribute information, and the extended *Louvain* method, which considers only the layer information, has highlighted that *ML@NetDE* obtains network divisions with higher density and lower entropy values, meaning that the nodes of the communities are well connected and with high feature homogeneity. It is worth pointing out that, since the research in this field is rather recent, a comparison with other methods is difficult because of two main problems. The first is the lack of real-world networks for which the ground-truth division is known and a generator of synthetic networks, like *LFR*, with multiple layers and attributes. The second one is that the main proposals of the last years either do not consider both the multilayer aspect and the attribute information, or they transform the network in a form apt for single layer methods. Our approach is a first proposal in this direction. Future work aims to evaluate the performance of the method on networks of larger size and with different mutation operators.

References

1. Blondel, V.D., Guillaume, J.L., Lambiotte, R., Lefevre, E.: Fast unfolding of communities in large networks. J. Stat. Mech. Theor. Exp. **2008**, 10008 (2008)
2. Boccaletti, S., et al.: The structure and dynamics of multilayer networks. Phys. Rep. **544**(1), 1–122 (2014)
3. Boden, B., Günnemann, S., Hoffmann, H., Seidl, T.: MiMAG: mining coherent subgraphs in multi-layer graphs with edge labels. Knowl. Inf. Syst. **50**(2), 417–446 (2017)

4. Bothorel, C., Cruz, J.D., Magnani, M., Micenkova, B.: Clustering attributed graphs: models, measures and methods. Netw. Sci. **3**(03), 408–444 (2015)
5. Boutemine, O., Bouguessa, M.: Mining community structures in multidimensional networks. TKDD **11**(4), 51:1–51:36 (2017)
6. Bródka, P., Grecki, T.: mLFR benchmark: Testing community detection algorithms in multi-layered, multiplex and multiple social networks (to appear)
7. Bródka, P., Filipowski, T., Kazienko, P.: An introduction to community detection in multi-layered social network. In: Lytras, M.D., Ruan, D., Tennyson, R.D., Ordonez De Pablos, P., García Peñalvo, F.J., Rusu, L. (eds.) WSKS 2011. CCIS, vol. 278, pp. 185–190. Springer, Heidelberg (2013). https://doi.org/10.1007/978-3-642-35879-1_23
8. Chen, P., Hero, A.O.: Multilayer spectral graph clustering via convex layer aggregation: theory and algorithms. IEEE Trans. Sig. Inf. Process. Netw. **3**(3), 553–567 (2017)
9. Chunaev, P.: Community detection in node-attributed social networks: a survey. arXiv:1912.09816v1 (2019)
10. Combe, D., Largeron, C., Egyed-Zsigmond, E., Géry, M.: Combining relations and text in scientific network clustering. In: 2012 IEEE/ACM International Conference on Advances in Social Networks Analysis and Mining (ASONAM), pp. 1248–1253. IEEE (2012)
11. Cruz, J.D., Bothorel, C., Poulet, F.: Semantic clustering of social networks using points of view. In: CORIA, pp. 175–182 (2011)
12. Dang, T., Viennet, E.: Community detection based on structural and attribute similarities. In: International Conference on Digital Society (ICDS), pp. 7–12 (2012)
13. Danon, L., Duch, J., Arenas, A., Díaz-Guilera, A.: Comparing community structure identification. J. Stat. Mech., P09008 (2005)
14. Das, S., Suganthan, P.N.: Differential evolution: a survey of the state-of-the-art. IEEE Trans. Evol. Comput. **15**(1), 4–31 (2011)
15. Das, S., Mullick, S.S., Suganthan, P.N.: Recent advances in differential evolution - an updated survey. Swarm Evol. Comput. **27**, 1–30 (2016)
16. Elhadi, H., Agam, G.: Structure and attributes community detection: comparative analysis of composite, ensemble and selection methods. In: Proceedings of the 7th Workshop on Social Network Mining and Analysis, p. 10. ACM (2013)
17. Fortunato, S.: Community detection in graphs. Phys. Rep. **486**, 75–174 (2010)
18. Girvan, M., Newman, M.E.J.: Community structure in social and biological networks. Proc. Nat. Acad. Sci. USA **99**, 7821–7826 (2002)
19. Kim, J., Lee, J.: Community detection in multi-layer graphs: a survey. SIGMOD Rec. **44**(3), 37–48 (2015)
20. Kivelä, M., Arenas, A., Barthelemy, M., Gleeson, J.P., Moreno, Y., Porter, M.A.: Multilayer networks. J. Complex Netw. **2**(3), 203–271 (2014)
21. Lancichinetti, A., Fortunato, S., Radicchi, F.: Benchmark graphs for testing community detection algorithms. Phys. Rev. E **78**(4), 046110 (2008)
22. Li, Z., Liu, J., Wu, K.: A multiobjective evolutionary algorithm based on structural and attribute similarities for community detection in attributed networks. IEEE Trans. Cybern. **48**(7), 1963–1976 (2018)
23. Loe, C.W., Jensen, H.J.: Comparison of communities detection for multiplex. Phys. A **431**, 29–45 (2015)
24. Mucha, P.J., Richardson, T., Macon, K., Porter, M.A., Onnela, J.P.: Community structure in time-dependent, multiscale, and multiplex networks. Science **328**(5980), 876–878 (2010)

25. Neville, J., Adler, M., Jensen, D.: Clustering relational data using attribute and link information. In: Proceedings of the Text Mining and Link Analysis Workshop, 18th International Joint Conference on Artificial Intelligence, pp. 9–15 (2003)
26. Newman, M.E.J., Girvan, M.: Finding and evaluating community structure in networks. Phys. Rev. **E69**, 026113 (2004). http://www.citebase.org/abstract?id=oai:arXiv.org:cond-mat/0308217
27. Onwubolu, G., Davendra, D.: Differential evolution for permutation—based combinatorial problems. In: Onwubolu, G.C., Davendra, D. (eds.) Differential Evolution: A Handbook for Global Permutation-Based Combinatorial Optimization. Studies in Computational Intelligence, vol. 175. Springer, Heidelberg (2009)
28. Papadopoulos, A., Pallis, G., Dikaiakos, M.D.: Weighted clustering of attributed multi-graphs. Computing **99**(9), 813–840 (2016). https://doi.org/10.1007/s00607-016-0526-5
29. Park, Y., Song, M.: A genetic algorithm for clustering problems. In: Proceedings of 3rd Annual Conference on Genetic Algorithms, pp. 2–9. Morgan Kaufmann Publishers (1989)
30. Pizzuti, C., Socievole, A.: Many-objective optimization for community detection in multi-layer networks. In: 2017 IEEE Congress on Evolutionary Computation, CEC 2017, Donostia, San Sebastián, Spain, 5–8 June 2017, pp. 411–418 (2017)
31. Pizzuti, C., Socievole, A.: Multiobjective optimization and local merge for clustering attributed graphs. IEEE Trans. Cybern. **PP**(99), 1–13 (2019). https://doi.org/10.1109/TCYB.2018.2889413
32. Pizzuti, C., Socievole, A.: Self-adaptive differential evolution for community detection. In: Sixth International Conference on Social Networks Analysis, Management and Security, SNAMS 2019, Granada, Spain, 22–25 October 2019, pp. 110–117 (2019)
33. Ruan, Y., Fuhry, D., Parthasarathy, S.: Efficient community detection in large networks using content and links. In: 22nd International World Wide Web Conference, WWW 2013, Rio de Janeiro, Brazil, 13–17 May 2013, pp. 1089–1098 (2013)
34. Tang, L., Wang, X., Liu, H.: Uncovering groups via heterogeneous interaction analysis. In: The Ninth IEEE International Conference on Data Mining, ICDM 2009, pp. 503–512 (2009)
35. Tang, L., Wang, X., Liu, H.: Community detection via heterogeneous interaction analysis. Data Min. Knowl. Disc. **25**(1), 1–33 (2012)
36. Zhou, Y., Cheng, H., Yu, J.X.: Graph clustering based on structural/attribute similarities. Proc. VLDB Endow. **2**(1), 718–729 (2009)

Databases and Data Management

Modeling and Enforcing Integrity Constraints on Graph Databases

Fábio Reina$^{(\boxtimes)}$ ⓘ, Alexis Huf$^{(\boxtimes)}$ ⓘ, Daniel Presser$^{(\boxtimes)}$ ⓘ,
and Frank Siqueira$^{(\boxtimes)}$ ⓘ

Graduate Program in Computer Science, Department of Informatics and Statistics,
Federal University of Santa Catarina, Florianópolis, Brazil
{f.reina,alexis.huf,daniel.presser}@posgrad.ufsc.br, fmreinaa@gmail.com,
frank.siqueira@ufsc.br

Abstract. The enormous volume and high variety of information that is constantly produced by computing systems requires storage technologies able to provide high processing velocity and data quality. The suitability for modeling complex data and for delivering performance are characteristics that are making graph databases become very popular. However, existing limitations still prevent database management systems that adopt the graph model to fully ensure data consistency, given that the means for ensuring data consistency are usually nonexistent or at most very simple. This work intends to overcome this limitation by extending the support for defining and enforcing integrity constraints on graph databases, in order to prevent the graph to reach an inconsistent state and compromise the correctness of applications. The proposed integrity constraints are implemented on OrientDB. Experimental results show that the prototype implementation can improve the performance in comparison to verification of constraints on a client application.

Keywords: Graph databases · Data consistency · Integrity constraints · Data integrity · OrientDB

1 Introduction

A well-known fact regarding the present state of information technology is that the amount of data produced and stored by computer systems is in constant growth. This phenomenon, known as Big Data, is nowadays the subject of a substantial amount of research work. Not only data is produced in larger *volumes*, but it is generated by multiples sources in different formats (*variety*), and must be stored and processed quickly (*velocity*) without compromising its *validity*.

Along with the volume of data, the need for better performance and for more efficient management became very relevant issues [13]. However, the traditional

Financed in part by CAPES ("Coordenação de Aperfeiçoamento de Pessoal de Nível Superior") - Brasil - Finance Code 001 and by FAPESC ("Fundação de Amparo à Pesquisa e Inovação do Estado de Santa Catarina").

© Springer Nature Switzerland AG 2020
S. Hartmann et al. (Eds.): DEXA 2020, LNCS 12391, pp. 269–284, 2020.
https://doi.org/10.1007/978-3-030-59003-1_18

databases (DBs) are not always able to handle all the requirements of such large volumes [17]. As a result, the category of Database Management Systems (DBMSs) known as Not only SQL (NoSQL) has emerged aiming to fulfill these requirements. The main characteristic of NoSQL DBMSs is that data consistency is relaxed with the aim of improving performance. It allows data to be inconsistent for some time, which may be prohibitive for some applications. Among the existing categories of NoSQL systems, graph DBMSs have been increasing in popularity, and are the subject of this work.

In general, when graph DBMSs offer support for specifying and enforcing Integrity Constraints (ICs), the supported ICs are usually very limited. The most common constraints allow database administrators to enforce attributes to have unique values, limit minimum and maximum values, and define attribute types. Due to the lack of support for more complex ICs, data validation often becomes the responsibility of the application that uses the graph DB, resulting in an increase of development effort.

The work presented in this paper aims to tackle the lack of support for complex ICs in graph DBMSs. Thus, it is possible to create rules using two or more elements (attribute, node or edge) at the same time. Therefore, we propose a specification syntax for ICs as well as a mechanism for their enforcement during operations that modify the graph. To achieve this goal we define six new ICs and implement them on the OrientDB, allowing the definition of: (1) conditions on node attributes, (2) required edges, (3) type of in/out nodes of an edge, (4) edge cardinality, (5) bidirectionality of edges, and (6) conditions on attributes of nodes linked by an edge. We evaluate the impact of these constraints over the performance of OrientDB. With these extensions, we intend to transfer the responsibility for validating data constraints from the client application to the DBMS, enforcing data integrity with less effort when developing applications.

The remainder of this paper is organized as follows. Section 2 presents the most relevant concepts used in this work. Section 3 explains our proposal, specifying in more detail the ICs that are supported by our extended version of OrientDB. Then, Sect. 4 describes the evaluation study performed over our implementation. Next, Sect. 5 identifies some similar proposals described in the literature and compares them to the solution proposed in this paper. Finally, Sect. 6 presents the conclusions reached with the development of this work and singles out some open issues that require further research.

2   Fundamental Concepts

A Graph DB is essentially a DB that uses the explicit structure of a graph to store, query and manipulate data. Vertices, also called nodes, represent database records, while edges represent relationships between data [3]. In general, every edge has a label that identifies the relationship it represents. Vertices and edges may also have properties. Therefore, edges are as important as vertices, due to the potentially relevant information they carry. Due to its composition of vertices, edges and properties, this structure is said to be a property graph [19].

This work considers property graphs as defined in Definition 1. This definition was adapted from [2] to remove multiple labels and multiple attribute values, aiming to better align with current graph DBMSs implementations, including Neo4J, OrientDB, InfinityGraph, Trinity, Titan and ArangoDB.

Definition 1. *A property graph is a tuple $G = (N, E, \rho, \lambda_N, \lambda_E, \sigma)$, such that:*

1. *N is a set of nodes;*
2. *E is a set of edges;*
3. *$\rho : E \to (N \times N)$ is a function that associates an edge in E with a pair (o, t) of origin and target nodes in N;*
4. *$\lambda_N : N \to L_N$ is a function that associates a node with a node label;*
5. *$\lambda_E : E \to L_E$ is a function that associates a edge with a edge label;*
6. *$\sigma : (N \cup E) \times P \to K$ is a function that given a node or edge together with a property P, associates the pair to a value from K.*

The representation of highly connected data in a relational model is possible, however it results in several many-to-many relationships. A query under this conditions may require a big number of joins, which can degrade the performance [18]. In contrast, the graph structure allows a better management of this type of data, resulting in faster query processing [5]. The most recurring example of application that benefit from using graph DBMSs are social networks. Nevertheless, graph DBs are very useful in financial systems, for fraud detection and transaction monitoring [6,18]; on document analysis to analyze speech data and identify stakeholders' intention [15]; on the retail sector, helping with decision making and product recommendation [1]; among several other applications.

Differently from relational DBs, which have a well-defined schema, graph DBs do not have a rigid structure. It means that while in a relational DB an insert operation can only set properties defined in the DB schema, in a graph DB it can add new properties that were not defined on the schema. This characteristic results in faster execution of operations. However, it becomes harder to impose ICs, given that there is no schema to follow. As a result, graph DBs in general either do not provide tools to ensure consistency, or only support very basic ICs.

Graph DBs are categorized under a larger category of DBMSs, named NoSQL DBs. Unlike the acronym suggests, this category does not preclude the use of SQL, but indicates the use of alternatives to the relational model, which backs SQL. They also follow Basically Available, Soft state, and Eventual consistency (BASE) properties, which imply that the DB has to be available most of the time for read and write operations. It also indicates that data may be inconsistent during some time, but will become accurate in a future moment [16].

In DBs with BASE properties, consistency refers to transactions performing reads on up-to-date and committed data. Correctness of data is not a concern of consistency, but rather a concern of data integrity. Thus, data integrity can be defined as the maintenance and assurance of the data correctness during all its life cycle [14]. It is a major concern of many systems, especially those that mange real world data, and can be achieved by the use of a set of rules that specify all the allowed update over the data. In this scenario comes the concept of ICs, which can be described as a set of general rules that define a consistent state of the DB, as well as allowed modifications [3]. These rules must be applied on every inserted data to avoid inconsistency. If the integrity cannot be secured by the DBMS, its assurance must be implemented at the application level.

In relational DBMSs, the constraints are linked to one of the following categories: entity integrity, referential integrity, and domain integrity [9]. The use of ICs ensures that the information stored on the DB conforms to these rules and, as a result, enforces integrity. However, most of the graph DBMSs available on the market nowadays do not support ICs, or only allow elementary rules to be defined, that are unable to provide the level of integrity required by most of the applications that store and manipulate graph data.

3   Extending Support for Integrity Constraints in Graphs

This work proposes an extension of a graph DBMS to allow the specification of complex ICs, which are useful for applications that need to store data in the form of a graph database with a high level of integrity. The graph DBMS chosen to receive this extension was OrientDB[1]. One of the features that motivated its choice is the fact that OrientDB already provides support for the definition of simple constraints. Besides that, it is ranked among the most popular graph DBMSs, according to DBEngines[2].

In OrientDB, every class, also known as node type, is associated to a set of metadata. The proposed extension aims to store the constraint definitions into these metadata. Each IC is defined by the OrientDB client application using our extension of the OrientDB Data Definition Language (DDL). After an IC is stored on the metadata, any modification to the set of instances of that class will be validated against the IC before the corresponding transaction commits.

All ICs already supported by OrientDB are restricted to comparing values between the data being inserted or updated against threshold values defined in OrientDB metadata for a given node type. One approach to implement more complex ICs not limited to node attributes is to dynamically compute node attributes and compile definitions of such complex ICs into simpler ICs that are limited to checking node attribute values. However, this strategy incurs a large overhead in the form of such dynamic attributes and the potentially large amount of metadata required to implement the ICs. To avoid such drawbacks, the extension adds a new first-class component, the constraint manager, to the internal architecture of OrientDB. This component provides its own constraint validation mechanisms that are used by the new IC types, without relying on the constraints already supported by OrientDB. The use of a dedicated constraint manager allows for ICs that also involve edges, in addition to nodes and properties. Its existence also eases the introduction of new constraint types by providing a single interception point for their validation. The extension proposed

---

[1] https://orientdb.com/.
[2] https://db-engines.com/en/ranking/graph+dbms.

in this work introduces support for six new IC types: node condition, required edge, in/out, edge cardinality, bidirectional edge and edge condition.

The constraint manager is initialized together with the OrientDB server and uses an SB-Tree [8] to store and manage *Constraint* objects. When the user defines an IC using the `CREATE CONSTRAINT` command, the parser extracts the constraint type, its target and constraint-specific arguments. The result of the parsing is fed into a constraint factory and the resulting Constraint object is serialized into the SB-Tree. In addition to adding the constraint manager, the OrientDB DDL grammar was extended to support the new constraint types.

Once the constraint is declared and stored on the SB-Tree, every write operation on the graph DB triggers the validation of the relevant *Constraint* objects. After locating all relevant *Constraint* objects, the constraint manager collects all relevant data for the write operation and invokes the validation procedure of each *Constraint* object. If any *Constraint* validation fails, an exception is thrown, leading to the abortion of the whole transaction.

The syntax of each new IC supported is presented in the remainder of this section, together with a formal definition. All definitions use the property graph defined by [2] and presented in Definition 1.

Some constraints allow for relational operators, which are denoted by $\alpha \in \mathcal{O}$, where $\mathcal{O} = \{<, \leq, =, \neq, \geq, >\}$. Node and edge types are denoted using the notation $\mathcal{T}_\beta$, where $\beta \in L$ is a label that identifies the type unambiguously. Formally, types are defined by the sets of their instances: $\mathcal{T}_\beta = \{x \mid \lambda_N(x) = \beta \wedge \lambda_E(x) = \beta\}$. Given that, labels for node and edges are disjoint, $\beta \in L_N \iff \mathcal{T}_\beta \subseteq N$ and $\beta \in L_E \iff \mathcal{T}_\beta \subseteq E$.

Node Condition Constraint. The goal of this constraint is to compare properties of a node and to validate values assigned to them according to a previously defined condition. Therefore, this IC is applied to the node class. It is formally specified by Definition 2 and the general syntax is shown in Fig. 1.

Definition 2. *Given a property graph* $G = (N, E, \rho, \lambda_N, \lambda_E, \sigma)$, *a Node Condition constraint is a tuple* $Cond = (\mathcal{T}_o, p_1 \in P, \alpha_1 \in \mathcal{O}, k_1 \in K, p_2 \in P, \alpha_2 \in \mathcal{O}, k_2 \in K, p_3 \in P, \alpha_3 \in \mathcal{O}, k_3 \in K)$, *such that:*

$$\forall o \in \mathcal{T}_o \ . \ \begin{cases} \sigma(o, p_2) \ \alpha_2 \ k_2, & \text{if } k_1 \in \sigma(o, p_1) \\ \sigma(o, p_3) \ \alpha_3 \ k_3, & \text{otherwise} \end{cases}$$

```
1    <CREATE> <CONSTRAINT> name <ON> class <(> attribute <)> <CONDITIONAL>
2    <(>   <IF> property (>|<|>=|<=|=|!=) expression
3          <THEN> property (>|<|>=|<=|=|!=) expression
4          [ <ELSE> property (>|<|>=|<=|=|!=) expression ]  <)>
```

Fig. 1. General syntax of the Node Condition constraint.

Required Edge Constraint. This IC defines that a node class has a mandatory outgoing edge of a given type, that will point to an instance of a target node class. Therefore, if there is a node whose class appears as origin class in the constraint, then there must be at least one edge leaving this node and arriving at a node with the given target class. The constraint is associated with the origin node class metadata. This IC is formally described by Definition 3 and the general syntax is shown in Fig. 2.

Definition 3. *Given a property graph* $G = (N, E, \rho, \lambda_N, \lambda_E, \sigma)$, *a Required Edge constraint is a tuple* $Req = (\mathcal{T}_o, \mathcal{T}_e, \mathcal{T}_t)$ *such that:* $\forall o \in \mathcal{T}_o : (\exists e, t : e \in \mathcal{T}_e \wedge t \in \mathcal{T}_t \wedge \rho(e) = (o, t))$.

```
1    <CREATE> <CONSTRAINT> name <ON> origin_class
2          <REQUIRED_EDGE> [edge_type] <TO> target_class
```

Fig. 2. General syntax of the required edge constraint.

In/out Constraint. This constraint type restricts the classes of nodes that are connected by an edge class. At the same time, this constraint is also useful to enforce the direction of the represented relationship. For example, a person authors a document, and not the other way around. Unlike the previous constraints, this one is associated to the edge class metadata instead of the node class. This IC can be formally specified as shown by Definition 4 and the general syntax is shown in Fig. 3.

Definition 4. *Given a property graph* $G = (N, E, \rho, \lambda_N, \lambda_E, \sigma)$, *an In/Out constraint is a tuple* $IO = (\mathcal{T}_o, \mathcal{T}_e, \mathcal{T}_t)$ *such that:* $\forall e \in \mathcal{T}_e : o \in \mathcal{T}_o \wedge t \in \mathcal{T}_t$, *where* $\rho(e) = (o, t)$.

```
1    <CREATE> <CONSTRAINT> name <ON> edge_type <IN_OUT_EDGE> (
2        <FROM> origin_class <TO> target_class | <FROM> origin_class
3        | <TO> target_class )
```

Fig. 3. General syntax of the in/out constraint.

Cardinality Constraint. The goal of this IC is to restrict the number of edges of a given class that connect an origin node to target nodes. The cardinality specification consists of two integer numbers, with N serving as a placeholder for "unspecified", i.e., any number is allowed. Unlike in/out constraints, cardinality constraints are associated with the origin node class and therefore do not allow an unspecified origin node class. Another important difference between both constraints is the form of validation. Since OrientDB associates node identification numbers with classes, in/out validation can be implemented performing no reads on the DB. Figure 4 shows the general syntax of this constraint type.

The constraint is always associated with the origin node class and to the edge class. After specification of the cardinality, one may optionally specify the destination node class. This constraint is formally specified by Definition 5.

Definition 5. *Given a property graph* $G = (N, E, \rho, \lambda_N, \lambda_E, \sigma)$, *a cardinality constraint is a tuple* $Card = (T_o, T_e, T_t, \alpha \in \mathcal{O}, k_o \in K \wedge k_t \in K)$ *such that:* $\forall o \in T_o : ||\{(e,t) \mid (e,t) \in (T_e \times T_t) \wedge \rho(e) = (o,t)\}|| \; \alpha \; k_o \wedge ||\{(e,o) \mid (e,o) \in (T_e \times T_o) \wedge \rho(e) = (t,o)\}|| \; \alpha \; k_t$.

```
1    <CREATE> <CONSTRAINT> name <ON> origin_class <CARDINALITY>
2      [edge_type] <(> <INT | N> <..> <INT | N> <)> [<TO> target_class ]
```

Fig. 4. General syntax of the cardinality constraint.

Bidirectional Edge Constraint. The purpose of this IC is to ensure bidirectionality of the direction of a relationship. That is, given two nodes, if there is an edge from A to B, there must be another edge of the same type linking B to A. The constraint is associated with the edge metadata and its validation verifies if the specified nodes are linked by an incoming and an outgoing edge of the required type. This constraint is formally described by Definition 6 and its general syntax is shown in Fig. 5.

Definition 6. *Given a property graph* $G = (N, E, \rho, \lambda_N, \lambda_E, \sigma)$, *a Bidirectional Edge constraint is a tuple* $BiDir = (T_o, T_e, T_t)$ *such that the following two conditions hold:*

$$\forall \, e \in T_e : \quad \exists \, \bar{e} : \quad \rho(e) = (o,t) \implies \rho(\bar{e}) = (t,o) \; \wedge \; t \in T_t \; \wedge \; o \in T_o$$
$$\forall \, \bar{e} \in T_e : \quad \exists \, e : \quad \rho(\bar{e}) = (t,o) \implies \rho(e) = (o,t) \; \wedge \; o \in T_o \; \wedge \; t \in T_t$$

```
1    <CREATE> <CONSTRAINT> name <ON> edge_type <BIDIRECTIONAL_EDGE>
2      <BETWEEN> origin_class <AND> target_class
```

Fig. 5. General syntax of the bidirectional edge constraint.

Edge Condition Constraint. The goal of this IC is to enforce a condition over attributes in both ends of an edge. In other words, it compares property values of the two nodes connected by one edge. It is associated with the edge metadata and also enforces the direction of the relationship. It can be formally defined as shown in Definition 7 and its syntax is presented in Fig. 6.

Definition 7. *Given a property graph* $G = (N, E, \rho, \lambda_N, \lambda_E, \sigma)$, *an edge condition constraint is a tuple* $ECond = (\mathcal{T}_e, p_o \in P, p_t \in P, \alpha \in O)$, *such that:*
$$\forall e \in \mathcal{T}_e \; : \; \rho e = (o, t) \implies \sigma(o, p_o) \; \alpha \; \sigma(t, p_t)$$

```
1    <CREATE> <CONSTRAINT> name <ON> edge_type <EDGE_CONDITION>
2        origin_class<.>property (>|<|>=|<=|=|!=) target_class<.>property
```

Fig. 6. General syntax of the edge condition constraint.

Since all constraints are built on top of the formalization in Definition 1, which is aligned with several other graph DBMSs, the constraints are applicable to those DBMSs as well. In general, applying the method to other DBMS involves modifying the host DBMS for three tasks: parsing the constraints, storing them and validating them. Parsing involves extending the DDL or creating one with the IC syntax presented in this work, so the DB recognizes the CREATE CONSTRAINT commands. The presented implementation of the constraint manager uses an SB-Tree to store and manage constraint objects because this is a general use index algorithm already provided by OrientDB. In other DBMSs, this structure could be replaced by similar indexing algorithms, such as B-Trees. In addition, a trigger to the validation routines must be implemented after procedures that recognize the create, delete and update commands of the DBMS.

4 Evaluation

The support for ICs proposed in this paper was implemented on OrientDB 3.1.0. To evaluate the impact of the modifications, the most relevant and quantifiable factor is the execution time of operations in the DBMS. Therefore, this evaluation focuses on the impact of IC validation on query execution time.

The strategy adopted assumes that ICs are originated from business modeling and not only from implementation aspects. As a result, when a developer selects a weakly consistent DBMS or one without support for enforcing ICs, constraint validation must be enforced by logic introduced in the application code. That way, the experiments are built to allow the comparison of three variants of the OrientDB. The first one is the *Original* OrientDB server, without any modification. The *Modified* variant consists in a version of OrientDB modified to incorporate the constraint manager and to accept the definition of the six new ICs described in the previous section. Finally, the *Application* variant corresponds to performing IC validation within the client application and sending the corresponding data manipulation commands to the original OrientDB server.

The source code for reproducing the experiments is available in a public repository[3]. Countermeasures were adopted to avoid spurious interference on the results provoked by the Java Virtual Machine (JVM). A new pair of JVMs is created for every measurement: one JVM executes the OrientDB server while the

---
[3] https://bitbucket.org/fmreina/orient-driver/src/master/.

other executes the test client code (`constraints-tests-client`), which sends the test transaction to the server and measures its execution time. Both JVMs are created on the same host – an Intel i7-4510U dual-core at 2.0 GHz, with 16 GB of RAM, running Ubuntu 18.04.3 LTS 64bit (kernel 5.0). A single measurement consists of two phases, each executed on a new pair of client and server JVMs. In the first phase, the client sets up the DB within OrientDB using administration commands. This setup includes basic schema information, including ICs, and population of the DB where applicable. On the second phase, that is executed on a new pair of client/server JVMs, 100 analogous transactions are executed 3 times. The number of transactions aims to avoid unreliable measurement of fast operations and the 3 executions aim to avoid interference from non-deterministic background tasks, such as the GC (Garbage Collector), disk caching and JIT (Just In Time) compilation. Between each of these 3 executions, disk caches are flushed (using the `sync()` system call) and the GC is requested to run. Only the third execution had its time recorded and was considered for analysis.

The first scenario evaluates the *Node Condition* constraint. In this experiment, each transactions tries to modify the values of properties of one instance of class *Person*, violating the constraint. The constraint imposed in this first experiment is shown in Fig. 7.

The next three scenarios evaluate the *Required Edge*, *In/out* and *Cardinality* constraints. Figure 8(a) presents the constraints created in these three scenarios. The DB was populated with 100 replicas of the structure shown in Fig. 8(b); then, 100 transactions are executed sequentially, resulting in the structure shown in Fig. 8(c). The *Required Edge* IC is created to enforce the existence of at least one edge of type *owns* between nodes *Person* and *Company*. In the *In/Out* case, the IC is created on the edge class *owns* to specify that edges of this type are

```
1  CREATE CONSTRAINT cond ON Person (attr1) CONDITIONAL
2    (IF attr2 < 3 THEN attr1 < 2 ELSE attr1 > 4);
```

Fig. 7. Node Condition constraint created in the first experiment.

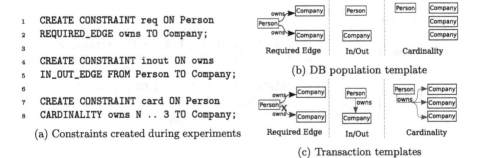

```
1  CREATE CONSTRAINT req ON Person
2    REQUIRED_EDGE owns TO Company;
3
4  CREATE CONSTRAINT inout ON owns
5    IN_OUT_EDGE FROM Person TO Company;
6
7  CREATE CONSTRAINT card ON Person
8    CARDINALITY owns N .. 3 TO Company;
```

(a) Constraints created during experiments

(b) DB population template

(c) Transaction templates

Fig. 8. Constraints (a), models for DB population (b) and transactions (c).

only valid if they have a node *Person* as source and a node *Company* as target. Finally, the *Cardinality* IC is created on the node class *Person* to limit the number of nodes of type *Company* the same person can own.

The transaction that tests the *Required Edge* scenario, due to the nature of this IC, performs an edge removal rather than an insertion. In the *Original* and *Modified* variants, a single command is sent to OrientDB per transaction. In the *Application* variant, this is not possible. Therefore, the application performs a *SELECT* operation for each node to validate the existing edge cardinality, and then sends a command that creates the edges between the nodes. The same approach is adopted for the *Application* variant in all scenarios, as it is necessary to validate the IC before actually performing the intended operation.

The last experiments evaluate the other two ICs proposed in this paper: bidirectional edge and edge condition. Figure 9(a) illustrates the transactions for each scenario where new edges are created meeting the constraint requirements. As with the first batch, this operations can be executed with a single command in the variants *Modified* and *Original*. However, in the variant *Application*, it is necessary to perform more operations to query the involved nodes and run the validation routine before the command to create the edges is sent to the DB.

The results obtained with the three variants (*i.e., Original, Modified* and *Application*) are illustrated in Fig. 10. In this figure, narrow boxes with white background represent the central quartiles, totaling 50% of measurements and are divided into the median. For the limits, from which outliers are found as points, we used $min(Maximum, Median \pm 1.5IQR)$, where IQR is the *Inter-Quartile Range*, to the height of the boxes. In addition to the classic elements of a box diagram, the average, represented by triangles, and the 95% confidence interval (CI), represented by wide filled rectangles, have been added to the illustration. The labels on the bars show the increase of the runtime as a proportion of the variation *Original*.

The first box diagram in Fig. 10 presents the results for the *Node Condition* constraint. There is a performance cost for using the *Modified* OrientBD to impose the constraint, but this cost is significantly lower than doing the required validation at application level.

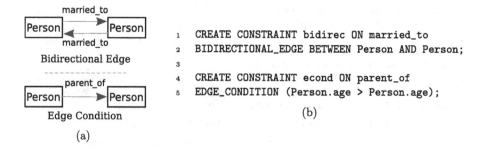

(a)

(b)

Fig. 9. Models for DB population and transactions (a), and ICs created (b).

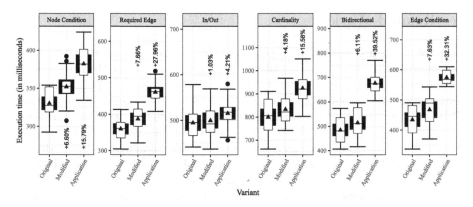

Fig. 10. Impact of constraint checking on the average time required to execute a batch of 100 write transactions.

The second, third and fourth plots show the runtime of the *Required Edge*, *In/out* and *Cardinality* scenarios using the three variants mentioned before. In all scenarios it is also observed that IC validation using the modified OrientDB is, on average, more efficient than validation at the application. Another important conclusion is that there is a large intersection between the performance observed with the *Original* OrientDB and the time with the *Modified* variant. This wide intersection prevents the differences in performance from being considered statistically significant.

The last two plots in Fig. 10 present the execution time of transactions in scenarios with the *Bidirectional Edge* and *Edge Condition* constraints. In these two scenarios, the same behavior of the previous ICs is also observed. The validation in the modified version of OrientDB is equally more efficient than the *Application* variant, since the CIs do not intersect. Comparing the other two variants, *Original* and *Modified*, the latter presents a higher positive skew, however the intersection between them is large as well.

Although the *Application* variant has, in all cases, shown worse results than the *Modified* OrientDB, there are scenarios where this difference stands out. In the *Required Edge, Cardinality, Bidirectional Edge* and *Edge Condition* scenarios, additional database access is required for the application to retrieve the information necessary to perform data validation. New tests with significantly larger graphs and more complex constraints are planned as future work.

The experiments assume that the client application is capable of ensuring adequate concurrency control in order to preserve IC consistency, which can be difficult to achieve in practical scenarios. Two instances of the application can concurrently validate an IC and then concurrently perform operations that together violate the IC. In such situations, where there are applications modifying the graph concurrently, distributed concurrency control techniques should be applied. As a result, there will be greater complexity of implementation, causing additional impacts on applications beyond those shown in Fig. 10.

5   Related Work

Some of the characteristics of graph DBs, such as being schema-less and follow-ing the BASE properties, make them more flexible and allow them to provide better performance. On the other hand, the same aspects are responsible for the lack of consistency that may be fundamental for some categories of application. However, it is important to remember that a strong schema definition may dete-riorate the performance of the DBMS. Thus, the challenge is to find a solution for the lack of consistency without impairing flexibility and performance. Most of the proposals found in the literature that address ICs are developed for relational DBs. Among the few that discuss the graph model, many of them only compare the available implementations, while others suggest supporting new constraints.

Pokorný [10] presents a general overview of graph DBs, covering storage, query, scalability, transaction processing, categories of graph DBs and their lim-itations. Among the limitations, Pokorný [10] lists features that are not entirely supported by current graph DBMS, such as data partitioning capacity, support for declarative queries, vector operations, and model restrictions that could make possible the definition of data schema. Within the topic of model constraints, ICs play a central role but are not well supported by graph DBMSs.

Barik et al. [4] employ graph DBs to analyze possible attack paths of net-worked applications. Most of the discussion in this paper centers on the anal-ysis of vulnerabilities and attacks employing graphs. In their analysis, the pre-conditions necessary to perform an attack form a dependency graph. Therefore, an attack is seen as a progression that satisfies the dependencies. The authors argue that the use of ICs assists the process of generating an attack graph. Thus, they propose an extension of Neo4j[4] in order to create a constraint layer that allows ICs of unique values, primary and foreign keys, value range, in/out (known as edge model) and edge cardinality.

In a comparison between relational and graph databases, Pokorný [11] lists characteristics of relational DBMSs that are not currently supported by graph DBMSs. One of such features is the explicit schema definition, including the specification of ICs. This absence makes verifying accuracy of graph DBs more difficult. Pokorný [11] argues that graph DBs are based on a logical model that has three components: i) a set of types and data structures; ii) a set of inference operators; iii) a set of ICs. Actual graph DBMSs typically lack at least one of these three components, with ICs usually being the missing component. Later, in [12], the authors suggest the definition of ICs in the conceptual or the DB level. For this, they considered property data types, property value ranges, class disjunction (i.e., a node cannot belong to two classes simultaneously), mandatory edges and unique values for a property composition. Support for these ICs is added to Neo4j through the extension of its Cypher language.

Roy-Hubara et al. [13] discuss data modeling and a schema definition. The authors present an approach based on the entity-relationship (ER) model of the application domain and create a mapping from the ER model for a graph DB,

---

[4] https://neo4j.com/.

along with using a DDL. They argue that the approach could be applied to any graph DB, but do not show nor refer to any implementation.

Lastly, Angles [2] attempts to find common theoretical grounds for the plethora of graph models implemented by several graph DBMSs. The author adopts property graphs as the starting point and provides a logical formalization, including the notion of a schema. This basic notion of schema is extended with ICs and the syntax and semantics of a unified query language are described. All ICs are defined and discussed from a theoretical standpoint, without discussing their support in existing graph DBMSs.

Table 1 shows which ICs are natively supported by Neo4j and which are added to it by the extensions proposed in [4] and [12]. In the case of OrientDB, the table only shows those natively supported by OrientDB and those added by this work, since it has not received such extensions before. In the table, empty circles represent ICs that are natively supported by the DBMS, while filled circles denote the new ICs proposed by the work referenced on the table header. Those that are natively supported but are also repeated with filled circles were re-implemented or improved by the work referenced in the table header.

Some of the ICs that appear in Table 1 require further discussion. In an relational DB, foreign keys are used to represent relationships between tuples from distinct tables. In a graph DB, this can be considered an anti-pattern, since instead of using a foreign key property, one should use edges to represent relationships between nodes, which correspond to tuples in the relational model. Using foreign keys with properties to maintain relationships will yield more complex queries and lead to inefficient query processing, since graph DBMSs are not designed to handle such kind of property joins. The primary key constraint, mentioned by some authors, can be replaced by the simultaneous definition of unique and required ICs. If primary keys are used to maintain referential integrity, one falls into the same anti-pattern of using foreign key constraints. Similarly, composite primary keys can be obtained by combining composite unique constraints with a required property constraint for each component property.

Class disjointness ICs disallow membership of a single node to two or more classes. In the case of OrientDB, this IC is a design decision of the DBMS itself and every node must belong to a single node class. Therefore, this IC does not apply to OrientDB in its literal sense. In contrast, one may use a node property to store the "category" or "class" of a node. If multiple classifications are desired, one may model the classes as nodes and model membership as an edge from the instance to the class. If a single extra classification is allowed, one may use a single property and limit its value range using maximum/minimum constraints or using RegEx ICs. A RegEx (short for Regular Expression) is a string that compactly describes a whole set of allowed values. RegExes can be used to describe also sets of allowed values, such as the expression `adult|minor` which accepts only two possible values: "adult" and "minor".

Table 1. Comparison between native ICs and proposed extensions.

Constraint types	Neo4J	Barik et al.	Pokorny et al.	OrientDB	This work
Bidirecional edge					●
Cardinality of Edges		●			●
Class Disjunction		●	●	○ (1)	○ (1)
Edge Condition					●
Node Condition					●
Foreign Keys		●			
In/Out (EndPoint)		●			●
Primary Keys	○	○	○	○ (2)	○ (2)
Property Type			●	○	○
Range of property		●	●	○	○
RegEx				○	○
Required Edge	○	○	●		●
Required Property	○	○	○	○	○
Unique	○	●	○	○	○
Unique - Composed			●	○	○

(1): OrientDB enforces a single label (class) per node.
(2): The primary key constraint is equivalent to a composition of the required property and unique property.

6   Conclusions

Due to the large volume of data produced continuously by computing systems, that resulted in the phenomena called *Big data*, new solutions for managing and storing data have been developed. The greatest motivation for such development comes from the fact that traditional technologies for data storage and management are unable to meet the performance and scalability requirements demanded by most of the novel data-intensive applications that have emerged recently. New data models have been adopted, allowing large volumes of data to be handled, resulting in a fast adoption of these technologies in the software market. In this context, graph DBs gained popularity due to their ability to easily represent data used by several applications, for which the graph data model suits perfectly. However, this category of DBMS still lacks effective mechanisms to enforce the integrity of the stored data. Therefore, this work proposed extending a graph DBMS, adding support for 6 new ICs: node condition, required edge, in/out, edge cardinality, bidirectional edge and edge condition. As future work, we plan to extend the number of supported constraints, including ICs that appear in the related work and in [7], and also to evaluate their efficiency and relevance in comparison with the constraints that are implemented by other works and in the current DBMSs.

An evaluation study compared the original OrientDB version, a modified version with added support for the new ICs and a third case in which data validation is done by the client application. Experiments demonstrated that the modified OrientDB presented a small increase in the execution time of data manipulation operations, having the original version as baseline. This increase, though, is not big enough to be considered statistically significant for five of the new ICs. Only in one case – the node condition constraint – there is a noticeable, but still small, performance loss. However, the modified version of OrientDB is significantly faster than performing validations at the application level. In addition, the resulting implementation also simplifies the development of client applications, which can skip data validation checks and leave them to be done by the DBMS.

The same strategy could also be adopted to add support for new ICs to other graph DBs. Furthermore, despite adding support for only six ICs, the proposed solution allows the easy addition of other constraints, with the aim of further extending the mechanisms that can guarantee the integrity of graph DBs.

References

1. Amin, M.S., Yan, B., Sriram, S., Bhasin, A., Posse, C.: Leveraging a social graph to deliver relevant recommendations, August 2019. https://patents.google.com/patent/US10380629B2/en
2. Angles, R.: The property graph database model. In: Proceedings of the 12th Alberto Mendelzon International Workshop on Foundations of Data Management, Cali, Colombia, 21–25 May 2018 (2018). http://ceur-ws.org/Vol-2100/paper26.pdf
3. Angles, R., Gutierrez, C.: Survey of graph database models. ACM Comput. Surv. **40**(1), 1:1–1:39 (2008). https://doi.org/10.1145/1322432.1322433
4. Barik, M.S., Mazumdar, C., Gupta, A.: Network vulnerability analysis using a constrained graph data model. In: Ray, I., Gaur, M.S., Conti, M., Sanghi, D., Kamakoti, V. (eds.) ICISS 2016. LNCS, vol. 10063, pp. 263–282. Springer, Cham (2016). https://doi.org/10.1007/978-3-319-49806-5_14
5. Corbellini, A., Mateos, C., Zunino, A., Godoy, D., Schiaffino, S.: Persisting big-data: the NoSQL landscape. Inf. Syst. **63**, 1–23 (2017). http://www.sciencedirect.com/science/article/pii/S0306437916303210
6. van Erven, G.C.G., Holanda, M., Carvalho, R.N.: Detecting evidence of fraud in the Brazilian government using graph databases. In: Rocha, Á., Correia, A.M., Adeli, H., Reis, L.P., Costanzo, S. (eds.) Recent Advances in Information Systems and Technologies, pp. 464–473. Springer, Cham (2017)
7. Ghrab, A., Romero, O., Skhiri, S., Vaisman, A., Zimányi, E.: Grad: on graph database modeling (2016). https://arxiv.org/ftp/arxiv/papers/1602/1602.00503.pdf
8. Ip, H., Tang, H.: Parallel evidence combination on a SB-tree architecture. In: 1996 Australian New Zealand Conference on Intelligent Information Systems. Proceedings. ANZIIS 96. IEEE (1996). https://doi.org/10.1109/anziis.1996.573882
9. Navathe, S.B., Elmasri, R.: Fundamentals of Database Systems, 7th edn. Pearson (2016)
10. Pokorný, J.: Graph databases: their power and limitations. In: Saeed, K., Homenda, W. (eds.) CISIM 2015. LNCS, vol. 9339, pp. 58–69. Springer, Cham (2015). https://doi.org/10.1007/978-3-319-24369-6_5

11. Pokorný, J.: Conceptual and database modelling of graph databases. In: Proceedings of the 20th International Database Engineering Applications Symposium, IDEAS 2016, pp. 370–377. ACM, New York (2016). https://doi.org/10.1145/2938503.2938547

12. Pokorný, J., Valenta, M., Kovačič, J.: Integrity constraints in graph databases. Procedia Comput. Sci. **109**, 975–981 (2017). http://www.sciencedirect.com/science/article/pii/S1877050917311390, 8th International Conference on Ambient Systems, Networks and Technologies, ANT-2017 and the 7th International Conference on Sustainable Energy Information Technology, SEIT: 16–19 May 2017. Madeira, Portugal (2017)

13. Roy-Hubara, N., Rokach, L., Shapira, B., Shoval, P.: Modeling graph database schema. IT Professional **19**(6), 34–43 (2017). https://doi.org/10.1109/MITP.2017.4241458

14. Sandhu, R.S.: On five definitions of data integrity. In: Proceedings of the IFIP WG11.3 Working Conference on Database Security VII. pp. 257–267. North-Holland Publ., Amsterdam, The Netherlands (1994)

15. Shirasaki, Y., Kobayashi, Y., Aoyama, M.: A speech data-driven stakeholder analysis methodology based on the stakeholder graph models. In: 2019 IEEE 43rd Annual Computer Software and Applications Conference (COMPSAC), vol. 2, pp. 213–220, July 2019. https://doi.org/10.1109/COMPSAC.2019.10209

16. de Souza, V.C.O., dos Santos, M.V.C.: Maturing, consolidation and performance of NOSQL databases: Comparative study. In: Proceedings of the Annual Conference on Brazilian Symposium on Information Systems: Information Systems: A Computer Socio-Technical Perspective, vol. 1, pp. 32:235–32:242. SBSI 2015, Brazilian Computer Society, Porto Alegre, Brazil, Brazil (2015), http://dl.acm.org/citation.cfm?id=2814058.2814097

17. Stokebraker, M.: SQL databases v. NoSQL databasesd. Commun. ACM **53**, 10–11 (2010)

18. Vaz, R.V., de Oliveira, J.D.Q., Ribeiro, L.A.: Duplicate management using graph database systems: a case study. In: Proceedings of the XV Brazilian Symposium on Information Systems, SBSI 2019, pp. 50:1–50:8. ACM, New York (2019). https://doi.org/10.1145/3330204.3330260

19. Yamaguchi, F., Golde, N., Arp, D., Rieck, K.: Modeling and discovering vulnerabilities with code property graphs. In: 2014 IEEE Symposium on Security and Privacy, pp. 590–604, May 2014. https://doi.org/10.1109/SP.2014.44

Bounded Pattern Matching Using Views

Xin Wang[1(✉)], Yang Wang[1], Ji Zhang[2], and Yan Zhu[3]

[1] Southwest Petroleum University, Chengdu, China
xinwang.ed@gmail.com, wangyang@swpu.edu.cn
[2] University of Southern Queensland, Toowoomba, Australia
Ji.Zhang@usq.edu.au
[3] Southwest Jiaotong University, Chengdu, China
yzhu@swjtu.edu.cn

Abstract. Bounded evaluation using views is to compute the answers $Q(\mathcal{D})$ to a query Q in a dataset $\mathcal{D}$ by accessing only cached views and a small fraction D_Q of $\mathcal{D}$ such that the size $|D_Q|$ of D_Q and the time to identify D_Q are independent of $|\mathcal{D}|$, no matter how big $\mathcal{D}$ is. Though proven effective for relational data, it has yet been investigated for graph data. In light of this, we study the problem of *bounded pattern matching using views*. We first introduce *access schema* $\mathcal{C}$ for graphs and propose a notion of *joint containment* to characterize *bounded pattern matching using views*. We show that a pattern query Q can be boundedly evaluated using views $\mathcal{V}(G)$ and a fraction G_Q of G *if and only if* the query Q is *jointly contained* by $\mathcal{V}$ and $\mathcal{C}$. Based on the characterization, we develop an efficient algorithm as well as an optimization strategy to compute matches by using $\mathcal{V}(G)$ and G_Q. Using real-life and synthetic data, we experimentally verify the performance of these algorithms, and show that (a) our algorithm for joint containment determination is not only effective but also efficient; and (b) our matching algorithm significantly outperforms its counterpart, and the optimization technique can further improve performance by eliminating unnecessary input..

1 Introduction

With the advent of massive scale data, it is very urgent to have effective methods for query evaluation on large scale data. One typical solution is by means of *scale independence* [8,9], whose idea is to compute the answers $Q(\mathcal{D})$ to a query Q in a dataset $\mathcal{D}$ by accessing a small fraction D_Q of $\mathcal{D}$ with bounded size, no matter how big the underlying $\mathcal{D}$ is. Following the idea, [13,14] show that nontrivial queries can be *scale independent* under a set $\mathcal{C}$ of *access constraints*, that are a form of cardinality constraints with associated indices; [13,14] also refer to a query Q as *boundedly evaluable* if for all datasets $\mathcal{D}$ that satisfy $\mathcal{C}$, $Q(\mathcal{D})$ can be evaluated from a fraction D_Q of $\mathcal{D}$, such that the time for identifying and fetching D_Q and the size $|D_Q|$ of D_Q are independent of $|\mathcal{D}|$. Still, many queries are not *boundedly evaluable*, hence *bounded evaluation with views* was proposed by [10], which is to select and materialize a set $\mathcal{V}$ of small views, and answer

© Springer Nature Switzerland AG 2020
S. Hartmann et al. (Eds.): DEXA 2020, LNCS 12391, pp. 285–303, 2020.
https://doi.org/10.1007/978-3-030-59003-1_19

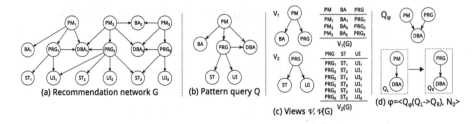

Fig. 1. Graph G, pattern Q, views $\mathcal{V}$, $\mathcal{V}(G)$ and access constraint φ

Q on $\mathcal{D}$ by using cached views $\mathcal{V}(\mathcal{D})$ and an additional small fraction D_Q of $\mathcal{D}$. Then, the queries that are not *boundedly evaluable* can be efficiently answered with views and a small fraction of original data with bounded size.

Bounded evaluation with views have proven effective for querying relational data [11], but the need for studying the problem is even more evident for graph pattern matching (GPM), since (a) GPM has been widely used in social analysis [28] which is becoming increasingly important nowadays; (b) it is a challenging task to perform graph pattern matching on real-life graphs due to their sheer size; and (c) view-based matching technique is often too restrictive. Fortunately, *bounded pattern matching using views* fills this critical void. Indeed, cardinality constraints are imposed by social graphs, *e.g.*, on LinkedIn, a person can have at most 30000 connections and most of people have friends less than 1000 [3]; on Facebook, a person can have no more than 5000 friends [6], etc. Given the constraints and a set of well-chosen views $\mathcal{V}$ along with their caches $\mathcal{V}(G)$ on graphs G, GPM can be evaluated by using the views plus a small fraction G_Q of G of bounded size, no matter how large G is.

Example 1. A fraction of a recommendation network G is shown in Fig. 1(a), where each node denotes a person with job title (*e.g.*, project manager (PM), business analyst (BA), database administrator (DBA), programmer (PRG), user interface designer (UI) and software tester (ST)); and each edge indicates collaboration, *e.g.*, (PM_1, PRG_1) indicates that PRG_1 worked well with PM_1 on a project led by PM_1.

To build a team for software development, one issues a pattern query Q depicted in Fig. 1(b). The team members need to satisfy the following requirements: (1) with expertise: PM, BA, DBA, PRG, UI and ST; (2) meeting the following collaborative experience: (i) BA, PRG and DBA worked well under the project manager PM; and (ii) DBA, ST and UI have been supervised by PRG, and collaborated well with PRG. It is then a daunting task to perform graph pattern matching since it takes $O(|G|!|G|)$ time to identify all the isomorphic matches of Q in G [12], where $|G| = |V| + |E|$ indicates the size of G.

While one can do better by *bounded pattern matching using views*. Suppose that (1) a set of views $\mathcal{V} = \{V_1, V_2\}$ is defined and cached ($\mathcal{V}(G) = \{V_1(G), V_2(G)\}$) as shown in Fig. 1(c), and (2) there exists an *access constraint* $\varphi = \langle Q_\varphi(Q_L \rightarrow Q_R), N_0 \rangle$ (Fig. 1(d)) that G satisfies. Here φ states that for each DBA that has been supervised by a PM, he can be supervised by at most N_0 distinct PRG. One may associate φ with an index $\mathcal{I}_\varphi$ for fast access, then a fraction G_Q of G of bounded size can be efficiently constructed and $Q(G)$ can be answered by using $\mathcal{V}(G)$ and G_Q (will be elaborated). Since (a) $\mathcal{V}(G)$ already contains partial answers to Q in G, and (b) $\mathcal{V}(G)$, G_Q are often much smaller than G, thus the cost for computing $Q(G)$ can be substantially reduced.   □

This example suggests that we perform pattern matching by using views $\mathcal{V}$, $\mathcal{V}(G)$ and a fraction G_D of original graph G of bounded size. In doing so, two key issues have to be settled. (1) How to decide whether a pattern query Q is *boundedly evaluable with views*? (2) How to efficiently compute $Q(G)$ with $\mathcal{V}(G)$ and G_Q?

Contributions. This paper investigates the aforementioned questions. We focus on graph pattern matching with *subgraph isomorphism* [12].

(1) We introduce *access schema* defined on graph data, and formalize the problem of *bounded pattern matching using views* (Sect. 2).

(2) We propose a notion of *joint containment* for determining whether a pattern query is *boundedly evaluable with views*. Given a pattern query Q, a set of views $\mathcal{V}$ and access schema $\mathcal{C}$ with indexes associated, we show that Q is *boundedly evaluable with views* if and only if Q is *jointly contained* by $\mathcal{V}$, $\mathcal{C}$ and provide an algorithm, that works in $O((||\mathcal{V}|| + ||\mathcal{C}||)|Q|!|Q|)$ time to determine *joint containment*, where $||\mathcal{V}||$ and $||\mathcal{C}||$ refer to the cardinality of $\mathcal{V}$ and $\mathcal{C}$, respectively, and $|Q|$ indicates the size of Q. As the cost of the algorithm is dominated by $||\mathcal{V}||$, $||\mathcal{C}||$ and $|Q|$, which are often small in practice, the algorithm hence very efficiently performs (Sect. 3).

(3) Based on *joint containment*, we develop an algorithm to evaluate graph pattern matching by using $\mathcal{V}(G)$ and G_Q (Sect. 4). Given a pattern query Q, a set of views $\mathcal{V}$ and its extension $\mathcal{V}(G)$ on a graph G, and an access schema $\mathcal{C}$ that G satisfies, the algorithm computes $Q(G)$ in $O((|\mathcal{V}|^{|Q|}|\mathcal{V}(G)|)^{||\mathcal{V}||}(|Q|^{|Q|} \cdot N_m)^{||\mathcal{C}||})$ time, *without accessing G at all*, when Q is *jointly contained* in $\mathcal{V}$ and $\mathcal{C}$. It is far less costly than the algorithm [12] that takes $O(|G|!|G|)$ time to evaluate Q directly on G, since $|Q|$, $|\mathcal{V}|$, $||\mathcal{V}||$, $||\mathcal{C}||$ are very small, and $|\mathcal{V}(G)|$ is typically *much smaller* than $|G|$ in practice. We also study the *minimum containment problem*, which is to find a pair of subsets $\langle \mathcal{V}', \mathcal{C}' \rangle$ of $\mathcal{V}$ and $\mathcal{C}$ such that Q is jointly contained by $\langle \mathcal{V}', \mathcal{C}' \rangle$ and moreover, the input used by the matching algorithm can be dramatically reduced.

(4) Using real-life and synthetic graphs, we experimentally verify the performances of our algorithms (Sect. 5). We find that (a) our algorithm for joint containment checking is very efficient, *e.g.*, taking only 145.5 ms to determine whether a pattern query is contained by a set of views; (b) our view-based matching algorithm is efficient: it is 9.7 times faster than conventional method on *Youtube* [5] with 1.6 million nodes and 4.5 million edges; (c) our optimization technique is effective: it can reduce the size of input by 75% and improve the efficiency by 143%, on average, over real-life graphs; and (d) our matching algorithm scales well with the data size.

In a summary of the scientific contributions, this work gives a full treatment for *bounded pattern matching using views*, which fills one critical void for graph pattern matching on big graphs, and yields a promising approach to querying "big" social data. All the proofs, algorithms and complexity analyses can be found in [2].

Related Work. We next categorize related work as follows.

Query Answering Using Views. Answering queries using views has been well studied for relational data (see [7,18,22] for surveys), XML data [17,24,26], and [16]. This work differs from them in the following aspects: (i) we adopt subgraph isomorphism as the semantic of pattern matching, instead of graph simulation [20] and bounded simulation [15], that are applied by [16]; (ii) we study a more practical problem to answer graph pattern matching using available views and a small fraction of graph G with bounded size; and (iii) we also investigate the problem of view selection, and provide effective technique for this problem.

Scale Independence. The idea of scale independence, *i.e.*, querying dataset $\mathcal{D}$ by accessing only a bounded amount of data in $\mathcal{D}$, is proposed by [8–10]. Extending the idea with *access schema*, [13,14] introduced *bounded evaluation*. To cope with nontrivial queries, [10] proposed *bounded evaluation with views*, *i.e.*, evaluating queries that are not boundedly evaluable on a dataset $\mathcal{D}$ by accessing not only cached views $\mathcal{V}(\mathcal{D})$ but also a small fraction of $\mathcal{D}$ with bounded size. Furthermore, [11] explored fundamental problems of *bounded evaluation with views*. This work differs from [10,11] in that the query semantics are different, and we not only conduct static analysis for fundamental problems, but also provide effective technique for matching evaluation.

2 Preliminaries

In this section, we first review data graphs, pattern queries and graph pattern matching. We then introduce the problem of *bounded pattern matching using views*.

2.1 Basic Definitions

We start with basic notations: data graphs, pattern queries and graph pattern matching.

Data Graphs. A *data graph* is a node-labeled, directed graph $G = (V, E, L)$, where (1) V is a finite set of data nodes; (2) $E \subseteq V \times V$, where $(v, v') \in E$ denotes a *directed* edge from node v to v'; and (3) $L(\cdot)$ is a function such that for each node v in V, $L(v)$ is a label from an alphabet Σ. Intuitively, $L(\cdot)$ specifies *e.g.*, job titles, social roles, ratings, etc [21].

Pattern Queries. A *pattern query* (or shortened as pattern) is a directed graph $Q = (V_p, E_p, f_v)$, where (1) V_p is the set of *pattern nodes*, (2) E_p is the set of *pattern edges*, and (3) $f_v(\cdot)$ is a function defined on V_p such that for each node $u \in V_p$, $f_v(u)$ is a label in Σ.

Subgraphs and Sub-patterns. A graph $G_s = (V_s, E_s, L_s)$ is a subgraph of $G = (V, E, L)$, denoted by $G_s \subseteq G$, if $V_s \subseteq V$, $E_s \subseteq E$, and moreover, for each $v \in V_s$, $L_s(v) = L(v)$. Similarly, a pattern $Q_s = (V_{p_s}, E_{p_s}, f_{v_s})$ is *subsumed by* another pattern $Q = (V_p, E_p, f_v)$, denoted by $Q_s \subseteq Q$, if Q_s is a subgraph of Q, *i.e.*, $V_{p_s} \subseteq V_p$, $E_{p_s} \subseteq E_p$ and for each $u \in V_p$, $f_{v_s}(u) = f_v(u)$. We say Q_s a sub-pattern of Q when $Q_s \subseteq Q$.

Graph Pattern Matching [12]. A *match* of Q in G via *subgraph isomorphism* is a subgraph G_s of G that is isomorphic to Q, *i.e.*, there exists a *bijective function* h from V_p to the node set V_s of G_s such that (1) for each node $u \in V_p$, $f_v(u) = L(h(u))$ $(h(u) \in V_s)$; and (2) (u, u') is an edge in Q if and only if $(h(u), h(u'))$ is an edge in G_s.

We also use the following notations. (1) The match result of Q in G, denoted as $Q(G)$, is a set consisting of all the matches G_s of Q in G. (2) For a pattern edge $e = (u, u')$, we derive a set $S(e)$ from $Q(G)$ by letting $S(e) = \{(v, v')|v = h(u), v' = h(u'), h \in Q(G), (v, v') \in E\}$, and denote $S(e)$ as the match set of pattern edge e. (3) We use $Q \sim G_s$ to denote that G_s is a match of Q. (4) We denote $|V_p| + |E_p|$ as the size $|Q|$ of Q and $|V| + |E|$ as the size $|G|$ of G.

2.2 Problem Formulation

We next formulate the problem of *bounded pattern matching using views*. We start from notions of views, followed by access schema and problem statement.

Views. A *view* (*a.k.a.* view definition) V is also a pattern query. Its match result $V(G)$ in a data graph G is denoted as *view extension*, or *extension* when it is clear from the context [19]. As shown in Fig. 1(c), a set of views $\mathcal{V} = \{V_1, V_2\}$ are defined, with extensions $\mathcal{V}(G) = \{V_1(G), V_2(G)\}$ on G cached.

Access Schema. Extended from [11], we define *access schema* on graphs as follows. An *access schema* $\mathcal{C}$ is defined as a set of *access constraints* $\varphi = \langle Q_\varphi(Q_L \to Q_R), N \rangle$, where Q_φ is a pattern query, Q_L and Q_R are sub-patterns of Q_φ such that the union of edge sets of Q_L and Q_R equals to the edge set of Q_φ, and N is a natural number.

Given a graph G, a pattern Q_φ and its sub-pattern Q_L, a match G_L (resp. G_R) of Q_L (resp. Q_R) is denoted as a Q_L-value (resp. Q_R-value) of Q_φ in G. Then, we denote by $G_{[Q_\varphi:Q_R]}(Q_L \sim G_L)$ the set $\{G_R|G_R \subseteq G_s, G_R \in Q_R(G), G_L \subseteq G_s, G_s \in Q_\varphi(G)\}$, and write it as $G_{Q_R}(Q_L \sim G_L)$, when Q_φ is clear from the context.

A graph G *satisfies* the access constraint φ, if

○ for any Q_L-value G_L, $|G_{Q_R}(Q_L \sim G_L)| \leq N$; and
○ there exists a function (referred to as an *index*) that given a Q_L-value G_L, returns $\{G_s|G_s \in Q_\varphi(G), G_L \subseteq G_s\}$ from G in $O(N)$ time.

Table 1. A summary of notations

Symbols	Notations	Symbols	Notations
$G_s \subseteq G$ (resp. $Q_s \subseteq Q$)	G_s (resp. Q_s) is a subgraph (resp. sub-pattern) of G (resp. Q)	$Q \sqsubseteq \mathcal{V}$	Q is contained in $\mathcal{V}$
$Q(G)$	match result of Q in G	$Q \sqsubseteq_J [\mathcal{V}, \mathcal{C}]$	Q is jointly contained in $\mathcal{V}$ and $\mathcal{C}$
$S(e)$	match set of pattern edge e in G	Q_g	a containing rewriting of Q
$\mathcal{V} = \{V_1, \ldots, V_n\}$	a set of view definitions	$\bar{E}_g$	"uncovered edges" of Q_g in Q
$\mathcal{V}(G) = \{V_1(G), \ldots, V_n(G)\}$	a set $\mathcal{V}(G)$ of view extensions $V_i(G)$	$\|Q\|$ (resp. $\|V\|$)	size (total number of nodes and edges) of a pattern Q (resp. view definition V)
$\varphi = \langle Q_\varphi(Q_L \to Q_R), N \rangle$	access constraint	$\|Q(G)\|$ (resp. $\|V(G)\|$)	total size of matches of Q (resp. V) in G
$\mathcal{C} = \{\varphi_1, \cdots, \varphi_n\}$	access schema	$\|\mathcal{V}\|$ (resp. $\|\mathcal{Q}\|$)	total size of view definitions in $\mathcal{V}$ (resp. pattern queries in $\mathcal{Q}$)
N_m	maximum cardinality of access constraints in $\mathcal{C}$	$\|\mathcal{V}\|$ (resp. $\|\mathcal{C}\|$, $\|\mathcal{Q}\|$)	total number of V in $\mathcal{V}$ (resp. φ in $\mathcal{C}$, Q in $\mathcal{Q}$)
H_V^Q	the shadow of a view V in Q	$\|\mathcal{V}(G)\|$	total size of matches in $\mathcal{V}(G)$

Intuitively, an *access constraint* φ is a combination of a cardinality constraint and an index $\mathcal{I}_\varphi$ on Q_L for Q_R. It tells us that given any Q_L-value, there exist at most N distinct Q_R-values, and these Q_R-values can be efficiently fetched by using $\mathcal{I}_\varphi$. By using indices, we can also construct a fraction G_Q of G, whose size is bounded by $\Sigma_{\varphi_i \in \mathcal{C}} N_i \cdot |Q_{\varphi_i}|$. We refer to the maximum cardinality of access constraints in an access schema $\mathcal{C}$ as N_m. A graph G satisfies *access schema* $\mathcal{C}$, denoted by $G \models \mathcal{C}$, if G satisfies all the access constraints φ in $\mathcal{C}$.

Example 2. An *access constraint* φ is shown in Fig. 1(d). By definition, pattern Q_φ takes two edges that are from its sub-patterns Q_L and Q_R, respectively. Assume that an index $\mathcal{I}_\varphi$ is constructed on G (Fig. 1(a)), then given a Q_L-value (PM$_2$, DBA$_1$), one can fetch from $\mathcal{I}_\varphi$ a set of matches of Q_φ, *i.e.*, $\{(\mathsf{PM}_2, \mathsf{DBA}_1, \mathsf{PRG}_1), (\mathsf{PM}_2, \mathsf{DBA}_1, \mathsf{PRG}_2)\}$. One may further verify that $|G_{Q_R}(Q_L \sim G_L)| = 2$. □

Bounded Pattern Matching Using Views. Given a pattern query Q, a set $\mathcal{V}$ of view definitions and an *access schema* $\mathcal{C}$, *bounded pattern matching using views* is to find another query $\mathcal{A}$ such that for any graph G that satisfies $\mathcal{C}$,

- $\mathcal{A}$ is equivalent to Q, *i.e.*, $Q(G) = \mathcal{A}(G)$; and
- $\mathcal{A}$ only refers to views $\mathcal{V}$, their extensions $\mathcal{V}(G)$ in G and G_Q only, without accessing original graph G. Here G_Q is a fraction of G and can only be constructed with indexes $\mathcal{I}_\varphi$, that are associated with access constraints $\varphi \in \mathcal{C}$, such that the time for generating G_Q is in $O(\Sigma_{\varphi_i \in \mathcal{C}} N_i)$, and the size $|G_Q|$ of G_Q is bounded by $O(\Sigma_{\varphi_i \in \mathcal{C}} N_i \cdot |Q_{\varphi_i}|)$.

If such an algorithm $\mathcal{A}$ exists, we say that pattern Q is *boundedly evaluable with views*, and can be evaluated using $\mathcal{V}(G)$ and a fraction G_Q of G of bounded size, no matter how big G is (Table 1).

3   Characterization for Bounded Pattern Matching Using Views

We propose a characterization for *bounded pattern matching using views, i.e.*, a sufficient and necessary condition for deciding whether a pattern query is *boundedly evaluable with views*.

3.1   Joint Containment Problem

We first introduce the notion of *joint containment*.

Joint Containment. A pattern query Q with edge set E_p is *jointly contained* by a set of views $\mathcal{V} = \{\mathsf{V}_1, \cdots, \mathsf{V}_n\}$, and a set of access constraints $\mathcal{C} = \{\varphi_1, \cdots, \varphi_k\}$, denoted by $Q \sqsubseteq_J [\mathcal{V}, \mathcal{C}]$, if for any graph G that satisfies $\mathcal{C}$, there exist a pair of mappings $\langle \lambda, \rho \rangle$, such that

- E_p is divided into two disjoint parts E_c and E_u;
- E_c is mapped via λ to powerset $\mathcal{P}(\bigcup_{i \in [1,n]} E_{\mathsf{V}_i})$, and $S(e) \subseteq \bigcup_{e' \in \lambda(e)} S(e')$ for any $e \in E_c$; and moreover,
- E_u is mapped via ρ to powerset $\mathcal{P}(\bigcup_{j \in [1,k]} E_{\varphi_j})$, and $S(e) \subseteq \bigcup_{e' \in \rho(e)} S(e')$ for any edge $e \in E_u$,

where E_{V_i} refers to the edge set of the i-th view definition V_i in $\mathcal{V}$ and E_{φ_j} indicates the edge set of Q_{φ_j} of the j-th access constraint φ_j in $\mathcal{C}$.

Intuitively, $Q \sqsubseteq_J [\mathcal{V}, \mathcal{C}]$ indicates that Q can be divided into two disjoint parts, that take edge sets E_c and E_u, respectively; and moreover, there exist mappings λ and ρ, that map E_c to edges in $\mathcal{V}$ and E_u to edges in $\mathcal{C}$, respectively, such that match set $S(e)$ can be derived from either $\mathcal{V}(G)$ or G_Q, for any e in Q, without accessing original graph G.

Example 3. Recall Example 1. One may verify that $Q \sqsubseteq_J [\mathcal{V}, \mathcal{C}]$, since the edge set of Q can be divided into two parts $E_c = \{(PM, BA), (PM, PRG), (PRG, ST), (PRG, UI)\}$ and $E_u = \{(PM, DBA), (PRG, DBA)\}$, that are mapped via mappings λ and ρ to the sets of edges in $\mathcal{V}$ and $\mathcal{C}$, respectively. For any graph G and any edge e of Q, the match set $S(e)$ must be a subset of the union of the match sets of the edges in $\lambda(e)$ or $\rho(e)$, *e.g.*, $S(PM, DBA)$ in G is $\{(PM_1, DBA_1), (PM_2, DBA_1), (PM_3, DBA_2)\}$, that is contained in the match set of Q_L of Q_φ.   □

Theorem 1. *Given a set of views $\mathcal{V}$ and an access schema $\mathcal{C}$, a pattern query Q is boundedly evaluable with views if and only if $Q \sqsubseteq_J [\mathcal{V}, \mathcal{C}]$.*

As introduced earlier, if a pattern Q is *boundedly evaluable with views*, Q can be answered by using $\mathcal{V}(G)$ and a small fraction G_Q of G with bounded size, no matter how big G is. Theorem 1 indeed shows that *joint containment* characterizes whether a pattern Q is *boundedly evaluable with views*. This motivates us to study the *joint containment* (JPC) problem, which is to determine, given a pattern Q, a view set $\mathcal{V}$ and an access schema $\mathcal{C}$, whether $Q \sqsubseteq_J [\mathcal{V}, \mathcal{C}]$.

Remarks. (1) A special case of *joint containment* is the pattern containment [27]. Indeed, when access schema $\mathcal{C}$ is an empty set, *joint containment* problem becomes pattern containment problem. Techniques in [27] is in place for answering graph pattern matching using views only. (2) When Q is not contained in $\mathcal{V}$ (denoted as $Q \not\sqsubseteq \mathcal{V}$), one may find another pattern query $Q_g = (V_g, E_g)$, referred to as a *containing rewriting* of Q w.r.t. $\mathcal{V}$, such that $Q_g \subseteq Q$ and $Q_g \sqsubseteq \mathcal{V}$, compute matches $Q_g(G)$ of Q_g from $\mathcal{V}(G)$ and treat $Q_g(G)$ as "approximate matches" [27]. While, we investigate more practical but nontrivial cases, *i.e.*, $Q \not\sqsubseteq \mathcal{V}$, and advocate to integrate access schema into view-based pattern matching such that exact matches can be identified by using a small portion of additional data of bounded size.

3.2   Determination of Joint Containment

To characterize *joint containment*, a notion of *shadow*, which is introduced in [27] is required. To make the paper self-contained, we cite it as follows (rephrased).

   Given a pattern query Q and a view definition V, one can compute $V(Q)$ by treating Q as data graph, and V as pattern query. Then the *shadow* from V to Q, denoted by H_V^Q, is defined to be the union of edge sets of matches of V in Q.

   We denote by $\bar{E}_g = E_p \setminus E_g$ as the "uncovered edges" of Q_g (a *containing rewriting* of Q) in Q, where E_p and E_g are the edge sets of Q and Q_g, respectively.

   The result below shows that *shadow* yields a characterization of *joint containment* (see [2] for the proof). Based on the characterization, we provide an efficient algorithm for the determination of *joint containment*.

Proposition 1. *For a pattern Q, a set of views $\mathcal{V}$ and an access schema $\mathcal{C}$, $Q \sqsubseteq_J [\mathcal{V}, \mathcal{C}]$ if and only if there exists a containing rewriting Q_g of Q such that $\bar{E}_g \subseteq \bigcup_{\varphi \in \mathcal{C}} H_{Q_\varphi}^Q$.*

Input: A pattern $Q = (V_p, E_p)$, views $\mathcal{V}$ and access schema $\mathcal{C}$.
Output: A boolean value ans that is true if and only if $Q \sqsubseteq_J [\mathcal{V}, \mathcal{C}]$.

1. boolean ans := false; set $E_c := \emptyset$, $E_u := \emptyset$, $\mathcal{C}' := \emptyset$;
2. **for each** view definition $V_i \in \mathcal{V}$ **do**
3.   compute $H_{V_i}^Q$; $E_c := E_c \bigcup H_{V_i}^Q$;
4. $E_u := E_p \setminus E_c$;
5. **for each** access constraint $\varphi = \langle Q_\varphi(Q_L \to Q_R), N \rangle$ in $\mathcal{C}$ **do**
6.   **if** $H_{Q_\varphi}^Q \cap E_u \neq \emptyset$ **then**
7. $\mathcal{C}' := \mathcal{C}' \cup \{\varphi\}$; $E_u := E_u \setminus H_{Q_\varphi}^Q$;
8.   **if** $E_u = \emptyset$ **then**
9.   ans := true; **break** ;
10. **return** ans;

Fig. 2. Algorithm JCont

Theorem 2. *It is in $O((\|\mathcal{V}\| + \|\mathcal{C}\|)|Q|!|Q|)$ time to decide whether $Q \sqsubseteq_J [\mathcal{V}, \mathcal{C}]$, and if so, to compute associated mappings from Q to $\mathcal{V}, \mathcal{C}$.*

Proof. We show Theorem 2 by presenting an algorithm as a constructive proof (Fig. 2).

Algorithm. The algorithm, denoted as JCont, takes Q, $\mathcal{V}$ and $\mathcal{C}$ as input, and returns true if and only if $Q \sqsubseteq_J [\mathcal{V}, \mathcal{C}]$. The algorithm works in three stages. In the first stage, it initializes a boolean variable ans, and three empty sets E_c, E_u and $\mathcal{C}'$, to keep track of "covered edges", "uncovered edges", and selected access constraints, respectively (line 1). In the second stage, it identifies an edge set E_c such that the sub-pattern of Q induced with E_c is *contained by* $\mathcal{V}$. Specifically, it (1) computes shadow $H_{V_i}^Q$ for each V_i in $\mathcal{V}$, by invoking the revised subgraph isomorphism algorithm, which finds all the matches of V_i in Q with algorithm in [12], and then merges them together; (2) extends E_c with $H_{V_i}^Q$ (lines 2–3). After all the shadows are merged, JCont generates an edge set $E_u = E_p \setminus E_c$ (line 4). In the last stage, JCont verifies the condition for *joint containment* as follows. It checks for each access constraint φ whether its Q_φ can *cover* a part of E_u, *i.e.*, $H_{Q_\varphi}^Q \cap E_u \neq \emptyset$ (line 6). If so, JCont enlarges $\mathcal{C}'$ with φ and updates E_u with $E_u \setminus H_{Q_\varphi}^Q$ (line 7). When the condition $E_u = \emptyset$ is encountered, the variable ans is changed to true, and the **for** loop (line 5) immediately terminates (line 9). JCont finally returns ans as result (line 10).

Example 4. Consider Q, $\mathcal{V} = \{V_1, V_2\}$ and $\mathcal{C} = \{\varphi\}$ in Fig. 1. JCont first computes shadows for each $V_i \in \mathcal{V}$ and obtains $H_{V_1}^Q = \{(PM, BA), (PM, PRG)\}$ and $H_{V_2}^Q = \{(PRG, ST), (PRG, UI)\}$. Then E_u includes $\{(PM, DBA), (PRG, DBA)\}$. It next computes $H_{Q_\varphi}^Q = \{(PM, DBA), (PRG, DBA)\}$, which exactly covers E_u. Finally, JCont returns true indicating that Q is *boundedly evaluable with views*. $\square$

Input: Pattern Q, views $\mathcal{V}$, $\mathcal{V}(G)$, access schema $\mathcal{C}$, mappings λ, ρ.
Output: The query result M as Q(G).

1. initialize an empty pattern Q_o; set M := $\emptyset$, M_1 := $\emptyset$;
2. **for each** view V_i that is mapped via λ from E_p **do**
3.   **for each** $G_s \in \lambda^{-1}(V_i)$ **do**
4. $Q_o \oplus G_s$;
5.   **for each** $m_1 \in M$ and **each** $m_2 \in V_i(G)$ **do**
6.   **if** m_1 and m_2 can be merged **then**
7. $m := m_1 \oplus m_2$; $M_1 := M_1 \cup \{m\}$;
8.   update M, M_1;
9. **for each** access constraint φ in $\rho(E_p)$ **do**
10. $\langle Q_o, M \rangle$:= Expand(Q_o, M, φ, ρ);
11. **return** M;

Fig. 3. Algorithm BMatch

Correctness and Complexity. The correctness is ensured by that when JCont terminates, JCont correctly identifies a part Q_c (with edge set E_c) of Q that can be answered by using $\mathcal{V}$; and in the meanwhile, the remaining part Q_u (with edge set E_u) of Q can also be *covered* by $\mathcal{C}$. To see the complexity, observe the following. The initialization of JCont is in constant time. For the second stage, JCont iteratively computes shadow $H_{V_i}^Q$ for each $V_i \in \mathcal{V}$. As it takes $O(|Q|!|Q|)$ time to compute shadow from V_i to Q for a single iteration, and the **for** loop repeats $||\mathcal{V}||$ times, thus, it is in $O(||\mathcal{V}|||Q|!|Q|)$ time for the second stage. In the last stage, JCont computes shadow from Q_φ to Q for each access constraint φ in $\mathcal{C}$, which is in $O(|Q|!|Q|)$ time for a single iteration as well. As the iteration executes $||\mathcal{C}||$ times, it hence takes JCont $O(||\mathcal{C}|||Q|!|Q|)$ time. Putting these together, JCont is in $O((||\mathcal{V}|| + ||\mathcal{C}||)|Q|!|Q|)$ time.   □

Remarks. Algorithm JCont can be easily adapted to return a pair of mappings $\langle \lambda, \rho \rangle$ that serve as input for the matching algorithm (shown in Sect. 4).

4   Matching Evaluation

We study how to evaluate pattern matching using views $\mathcal{V}(G)$ and a fraction G_Q of G.

4.1   A Matching Algorithm

Along the same line as pattern matching using views, on a graph G that satisfies access schema $\mathcal{C}$, a pattern Q can be answered with $\mathcal{V}(G)$ and G_Q with below technique: (1) determine whether $Q \sqsubseteq_J [\mathcal{V}, \mathcal{C}]$ and compute a pair of mappings $\langle \lambda, \rho \rangle$ with revised algorithm of JCont; and (2) compute Q(G) with a matching algorithm that takes λ, ρ, $\mathcal{V}$, $\mathcal{V}(G)$ and $\mathcal{C}$ as input, if $Q \sqsubseteq_J [\mathcal{V}, \mathcal{C}]$. We next show such an algorithm.

Theorem 3. *For any graph G that satisfies $\mathcal{C}$, a pattern Q can be answered by using $\mathcal{V}$, $\mathcal{V}(G)$ and G_Q in $O((|\mathcal{V}|^{|Q|}|\mathcal{V}(G)|)^{||\mathcal{V}||}(|Q|^{|Q|} \cdot N_m)^{||\mathcal{C}||})$ time, if $Q \sqsubseteq_J [\mathcal{V}, \mathcal{C}]$.*

Proof. We provide an algorithm with analyses as a constructive proof of Theorem 3.

Algorithm. The algorithm BMatch is shown in Fig. 3. It takes a pattern Q, a set of views $\mathcal{V}$, $\mathcal{V}(G)$ and mappings λ and ρ, as input, and works in two stages: "merging" views $V_i(G)$ following λ; and "expanding" partial matches with G_Q under the guidance of ρ.

More specifically, BMatch starts with an empty pattern query Q_o, an empty set M to keep track of matches of Q_o and another empty set M_1 for maintaining intermediate results (line 1). It then iteratively "merges" $V_i(G)$ following mapping λ (lines 2–8). Specifically, for each view definition V_i that is mapped via λ from edge set E_p (line 2) and each match G_s of V_i in Q (line 3), BMatch expands Q_o with G_s (line 4), and iteratively expands each match m_1 of Q_o with each match m_2 of V_i if they can be merged in the same way as the merging process of Q_o and G_s, and includes the new match m in M_1 (lines 5–7). When a round of merging process for matches of Q_o and V_i finished, BMatch updates M and M_1 by letting M := M_1 and M_1 := $\emptyset$ (line 8). When the first stage finished, Q_o turns to a containing rewriting Q_g of Q and M includes all the matches of Q_g in G. In the following stage, BMatch iteratively invokes Procedure Expand to "expand" Q_o and its matches under the guidance of mapping ρ (lines 9–10). It finally returns M as matches of $Q(G)$ (line 11).

Procedure Expand. Given mapping ρ, access constraint φ, pattern Q_o and its match set M, Expand (not shown) expands Q_o and M as follows. It first initializes an empty set M_1. For each match G_s of Q_φ in Q, Expand first extends Q_o with G_s; it next expands a match m_1 of Q_o with a match G_L of Q_L if m_1 has common nodes or edges with G_L and further expands m_1 with each Q_R-value G_R of G_L, for each match m_1 of Q_o and each match G_L of Q_L. The new matches are maintained by the set M_1. After all the G_s are processed, Expand returns updated Q_o and its match set M_1 as final result.

Example 5. Consider Q, $\mathcal{V}$, $\mathcal{V}(G)$ and φ shown in Fig. 1. BMatch first merges "partial matches" in $\mathcal{V}(G)$ following the guidance of mapping λ. Specifically, it first initializes Q_o and M with V_1 and $V_1(G)$ (as shown in Fig. 1(c)); it next expands Q_o with V_2, and merges each match in M with each match in $V_2(G)$. After "partial matches" are merged, Q_o includes all the edges of V_1 and V_2, and set M includes below matches: $m_1 = \{\mathsf{PM_1, BA_1, PRG_1, ST_1, UI_1}\}$, $m_2 = \{\mathsf{PM_2, BA_2,}$ $\mathsf{PRG_2, ST_2, UI_1}\}$, $m_3 = \{\mathsf{PM_2, BA_2, PRG_2, ST_2, UI_2}\}$, $m_4 = \{\mathsf{PM_2, BA_2, PRG_2, ST_3,}$ $\mathsf{UI_1}\}$, $m_5 = \{\mathsf{PM_2, BA_2, PRG_2, ST_3, UI_2}\}$, and $m_6 = \{\mathsf{PM_3, BA_2, PRG_3, ST_3, UI_3}\}$.

Guided by mapping ρ, BMatch expands Q_o and its matches via Procedure Expand. Expand first merges Q_o with Q_φ. Then, it first expands m_1 with G_L (with edge set $\{(\mathsf{PM_1, DBA_1})\}$), and then merges m_1 with G_R (with edge set $\{(\mathsf{PRG_1, DBA_1})\}$). The above merge process repeats another 5 times for each

m_i ($i \in [2,6]$). Finally, BMatch returns a set of 6 matches that are grown from $m_1 - m_6$, respectively. □

Correctness and Complexity. The correctness is guaranteed by the following three invariants: (1) BMatch correctly merges Q_o (resp. M) with views V_i (resp. $V_i(G)$); (2) procedure Expand correctly expands Q_o and M using indexes associated with access schema; and (3) when BMatch terminates, Q_o (resp. M) is equivalent to Q (resp. Q(G)). Interested readers may refer to [2] for more details.

   We give a detailed complexity analysis as below.

(I) BMatch iteratively merges Q_o and M with view V_i and $V_i(G)$, respectively. For a single iteration, it takes BMatch $|\lambda^{-1}(V_i)||M||V_i(G)|$ time for the "merge" task. As in the worst case, (1) there may exist $|V_{V_i}|^{|Q_c|}$ matches of V_i in Q_c, where V_{V_i} denotes the node set of V_i and $|Q_c|$ is bounded by $|Q|$, hence $|\lambda^{-1}(V_i)|$ is bounded by $|V_{V_i}|^{|Q|}$; (2) $|M|$ is bounded by $\prod_{i \in [1,k-1]} |V_{V_i}|^{|Q|} |V_i(G)|$ before the k-th iteration; and (3) the iteration repeats at most $||\mathcal{V}||$ times, hence the first stage is bounded by $\prod_{i \in [1,||\mathcal{V}||]} |V_{V_i}|^{|Q|} |V_i(G)|$, which is in $O((|\mathcal{V}|^{|Q|}|\mathcal{V}(G)|)^{||\mathcal{V}||})$ time.

(II) BMatch repeatedly invokes Procedure Expand to process expansion of Q_o and M with access constraint φ. For a single process, it takes Expand $|\rho^{-1}(Q_R)||M||\mathcal{I}_\varphi(G_L)|$ time. Note that (1) $|\rho^{-1}(Q_R)|$ is bounded by $|V_R|^{|Q_u|}$ (V_R refers to the node set of Q_R), which is further bounded by $|V_R|^{|Q|}$, as $|Q_u|$ is bounded by $|Q|$; (2) $|M|$ is bounded by $(|\mathcal{V}|^{|Q|}|\mathcal{V}(G)|)^{||\mathcal{V}||} \cdot \prod_{i \in [1,k-1]} |V_{R_i}|^{|Q|} \cdot N_i$, before the k-th iteration; and (3) $|\mathcal{I}_\varphi(G_L)| \leq N_k$ at the k-th iteration, thus, Expand is in $O(|V_R|^{|Q|} \cdot (|\mathcal{V}|^{|Q|}|\mathcal{V}(G)|)^{||\mathcal{V}||} \prod_{i \in [1,k]} (|V_{R_i}|^{|Q|} \cdot N_i))$ time. As the iteration repeats at most $||\mathcal{C}||$ times, the second stage is hence in $O(|Q|^{|Q|}(|\mathcal{V}|^{|Q|}|\mathcal{V}(G)|)^{||\mathcal{V}||}(|Q|^{|Q|} \cdot N_m)^{||\mathcal{C}||})$ time. Putting these together, BMatch is in $O((|\mathcal{V}|^{|Q|}|\mathcal{V}(G)|)^{||\mathcal{V}||}(|Q|^{|Q|} \cdot N_m)^{||\mathcal{C}||})$ time, where N_m is the maximum cardinality of access schema $\mathcal{C}$. □

4.2   Optimization Strategy

As the cost of BMatch is partially determined by $|\mathcal{V}(G)|$ and $|G_Q|$, it is beneficial to reduce their sizes. This motivates us to study the *minimum containment problem*.

Minimum Containment Problem. Given a pattern query Q, a set of view definitions $\mathcal{V}$ with each V_i associated with weight $|V_i(G)|$, and an access schema $\mathcal{C}$, the problem, denoted as MCP, is to find a subset $\mathcal{V}'$ of $\mathcal{V}$ and a subset $\mathcal{C}'$ of $\mathcal{C}$, such that (1) Q $\sqsubseteq_J [\mathcal{V}', \mathcal{C}']$, and (2) for any subset $\mathcal{V}''$ of $\mathcal{V}$ and any subset $\mathcal{C}''$ of $\mathcal{C}$, if Q $\sqsubseteq_J [\mathcal{V}'', \mathcal{C}'']$, then $|\mathcal{V}'(G)| + \Sigma_{\varphi_i \in \mathcal{C}'} N_i \cdot |Q_{\varphi_i}| \leq |\mathcal{V}''(G)| + \Sigma_{\varphi_j \in \mathcal{C}''} N_j \cdot |Q_{\varphi_j}|$.

   As will be seen in Sect. 5, MCP is effective: it can eliminate redundant views (as well as their corresponding extensions), and reduce the size of G_Q thereby improving the efficiently of BMatch. However, MCP is nontrivial, its decision problem is NP-hard. Despite of this, we develop an algorithm for MCP, which is approximable within $O(\log |Q|)$. That's, the algorithm can identify a subset

$\mathcal{V}'$ of $\mathcal{V}$ and a subset $\mathcal{C}'$ of $\mathcal{C}$ when $Q \sqsubseteq_J [\mathcal{V}, \mathcal{C}]$, such that $Q \sqsubseteq [\mathcal{V}', \mathcal{C}']$ and $|\mathcal{V}'(G)| + \Sigma_{\varphi_i \in \mathcal{C}'} N_i \cdot |Q_{\varphi_i}|$ is guaranteed to be no more than $\log(|Q|) \cdot (|\mathcal{V}_{\mathsf{OPT}}(G)| + \Sigma_{\varphi_j \in \mathcal{C}_{\mathsf{OPT}}} N_j \cdot |Q_{\varphi_j}|)$, where $\mathcal{V}_{\mathsf{OPT}}$ and $\mathcal{C}_{\mathsf{OPT}}$ are the subsets of $\mathcal{V}$ and $\mathcal{C}$, respectively, and moreover, $Q \sqsubseteq_J [\mathcal{V}_{\mathsf{OPT}}, \mathcal{C}_{\mathsf{OPT}}]$ and $|\mathcal{V}_{\mathsf{OPT}}(G)| + \Sigma_{\varphi_j \in \mathcal{C}_{\mathsf{OPT}}} N_j \cdot |Q_{\varphi_j}|$ is minimum, among all possible subset pairs of $\mathcal{V}$ and $\mathcal{C}$.

Theorem 4. *There exists an algorithm for* MCP *that finds a subset* $\mathcal{V}'$ *of* $\mathcal{V}$ *and a subset* $\mathcal{C}'$ *of* $\mathcal{C}$ *with* $Q \sqsubseteq_J [\mathcal{V}', \mathcal{C}']$ *and* $|\mathcal{V}'(G)| + \Sigma_{\varphi_i \in \mathcal{C}'} N_i \cdot |Q_{\varphi_i}| \leq \log(|Q|) \cdot (|\mathcal{V}_{\mathsf{OPT}}(G)| + \Sigma_{\varphi_j \in \mathcal{C}_{\mathsf{OPT}}} N_j \cdot |Q_{\varphi_j}|)$ *in* $O(||\mathcal{V}|||Q|!|Q| + (||\mathcal{V}|||Q|)^{3/2})$ *time.*

Proof. We next show Theorem 4 by providing an algorithm as a constructive proof.

Algorithm. The algorithm, denoted as Minimum (not shown), takes a pattern query Q, a set of view definitions $\mathcal{V}$ with each V_i in $\mathcal{V}$ taking a weight $|\mathsf{V}_i(G)|$, and an access schema $\mathcal{C}$ as input, identifies a pair $\langle \mathcal{V}', \mathcal{C}' \rangle$ of subsets of $\mathcal{V}$ and $\mathcal{C}$ such that (1) $Q \sqsubseteq_J [\mathcal{V}', \mathcal{C}']$ if $Q \sqsubseteq_J [\mathcal{V}, \mathcal{C}]$ and (2) $|\mathcal{V}'(G)| + \Sigma_{\varphi_i \in \mathcal{C}'}(N_i \cdot |Q_{\varphi_i}|) \leq \log(|Q|) \cdot (|\mathcal{V}_{\mathsf{OPT}}(G)| + \Sigma_{\varphi_j \in \mathcal{C}_{\mathsf{OPT}}}(N_j \cdot |Q_{\varphi_j}|))$, where (a) $\log(|Q|)$ is the approximation ratio, and (b) $\mathcal{V}_{\mathsf{OPT}}, \mathcal{C}_{\mathsf{OPT}}$ are the subsets of $\mathcal{V}, \mathcal{C}$, respectively, and moreover, $Q \sqsubseteq_J [\mathcal{V}_{\mathsf{OPT}}, \mathcal{C}_{\mathsf{OPT}}]$ and $|\mathcal{V}_{\mathsf{OPT}}(G)| + \Sigma_{\varphi_j \in \mathcal{C}_{\mathsf{OPT}}}(N_j \cdot |Q_{\varphi_j}|)$ is minimum, among all possible subset pairs of $\mathcal{V}$ and $\mathcal{C}$.

In a nutshell, the algorithm applies a greedy strategy to find a views V_i in $\mathcal{V}$ or an access constraint φ_j from $\mathcal{C}$ that is considered "best" during the iteration. To measure the goodness of the views and access constraints, we define a metric $\alpha(\mathsf{V}_i) = \frac{|\mathsf{V}_i(G)|}{|\mathsf{H}^Q_{\mathsf{V}_i} \setminus Q_o|}$ for a view V_i, and $\alpha(\varphi_j) = \frac{N_j \cdot |Q_{\varphi_j}|}{|\mathsf{H}^Q_{Q_{\varphi_j}} \setminus Q_o|}$ for an access constraint φ_j. Here, Q_o takes edges from shadows whose corresponding views (resp. access constraints) are chosen in $\mathcal{V}'$ (resp. $\mathcal{C}'$). Intuitively, $\alpha(\mathsf{V}_i)$ (resp. $\alpha(\varphi_j)$) indicates how costly it is to "cover" the remaining part $Q \setminus Q_o$ of Q with $\mathsf{H}^Q_{\mathsf{V}_i}$ (resp. $\mathsf{H}^Q_{Q_{\varphi_j}}$), hence a V_i or φ_j with the least $\alpha(\cdot)$ is favored in each round iteration. The algorithm works in two stages.

(I) It initializes three empty sets $\mathcal{V}'$, $\mathcal{C}'$ and $\mathcal{F}$, and computes the shadow $\mathsf{H}^Q_{\mathsf{V}_i}$ for each $\mathsf{V}_i \in \mathcal{V}$ (resp. $\mathsf{H}^Q_{Q_{\varphi_j}}$ for each $\varphi_j \in \mathcal{C}$) and maintains a pair $\langle \mathsf{H}^Q_{\mathsf{V}_i}, |\mathsf{V}_i(G)| \rangle$ (resp. $\langle \mathsf{H}^Q_{Q_{\varphi_j}}, N_j \cdot |Q_{\varphi_j}| \rangle$) in $\mathcal{F}$. Intuitively, $|\mathsf{V}_i(G)|$ (resp. $N_j \cdot |Q_{\varphi_j}|$) can be viewed as the "weight" of its corresponding V_i (resp. φ_j).

(II) Minimum invokes procedure PtnFinder to compute a pair $\langle \mathcal{V}', \mathcal{C}' \rangle$ of subsets of $\mathcal{V}$ and $\mathcal{C}$. Specifically, it first initializes an empty pattern Q_o and two empty sets $\mathcal{V}'$, $\mathcal{C}'$. It then iteratively selects an object obj, whose corresponding $\alpha(\cdot)$ is the least. Here, the chosen object obj is either a view definition V_i or an access constraint φ_j. Once there does not exist any obj whose shadow can expand Q_o, i.e., $(\mathsf{H}^Q_{\mathsf{obj}} \setminus Q_o) = \emptyset$, PtnFinder breaks the loop. Otherwise, it updates $\mathcal{F}$ by removing $\langle \mathsf{H}^Q_{\mathsf{V}_i}, |\mathsf{V}_i(G)| \rangle$ (resp. $\langle \mathsf{H}^Q_{Q_{\varphi_j}}, N_j \cdot |Q_{\varphi_j}| \rangle$), and expands Q_o with shadow $\mathsf{H}^Q_{\mathsf{obj}}$. If obj is a view V_i, PtnFinder includes it in $\mathcal{V}'$, otherwise, PtnFinder enriches $\mathcal{C}'$ with φ_j. After **while** loop terminates, if E_o is not equivalent to E_p, PtnFinder returns an empty set, since Q $\not\sqsubseteq_j$ $[\mathcal{V}, \mathcal{C}]$ and hence no subset pair exists. Otherwise, PtnFinder returns $\langle \mathcal{V}', \mathcal{C}' \rangle$ as final result.

Correctness and Complexity. Observe that Minimum either finds a pair $\langle \mathcal{V}', \mathcal{C}' \rangle$ of subsets of $\mathcal{V}$ and $\mathcal{C}$ such that Q $\sqsubseteq_J$ $[\mathcal{V}', \mathcal{C}']$ or an empty set indicating Q $\not\sqsubseteq_J$ $[\mathcal{V}, \mathcal{C}]$. This is ensured by joint containment checking of the algorithm, by following Proposition 1. Moreover, PtnFinder identifies $\langle \mathcal{V}', \mathcal{C}' \rangle$ with a greedy strategy, which is verified to guarantee $\log(|Q|)$ approximation ratio for *weighted set cover problem* [25]. Algorithm Minimum computes "shadows" for each V_i in $\mathcal{V}$ and each φ_j in $\mathcal{C}$ in $O((\|\mathcal{V}\| + \|\mathcal{C}\|)|Q|!|Q|)$ time. The procedure PtnFinder is in $O(((\|\mathcal{V}\| + \|\mathcal{C}\|)|Q|)^{3/2})$ time, as the while loop is executed $\min\{(\|\mathcal{V}\| + \|\mathcal{C}\|), |Q|\}$ times, which is bounded by $O(((\|\mathcal{V}\| + \|\mathcal{C}\|)|Q|)^{1/2})$ time, and each iteration takes $O((\|\mathcal{V}\| + \|\mathcal{C}\|)|Q|)$ time to find a view with least $\alpha(\cdot)$ [25]. Thus, Minimum is in $O((\|\mathcal{V}\| + \|\mathcal{C}\|)|Q|!|Q| + ((\|\mathcal{V}\| + \|\mathcal{C}\|)|Q|)^{3/2})$ time.

The analysis above completes the proof of Theorem 4. □

5 Experimental Evaluation

We conducted two sets of tests to evaluate performance of algorithms for (1) *joint containment* checking and (2) *bounded pattern matching using views*.

Experimental Setting. We used the following data.

(1) Real-life graphs. We used three real-life graphs: (a) *Amazon* [4], a product co-purchasing network with 548K nodes and 1.78M edges; (b) *Citation* [1], a collaboration network with 1.4M nodes and 3M edges; (c) *YouTube* [5], a recommendation network with 1.6M nodes and 4.5M edges.

(2) Synthetic graphs. We designed a generator to produce random graphs, controlled by the number $|V|$ of nodes, the number $|E|$ of edges, and an alphabet Σ for node labels. We enforced a set of access constraints during random generation.

(3) Pattern queries. We implemented a generator for pattern queries controlled by: the number $|V_p|$ (resp. $|E_p|$) of pattern nodes (resp. edges), and node label f_v from an alphabet Σ of labels drawn from corresponding real-life graphs. We denote $(|V_p|, |E_p|)$ as the size of pattern queries, and generated a set of 30 pattern queries with size $(|V_p|, |E_p|)$ ranging from $(3, 2)$ to $(8, 16)$, for each data graph.

(4) Views. We generated views for *Amazon* following [23], designed views to search for papers and authors in computer science for *Citation*, and generated views for *Youtube* following [16]. For each real-life graph, a set $\mathcal{V}$ of 50 view definitions with different sizes *e.g.*, $(2,1)$, $(3,2)$, $(4,3)$, $(4,4)$ and structures are generated. For synthetic graphs, we randomly generated a set of 50 views whose node labels are drawn from a set Σ of 10 labels and with size of $(2,1)$, $(3,2)$, $(4,3)$ and $(4,4)$. For each view set $\mathcal{V}$, we force that Q $\not\sqsubseteq \mathcal{V}$, for each Q used for testing.

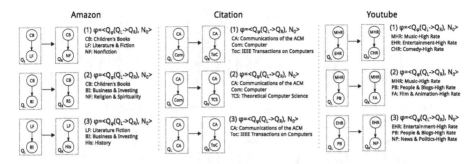

Fig. 4. Typical access constraints on real-life graphs

(5) Access schema. We investigated real-life graphs, and extracted an access schema $\mathcal{C}$ with a set of access constraints φ for each of them. A set of typical access constraints are shown in Fig. 4. For each access constraint $\varphi = \langle Q_\varphi(Q_L \rightarrow Q_R), N \rangle$, we computed an index $\mathcal{I}_\varphi$ for it. Since N is no more than 100 for each chosen φ, hence the space cost of an index $\mathcal{I}_\varphi$ ranges from a few megabytes to dozens of megabytes, for each φ. In addition, the index $\mathcal{I}_\varphi$ is built upon a hashtable, with a distinct match G_L of Q_L as the key, and a set of matches G_s of Q_φ as values. Thus, the fetch time on $\mathcal{I}_\varphi$ for any G_L is very fast, and can be viewed as a constant. On synthetic graphs, we manually generated a set of access constraints and computed indexes for these access constraints, along the same line as performed on real-life graphs. As index construction is a one-off task and can be performed off-line, we do not report its computational time.

(6) Implementation. We implemented below algorithms, in Java: (1) JCont for joint containment checking; (2) VF2 [12] for matching evaluation on G, BMatch for matching with $\mathcal{V}(G)$ and G_Q; and (3) Minimum for identifying a pair $\langle \mathcal{V}', \mathcal{C}' \rangle$ from $\mathcal{V}$ and $\mathcal{C}$, and BMatch$_{min}$ which revises BMatch by using $\langle \mathcal{V}', \mathcal{C}' \rangle$ identified by Minimum.

All the tests were run on a machine with an Intel Core(TM)2 Duo 3.00 GHz CPU and 4 GB memory, using Ubuntu. Each test was run 10 times and the average is reported.

Experimental Results. We next present our findings.

Exp-1: Joint Containment Checking. We evaluate performance of JCont vs. Minimum.

Performance of JCont *vs.* Minimum. We evaluate the efficiency of JCont vs. Minimum. Fixing $\mathcal{V}$ and $\mathcal{C}$ for real-life graphs, we varied the pattern size from $(4,4)$ to $(8,16)$, where each size corresponds to a set of pattern queries with different structures and node labels. We find the following. (1) JCont and Minimum both are efficient, *e.g.*, it takes JCont on average 145.5 ms to decide whether a pattern with size $(8,16)$ is jointly contained in $\mathcal{V}$ and $\mathcal{C}$. (2) Both two algorithms spend more time over larger patterns, which are consistent with their computational complexities. Due to space constraint, we do not report detailed

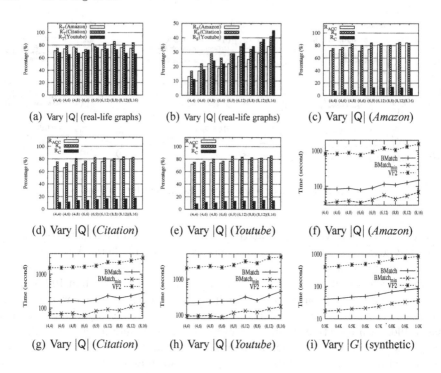

Fig. 5. Performance evaluation

figures here. (3) We compute a ratio $R_T = \frac{T_{\mathsf{JCont}}}{T_{\mathsf{Minimum}}}$, where T_{JCont} and T_{Minimum} are the time used by JCont and Minimum, respectively, to evaluate performance gap between JCont and Minimum. As shown in Fig. 5(a), JCont accounts for about 75.7% of the time of Minimum, on average, since it takes Minimum more time to pick a V_i from V (resp. φ_j from C).

To investigate the effectiveness of Minimum, we defined a ratio $R_S = \frac{S_a}{S_b}$, where $S_a = |V'(G)| + \Sigma_{\varphi_j \in C'} N_j \cdot |Q_{\varphi_j}|$, $S_b = |V(G)| + \Sigma_{\varphi_j \in C} N_j \cdot |Q_{\varphi_j}|$, as the ratio of the total size of view extensions $V'(G)$ and G_Q identified by Minimum to the total size of whole set of $V(G)$ and indexes in C. Fixing V and C over real-life graphs, we varied $(|V_p|, |E_p|)$ from $(4, 4)$ to $(8, 16)$ and evaluated the ratio R_S. As shown in Fig. 5(b), Minimum is effective, it finds a pair $\langle V', C' \rangle$ with total size S_a *substantially smaller* than S_b, *i.e.*, taking only about 25.2% of S_b for all real-life graphs, on average. As will be shown, using $\langle V', C' \rangle$ can substantially improve efficiency of matching computation.

Exp-2: Bounded Pattern Matching Using Views. We study the effectiveness, efficiency and scalability of BMatch, BMatch_min, compared to VF2.

Effectiveness. We define following three metrics and evaluate effectiveness of *bounded pattern matching using views* with real-life graphs.

(1) We defined a ratio R_{ACC} as accuracy, to measure the result quality when access schema is absent, by following F-measure [29]. Here, $R_{ACC} = \frac{2 \cdot \text{recall} \cdot \text{precision}}{\text{recall} + \text{precision}}$, where $\text{recall} = \frac{|S|}{|S_t|}$, $\text{precision} = \frac{|S|}{|S_m|}$, S_t consists of matches in $Q(G)$, S_m is the set of matches in $Q_g(G)$, and S consists of "true" matches that can be identified from $Q_g(G)$, where Q_g refers to the maximally containing rewriting of Q $w.r.t.$ $\mathcal{V}$.

(2) We used $R_{\mathcal{C}} = \frac{\Sigma_{\varphi_i \in \mathcal{C}'} N_i \cdot |Q_{\varphi_i}|}{|Q(G)|}$ to show the propotion of $|G_Q|$ in $|Q(G)|$. Here, G_Q is constructed from $\mathcal{C}'$, which is a subset of $\mathcal{C}$ and takes a set of access constraints φ_i that are used by BMatch when matching evaluation, and $|Q(G)|$ is the total size of matches of Q in G.

(3) We used $R_e = \frac{|\bar{E}_g|}{|E_p|}$ to show how large a pattern query needs to be "covered" by access constraints.

Figures 5(c)–5(e) report three ratios on real-life graphs, which tell us the following. The ratio R_{ACC} is, on average, 77.3%, 73.6%, and 76.8% on *Amazon*, *Citation* and *Youtube*, respectively. In the meanwhile, the average ratios of $R_{\mathcal{C}}$ and R_e reach 10.4% and 18.8%, 14% and 20.6%, 11.7% and 19.4% on *Amazon*, *Citation* and *Youtube*, respectively. These together show that access constraints often cover a small but critical part of pattern queries, *e.g.*, 16.11% of total edges of Q on average, and provide limited but key information, *e.g.*, 10.89% of the size of match result to improve accuracy by more than 20%, on average.

Efficiency. Figures 5(f), 5(g) and 5(h) show the efficiency on *Amazon*, *Citation* and *YouTube*, respectively. The x-axis represents pattern size ($|V_p|, |E_p|$). The results tell us the following. (1) BMatch and BMatch$_{min}$ substantially outperform VF2, taking only 9.5% and 3.8% of its running time on average over all real-life graphs. (2) All the algorithms spend more time on larger patterns. Nonetheless, BMatch and BMatch$_{min}$ are less sensitive to the increase of $|Q|$ than VF2, as they reuse earlier computation cached in view extensions and hence save computational cost. (3) BMatch$_{min}$ is more efficient than BMatch, taking only 41.2% time on average over real-life graphs, as it uses smaller $\mathcal{V}(G)$ and G_Q.

Scalability. Using synthetic graphs, we evaluated the scalability of BMatch, BMatch$_{min}$ and VF2. Fixing $|Q| = (4,6)$, we varied the node number $|V|$ of data graphs from $0.3M$ to $1M$, in $0.1M$ increments, and set $|E| = 2|V|$. As shown in Fig. 5, BMatch$_{min}$ scales best with $|G|$ and is on average 1.4 and 22.1 times faster than BMatch and VF2, which is consistent with the complexity analysis, and the observations in Figs. 5(f), 5(g) and 5(h).

6   Conclusion

We have studied *bounded pattern matching using views*, for pattern queries defined in terms of subgraph isomorphism, from theory to algorithms. We have introduced *access schema* for graphs, proposed a notion of joint containment for characterizing bounded pattern matching using views, and provided an efficient algorithm for joint containment checking. Based on the characterization, we have developed an matching algorithm by using views and a size-bounded fraction, and moreover, we have also provided optimization strategy to improve efficiency of matching compuation. Our experimental results have verified the effectiveness, efficiency and scalability of our algorithms, using real-life and synthetic data. The study of bounded pattern matching using views is still in its infancy. One issue is the selection of views and *access schema*. Another problem concerns scale-inpendence for GPM.

References

1. Citation. http://www.arnetminer.org/citation/
2. Full version. https://github.com/xgnaw/sun/raw/master/boundedPM.pdf
3. Linkedin statistics. https://www.omnicoreagency.com/linkedin-statistics/
4. Stanford large network dataset collection. http://snap.stanford.edu/data/index.html
5. Youtube dataset. http://netsg.cs.sfu.ca/youtubedata/
6. Facebook (2013). http://newsroom.fb.com
7. Afrati, F.N., Chirkova, R.: Answering Queries Using Views, Second Edition. Synthesis Lectures on Data Management. Morgan & Claypool Publishers (2019)
8. Armbrust, M., Curtis, K., Kraska, T., Fox, A., Franklin, M.J., Patterson, D.A.: PIQL: success-tolerant query processing in the cloud. PVLDB **5**(3), 181–192 (2011)
9. Armbrust, M., et al.: SCADS: scale-independent storage for social computing applications. In: CIDR (2009)
10. Armbrust, M., Liang, E., Kraska, T., Fox, A., Franklin, M.J., Patterson, D.A.: Generalized scale independence through incremental precomputation. In: SIGMOD, pp. 625–636 (2013)
11. Cao, Y., Fan, W., Geerts, F., Lu, P.: Bounded query rewriting using views. ACM Trans. Database Syst. **43**(1), 6:1–6:46 (2018)
12. Cordella, L.P., Foggia, P., Sansone, C., Vento, M.: A (sub)graph isomorphism algorithm for matching large graphs (2004)
13. Fan, W., Geerts, F., Cao, Y., Deng, T., Lu, P.: Querying big data by accessing small data. In: PODS, pp. 173–184 (2015)
14. Fan, W., Geerts, F., Libkin, L.: On scale independence for querying big data. In: PODS, pp. 51–62 (2014)
15. Fan. W., Li, J., Ma, S., Tang, N., Wu, Y., Wu, Y.: Graph pattern matching: from intractability to polynomial time. In: PVLDB (2010)
16. Fan, W., Wang, X., Wu, Y.: Answering pattern queries using views. IEEE Trans. Knowl. Data Eng. **28**(2), 326–341 (2016)
17. Gerome, M., Suciu, D.: Containment and equivalence for an xpath fragment. In: PODS (2002)
18. Halevy, A.Y.: Answering queries using views: a survey. VLDB J. **10**(4), 270–294 (2001)
19. Halevy, A.Y.: Theory of answering queries using views. SIGMOD Record **29**(4), 56 (2001)
20. Henzinger, M.R., Henzinger, T., Kopke, P.: Computing simulations on finite and infinite graphs. In: FOCS (1995)
21. Kumar, R., Novak, J., Tomkins, A.: Structure and evolution of online social networks. In: KDD (2006)

22. Lenzerini, M.: Data integration: a theoretical perspective. In: PODS (2002)
23. Leskovec, J., Singh, A., Kleinberg, J.: Patterns of influence in a recommendation network. In: Advances in Knowledge Discovery and Data Mining (2006)
24. Neven, F., Schwentick, T.: XPath containment in the presence of disjunction, DTDs, and variables. In: ICDT (2003)
25. Papadimitriou, C.H.: Computational Complexity. Addison-Wesley (1994)
26. Wang, J., Li, J., Yu, J.X.: Answering tree pattern queries using views: a revisit. In: EDBT (2011)
27. Wang, X.: Answering graph pattern matching using views: a revisit. In: DEXA (2017)
28. Wang, X., Zhan, H.: Approximating diversified top-k graph pattern matching. In: DEXA, pp. 407–423 (2018)
29. Wikipedia. F-measure. http://en.wikipedia.org/wiki/F-measure

A Framework to Reverse Engineer Database Memory by Abstracting Memory Areas

James Wagner[1]($\boxtimes$) and Alexander Rasin[2]

[1] University of New Orleans, New Orleans, LA 70148, USA
jay.wagner88@gmail.com
[2] DePaul University, Chicago, IL 60604, USA
arasin@cdm.depaul.edu

Abstract. The contents of RAM in an operating system (OS) are a critical source of evidence for malware detection or system performance profiling. Digital forensics focused on reconstructing OS RAM structures to detect malware patterns at runtime. In an ongoing arms race, these RAM reconstruction approaches must be designed for the attack they are trying to detect. Even though database management systems (DBMS) are collectively responsible for storing and processing most data in organizations, the equivalent problem of memory reconstruction has not been considered for DBMS-managed RAM.

In this paper, we propose and evaluate a systematic approach to reverse engineer data structures and access patterns in DBMS RAM. Rather than develop a solution for specific scenarios, we describe an approach to detect and track any RAM area in a DBMS. We evaluate our approach with the four most common RAM areas in well-known DBMSes; this paper describes the design of each area-specific query workload and the process to capture and quantify that area at runtime. We further evaluate our approach by observing the RAM data flow in presence of built-in DBMS encryption. We present an overview of available DBMS encryption mechanisms, their relative advantages and disadvantages, and then illustrate the practical implications for the four memory areas.

1 Introduction

Database managements systems (DBMS) serve as the main data repositories for applications ranging from personal use (e.g., text messaging, web browsers) to enterprise data warehouses (e.g., airlines, merchants). In order to perform "live" (i.e., runtime) forensic, security, or performance analysis in a DBMS, an understanding of its RAM layout is necessary. There are currently no approaches or tools that can reverse engineer RAM contents of a DBMS. Current work in OS RAM analysis (see Sect. 2) seeks to detect specific malware patterns, offering no generalized solution. Although OS RAM may be too general, DBMS memory can be abstracted by identifying and quantifying each type of its memory area.

© Springer Nature Switzerland AG 2020
S. Hartmann et al. (Eds.): DEXA 2020, LNCS 12391, pp. 304–319, 2020.
https://doi.org/10.1007/978-3-030-59003-1_20

In this paper, we describe our approach and validate its generality on four major RAM areas across several representative DBMSes.

DBMSes allocate multiple RAM areas within their process memory to serve a particular purpose. For example, I/O buffer area caches pages accessed from disk (and some other operations); in Sect. 5 we describe four ubiquitous memory areas and other special-purpose RAM areas. A memory area can be detected and quantified by executing a customized synthetic workload and capturing the resulting RAM snapshots. In Sect. 7, we illustrate how to capture any memory area, describing our process and include the link to our query workloads.

A significant contribution of our approach to reverse engineering memory is assigning context to data. We demonstrate this with DBMS encryption as a use case (Sect. 9 outlines other use cases). While data can be encrypted outside of a DBMS, all major DBMSes (e.g., IBM DB2, Microsoft SQL Server, Oracle, MySQL, SQLite) manage their own encryption. "Foreign" encryption imposes trade-offs between protection guarantees and limiting DBMS functionality. Section 2 summarizes two encryption types: disk encryption and client-side encryption. Disk encryption protects data at rest (i.e., in persistent storage) with software between the I/O subsystem and DBMS. While this approach is transparent to a DBMS and, thus, does not interfere with DBMS functionality, it offers little control over encryption granularity; a malicious system administrator (or an attacker who gained similar privileges) can access DBMS files at byte-level. Alternatively, data encrypted and decrypted by a client application protects data both in-motion and at-rest. A major trade-off for this approach is a loss of DBMS functionality. The built-in encryption DBMS mechanisms offer a balanced solution between disk and client-side encryption. Section 8 demonstrates how to assess encryption vulnerabilities based on the purpose of each memory area. For example, a decrypted credit card number could appear in memory as part of an INSERT or SELECT query, as an internal copy in buffer cache, or an intermediate computation. The major contributions of the paper are:

- A survey of encryption mechanisms supported by popular DBMSes (Sect. 4). We review encryption options in IBM DB2, Microsoft SQL Server, Oracle, MySQL, PostgreSQL, SQLite, Firebird, and Apache Derby.
- A taxonomy that abstracts four ubiquitous categories of DBMS memory architecture: the I/O buffer, sort area, transaction buffer, and query buffer.
- A framework for isolating and identifying DBMS memory areas (Sect. 6).
- An evaluation of our framework (Sect. 7) demonstrating successful RAM analysis for three representative DBMSes: MySQL, Oracle, and PostgreSQL.
- A use-case study demonstrating how to assign context to encrypted data in RAM (Sect. 8) using a MySQL DBMS instance.

2   Related Work

Assigning Context to Forensic Data. Foundational digital forensic analysis applies file carving techniques, which reconstruct data without using file system metadata. The work in [7,20] presented some of the earliest research around file carving performed as a "dead analysis" on disk images. As the field of digital forensics matured, memory forensics "live analysis" has emerged [6]. An important application for memory forensic investigation is inspecting runtime code to detect malware (e.g., [5]). Such work requires not only carving but a complicated analysis of application and kernel data structures.

Since DBMSes manage their own internal storage separately from the OS and DBMS files are not standalone (unlike PDFs or JPEGs), file carving cannot be applied to DBMS data. Carving DBMS storage was explored in [25,27]. However, database carving has only been part of a "dead analysis." Combining the work in this paper with database carving would enable a "live analysis", such as detecting unusual DBMS access patterns similar to malware detection.

Query Processing for Encrypted Data. Client-side encryption (i.e., encrypting data before loading it into a database) prevents the DBMS from processing data unless data properties are preserved. Deterministic encryption always produces the same ciphertext for a given plaintext and thus supports equality predicates (e.g., WHERE Name = 'Alice'), equality based joins, and DISTINCT operations. GROUP BY operations can be used, but beyond the columns in the GROUP BY clause, deterministic encryption is essentially limited to the COUNT function (e.g., SELECT City, COUNT(*) FROM Customer GROUP BY City).

Order preserving encryption (OPE) produces ciphertext that preserves the plaintext value ordering [1,4]. OPE supports sorting (i.e., ORDER BY), range scans (e.g., WHERE Salary BETWEEN 50K AND 80K), and covering indexes. Homomorphic encryption (e.g., [12]) supports computations on ciphertext, returning an encrypted result. Fully homomorphic encryption supports unbounded computations, but research identified major trade-offs [13,17]. Partially homomorphic encryption offers a more balanced solution by supporting only bounded computations [10]. Support for the standard SQL string wildcard operators (i.e., % and _) on encrypted data was explored in [23]. However, it is only suitable for strings with known patterns. There are no solutions that support query processing on ciphertext with arbitrary wildcard expressions or regular expressions.

Systems such as CryptDB [19], Cipherbase [3], and Microsoft SQL Server's Always Encrypted [28] extend SQL and relational DBMSes to support query processing on encrypted data. However, these systems still sacrifice important functionality, such as nontrivial computations (e.g., multiplication and addition in the same expression) and regular expressions. More importantly, the encryption schemes should be designed with knowledge of the query workload. For example, homomorphic encryption does not support a workload that requires sorting. These systems also remain vulnerable to inference attacks since the ciphertext still preserves data properties [2,14]. Alternatively, the use case in this paper considers encryption that is natively supported by DBMSes. These mechanisms do not sacrifice DBMS functionality and provide access granularity.

3   Background

Global vs. Local DBMS Memory. DBMSes divide memory into either a global or local context. Global memory stores data and objects shared by all users, sessions, or all DBMS processes. Local memory stores data for an individual DBMS process, session, SQL statement, or an operation (e.g., sorting) within a SQL statement. All components in global memory remain active once the DBMS instance is started. Components in local memory may be allocated when the process, session, SQL statement, or operation starts and de-allocated when it ends. Data that is loaded into local DBMS memory is therefore likely to leak into the OS RAM after it has been de-allocated.

Temporary Table. Temporary tables are used to simplify DBMS procedures and improve performance of processing intermediate query results. Temporary tables are only visible to their user session. They are automatically dropped when the session ends; most DBMSes support the option to drop a temporary table on COMMIT. Temporary tables are typically stored in local memory.

Table 1. Encryption features supported by major DBMSes.

DBMS	Instance level	Column level
Apache Derby	✓	✗
DB2	TDE	Pre-built functions, masking
Firebird	✓	✗
MySQL	TDE	Pre-built functions, masking
Oracle	TDE	TDE, masking, pre-built functions
PostgreSQL	✗	Pre-built functions
SQLite	✓	✗
SQL Server	TDE	Client-side TDE, masking, pre-built functions

4   Native DBMS Encryption

Table 1 summarizes the 8 popular DBMSes investigated in this paper their encryption mechanisms. At a high-level we partition all native database encryption into two categories: instance-level and column-level.

Instance-Level Encryption. Instance-level encryption supports encrypting DBMS storage at the granularity of individual files, or other storage structures (e.g., tablespaces). We further categorize the instance-level encryption mechanisms into the standard encryption and transparent data encryption (TDE).

Standard instance-level encryption encrypts all reads and writes to and from DBMS storage; the encryption key is only provided when the DBMS is started or during a new session login. This mechanism works by encrypting entire pages (e.g., table data, binary large objects, and indexes) that make up the DBMS files. Encrypted data typically includes not only user data in tables and indexes, but also WAL files and temporary files created by the DBMS. Standard instance-level encryption is supported by Apache Derby [26], Firebird (user-customized crypt plug-ins [18]), and SQLite (SQLite Encryption Extension, SEE [24]).

TDE is a more advanced version of the standard instance-level encryption, offered primarily by enterprise DBMSes. The major difference between TDE and standard instance-level encryption is a two-tier encryption key architecture. To implement TDE, a DBMS explicitly manages data encryption key(s) to encrypt/decrypt data. The data encryption key(s) themselves are stored in DBMS storage and are further encrypted with a master key(s) created by the user. The master key is stored in a key store that exists externally and independently from the DBMS files. Two-tier key management creates a further diffusion of privilege required to decrypt the data; the master key can remain hidden from the database administrator. TDE is supported by DB2 (Native Encryption [9]), MySQL Enterprise [16], Oracle [15], and SQL Server [11].

Column-Level Encryption. Column-level encryption refers to the DBMS ability to encrypt individual columns or values in a column. The most common form of column-level encryption is pre-built functions in the DBMS engine (implemented in DB2, MySQL, PostgreSQL, Oracle, and SQL Server). To encrypt new data with pre-built encryption functions, the user must include both the plaintext to encrypt and the encryption key with INSERT or UPDATE statements. Similarly, to query encrypted data the user must provide the encryption algorithm, the encrypted value, and the encryption key with SELECT, DELETE, or UPDATE statements. The following query illustrates how these functions are used:

```
SELECT Decrypt(Name, key1) FROM Employee
WHERE SSN = Encrypt('123-45-6789', key2);
```

Another form of pre-built encryption functions offered by enterprise DBMSes is masking (or redaction). Masking allows users to specify a function describing which parts of a value must be hidden. Common examples of masking include revealing only the last four digits of a social security number or the last digits of a credit card number. Masking is supported by DB2, MySQL Enterprise, Oracle, and SQL Server.

In addition to the instance-level TDE, Oracle also supports a TDE mechanism for columns-level encryption. Column-level TDE still uses the two-tier encryption key architecture. SQL Server also supports a form of column-level TDE with Always Encrypted. The main difference with Always Encrypted is that the master key(s) is designed to be stored on the client-side application.

5   Abstracting DBMS Memory Structures

This section describes the abstraction of DBMS memory areas, based on the type of runtime operations each area supports. Area categories can be identified

Table 2. DBMS-specific names for major memory areas.

DBMS	I/O buffer	Sort area	TXN buffer	Query buffer
Apache Derby	Page Cache	JVM Sort Heap	Write Cache	Statement Cache
DB2	Buffer pool	Sort heap	Log buffer	Query heap
Firebird	Page cache	TempCache	Undo log buffer	Metadata Cache
MySQL	Buffer pool	Sort buffer	Redo log buffer	Query Cache
Oracle	Buffer cache	SQL work areas	Redo log buffer	Result Cache
PostgreSQL	Buffer pool	Work_mem	WAL buffer	Query plan Cache
SQLite	Page cache	Transient index[a]	Journal buffer	Tokenizer
SQL Server	Page cache	Work table[a]	Log cache	Procedure Cache

[a]Stored in the I/O buffer

with the help of DBMS documentation and database textbooks; each type of DBMS operation can be consistently mapped to an area. For example, regular table access (e.g., table scan or index-based access) uses the I/O buffer in RAM to cache pages; hash-join execution uses a memory-intensive operation area.

We chose four areas that best represent the power of area-based memory abstraction: I/O buffer, the area for memory-intensive operations (or sort area), transaction (TXN) buffer, and query cache. Each DBMS uses some variant of these four areas – Table 2 lists their DBMS-specific names. An area may exhibit DBMS-specific configuration properties (e.g., sort area is allocated at a different granularity across DBMSes). DBMS can include other specialized memory areas, which can similarly be abstracted through the process described in this paper.

I/O Buffer. The I/O buffer caches table, index, and materialized view pages recently accessed from files on disk. While each DBMS uses a custom algorithm to decide when to store or evict data from the I/O buffer, some variation of the least recently used (LRU) policy is typically used. When at least one page record is accessed by a query, the entire page is cached in RAM and possibly decrypted. In most DBMSes, the I/O buffer contains a significant number of index pages, including the intermediate nodes and leaf pages of B-Tree indexes.

Sort Area. DBMSes reserve a separate area(s) for memory-intensive operations, which we refer to as the sort area. Sorting-like operations include the straight-forward ORDER BY and DISTINCT clauses along with certain types of JOINs, such as merge-join or hash-join. Nested loop join does not require as much memory (for sorting or hashing) and is typically performed in the I/O buffer. Our experiments illustrate the variations in sort area implementation. Oracle creates a sort area per session (i.e., per user connection); MySQL allocates a sort area for each query, even for the same session; PostgreSQL allocates a sort area for each operation (potentially allocating multiple sort areas for a single query). Once the operation associated with the sort area concludes, the sort area is de-allocated.

DBMSes almost always use temporary tables for sorting. This allows the DBMS to process data-intensive operations in parts, while storing the rest of the data in temporary files in persistent storage. The temporary tables are created in a dedicated sort area. Two DBMSes are an exception to that rule. SQL Server also uses temporary tables for sorting (called Work Tables) but actually stores them in the I/O buffer rather than in a dedicated local memory area. SQLite sorts data using temporary indexes rather than temporary tables. This is a consistent approach for SQLite since their tables are in the form of index organized tables.

Transaction Buffer. DBMSes use a TXN buffer to store write-ahead log (WAL) entries, sometimes referred to as redo or journal log entries. These log entries describe the transactional change history to data, including information needed to rollback or recover from the changes made to the database through DML operations (e.g., DELETE or UPDATE). DBMSes typically write to the TXN buffer in a circular pattern while a background process writes the entries to the WAL (or redo) log files on disk. The TXN buffer must be a part of global DBMS memory to avoid conflicting modifications among different users.

Query Cache. The query cache corresponds to operations that store raw SQL code in RAM as well as query execution plans. DBMS can subsequently reuse cached query execution plans in optimizing similar queries. Query cache area may also contain the DBMS-specific general programming code, e.g., PL/SQL (Oracle), PL/pgSQL (PostgreSQL), or T-SQL (SQL Server). Prepared statements and bind variable values are also stored in this area. Depending on the DBMS, the query cache can be part of global memory or local memory.

Other Areas. DBMSes reserve memory areas for background or user-issued maintenance operations. For example, the MySQL Change Buffer maintains indexes in the background, and the PostgreSQL maintenance_work_mem is reserved for the user-issued VACUUM and REINDEX operations. DBMSes often maintain custom resource scheduling information. Examples include the PostgreSQL commit log which stores the current state of each transaction (i.e., in-progress, committed, or aborted), the Firebird LockMem and Oracle Library Cache acquire locks for database objects, and the DB2 locklist that maintains a list of currently locked objects.

6   Experiment Overview

This section describes using our framework to isolate and identify the four memory areas from Sect. 5. Section 7 demonstrates the effectiveness of this framework on MySQL, Oracle, and PostgreSQL (chosen as representative of different internal storage implementations in a DBMS). Section 8 further shows how to apply this framework to assign context to decrypted data in memory for MySQL.

Table 3. SSBM Scale 4 table sizes used for experiments.

	DWDate	Supplier	Customer	Part	Lineorder
Size	200 KB	700 KB	10 MB	50 MB	2.3 GB
Records	2556	8000	120 K	600 K	24M

Our experimental analysis does not consider an exhaustive list of DBMSes and possible configurations; rather, our framework is designed to be independent of such variables. For example, Sect. 7 considers three representative DBMSes of the eight DBMSes listed in Sect. 5, but the same process can be applied to any relational DBMS. Similarly, Sect. 8 only considers MySQL, although the same analysis could be performed for the other DBMSes. This framework focuses on how DBMSes manage their internal process memory. Although we consider default implementations, a researcher could further explore a specific environment (e.g., compare DBMS memory behavior for ptmalloc2 vs. tcmalloc).

6.1 Setup

Dataset. In our experiments we used the Star Schema Benchmark (SSBM) Scale 4 (~2.4 GB or ~25M records). SSBM is widely used in database research community to represent a data warehouse evaluation. It combines a realistic distribution of data (maintaining data types and cross-column correlations) with a synthetic data generator that can create datasets at different scale. Table 3 summarizes the sizes of the SSBM tables used throughout the experiments.

Table 4. DBMS memory area configurations used for experiments

DBMS	I/O buffer	Sort area	TXN buffer	Query buffer	Proc Mem
MySQL	100 MB	256 KB	1 MB	10 MB	383 MB
Oracle	1.6 GB	262 MB	7 MB	200 MB	3.6 GB
PostgreSQL	128 MB	4 MB	4 MB	12 MB	248 MB

DBMS Configuration. Table 4 lists the DBMSes we chose for an evaluation as well as their memory area parameter settings. We chose the settings for memory size in consultation with each DBMS' documentation and the established best practices. For example, MySQL and PostgreSQL are relatively lightweight engines, while Oracle requires significantly more memory. Furthermore, although the memory area serves the same function across DBMSes, the setting depends on DBMS engine implementation. For example, 4 MB for PostgreSQL vs 262 MB for Oracle is not as different as it appears: PostgreSQL initializes a sort area per operation (thereby creating multiple 4 MB buffers per query in many cases), while Oracle uses a shared sort area.

Oracle 12c and MySQL 5.7 were deployed on a Windows 10 server. PostgreSQL 9.6 was deployed on a CentOS 6.5 server. Based on our experimental analysis, DBMS behavior remains similar between Windows and Linux servers.

6.2 Workload

We designed a SQL workload to populate each memory area with data. This includes three specialized sets of queries: 1) for filling the I/O buffer, 2) for filling the sort area with data, and 3) for filling the TXN buffer. For evaluation of the query cache area, we used the queries from the other three custom workloads. We next discuss the workload design in the context of each memory area. The workloads and workload generators can be downloaded from our research group website: http://dbgroup.cdm.depaul.edu/downloads/DB_Mem_Workloads/Workloads.zip These queries are designed sepcifically to highlight the different memory areas. While randomized queries would populate the same memory areas, they do not contribute to the goal of identifying the different memory areas, thus we do not include any.

I/O Buffer. We generated a total of 300,000 SELECT queries: 290,000 for Lineorder, 8,000 for Part, 1,500 for Customer, 400 for Supplier, and 100 for DWDate. All queries included a predicate that accessed equality on a value from an indexed column (to produce query execution with index-based access). An index-based access caches and retains all accessed data pages. Alternatively, full table scan may only cache a small portion of the table in memory and the DBMS is likely to immediately free-list that data. Since the primary key column contains all unique value and an index is automatically created on a primary key column, random values were accessed based on the primary key column. The following query template was used to generate this workload; '?' is a placeholder that was replaced by a (uniformly distributed) random value.

```
SELECT * FROM [Lineorder/Part/Supplier/Customer/DWDate]
WHERE [LO_Orderkey/P_Partkey/S_Suppkey/C_Custkey/D_Datekey] = ?;
```

Sort Area. We designed a memory-intensive query to perform a JOIN on all five tables in SSBM. To force result sorting, the query used a four column composite ORDER BY clause. The SELECT clause used 8 columns (as this is what is sorted in memory); these columns were arranged to uniquely identify them as sorted result among any records found in the SSBM tables and thus in the RAM snapshot. We experimentally chose the number of columns to be sufficiently large to fill each DBMS respective sort area.

```
SELECT S_Name,C_Name,P_Name,D_Day,S_City,S_Nation,S_Phone,C_Nation
FROM Lineorder JOIN Part JOIN Customer JOIN Supplier JOIN DWDate
ORDER BY S_Name, C_Name, P_Name, D_Dayofweek;
```

Transaction Buffer. We issued 10 UPDATE queries against the Part table. Each query updated 150,000 different records. To definitively detect entries in transaction buffer area, every query updated the container column to a (string + a unique ID) value not already used in the table. We used the following template for our update queries. The first question mark was replaced by a unique ID, the second question mark was replaced by a value from the P_Container column.

```
UPDATE Part SET P_Container = 'DEXA'+ ? WHERE P_Container = ?;
```

6.3   Experimental Procedure

We performed the experiments in the following sequence of steps for each DBMS:
1) Set up a new DBMS instance, 2) Load the SSBM tables into the DBMS,
3) Run the I/O cache query workload, 4) Run the transaction buffer query
workload, 5) Run the sort area query workload. The RAM snapshot was generated *during* step #5 while the sort area workload was still running. Since sort
area is part of local memory, it would become de-allocated after the sort area
workload was completed. Therefore, the memory had to be captured while this
local area was still allocated to the DBMS process. We verified that, if taken
after step #5, the sort area was no longer a part of the captured DBMS process
memory for all three evaluated DBMS. We used procdump v9.0 [21] to collected
DBMS process snapshot on the Windows server, and read the process snapshot
data under /proc/$pid/mem on the Linux server.

To evaluate the contents of the memory snapshots, used regular expressions
with Python 2.7 to locate matching data values and their offsets. We designed the
regular expression to search for known string values introducing enough slack for
metadata content (varies by DBMS). For example, we used the following regular
expression to detect customer records. Each string represents possible values
(e.g., 'Customer#000000042', 'EUROPE', '85-234-621-3704') plus the additional
wildcards for numeric columns and metadata characters.

```
''Customer#[0-9]{9}.{5,60}((EUROPE)|(AFRICA)|(AMERICA)|(MIDDLE EAST)|(ASI
 ↪ A)).{1,10}[0-9]{2}-[0-9]{3}-[0-9]{3}-[0-9]{4}''
```

7   Memory Experiments

For each experiment, we performed at least five evaluations and chose a representative snapshot (snapshots were always consistent with minor variations).

7.1   RAM Spectroscopy Graphs

Figure 1 summarizes the memory contents; each DBMS is represented by a separate graph to describe and quantify contents of its process memory. The four
memory areas from Sect. 5 are annotated with the following legend: I/O cache
line is highlighted by square points, query cache line is denoted by diamonds, sort
area is identified by triangles, and the TXN buffer is marked by circles. We term
these graphs as *RAM spectroscopy*, which was inspired by infrared (IR) spectroscopy commonly used in analytical chemistry [22]. IR spectroscopy graphs
measure the amount of infrared light absorbed by a chemical sample at different
wavelengths. In an analogous manner, the purpose of our RAM spectroscopy
graphs is to visualize the amount of data found at different memory offsets. We
observed that each DBMS maintained a consistent shape throughout multiple
session connections and system restarts. RAM spectroscopy cannot be applied
to full OS RAM snapshots due to heavy fragmentation of the DBMS data.

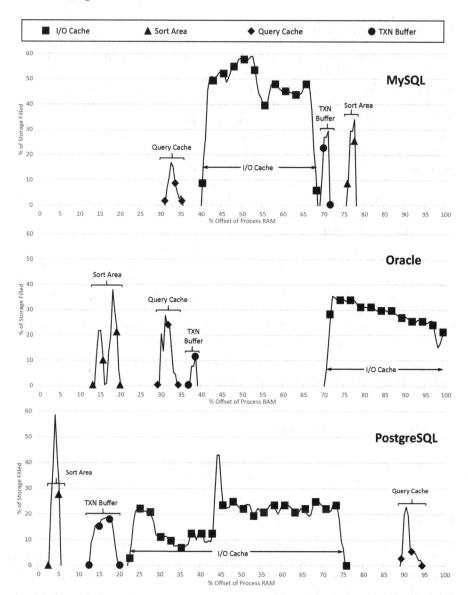

Fig. 1. Process memory representation for MySQL, Oracle, and PostgreSQL

For each RAM spectroscopy graph in Fig. 1, the x-axis represents the byte offset within the DBMS process snapshot, normalized as a percentage. For example, 50% represents 50 MB in a 100 MB process snapshot or 800 MB in a 1.6 GB process snapshot. We summarized the data to 200 points (i.e., a point at every 0.5%) to normalize the snapshots for DBMSes across different RAM sizes. The y-axis represents an estimated amount of memory storage filled at a given offset.

To estimate the percent of the storage filled by our data values, we assumed an additional 20% overhead to the data found. That is, for 'Customer#000000042' we accounted for (a total of 18×1.20) 21.6 bytes. This overhead is based on a generally accepted estimate of metadata associated with a DBMS page. While metadata varies between DBMSes, we chose a constant estimate to simplify our measurements. We also note that not all memory areas use pages (e.g., I/O buffer uses pages but sort area buffer does not). However, we only consider the relative heights of the peaks and we do not compare across areas (e.g., we do not compare I/O buffer peaks vs sort area buffer peaks).

7.2   Memory Observations

Memory Area Data. Figures 1 and 2 reflect only the SSBM table data distribution. Each area contains other data that we do not consider; we therefore never expect to observe values close to 100%. It is likely that memory areas are not densely packed or contain data from DBMS system tables (we did not load other data tables, but all DBMSes use internal "system" tables). Moreover, the memory areas typically contain auxiliary data or metadata in addition to raw table data. For example, the I/O cache includes index pages, which we did not measure in our report (I/O buffer regular expressions search for table rows and not index entries). The indexes used integer columns, and integers have their own DBMS-specific encodings that vary both in format and in size. Although index access and caching behavior would share similarities across DBMSes, we measured cached table rows (or SQL query result rows for query cache area) as the most consistent and representative way to detect the relevant memory areas.

Identifying Memory Area Regions. For all DBMSes the size of each memory area was consistent with sizes in the configuration files (see Table 4). When repeating and verifying these results, we observed the memory areas maintained the same order with slight shifts within the process memory snapshot. Therefore, we concluded that when a process memory snapshot is taken, data found in those offset regions belongs to the respective memory region. Each snapshot is a chosen representative of at least five independent snapshots we recorded. However, all of the snapshot were similar enough that any one of them could have been chosen for the spectroscopy figure report. Figure 1 also indicates how much of the overall DBMS memory process is occupied by the four memory areas. PostgreSQL snapshot uses relatively little space outside of these areas, while both Oracle and MySQL allocate a significant quantity of other RAM.

Local Memory. When the sort area query finished executing and the user session was disconnected, the sort area was no longer present in the process memory snapshot. This is consistent with the behavior of a local memory buffer, which stores the sort area. After de-allocation, the sort area data values (the output of the sort area SQL workload) could still be found in the full OS RAM snapshot, outside of the DBMS process. However, the sort area contents were now fragmented across OS RAM. Therefore, we concluded that when local DBMS memory is deallocated, its contents are effectively leaked into global OS RAM.

Memory Area Shapes. In Figure 1, MySQL I/O cache buffer fills by approximately 50%, in contrast with Oracle (approximately 25%) and PostgreSQL (approximately 20%). This is consistent with our expectations because MySQL uses index-organized tables. As a result, query access does not fetch index pages independently of the data pages (as there is no separate index structure). Specifically, B-Tree leaf pages with value-pointer pairs do not exist because data is in the leaf page of the B-Tree. Alternatively, both Oracle and PostgreSQL fetch a significant number of index pages, filling the I/O buffer cache with non-table pages. As a result, while the number of pages in the I/O buffer is similar, there are fewer table pages in Oracle and PostgreSQL compared to MySQL.

Oracle sort area in Fig. 1 exhibits two distinct peaks for the single query we executed. This is also consistent with our expectations because Oracle uses hash-join which is a memory-intensive operation that targets the sort area buffer. We therefore observed data originating from two different operations in Oracle's sort area: the results sorting and the hash-joins. PostgreSQL sort area in Fig. 1 exhibits only one peak. While PostgreSQL also uses a hash-join, it allocates a separate sort area for each operation. Therefore, the PostgreSQL hash-join operations use a different sort area that was de-allocated at the time the process snapshot was taken. MySQL uses nested loop join which will execute in the I/O buffer. Therefore, the MySQL sort area is dedicated to the result sorting.

8   Encryption Experiments

The purpose of this experiment is to demonstrate the importance of assigning context to data. We extend the Sect. 7 experiments using a new MySQL 5.7 instance with encryption enabled. The same setup and procedure described in Sect. 6 were used except TDE was enabled for all five SSBM tables. Since finding decrypted data in memory is an expected result, we emphasize that assigning context to this data can anticipate vulnerabilities.

Figure 2 displays the resulting RAM spectroscopy graph for the encrypted MySQL instance combined with the MySQL instance data from Fig. 1. The old unencrypted instance is represented with the gray line and the encrypted instance is represented with a black line.

All memory area peaks were observed equivalent, confirming that all data read into memory with TDE is decrypted and accessible in RAM. As a result, TDE has no significant impact on protecting the data from RAM perspective. However, it does not exhibit new vulnerabilities as does column-level encryption. Figure 2 also illustrates the consistency of the peak detection by superimposing results from two different snapshots. We note that the sort area buffer exhibited the same de-allocation behavior; as a result the decrypted data was released into global OS RAM. This data is particularly vulnerable because it could be observed in RAM and potentially captured with `malloc` from another process.

The experiment in Fig. 2 measured the data cached by MySQL using instance-level TDE. The column-level encryption that relies on pre-built functions can manifest additional data vulnerabilities, depending on the memory

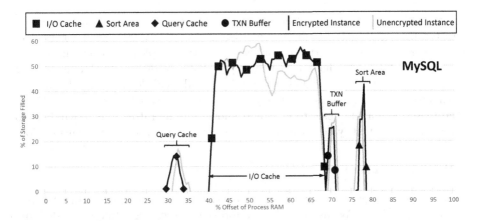

Fig. 2. Process memory representation for encrypted MySQL instance superimposed over the unencrypted MySQL instance from Fig. 1

area. The I/O buffer will expose less data with column-level encryption compared TDE. While the column-level encryption pages are visible in the I/O buffer, individual values in pages will remain encrypted in RAM. In contrast, both the query cache area and transaction buffer area will expose the encryption key in column-level encryption schemes. Pre-built encryption queries explicitly specify the encryption key in SQL commands which are cached in query cache and transaction buffer. The sort area will expose a similar amount of data for both column-level and TDE encryption because both TDE page requests and column-level encryption **SELECT** clause decrypts the queried values.

9   Future Work

The work in this paper supports future directions for third-party tools to assign context to data in addition to carving raw content from DBMS memory and providing detailed data flow tracking. Current DBMS APIs do not support data flow tracking and offer few limited system analysis features. For example, Oracle allows users to query the number of pages associated with table in the I/O buffer, but not the information about specific pages or records. Most DBMSes do not even offer the features provided by Oracle. We believe that data flow tracking has two primary application: security monitoring and performance analysis.

Current work in memory forensics detects activity patterns indicative of malware. The equivalent for DBMSes is detecting unusual data access patterns in RAM. Tools such as IBM Guardium [8] detect unusual patterns by observing SQL queries. While useful, this approach is limited – an obfuscated SQL query or a query that bypassed the monitoring proxy will escape detection. However, the approaches discussed here would allow monitoring memory operations in the event that an attacker circumvents current detection mechanisms.

DBMSes use a complex set of configuration settings. Our experiments demonstrated that these settings are not consistent across DBMSes; even for a corresponding setting (e.g., sort area buffer) the actual implementation can lead to a radically different behavior. For example, it is a known issue that increasing PostgreSQL area buffer setting (seemingly a good idea!) leads to significant performance deterioration as too many buffers are allocated in some workloads. Database memory forensic tools would allow administrators and researchers to more precisely identify performance bottlenecks and monitor memory utilization.

10   Conclusion

This paper presented a systematic approach to reverse engineering DBMS-controlled memory. We evaluated our approach by creating a taxonomy defining several common memory areas. Experiments demonstrated how to identify and isolate DBMS memory areas through design and evaluation of custom query workloads. We validated our approach on four memory areas using three representative DBMSes (PostgreSQL, Oracle, and MySQL). Finally, experiments showed the significance of assigning context to data in memory, an inherent feature of our reverse engineering approach.

References

1. Agrawal, R., Kiernan, J., Srikant, R., Xu, Y.: Order preserving encryption for numeric data. In: SIGMOD conference, pp. 563–574 (2004)
2. Akin, I.H., Sunar, B.: On the difficulty of securing web applications using cryptDB. In: Conference on Big Data and Cloud Computing, pp. 745–752. IEEE (2014)
3. Arasu, A., et al.: Orthogonal security with cipherbase. In: CIDR. Citeseer (2013)
4. Boldyreva, A., Chenette, N., Lee, Y., O'Neill, A.: Order-preserving symmetric encryption. In: Joux, A. (ed.) EUROCRYPT 2009. LNCS, vol. 5479, pp. 224–241. Springer, Heidelberg (2009). https://doi.org/10.1007/978-3-642-01001-9_13
5. Case, A., Richard III, G.G.: Detecting objective-C malware through memory forensics. Digit. Invest. **18**, S3–S10 (2016)
6. Case, A., Richard III, G.G.: Memory forensics: the path forward. Digit. Invest. **20**, 23–33 (2017)
7. Garfinkel, S.L.: Carving contiguous and fragmented files with fast object validation. Digit. Invest. **4**, 2–12 (2007)
8. IBM: Security guardium (2017). http://www-03.ibm.com/software/products/en/ibm-security-guardium-express-activity-monitor-for-databases
9. IBM: Db2 native encryption (2019). https://www.ibm.com/support/knowledgecenter/SSEPGG_11.1.0/com.ibm.db2.luw.admin.sec.doc/doc/c0061758.html
10. Liu, J., Mesnager, S., Chen, L.: Partially homomorphic encryption schemes over finite fields. In: Carlet, C., Hasan, M.A., Saraswat, V. (eds.) SPACE 2016. LNCS, vol. 10076, pp. 109–123. Springer, Cham (2016). https://doi.org/10.1007/978-3-319-49445-6_6
11. Microsoft: Transparent data encryption (2019). https://docs.microsoft.com/en-us/sql/relational-databases/security/encryption/transparent-data-encryption?view=sql-server-ver15

12. Microsoft: Microsoft seal (2020). https://www.microsoft.com/en-us/research/project/microsoft-seal/
13. Naehrig, M., Lauter, K., Vaikuntanathan, V.: Can homomorphic encryption be practical? In: Workshop on Cloud computing security, pp. 113–124 (2011)
14. Naveed, M., Kamara, S., Wright, C.V.: Inference attacks on property-preserving encrypted databases. In: SIGSAC Conference, pp. 644–655 (2015)
15. Oracle: Database advance security guide (2017). https://docs.oracle.com/database/121/ASOAG/toc.htm
16. Oracle Corporation: Innodb data-at-rest encryption (2020). https://dev.mysql.com/doc/refman/5.7/en/innodb-data-encryption.html
17. Peng, Z.: Danger of using fully homomorphic encryption: a look at microsoft seal. arXiv preprint arXiv:1906.07127 (2019)
18. Peshkov, A., Firebird foundation: encrypting firebird databases (2016). https://firebirdsql.org/file/documentation/release_notes/html/en/3_0/rnfb30-security-encryption.html
19. Popa, R.A., Redfield, C.M., Zeldovich, N., Balakrishnan, H.: CryptDB: protecting confidentiality with encrypted query processing. In: SOSP, pp. 85–100 (2011)
20. Richard III, G.G., Roussev, V.: Scalpel: a frugal, high performance file carver. In: DFRWS (2005)
21. Russinovich, M., Richards, A.: Procdump v9.0 (2017). https://docs.microsoft.com/en-us/sysinternals/downloads/procdump
22. Skoog, D., West, D., Holler, J., Crouch, S.: Fundamentals of analytical chemistry. Brooks-Cole, Molecular Absorption Spectroscopy (2014)
23. Song, D.X., Wagner, D., Perrig, A.: Practical techniques for searches on encrypted data. In: IEEE S&P conference, pp. 44–55. IEEE (2000)
24. SQLite: Sqlite encryption extension (2019). https://www.sqlite.org/see
25. Stahlberg, P., Miklau, G., Levine, B.N.: Threats to privacy in the forensic analysis of database systems. In: SIGMOD Conference, pp. 91–102 (2007)
26. The Apache Software Foundation: Configuring database encryption (2016). http://db.apache.org/derby/docs/10.13/security/cseccsecure24366.html
27. Wagner, J., Rasin, A., Malik, T., Heart, K., Jehle, H., Grier, J.: Database forensic analysis with DBcarver. In: CIDR Conference (2017)
28. Ward, B.: SQL Server 2019 Revealed. Apress, Berkeley, CA (2019). https://doi.org/10.1007/978-1-4842-5419-6

Collaborative SPARQL Query Processing for Decentralized Semantic Data

Arnaud Grall[1,2], Hala Skaf-Molli[1]([⊠]) [iD], Pascal Molli[1] [iD],
and Matthieu Perrin[1] [iD]

[1] LS2N – University of Nantes, Nantes, France
{arnaud.grall,hala.skaf,pascal.molli,matthieu.perrin}@univ-nantes.fr
[2] GFI Informatique - IS/CIE, Nantes, France
arnaud.grall@gfi.fr

Abstract. Decentralization allows users to regain freedom and control over their digital life. As a global shared data space, the Linked Data already supports decentralization. Data providers are free to publish their data on their web domains and users can execute decentralized SPARQL queries over multiple data sources. However, decentralization makes query processing challenging, raising well-known problems of source discovery, answer completeness and performance. Existing approaches for decentralized SPARQL query processing raise issues related to autonomy and answer completeness. In this paper, we propose Qasino, an original approach for querying decentralized RDF data that targets both answer completeness, and source autonomy. Qasino is based on a decentralized random service that allows for discovering all relevant data sources. To speed up query processing, sources executing similar queries cooperate by sharing their intermediate results. Our experimental results demonstrate that collaborative query processing can significantly speedup query processing in a decentralized setup.

Keywords: Decentralized data management · SPARQL query processing · Sources discovery

1 Introduction

Decentralization is a common way to give users back control over their digital life. As a global shared data space, the Linked Data already supports decentralization. Data providers are free to publish their data on their web domains and users can execute decentralized SPARQL queries over multiple data sources. However, decentralization introduces challenging problems for query processing related to well-known problems of source discovery, completeness and performance. Discovering all relevant sources for a query remains an issue. Existing federated query engines assume the existence of a catalog [3,27]. Link traversal [12] crawls links from a seed URI during query execution but cannot ensure that all relevant sources are reachable from the seed. Semantic P2P data management [5]

© Springer Nature Switzerland AG 2020
S. Hartmann et al. (Eds.): DEXA 2020, LNCS 12391, pp. 320–335, 2020.
https://doi.org/10.1007/978-3-030-59003-1_21

rebuild overlay networks on top of sources to allow efficient discovery. However, P2P data management raises issues on autonomy, i.e. participants must agree to participate in common tasks such as routing, indexing, and replication.

In this paper, we propose Qasino, an original approach for querying decentralized RDF data that targets both answer completeness and sources autonomy. Qasino relies on a decentralized random service ables to return a random participant. Thanks to the random service, Qasino crawls the set of participants while running a query. Crawling allows to discover both sources and similar running queries. Similar queries collaborate by sharing queries intermediate results and random draws. Collaboration allows to speed up queries termination while producing complete results. In this paper, we propose the following contributions:

- A new model for decentralized SPARQL query processing.
- A collaborative Monte-Carlo SPARQL query execution algorithm that allows collaborative discovery of datasources. If several participants execute similar queries, then they will eventually meet several times during query execution and share their results. Collaboration allows to speed up query termination and provides probabilistic guarantee on answers completeness.
- A simulator to run experimentations with thousands of participants, which goes beyond traditional decentralized query experimentations.

The paper is organized as follows. Section 2 describes the related works. Section 3 presents the Qasino approach. Section 4 presents Qasino algorithms. Section 5 presents experimental results. Section 6 concludes contributions and presents future works.

2   Related Works

Decentralization allows users to store their RDF data where they want on the Web. However, executing a SPARQL query over decentralized data requires to discover all relevant data sources. Solving the discovery problem at the scale of the Web while preserving sources autonomy is still an open issue.

The Solid project [19] promotes the vision of a decentralized web for social web applications. In Solid, users store their data in personal online datastores (pods). However, Solid does not describe how to run a query over a large-scale network of pods.

Link traversal [12] allows to execute SPARQL queries directly on the Linked Data resources, relying on URI dereferencing. Link traversal starts the query execution from a seed URI provided by users and considers every URI appearing in mappings as a new data source. Therefore, it is up to the user to know at least one seed, and the link traversal is able to discover new sources during query processing. Therefore, sources discovery is partially in charge of data consumers, moreover, query answer completeness is defined according to the reachability semantics [13].

Federated query engines [3, 27] allow data consumers to execute queries over a catalog of data sources. It is up to the data consumers to provide this catalog. Building this catalog traditionally implies to collect the description of all potential sources, which is again in charge of data consumers. The query answer completeness is defined according to the set of sources contained in the catalog.

Many systems rely on Distributed Hash Tables (DHT) such as P-Grid [1] or GridVine [2] for sources discovery. However, DHT systems do not allow users to choose where their data are stored, consistent hashing determine where data should be stored. DHT can be used also just as a distributed catalog and data remain located where users want as in the InterPlanetary File System (IPFS). In this case, participants have to agree to participate to keybase routing and store informations that do not belong to them.

Recently, unstructured P2P techniques have gained attention of Semantic Web community as potential decentralized architectures for Linked Data management [4, 10, 11, 20, 26]. Existing approaches maintain a neighborhood for each site. As a participant does not know where data are located, she floods the network with her query [4]. However, this approach does not scale and hardly delivers complete results. Flooding can be avoided with super-peer maintaining routing indices as in [24]. Having super-peers in a network of nodes is possible but they represent a point of failure which is a strong limitation to massive deployment of distributed applications in nodes. Flooding can also be reduced with spanning trees as in sensor networks [6]. A spanning tree reduces the flooding to the number of nodes in the network. However, spanning trees are costly to maintain on large networks with heavy churn. The network traffic can be significantly reduced using adapted replication strategies and random walks [18]. Flooding can also be limited by using multiple overlays as in Semantic Overlay Network (SON) or routing indices [5, 7]. Participants are clustered in communities according to their common interests. Queries are routed to the right community to be executed. SON restricts the number of sources for a query. This supposes that participants agree to compute this routing and maintain information that are not directly relevant for executing their queries.

In both structured or unstructured P2P networks, solving the discovery problem requires participants to loose autonomy. Participants have to route all messages, not only those they want and they cannot choose data to host or to replicate. In Qasino, we explore an original P2P approach that requires only that participants accept to be discovered.

3   Qasino Approach and Models

In Qasino, we consider a community of participants. We aim to preserve two properties: (i) *Completeness:* we aim to execute queries and get complete results over all data stored by the community. (ii) *Autonomy:* participants only host data they want and there is no routing. In this context, solving the discovery problem is to guarantee that each participant can discover all other participants.

We consider a random abstract service able to return a random participant among the community. Such random service is enough to enable discovery. This service can be implemented in a decentralized fashion relying on random peer sampling techniques [14]. Consequently, participants only collaborate to return a random node in the set of nodes. To build a query engine according to this model, we follow a bottom-up query evaluation strategy [16]; sources description are not available and the query engine discovers new sources incrementally and terminates when all sources or *a priori*-decided proportion of sources have been processed. The Qasino approach allows not only to discover new sources during queries evaluation but also to discover other participants running similar queries. In this case, queries can collaborate to speed up query termination.

3.1   Qasino Nodes Data Structures

We consider a community of participants as a set of n nodes, n is unknown to participants. A node has a local data structure and a shared one defined as follows.

Definition 1 (Local Structure). *A node N_i has access to:*

- *$\hat{n}$, a statistical estimator of the number of nodes.*
- **rand**, *a function generating independent and uniformly distributed random variables from the set of nodes, i.e. each time the node calls* **rand** *(), it gets a random node.*
- *$\mathcal{D}_i$, a local RDF dataset.*
- *Q_i, a SPARQL query. For each triple pattern tp of Q_i, $Q_i[tp]$ stores the set of mappings of the variables of tp.*

A set of nodes $\{N_1, \ldots N_n\}$ defines a virtual dataset $\mathcal{D}$ defined as $\mathcal{D} = \cup_{i=1}^{n} N_i . \mathcal{D}_i$, i.e. $\mathcal{D}$ is the union of local RDF datasets. For simplicity, in this work we suppose that $\mathcal{D}_i$ is immutable and a node executes only one SPARQL query, this can be easily extended to a set of queries.

Definition 2 (Remote Interface). *A node N_i exposes to other nodes:*

- *$[\![\cdot]\!]$, evaluation function as defined in [25], i.e. for a triple pattern tp, $N_i.[\![tp]\!]$ returns the set of mappings for the variables of tp that match the dataset $\mathcal{D}_i$.*
- *$\mathcal{E}(tp)$, a boolean function that returns true if N_i accepts to collaborate on the evaluation of tp, with $tp \in Q_i$.*

3.2   Qasino Query Processing Model

Each node maintains a local RDF data and can evaluate, at least, a triple pattern query. Query processing follows a bottom-up query evaluation strategy [16]. This strategy does not assume source descriptions to be available before query execution and computes results in a bottom-up fashion. A SPARQL query Q_i at a node N_i is processed in the following steps: (1) Built a left-tree query plan of Q_i. To determine the order of the evaluation of triple patterns, the cardinality of each triple pattern tp in Q is estimated using variable counting [29]. The most selective triple pattern is evaluated first. (2) Evaluate Q's triple patterns over D_i, the local dataset of N_i ($Q_i[tp] \leftarrow N_i.[\![tp]\!]$). The evaluation relies on the pushed-based symmetric hash join operator [16], i.e. results are produced as soon as input tuples are available and input tuples can arrive on all inputs in any order. (3) Discover a source randomly N_j, among the nodes. N_j evaluates Q's triple patterns, as detailed in the different algorithms in the next sections. (4) Receive partial results from N_j (only if N_j has results), add partial results to hash table of the corresponding triple pattern and produce results (if available). In Qasino, *source discovery* is an integral part of the query processing. Sources are discovered online, and query results are produced incrementally.

Problem Statement: Given a set of nodes $\{N_1, \ldots N_n\}$, our objective is to define a query execution function `execute` that ensures query termination and answer completeness. $\forall N_i \in \{N_1, \ldots N_n\}$, we expect:

(i) **Termination** N_i.`execute` eventually returns.

(ii) **Completeness** After a node N_i has terminated, $Q_i[tp] = [\![Q_i]\!]_{\mathcal{D}}$, i.e. Q_i is evaluated over the virtual dataset $\mathcal{D}$.

As nodes can only discover sources randomly, this makes the respect of both properties impossible. Among existing strategies for randomized algorithms are Las Vegas and Monte-Carlo algorithms. Las Vegas algorithms where the termination property is weakened to termination with probability 1, i.e. N_i.`execute` returns with probability 1. Monte-Carlo algorithms ensure termination but completeness is replaced by the following guarantee: N_i.`execute` returns $[\![Q_i]\!]_{\mathcal{D}}$ with some probability $\rho > 0$ independent of n. Consequently, two termination conditions are possible: (1) All nodes are discovered, (2) A proportion p of nodes is discovered. Termination conditions impact the query completeness, i.e. evaluating the query over all nodes ensures answer completeness, this is not always the case for the second condition. Moreover, termination conditions impact the complexity of steps 3 and 4 of query processing.

4   Algorithms

In the following, we detail existing strategies for randomized algorithms and we propose a new collaborative randomize algorithm for SPARQL query processing. The proposed algorithm allows to speed up query termination while preserving the proportion of discovered nodes.

4.1   Discover All Nodes: Las Vegas Algorithm

Algorithm 1: Las Vegas SPARQL engine

Data: $V_i \leftarrow \{N_i\}$: Set of visited nodes

1 **Function** $N_i.\texttt{execute}()$:
2   **while** $|V_i| < \hat{n}$ **do**
3   let $N_j \leftarrow \texttt{rand}()$
4   **if** $N_j \notin V_i$ **then**
5   **foreach** $tp \in Q_i$ **do**
6             $Q_i[tp] \leftarrow Q_i[tp] \cup N_j.[\![tp]\!]$
7           $V_i \leftarrow V_i \cup \{N_j\}$

Algorithm 1 presents a Las Vegas algorithm for evaluating a SPARQL query Q_i. For simplicity, we make the hypothesis that each node executes only one SPARQL query. Thanks to the random service, it may discover a new node at each iteration. If the discovered node has relevant data for the query, the query execution will produce new query results. Assuming the estimator $\hat{n}$ returns the exact number of nodes, consequently, it always produces correct and complete results, but its running time complexity is non-deterministic and it only terminates with probability 1. The main issue is to evaluate how many draws, in average, are necessary to get complete results. Such problem is similar to the coupon collector problem [22]. The average complexity is: $\sum_{i=1}^{n} \frac{n}{i} = n \times (\ln(n) + \gamma) + \mathcal{O}(1)$ iterations, where $\gamma \approx 0.577$ is the Euler–Mascheroni constant.

To illustrate, consider $n = 1000$ nodes, a node executing a query Q_i should try around 7484 random call to $\texttt{rand}$ () in order to discover all nodes.

This algorithm raises several issues: (i) It requires that the exact number of nodes is known and immutable, which is not realistic in a decentralized setting. If n is overestimated by $\hat{n}$, then the algorithm does not terminate. Conversely, if n is underestimated by $\hat{n}$, then the results may be incomplete. (ii) As illustrated, discovering all the nodes can be very costly, especially discovering the last missing nodes.

4.2 Discover a Proportion of Nodes: A Monte-Carlo Algorithm

Algorithm 2: Monte-Carlo SPARQL engine

Require: $p < 1$: Expected proportion of sources observed during a run
Data: $V_i \leftarrow \{N_i\}$: Set of visited nodes
1 **Function** $N_i.\texttt{execute}(p)$:
2   **for** k from 1 to $\hat{n} \times \ln\left(\frac{1}{1-p}\right)$ **do**
3   let $N_j \leftarrow \texttt{rand}()$
4   **if** $N_j \notin V_i$ **then**
5   **foreach** $tp \in Q_i$ **do**
6             $Q_i[tp] \leftarrow Q_i[tp] \cup N_j.[\![tp]\!]$
7           $V_i \leftarrow V_i \cup \{N_j\}$

Instead of discovering all the nodes, a user can decide to stop the exploration after a given number k of random draws, for example $2n$ draws, hoping to discover as many sources as possible. Algorithm 2 describes a Monte-Carlo algorithm for executing a query Q_i, based on this strategy.

Ideally, the user would decide to explore a proportion p of the nodes, for example only 99% of nodes to terminate. The main issue is to calibrate k such that, in average, the algorithm will discover $p \times n$ sources. Surprisingly, for a given p, the necessary number of draws is linear in n, as we will now detail.

Let us first compute the expected proportion $u_n(k)$ of the sources that have yet to be discovered after k draws, among n sources. Initially, no source has been discovered, so $u_n(0) = 1$. During the kth draw, a new source is discovered with probability $\frac{1}{n}u_n(k)$, so $u_n(k+1) = u_n(k) - \frac{1}{n}u_n(k) = \frac{n-1}{n}u_n(k)$. The solution to this geometric progression is $u_n(k) = \left(\frac{n-1}{n}\right)^k$.

The number k_{max} of random participants that must be drawn in average to see a proportion p of the sources verifies the equation $1 - u_n(k_{max}) = p$, that is $1 - p = \left(\frac{n-1}{n}\right)^{k_{max}}$. This equation can be rewritten as $k_{max} = \frac{\ln(1-p)}{\ln\left(\frac{n-1}{n}\right)} = \frac{\ln\left(\frac{1}{1-p}\right)}{\ln\left(\frac{n}{n-1}\right)}$. By the mean value theorem applied to function $\ln$, there is an $x \in [n-1; n]$ such that $\ln\left(\frac{n}{n-1}\right) = \frac{1}{x}$. In other words, $k_{max} = x \ln\left(\frac{1}{1-p}\right) \lesssim n \ln\left(\frac{1}{1-p}\right)$, which gives the number of iterations in Algorithm 2.

To illustrate, consider a set of $n = 1000$ nodes and $p = 99\%$, then Algorithm 2 requires $1000 * (\ln(1/1 - 0.99)) = 4605$ random draws to terminate. Compared to the Las Vegas algorithm, for a given p, the runtime complexity of Algorithm 2 is linear in n, compared to $\mathcal{O}(n \ln(n))$ for Algorithm 1. Moreover, the Monte-Carlo algorithm supports that n is approximated.

4.3 New Monte-Carlo Algorithm for Collaborative Query Processing

The random service allows to discover not only data but also other nodes running similar queries. Therefore, it is possible for queries to collaborate by sharing intermediate results and random draws. This allows to speed up query termination while preserving the proportion of discovered nodes.

Algorithm 3: Collaborative Monte-Carlo SPARQL engine

Require: $p < 1$: Expected proportion of sources observed during a run
Data: $V_i[tp] \leftarrow \{N_i\}$: set of visited nodes by a node searching tp
$k_i[tp][j]$: number of iterations by the node N_j searching tp.

1 **Function** $N_i.\text{execute}(p)$:
2   **while** $\exists tp \in Q_i : \sum_l k_i[tp][l] < \hat{n} \times \ln\left(\frac{1}{1-p}\right)$ **do**
3   **let** $N_j \leftarrow \text{rand}()$
4   **foreach** $tp \in Q_i$ **do atomically**
5        $k_i[tp][i]{+}{+}$
6   **if** $N_j.\mathcal{E}(tp)$ **then**
7          $N_i.\text{sync}(tp, N_j.\text{sync}(tp, \langle Q_i[tp], V_i[tp], k_i[tp]\rangle))$
8   **else if** $N_j \notin V_i[tp]$ **then**
9          $Q_i[tp] \leftarrow Q_i[tp] \cup N_j.[\![tp]\!]$
10          $V_i[tp] \leftarrow V_i[tp] \cup \{N_j\}$

11 **Function** $N_j.\text{sync}(tp, \langle \mathcal{M}_i, V_i, k_i\rangle)$:
12    $Q_j[tp] \leftarrow Q_j[tp] \cup \mathcal{M}_i$
13    $V_j[tp] \leftarrow V_j[tp] \cup V_i$
14    $k_j[tp] \leftarrow \max(k_j[tp], k_i)$
15   **return** $\langle Q_j[tp], V_j[tp], k_j[tp]\rangle$

Algorithm 3 extends Algorithm 2 to handle collaborative query processing. Compared to Algorithm 2, the local variables $Q_i[tp]$, V_i and k_i are replaced in Algorithm 3 by shared state-based Commutative Replicated Data Types (CRDT) data structures [28], i.e. two grow-only sets $Q_i[tp]$, $V_i[tp]$ and a counter $k_i[tp]$ per triple pattern $tp \in Q_i$. CRDT data structures allow shard data to *eventually converge* towards shared state without conflicts. In Algorithm 3, counters eventually converge towards the global number of increments, and CRDT grow-only sets eventually converge towards the union of sets. In order to collaborate on the computation of a triple pattern tp, each node exposes an additional function $\text{sync}(tp, \langle \mathcal{M}, k, V\rangle)$ that allows pairs of nodes to synchronize their mappings and counters.

Shared Counters. For each triple pattern $tp \in Q_i$, a node N_i maintains an associative array $k_i[tp][]$. $k_i[tp][j]$ represents the number of draws that N_j has participated for the computation of the triple tp, as known by N_i. $k_i[tp][i]$ represents the number of draws that N_i has participated for the computation of tp. Each time N_i draws a random source, it increments its own number of random draws. For instance, $k_1[tp] = [1 \mapsto 3; 4 \mapsto 42]$ means that N_1 has done 3 draws from the computation of tp and it knows that N_4 has participated to 42 draws for the computation of tp. If a random node N_j accepts to collaborate with N_i on the evaluation of tp, they merge their shared counter by taking the max value for each cell. For example, if $N_2.k_2[tp] = [1 \mapsto 1; 2 \mapsto 73]$, then $k_1[tp]$ and $k_2[tp]$

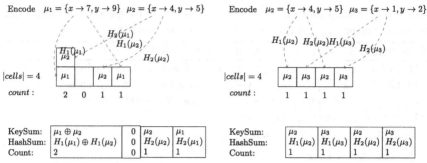

Encode $\mu_1 = \{x \to 7, y \to 9\}$ $\mu_2 = \{x \to 4, y \to 5\}$ Encode $\mu_2 = \{x \to 4, y \to 5\}$ $\mu_3 = \{x \to 1, y \to 2\}$

(a) Computing $IBLT_1$ for $N_1.Q_1[?x\ P1\ ?y]$ (b) Computing $IBLT_2$ for $N_2.Q_2[?x\ P1\ ?y]$

2) Difference: $IBLT_2$ - $IBLT_1$, where $W \oplus W = 0$

KeySum:	μ_1	μ_3	0	$\mu_3 \oplus \mu_1$
HashSum:	$H_1(\mu_1)$	$H_1(\mu_3)$	0	$H_2(\mu_3) \oplus H_2(\mu_1)$
Count:	-1	1	0	0

3) Decode($IBLT_2 - IBLT_1$): $D_{2-1} = \{\mu_1\}$ $D_{1-2} = \{\mu_3\}$

(c) Decoding all μ from the set difference $IBLT_2 \setminus IBLT_1$

Fig. 1. Computing $N_1.Q_1[?x\ P1\ ?y] \cup N_2.Q_2[?x\ P1\ ?y]$ with IBLTs.

are updated with $[1 \mapsto 3; 2 \mapsto 73; 4 \mapsto 42]$. The sum of this vector, $3 + 73 + 42$ is the number of random draws done by nodes N_1, N_2 and N_4 to obtain the triple pattern in $Q_1[tp]$. More precisely, $N_1.Q_1[tp]$ contains the results obtained by N_1 in 3 draws, N_2 in 73 draws and N_4 in 42 draw, i.e. N_1 takes advantages of visited nodes by N_2 and N_4. Consequently, the algorithm stops when this sum reaches k_{max}. If we consider q nodes running the same query Q, then the lower bound to terminate for one node is in average $n.\ln(1/(1 - p))/q$. As we can see, collaboration between q nodes can divide by q the number of random draws required to terminate.

Synchronizing Sets of Mappings. Synchronizing $Q_i[tp]$ (line 7) between nodes may become expansive when collaborative queries meet several times. Suppose two nodes N_i and N_j running queries that contain the same triple pattern tp. The sets of mappings of variables of tp $N_i.Q_i[tp]$ and $N_j.Q_j[tp]$ could be large as the query progresses, with potentially many duplicated mappings. This large number of mappings increases drastically the communication cost of sets synchronization. It is more efficient to exchange only missing mappings between nodes, especially, as nodes meet several time the set difference between the two sets of mappings gets smaller. Ideally, the communication between nodes should only depends on the size of the set difference rather than on the size of sets. In other words, considering two sets of mappings $\mathcal{M}_i$ and $\mathcal{M}_j$ where the set difference is $d = |\mathcal{M}_i \setminus \mathcal{M}_j| + |\mathcal{M}_j \setminus \mathcal{M}_i|$, computing $\mathcal{M}_i \cup \mathcal{M}_j$ depends only on $\mathcal{O}(d)$ elements. For efficient computation of set difference, we use a probabilistic data structure called Invertible Bloom Lookup Tables (IBLTs) [8,9]. Figure 1 illustrates how the set difference is computed using two IBLTs. Consider two nodes N_1 and N_2 running the same triple pattern $tp : ?x \; p1 \; ?y$. The evaluation of tp is the set $\{\mu_1, \mu_2\}$ on the node N_1 with $\{\mu_1 = \{x \to 7, y \to 9\}, \mu_2 = \{x \to 4, y \to 5\}\}$ and $\{\mu_2, \mu_3\}$ on the node N_2 with $\{\mu_2 = \{x \to 4, y \to 5\}, \mu_3 = \{x \to 1, y \to 2\}\}$. Let H_1 and H_2 be two different hash functions. Figure 1a shows how N_1 computes $IBLT_1$ designed to handle 4 differences $(d = 4)$. μ_1 (then μ_2) is hashed with H_1 and H_2, then assigned into two different cells. A cell is composed of three kinds of sum; keysum is the XOR sum of the keys $(\mu_1 \oplus \mu_2)$, HashSum is the XOR sum of the hashed keys $(H_1(\mu_1) \oplus H_2(\mu_2))$, and count is the number of elements assigned to the cell. When N_1 meets N_2, N_1 sends $IBLT_1$ to N_2. Then N_2 computes the set difference $IBLT_2 \setminus IBLT_1$ resulting in two different sets of mappings: (1) $D_{2-1} = \{\mu_3\}$ which is the missing set of mappings for node N_1 (2) and, $D_{1-2} = \{\mu_1\}$ which is the missing set of mappings for node N_2. Then, N_2 can send back the response to N_1 containing D_{2-1}.

Sending only the difference reduces considerably the traffic between collaborative nodes. However, as IBLTs' is a probabilistic structure, its accuracy depends on the number of cells in the IBLT compared to the real number of differences between the two sets. If the number of cells is too small, then the IBLT cannot compute the missing mappings. Concretely, the decode operation will fail, so the exchange of IBLTs is useless. In this case, the node sends its set of mappings. In our context, as for cooperative nodes the difference gets smaller while the query progresses, IBLTs are eventually efficient as demonstrated empirically in the experimental study.

5 Experimental Study

We want to empirically answer the following questions: (1) Does the random service generate independent and uniformly distributed random variables? (2) How does visiting only a proportion of the sources impact queries answer completeness? (3) What is the impact of the number of collaborating queries on the number of iterations? (4) What is the impact of Invertible Bloom Lookup Tables (IBLTs) on traffic?

We implemented different software to achieve the experimental study. The code and experiments are available in the companion website[1].

5.1 Implementations

Query Engines. The Qasino query engine is built on top of Apache Jena[2]. We implemented a new customized symmetric hash join operator that integrates IBLTs.

---

[1] https://github.com/folkvir/qasino-simulation.
[2] https://jena.apache.org.

Qasino Simulator. Decentralization raises the problem of running experiments with thousands of nodes. Deploying thousands of endpoints connecting, them with structured or unstructured network, measuring the traffic and the number of calls is intractable with a traditional experimental setup in federated query processing. To handle this issue, we deploy Qasino in PeerSim [21] to run experiments. PeerSim is configured to run in "cycles". In one cycle, each node executes synchronously several iterations of its `execute` function as described in the algorithms: Las Vegas, Monte-Carlo and Monte-Carlo collaborative. Therefore, we measure how many cycles (iterations) are necessary to get complete answers for queries and how many cycles are necessary to terminate, i.e. it is possible to get a query complete answers before the termination of the algorithms.

Random Service. Different approaches exist for implementing a decentralized random service [15] and for network size estimators [17]. We use Spray [23] because Spray integrates a network size estimator and implements the random service on an unstructured network. Each Spray node has a logarithmic subset of the whole network as direct neighbors.

5.2 Experimental Setup

Machine. We run experiment on a HPCS computer Xeon(R) CPU E5-2680v2@2.80 GHz with 160 cores and 130 GB RAM.

Queries and Dataset. As the size of the dataset does not impact the number of cycles necessary for terminating a query, we use the dataset Diseasome[3]. We generated 100 random queries from the dataset using PATH and STAR shaped templates with two to eight triple patterns instantiated with random values from the dataset. The triple patterns of these 100 queries selects 70417 triples over the 91182 triples of the whole Diseasome dataset. We distributed uniformly those 70417 triples over 1000 simulated nodes, each node stores 70 or 71 triples. We extracted five different queries presented in Table 2. These queries are varying in the number of triple patterns (from 2 to 7) and in the cardinality of triple patterns, evaluated over the dataset. Columns $NBIR$ and $Results$ present the number of intermediate results, and the number of final results, respectively.

	tp1	tp2	tp3	tp4	tp5	tp6	tp7	BN IR	Results
Q_{17}	1	1						2	1
Q_{22}	1	4213						4214	1
Q_{54}	1	2889	1284	1284				5458	1
Q_{73}	2	4213	2889	9670	1284	1284		19342	1
Q_{87}	1	1	1	1	1	1	4	10	4

Fig. 2. Five queries with the number of Triples, cardinalities and results per query

---

[3] https://old.datahub.io/dataset/fu-berlin-diseasome.

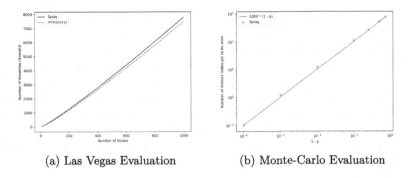

(a) Las Vegas Evaluation   (b) Monte-Carlo Evaluation

Fig. 3. Random evaluation

5.3 Experimental Results

Does the random service generating independent and uniformly distributed random variables? To answer this question, we compare the theoretical complexity (dashed line) with the empirical complexity (solid line). Figure 3a presents the number of random calls for varying number of nodes for the Las Vegas Algorithm. Compared to $n(ln(n) + \gamma)$, as computed in Sect. 4.1, the experimental values denote a slight deterioration of the complexity around 5%.

Figure 3b presents the proportion of visited nodes for different values of p and $n = 1000$. As expected the proportion of visited nodes is close to p.

Consequently, the experimentation confirms that the implementation of Qasino respects the theoretical model.

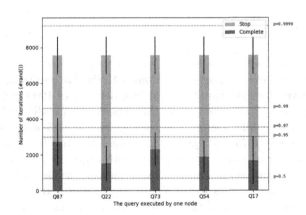

Fig. 4. The impact of the proportion of visited nodes p on queries answer completeness

How does visiting only a proportion of the sources impact queries answer completeness? We run several experimentations with the five queries of Table 2 with the Las Vegas algorithm. Only one node executes a query during an experimentation. Figure 4 presents the average of 100 executions per query. The *stop* bar chart represents the average number of iterations necessary to terminate the query with standard deviation. The *complete* bar chart represents the average number of iterations required to obtain complete result per query. As we can see, the number of iterations to terminate is higher than the number of iterations to get complete results. Moreover, for a proportion of visited sources equal to $p = 0.99$, the Monte-Carlo algorithm terminates with complete results in less than 4500 iterations.

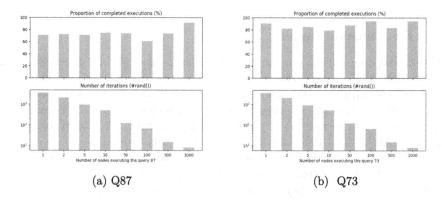

(a) Q87   (b) Q73

Fig. 5. Collaborative Monte-Carlo algorithm, $p = 0.97$ for different number of collaborative queries

What is the impact of the number of collaborating queries on the number of iterations? We run several experimentations with the five queries of Table 2 with the collaborative Monte-Carlo algorithm in a network of 1000 nodes, $p = 0.97$ with a perfect $n = 1000$. We repeat the experiment for different number of collaborative queries. All nodes run the same query. Figure 5 presents the average results for the query $Q73$ and $Q87$ for 100 runs. The top bar chart represents the average number of runs that terminate with complete results. The bottom bar chart presents the number of random draws to terminate. As we can see, the number of random draws to terminate decreases quickly as the number of similar queries increases, while the completeness of queries remains stable. This demonstrates the effectiveness of collaboration to speed up query execution.

What is the impact of Invertible Bloom Lookup Tables (IBLTs) on traffic? We analyze the traffic during query processing in terms of the number of transferred triples. We run the five queries 100 times with the collaborative Monte-Carlo algorithm in a network of 1000 nodes, $p = 097$ with a perfect $n = 1000$. All

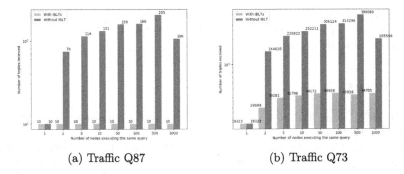

(a) Traffic Q87   (b) Traffic Q73

Fig. 6. The impact of IBLTs on traffic with the size difference of IBLTs are calibrated for $d = 500$ with 3 hash functions

node run the same query. We repeat the experiment for different number of collaborative queries. the same query. The IBLTs are configured for a number of differences < 500. Figure 6a shows the results for the query $Q87$. As the number of results per triple pattern is low (<500), the IBLT ensures optimal transfer, i.e. the number of transferred triples remains the same even if more queries collaborate. Figure 6b shows the results for the query $Q73$. As the query $Q73$ has much more intermediate results, IBLTs configured to handle only 500 differences can fail and trigger complete transfer between 2 collaborative queries. As collaborative queries eventually converge, IBLTs are eventually efficient and we can observe that the number of transferred triples remain stable after 50 collaboratives queries.

6   Conclusion

In this paper, we proposed Qasino, an original decentralized collaborative model for discovering RDF datasources and executing SPARQL queries. In contrast to traditional P2P models, Qasino respects the autonomy of participants. Qasino is based on P2P model where the cost of discovery is not shared by default and queries execution deliver complete results. Qasino approach allows to discover relevant sources, similar running queries and share intermediate results. With such collaborative query processing, participants only store data they want and therefore, preserve their autonomy. This work opens several perspectives. First, in the model, we relied on the network size estimator based on the random service. The knowledge of visited nodes and the number of random draws should allow to build a termination strategy that is independent of the size of the network. Second, we applied a simple strategy with IBLTs to synchronize queries, we can improve this strategy for better optimization of traffic. Finally, decentralization raises issues on completeness, autonomy and performance. We conjecture that only 2 of these 3 properties can be achieved in a system.

Acknowledgements. This work was partially funded by the French ANR projects O'Browser (ANR-16-CE25-0005-01) and DeKaloG (ANR-19-CE23-0014-01). Mr. Grall is funded by the GFI company, Nantes, France.

References

1. Aberer, K., et al.: P-grid: a self-organizing structured p2p system. ACM SIGMOD Record **32**(3), 29–33 (2003)
2. Aberer, K., Cudré-Mauroux, P., Hauswirth, M., Van Pelt, T.: GridVine: building internet-scale semantic overlay networks. In: McIlraith, S.A., Plexousakis, D., van Harmelen, F. (eds.) ISWC 2004. LNCS, vol. 3298, pp. 107–121. Springer, Heidelberg (2004). https://doi.org/10.1007/978-3-540-30475-3_9
3. Acosta, M., Vidal, M.-E., Lampo, T., Castillo, J., Ruckhaus, E.: ANAPSID: an adaptive query processing engine for SPARQL endpoints. In: Aroyo, L., et al. (eds.) ISWC 2011. LNCS, vol. 7031, pp. 18–34. Springer, Heidelberg (2011). https://doi.org/10.1007/978-3-642-25073-6_2
4. Aebeloe, C., Montoya, G., Hose, K.: A decentralized architecture for sharing and querying semantic data. In: Hitzler, P., et al. (eds.) ESWC 2019. LNCS, vol. 11503, pp. 3–18. Springer, Cham (2019). https://doi.org/10.1007/978-3-030-21348-0_1
5. Crespo, A., Garcia-Molina, H.: Semantic overlay networks for P2P systems. In: Moro, G., Bergamaschi, S., Aberer, K. (eds.) AP2PC 2004. LNCS (LNAI), vol. 3601, pp. 1–13. Springer, Heidelberg (2005). https://doi.org/10.1007/11574781_1
6. Diallo, O., Rodrigues, J.J., Sene, M., Lloret, J.: Distributed database management techniques for wireless sensor networks. IEEE Trans. Parallel Distrib. Syst. **26**(2), 604–620 (2015)
7. Doulkeridis, C., Vlachou, A., Nørvåg, K., Vazirgiannis, M.: Distributed semantic overlay networks. In: Shen, X., Yu, H., Buford, J., Akon, M. (eds.) Handbook of Peer-to-Peer Networking, pp. 463–494. Springer, Boston (2010)
8. Eppstein, D., Goodrich, M.T., Uyeda, F., Varghese, G.: What's the difference?: efficient set reconciliation without prior context. ACM SIGCOMM Comput. Commun. Rev. **41**(4), 218–229 (2011)
9. Goodrich, M.T., Mitzenmacher, M.: Invertible bloom lookup tables. arXiv preprint arXiv:1101.2245 (2011)
10. Grall, A., et al.: Ladda: SPARQL queries in the fog of browsers. In: Blomqvist, E., Hose, K., Paulheim, H., Lawrynowicz, A., Ciravegna, F., Hartig, O. (eds.) ESWC 2017. LNCS, vol. 10577, pp. 126–131. Springer, Cham (2017). https://doi.org/10.1007/978-3-319-70407-4_24
11. Grall, A., Molli, P., Skaf-Molli, H.: SPARQL query execution in networks of web browsers. In: Emerging Topics in Semantic Technologies - ISWC 2018 Satellite Events, Best Paper DeSemWeb@ISWC. pp. 55–68 (2018)
12. Hartig, O.: Zero-knowledge query planning for an iterator implementation of link traversal based query execution. In: Antoniou, G., Grobelnik, M., Simperl, E., Parsia, B., Plexousakis, D., De Leenheer, P., Pan, J. (eds.) ESWC 2011. LNCS, vol. 6643, pp. 154–169. Springer, Heidelberg (2011). https://doi.org/10.1007/978-3-642-21034-1_11
13. Hartig, O.: SPARQL for a web of linked data: semantics and computability. In: Simperl, E., Cimiano, P., Polleres, A., Corcho, O., Presutti, V. (eds.) ESWC 2012. LNCS, vol. 7295, pp. 8–23. Springer, Heidelberg (2012). https://doi.org/10.1007/978-3-642-30284-8_8

14. Kermarrec, A.M., Van Steen, M.: Gossiping in distributed systems. ACM SIGOPS Oper. Syst. Rev. **41**(5), 2–7 (2007)
15. King, V., Saia, J.: Choosing a random peer. In: Twenty-Third Annual ACM Symposium on Principles of Distributed Computing, PODC (2004)
16. Ladwig, G., Tran, T.: Linked data query processing strategies. In: Patel-Schneider, P.F., et al. (eds.) ISWC 2010. LNCS, vol. 6496, pp. 453–469. Springer, Heidelberg (2010). https://doi.org/10.1007/978-3-642-17746-0_29
17. Le Merrer, E., Kermarrec, A.M., Massoulié, L.: Peer to peer size estimation in large and dynamic networks: a comparative study. In: 15th IEEE International Conference on High Performance Distributed Computing, pp. 7–17. IEEE (2006)
18. Lv, Q., Cao, P., Cohen, E., Li, K., Shenker, S.: Search and replication in unstructured peer-to-peer networks. In: Proceedings of the 16th International Conference on Supercomputing, pp. 84–95. ACM (2002)
19. Mansour, E., et al.: A demonstration of the solid platform for social web applications. In: Proceedings of the 25th International Conference Companion on World Wide Web, pp. 223–226 (2016)
20. Marx, E., Saleem, M., Lytra, I., Ngomo, A.C.N.: A decentralized architecture for SPARQL query processing and RDF sharing: a position paper. In: 2th International Conference on Semantic Computing (ICSC), pp. 274–277 (2018)
21. Montresor, A., Jelasity, M.: PeerSim: a scalable P2P simulator. In: Proceedings of the 9th International Conference on Peer-to-Peer (P2P 2009), Seattle, WA, pp. 99–100, September 2009
22. Myers, A.N., Wilf, H.S.: Some new aspects of the coupon collector's problem. SIAM Rev. **48**(3), 549–565 (2006)
23. Nédelec, B., Tanke, J., Frey, D., Molli, P., Mostéfaoui, A.: An adaptive peer-sampling protocol for building networks of browsers. World Wide Web **21**(3), 629–661 (2017). https://doi.org/10.1007/s11280-017-0478-5
24. Nejdl, W., et al.: Super-peer-based routing and clustering strategies for RDF-based peer-to-peer networks. In: 12th international Conference on World Wide Web (2003)
25. Pérez, J., Arenas, M., Gutierrez, C.: Semantics and complexity of SPARQL. ACM Trans. Database Syst. **34**(3), 16 (2009)
26. Polleres, A., Kamdar, M.R., Fernández, J.D., Tudorache, T., Musen, M.A.: A more decentralized vision for linked data. In: DeSemWeb@ISWC (2018)
27. Schwarte, A., Haase, P., Hose, K., Schenkel, R., Schmidt, M.: FedX: optimization techniques for federated query processing on linked data. In: ISWC (2011)
28. Shapiro, M., Preguiça, N., Baquero, C., Zawirski, M.: A comprehensive study of convergent and commutative replicated data types. Research Report RR-7506, INRIA (2011)
29. Stocker, M., Seaborne, A., Bernstein, A., Kiefer, C., Reynolds, D.: SPARQL basic graph pattern optimization using selectivity estimation. In: 17th international conference on World Wide Web (2008)

Information Retrieval

Discovering Relational Intelligence
in Online Social Networks

Leonard Tan[1(✉)], Thuan Pham[1], Hang Kei Ho[2], and Tan Seng Kok[3]

[1] Engineering and Sciences, The University of Southern Queensland,
Toowoomba, Australia
{Leonard.Tan,Thuan.Pham}@usq.edu.au
[2] Faculty of Social Sciences, The University of Helsinki, Helsinki, Finland
hang.kei.ho@helsinki.fi
[3] Construction AI Research Labs, Applipro Services, Singapore, Singapore
aps_rudi@yahoo.com.sg

Abstract. Information networks are pivotal to the operational utility of key industries like medical, finance, governments, etc. However, applications in this area are not adequate in representing relationships between nodes [34]. Trending graph learning methodologies [9,16] like Graph Convolutional Networks (GCNs) [6] lack both representational power and accuracy to perform abstract computational tasks like prediction, classification, recommendation, etc. on real-time social networks. Furthermore, most such approaches known to date rely on learning temporal adjacency matrices to describe shallow attributes [9,16] like word co-occurance PMI [3] changes [6] and are unable to capture complex evolving entity relationships in real life for applications like event prediction, link prediction, topic tracking, etc. [34]. Importantly, such models ignore knowledge information geometry [1,24,32] completely, and sacrifices fidelity to speed of convergence. To address these challenges, a novel Relational Flux Turbulence (RFT) model was developed in this study - to identify relational turbulence in Online Social Networks (OSNs). Very good correlations between relational turbulence and sentiments exchanged within social transactions show promise in achieving these objectives.

Keywords: Relational turbulence · Social recognition · Deep learning

1 Introduction

Online Social Network (OSN) behavior has always been a topic of interest within various fields of social applications in artificial intelligence. These include: link detection, security threat identification, pattern recognition, recommendation, topic modeling and event prediction tasks, etc. Key relational behavior arises from manifolds of dynamic communication patterns which evolve over a temporal space of constant inceptions. Recent research include the use of directional dyads and signed reciprocity as a special representation of link "strength" [22].

© Springer Nature Switzerland AG 2020
S. Hartmann et al. (Eds.): DEXA 2020, LNCS 12391, pp. 339–353, 2020.
https://doi.org/10.1007/978-3-030-59003-1_22

Challenges. Many relational approaches used in this study however, lack depth and representative power [35]. The drawback of these techniques are that important correlational attributes shared between actors are ignored, resulting in shallow representations of relational states [35]. Methods based on feature similarities throughout studies in literature, have shown the lack of representational efficacy to model real life social structures effectively [2,28]. Generally speaking, there are several critical key questions in this field of study which remain unanswered. In an unstructured social network within an evolving construct of dynamic relationships [35]; how can we firstly, represent generalizations of evolutionary behavior within these social transactions accurately? Secondly, how can we recognize dynamic relational profiles which correlate to different social communication patterns? Finally, how can we quantify the dynamic errors arising from social disruptions (outliers) in our representations?

Data Models. We address these questions with the use of Fractal Neural Networks (FNNs). FNNs are used within the Relational Turbulence Model (RTM) framework to describe structures of chaos [25]. FNNs leverage on the dynamic structure of fractals as the lowest principle decompositions of never ending patterns. They are driven by a recursive process, and are adaptable enough to describe highly dynamic system representations [21]. In our approach, we define Relational Turbulence as probabilistic measures of Relational Intensity $P(\gamma_{rl})$, Relational Interference $P(\vartheta_{rl})$ and Relational Uncertainty $P(\varphi_{rl})$ [30]. RTM characterizes an artificial construct, which predicts communication behaviors. These behaviors are observed during relationship transitions in an environment of constant social disruptions [30]. We choose this model because alternative data models compromise accuracy and performance for simplicity in representation. Examples include node-based, neighbor-based, path-based, random walk-based, measures etc. [11]. These representations capture relational structures from a time static perspective and are not adaptable to real-life dynamic evolutions of relational states [31]. In this work, we focus on discovering relational intelligence through identifying relational profiles on three major social platforms: Twitter, Google and Enron email datasets.

Technical Model. In this paper, we introduce RFT to tackle the problem of misrepresentations as a time evolving flow of relational attributes. The model evolves into a multi-stage Deep Neural Network (DNN) from atomic fractal hybrid architectures [5]. The atom structure is morphed from standard concatenations of Restricted Boltzmann Machines (RBMs) and Recursive Neural Nets (RNNs). RFT accepts as inputs, key relational feature states f_i between actors a_j and global events E_ϵ from past and present social transactions to determine the likelihood of relational turbulence τ_{ij} within an identified social flux F_ϵ. Turbulence broadly corresponds to disruptive social communication patterns within various topic and event contexts. For example, passive negative sentiments transacted through discussions on major topics like trade wars, drive relational breakdowns in many aspects like trust, influence, status, etc. We develop a novel architecture from RTM to identify social disruptions by estimating relational turbulence profiles, within a given social context describing the state of flux. Then, we evaluate and demonstrate that our methods outperform similarity based feature and flat structural approaches in detecting social flux and turbulence.

Contributions. Our scientific contributions are presented as follows:

1. Our method adaptively learns from real-time online streaming data to identify key turbulent relationships within a given OSN.
2. An innovative RFT model was developed to capture key relational features which were used to detect and profile social communication patterns of eventful states within a given OSN.
3. Experiment results show that RFT is able to offer a good modeling of relational ground truths, while FNN efficiently and accurately represents evolving relational turbulence and flux profiles within a given OSN.

The remaining part of the paper is organized as follows: Sect. 2 presents a brief overview of related works drawn from social theories and relational structures. Section 3 introduces key concepts, theories and preliminaries of our proposed model. Section 4 discusses the methods and models we have developed for profiling relational turbulence in OSNs. Section 5 introduces our experimental design, implementation, results and presents our discussion. Section 6 leads to a conclusion and potential future directions.

2   Related Literature

Relational Turbulence. Relational Turbulence was first studied in [13]. It is characterized as a resultant state of conflicting interests between two or more actors. Conflict correlates to both a stimulus for communication and detrimental event occurrences [13]. Therefore, relational altering events are important discriminators to conflict detection and turbulence profiling. These events, if found to be in huge negative violations of expectancies between relational reciprocates of actors, can lead to instability in a relational flux [13]. The RTM [30] builds upon the core principles of relational state shifts and conflict management in an environment of continuous online social disruptions. The process of turbulent relationship development can be described as a continuous and communicative state of flux [30]. This state defines a consistent exchange of sentimental and affective information between the actor/s involved. Each transition to another state (e.g. professional colleagues to friendship) has the probability to cause friction (conflict), which may lead to a polarization of sentiments and affective communication flux in OSNs [30]. Two key features of the RTM are actor interferences and relational uncertainty [30]. They enable effective detection and prediction of conflicting events in sentimental and affective computing.

Neural Network Architectures. In [18], the authors present a minimalist neural network architecture for reliably and accurately estimating emotional states based on EEG captured data. Their model however, suffers from a lack of representation for more deeply complex emotional states (e.g. an in-betweenness in quantization across valance and arousal). Additionally, their reinforced gradient coefficient augments the errors calculated between expected-weighted and actual outputs which are then used to update the layered weights of their shallow Artificial Neural Network (ANN) model. This approach alleviates diminishing gradients at the expense of performance. In the same vein, [26] deals with social role recognition through the use of a Conditional Random Field (CRF) layered model architecture. However, for video image frames in which latent social role-based semantics exist, CRF architectures are ill-adapted to handle the complex representations of the depth to these roles in the identification process. This leads to poor performance output measures of their full model method. Building on the principles of Role Theory, the authors in [20] propose a deeper hierarchical model for human activity recognition based on identified actor roles within an eventful context. Their models performance suffer from scaling to larger event frameworks due to problems of overfitting and error gradient saddle points.

3   Preliminaries

Our RFT model leverages on two very important key concepts. The logical aspect is derived from the Relational Turbulence Model and the structural design is evolved from the Fractal Neural Network (FNN). The core idea of RFT is to iteratively adapt the structure of the neural network model to changing outputs (relational turbulence) at the inputs of the design. This is done in reference to the changing complexities of data at the inputs. A detailed architecture of the FNN used in our design is given in Fig. 1a.

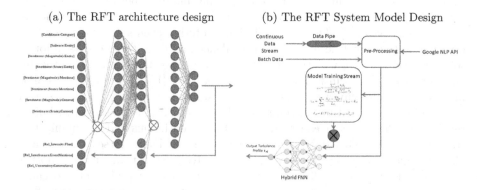

(a) The RFT architecture design   (b) The RFT System Model Design

Fig. 1. The RFT logical architecture

Relational Turbulence. From the RTM approach [29], we define Relational Intensity $P(\gamma_{rl})$, Relational Interference $P(\vartheta_{rl})$ and Relational Uncertainty $P(\varphi_{rl})$ to be three key probabilistic outputs of the RFT model which represent the relational turbulence $P(\tau_{rl})$ of a given link in an OSN. The key element types we have identified to be contributing features between the duration of the turning point and relationship development (as an unstable/turbulent process) are the Confidence ρ_{ij}, Salience ξ_{ij} and Sentiment λ_{ij} scores in an actor-actor relationship of a social transaction in question.

Expectancy Violation. It is noteworthy of mention that the ground truth reciprocities of these element types shared within a relational flux, violates expectancies - $E(\rho_{ij})$, $E(\xi_{ij})$ and $E(\lambda_{ij})$ respectively [29]. These violations, are a contributing factor to temporal representations of relational turbulence - γ_{rl}, ϑ_{rl} and φ_{rl}. Negative expectancy is defined has a polar mismatch between expected reciprocates against actual reciprocates (e.g. Actor i expecting a somewhat positive reciprocation of an egress sentiment stream, but instead, received a negative ingress sentiment stream from actor j). Positive expectancy is defined as the strong cosine similar vector alignment between these reciprocates. Both expectancy violation (EV) extremes, are characterized by sharp gradient changes of their weighted feature scores. This is given mathematically as:

$$\frac{\partial E_{rl}}{\partial \tau_{rl}} = \sum_{i,j=1}^{n} \prod_{\eta=\rho,\xi,\lambda} \frac{E(\eta_{ji})}{\partial \eta_{ji}} \times \frac{\partial \eta_{ij}}{\partial \tau_{ij}} \tag{1}$$

Where τ_{ij} is also known as the relational turbulence between node i and its surrounding neighbors j and η_{ij}, η_{ji} is the reciprocated sentiment from node i to j and j to i respectively.

Relational change or transition - also known as a turning point, defines some state-based critical threshold, beyond which relational turbulence and negative communication is irrevocable [19]. This critical threshold is specific to actor-actor relationships and learned through our model as a conflict escalation minimization function [27]. Conflict escalation is defined as the gradual increase in negative flux $\frac{-\nabla F_\epsilon}{\nabla t}$ over time within a classified context area L_{F_ϵ} of interest [27]. The critical threshold parameter is then driven mathematically as:

$$T_\epsilon = inf_{t\to\infty} \begin{cases} \frac{1}{2m}(-\frac{\nabla F_\epsilon}{\nabla t} \times \log_2(\frac{\nabla F_\epsilon}{\nabla L_\epsilon})) & \forall |\frac{\partial E_{rl}}{\partial \tau_{rl}}| > 1 \\ \log_2(|1 - \frac{\nabla F_\epsilon}{\nabla L_\epsilon}|) & \forall |\frac{\partial E_{rl}}{\partial \tau_{rl}}| < 1 \end{cases} \tag{2}$$

Where T_ϵ is the threshold of interest and m is the total number of training data over the time window t. The equation states simply that the relational transition threshold decreases drastically for strong EVs and gradually for weak EVs.

Problem Formulation. The problem statement which our work addresses can be summarised as follows: Given an OSN within an environment of constant social shocks, we wish to minimize inaccuracies in the representations from time evolving flow of relational attributes (time-realistic relationships) between actors. Furthermore, although DNNs are very powerful tools designed for use in both classification and recognition tasks, it is computationally abhorrent [14]. A drawback of a generative architectural approach involves the use of stochastic gradient decent methods during training which do not scale well to high dimensionalities [17,23]. Although still, generative DBNs offer many benefits like a supply of good initialization points, the efficient use of unlabeled data, etc.; thus, making its use in deep network architectures indispensable [5].

The Model Solution. To tackle the problem of computational efficiency and learning scalability to large data sets, we have adopted the DSN model framework for our study. Central to the concept of such an architecture is the relational use of stacking to learn complex distributions from simple core belief modules, functions and classifiers. Our approach leverages on the temporal transitions of stages in the relational evolution between nodes of an OSN [34]. It determines profiles of relational turbulence and encodes knowledge dimensionality into a highly volatile shallow fractal ANN architecture. This is used to either generate or collapse depth complexity during active learning - in response to random "anytime-sequenced" fluctuating data information.

4   Model and Methods

A high level system architecture of RFT is given in Fig. 1b. Specifically, in our design, data is fed into our model from two distinct sources. The first is batch processed from a repository of social data (Googles and Enron emails). The second is actively learned from live streaming tweet data (Twitter) pulled from multiple server sources using the twitter firehose API. It is then pushed through the model in stages. During pre-processing, data is first broken down into key relational features - Category confidence, Entity salience, Entity sentiment, Mentions sentiment and Context sentiment using the Googles NLP API. Then, in the next stage, these features are accepted as inputs into our RFT model (Fig. 1b) to estimate the output relational turbulence profiles. The input features of our RFT model is concatenated with the truth values of relational turbulence calculated from (6), (7) and (8) and synchronously fed back recursively into the intermediate confabulations of our FNN architecture (Fig. 1a). Errors in output expectations are backpropagated and corrected with inter-layer activity weight adjustments until they fall within pre-defined tolerance levels.

4.1   The Hybrid RFT Fractal Architecture

We begin with the definition of a soft kernel used to discover a markovian structure which we then encode into confabulations of fractal sub-structures. For a given set of data observables as inputs: $\chi \in X$ and outputs: $\Im \in \Xi$ we wish to loosely define a mapping such that the source space (X, α) maps onto a target space $(\Im, \omega)$. The conditional $P(\chi \vee \omega)$ assigns a probability from each source input χ to the final output space in ω. Each posterior state-space from in between input to output is generated and sampled through a random walk process. An indicator function which we have chosen to describe the state transition rule is:

$$\Theta_{t+1} = min \begin{cases} 0 \\ \prod_{c=1}^{n} \frac{\delta E_{t+1}^c}{\delta \chi_t^c} \end{cases} \tag{3}$$

Where δE_{t+1}^c is the error change from one hidden feature activity state $h_t \in H$ onto higher posterior confabulations. The objective function at each transition seeks to minimize error gradients. For a general finite state space markovian process, the markov kernel is thus defined as:

$$Kern(M) = \begin{cases} p : X \times \omega \rightarrow [0,1] \\ p(\chi|\omega) = \oint_\omega q(\chi, \Im)\nu(\delta\Im) \end{cases} \tag{4}$$

Once a unique markovian neural network has been discovered, a Single Layer Convolutional Perceptron (SLCP) is proposed as a baseline structure to learn the fractal sub-network from pre-existing posterior confabulations. The SLCP baseline structure changes as discovered knowledge is progressively encoded during the learning process.

4.2   The Fractal Neural Network

The model design we have chosen, with which to address the dynamic profiling of relational turbulence is the Fractal Neural Network (FNN) [21]. FNN adopts a hybrid architecture which incorporates the use of both generative and discriminative deep networks [5]. In our architecture, the generative DBN is used to initialize the DNN weights. Fine-tuning from the backpropagation process is then subsequently carried out sequentially layer by layer.

Generative Framework. In our learning model, the FNN generative framework is developed from the Restricted Boltzmann Machine (RBM) [12] layer stack. A Boltzmann Machine is architecturally defined as a stochastically coupled pair of binary units. These units contain a visible layer given as: $V \in 0, 1^D$ and a hidden layer vector: $H \in 0, 1^P$. The coupling between visible and hidden layers $V; H$ is driven by an energy state of layered interactivity; expressed as:

$$E(V, H, \theta) = -\frac{1}{2}V^T L V - \frac{1}{2}H^T J H - V^T W H \tag{5}$$

Where $\theta = W, J, L$ are Boltzmann Machine model weights between visible to hidden, visible to visible and hidden to hidden layers respectively. The discriminative architecture of the FNN model is built from the Tensorized Deep Stacking Recursive Neural Network (TDSN-RNN) model framework [5].

Discriminative Framework. The discriminative architecture of our FNN model is built from the Tensorized Deep Stacking Recursive Neural Network (TDSN-RNN) model framework. All deep architectures (Contrastive Divergence or per layer RBM to supervised backpropogation – perceptron golden architecture) rely on a back and forth recursive process through three core stages of their learning process. Stage 1 involves a forward pass which sequentially processes stacked training layers from input to output. Stage 2 backpropogates this layer-wise sequence from output to input using gradient descent. Stage 3 adjusts weights between layers to minimize output errors. This process is repeated in cycles until the final expectation is reached.

4.3   The Relational Turbulence Model

In our model, the probabilistic Relational Turbulence $P(\tau_{rl})$ of a given link in an OSN is determined by key features of an established relationship in any instance. They are the confidence ρ_{ij}, salience ξ_{ij} and sentiment λ_{ij} scores in a dyadic link. We define relational intensity as the continuous integration of sentimental transactions F_ϵ per context (event topic) L_{F_ϵ} area, the relational uncertainty as the likelihood from opposing sentiment mentions and relational interference as the probabilistic deviations in expectancies from predicted uncertainties and flux intensities. Mathematically, these are given as:

For **Relational Intensity:**

$$\gamma_{rl} = \sum_{i,j=1}^{n} \frac{\beta_{ij}| - \frac{\nabla F_{\epsilon j}}{\nabla t}|}{L_{F_\epsilon}} + \chi_{rl} + \dot{\theta}_{rl} \tag{6}$$

Where β_{ij} is defined as the temporal derivative of the latent topic (context) oscillation phase ϵ, χ_{rl} is the reciprocal bias and $\dot{\theta}_{rl}$ is the gradient of social influence from one actor to another across a relational link.

For **Relational Uncertainty:**

$$\varphi_{rl} = \frac{\sum_{i,j=1}^{n} S_i S_j}{\sqrt{\sum_{i=1}^{n} S_i}\sqrt{\sum_{j=1}^{n} S_j}} \tag{7}$$

Where S_i and S_j are sentiments transacted from nodes i to j and from nodes j to i respectively.

For **Relational Interference:**

$$\vartheta_{rl} = E(F(\gamma_{rl}, \varphi_{rl} : \mu_{\gamma\varphi}, \omega_{\gamma\varphi}^2))$$
$$= \frac{1}{2} + \frac{1}{\sqrt{2\pi}\omega} \sum_{\gamma_{rl}, \varphi_{rl}=0}^{n} \frac{1}{2} erf\left(\frac{\gamma_{rl},\varphi_{rl}-\mu}{\sqrt{2}\omega}\right) \exp^{-\frac{(\gamma_{rl},\varphi_{rl}-\mu)^2}{2\omega^2}} \tag{8}$$

Where,

$$F(\gamma_{rl}, \varphi_{rl} : \mu_{\gamma\varphi}, \omega_{\gamma\varphi}^2) = \frac{1}{\sqrt{2\pi}\omega} \sum_{t=-\infty}^{\gamma_{rl},\varphi_{rl}} \exp^{-\frac{(t-\mu)^2}{2\omega^2}} dt \tag{9}$$

Here, $F(\gamma_{rl}, \varphi_{rl} : \mu_{\gamma\varphi}, \omega_{\gamma\varphi}^2)$ is the Cumulative Distribution Function (CDF), and $erf(x)$ is the error function of the predicted outcomes γ_{rl} and φ_{rl}.

Finally **Relational Turbulence:** was calculated from conditional posteriors of γ_{rl}, ϑ_{rl} and φ_{rl} as the mathematical relation of:

$$P(\tau_{rl}) = \sum_{i=1}^{n} \frac{P(\gamma_i|\theta_i)P(\vartheta_i|\varphi_i)P(\varphi_i|\gamma_i)}{N_i P(\gamma_i)P(\vartheta_i)P(\varphi_i)} \tag{10}$$

Here, N_i is the conditional scaling factor. The inputs were tested across the RFT dynamically stacked Fractal Neural Network (FNN) and the chosen baseline models.

5 Experiments

5.1 Dataset

The experiments were conducted on three datasets using RFT and five different baseline algorithms. The datasets are: Twitter, Google and Enron emails. These three datasets were chosen because they are widely benchmarked throughout the academic circle for studies in sentimental computing and can be easily understood by the audience of this paper. They are detailed in Table 1.

Table 1. Statistics of datasets

Dataset	#Entities	#Dyads	Size	Avg. text len
Enron email[a]	162	1.5 mil	500000 emails	1000
Googles[b]	3566224	436994489	279 mil crawls	5000
TwitterAPI[c,d]	50 mil	1 bil	100 mil tweets	283

[a]https://archive.ics.uci.edu/ml/datasets/bag+of+words.
[b]http://commoncrawl.org/2014/07/april-2014-crawl-data-available/.
[c]https://developer.twitter.com/en/docs.html.
[d]http://help.sentiment140.com/api.

5.2 Baselines

Several state-of-the-art methods were considered for comparison with the proposed RFT model. Since the model is the first in line for this type of adaptive online active learning approach, modified versions of similar methods were used along with the baselines, developed earlier for comparison. Another notable point is although many prediction models exist, not all methods have the same goal or data features as this study. Therefore consideration is given only to the models which use similar data for comparison. It should be mentioned that not all the methods can both predict relational turbulence and profile communication patterns together. Therefore we compare only the profiles of relational turbulence outputs between each other. Descriptions of the competing methods are given in Table 2. The key difference between DCN and RFT is that in DCN the number of layers are fixed at 45 while in RFT, the layers are allowed to grow and collapse as new feature complexity representations are learned over time.

Tuning Parameters. In the experiments, system model parameters were chosen based on the combined effect of several factors - including errors in observational data, choices of calibration methods and Design Of Experiment (DOE) criterias [10]. A hybrid of both global and local Sensitivity Analysis (SA) approaches was used to determine and specify the best performing parameters for experimentation based on a predefined behavior threshold for the model. The experiments were conducted on the training model with a learning rate set to 1.1, a sliding window set to 3, an error tolerance set to 0.1 (10%), a data outlier

Table 2. Baseline models

Baseline	Class	Data	Modalities
SLFN [15]	ANN	Feature parameters	Parametric inputs
DCN [14]	DNN	Feature parameters	45-layer DCN
IMPALA [7]	RL	Feature parameters	Reinforcement Learning
MVVA [35]	VAR	Endogenous variables	Vector Auto-Regression
EnsemDT [8]	Ensemble	Group learners	SLP, DCN, IMPALA and MVVA
RTM (True Value)	RTT	Feature parameters	(6)–(10)

threshold set to 1.0, with scaling set to 10, a vanishing gradient error threshold at 0 and an exploding gradient error threshold set to 100. Finally, both trust region radius parameter was set to 5 and the softmax temperature regularization parameter was staged at 1.2.

5.3   Performance Measurements

Kendall Coefficient. The Kendall (tau-b coefficient) was used to measure the strength of associations between predicted and expected outputs of the learning models. It is given as:

$$\tau_b = \frac{N_c - N_d}{\sqrt{(N_0 - N_x)(N_0 - N_y)}} \tag{11}$$

Where N_c, N_d are the number of concordant and discordant pairs respectively, u_i is the number of tied values in the ith group of ties for the first quantity and v_j is the number of tied values in the jth group of ties for the second quantity.

Spearman Coefficient. The Spearman (rho coefficient) was used to measure the monotonic relationship between the independent variables (Category confidence $\mathfrak{C}_i$, Entity Sailence $\mathcal{J}_i$, Entity sentiments - magnitude and scores $(\mathfrak{S}_i, \beth_i)$, Mention sentiments -magnitude and scores $(\mathcal{L}_i, \lambda_i)$, Context sentiments - magnitude and scores $(\mathcal{O}_i, \mathsf{T}_i)$) and the dependent variables (Relational Intensity γ_{rl}, Relational Interference ϑ_{rl} and Relational Uncertainty φ_{rl}). It is calculated as:

$$\Gamma_S = 1 - \frac{6 \sum D_i^2}{N(N^2 - 1)} \tag{12}$$

Where $D_i = rank(X_i) - rank(Y_i)$ is the difference in ranks between the observed independent variable X_i and dependent variable Y_i and N is the number of predictions to input data sets for all three sources.

K-Fold Validation. Finally, during the experimentation, the full datasets obtained from the different sources (twitter, google and enron) were partitioned into k-subsamples. K-fold validation [33] was performed over all deep learning models across the Mean Absolute Percentage Error (MAPE) [4] measurement of each run. Mathematically, MAPE can be expressed as:

$$\delta_{MAPE} = \frac{1}{N} \sum_{i=1}^{N} |\frac{E_i(x) - Y_i(t)}{E_i(x)}| \tag{13}$$

Where $E_i(x)$ is the expectation at the output of data input set i and $Y_i(t)$ is the corresponding prediction over N total subsamples. δ_{MAPE} is the average measure of errors in expectations at the output.

5.4   Results

The tests were run across the baselines and our RFT model. For clarity and simplicity of explainations, only every 10th running data from a chosen output sample set is plotted on a graph and displayed for discussion purposes. The line of best fit was used to graph the curve through the points. Additionally because of space constraints, only the table on Kendall correlation experimented on the chosen datasets is displayed. The results are shown in Table 3 and Fig. 2a–c.

(a) Graph of Enron Relational Turbulence (b) Graph of Googles Relational Turbulence
Profile.   Profile.

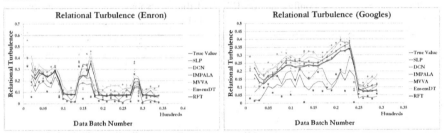

(c) Graph of Twitter Relational Turbulence
Profile.

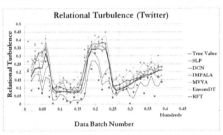

Fig. 2. Graph of relational turbulence across three datasets

5.5   Investigation

As can be seen from the graphs, SLP models consistently underperforms in ranking where prediction accuracy is concerned, the Kendall (tau-b coefficient) test

Table 3. Table of Spearman, Kendall and K-fold results

(a) Table of Kendall coefficient

Correlation Matrix	SLP (Enron)	DCN (Enron)	IMPALA (Enron)	MVVA (Enron)	EnsemDT (Enron)	RFT (Enron)
Kendall's Tau B	0.300	0.456	0.326	0.372	0.392	**0.487**
p-value	<.001	<0.001	<0.001	<0.001	<0.001	<0.001
	SLP (Googles)	DCN (Googles)	IMPALA (Googles)	MVVA (Googles)	EnsemDT (Googles)	RFT (Googles)
Kendall's Tau B	0.351	**0.764**	0.413	0.457	0.674	0.762
p-value	<.001	<0.001	<0.001	<0.001	<0.001	<0.001
	SLP (Twitter)	DCN (Twitter)	IMPALA (Twitter)	MVVA (Twitter)	EnsemDT (Twitter)	RFT (Twitter)
Kendall's Tau B	0.377	0.766	0.427	0.512	0.598	**0.810**
p-value	<.001	<0.001	<0.001	<0.001	<0.001	<0.001

(b) Table of K-fold MAPE

k	δ_{MAPE} (SLP)	δ_{MAPE} (DCN)	δ_{MAPE} (IMPALA)	δ_{MAPE} (MVVA)	δ_{MAPE} (EnsemDT)	δ_{MAPE} (RFT)
20	0.461	0.189	0.348	0.218	0.188	**0.127**
30	0.424	0.175	0.357	0.302	0.172	**0.131**
50	0.420	0.173	0.401	0.327	0.156	**0.112**
80	0.418	0.169	0.376	0.287	0.141	**0.110**
100	0.421	0.166	0.392	0.308	0.127	**0.107**

(c) Table of Spearman coefficient for RFT

	Spearman (rho) coefficient		
	$P(\gamma_{rl})$	$P(\vartheta_{rl})$	$P(\varphi_{rl})$
$\mathfrak{C}_i$	0.074	0.070	0.076
$\mathfrak{I}_i$	-0.738	0.825	0.007
$\mathfrak{R}_i$	0.842	0.847	0.787
$\mathfrak{Z}_i$	0.887	0.834	0.837
$\mathcal{L}_i$	0.846	0.884	0.901
λ_i	0.784	0.846	0.871
$\mathcal{O}_i$	-0.285	-0.292	-0.269
$\daleth_i$	-0.273	-0.287	-0.278

(d) Table of Spearman coefficient for DCN

	Spearman (rho) coefficient		
	$P(\gamma_{rl})$	$P(\vartheta_{rl})$	$P(\varphi_{rl})$
$\mathfrak{C}_i$	0.079	-0.074	-0.077
$\mathfrak{I}_i$	-0.783	0.776	0.001
$\mathfrak{R}_i$	0.874	0.843	0.767
$\mathfrak{Z}_i$	0.892	0.882	0.846
$\mathcal{L}_i$	0.864	0.891	0.921
λ_i	0.779	0.833	0.888
$\mathcal{O}_i$	-0.293	-0.289	-0.278
$\daleth_i$	-0.275	-0.276	-0.284

(e) Table of Spearman coefficient for IM-PALA

	Spearman (rho) coefficient		
	$P(\gamma_{rl})$	$P(\vartheta_{rl})$	$P(\varphi_{rl})$
$\mathfrak{C}_i$	0.022	-0.152	-0.035
$\mathfrak{I}_i$	-0.863	0.415	0.073
$\mathfrak{R}_i$	0.512	0.772	0.822
$\mathfrak{Z}_i$	0.446	0.563	0.851
$\mathcal{L}_i$	0.734	0.825	0.416
λ_i	0.378	0.715	0.781
$\mathcal{O}_i$	-0.745	-0.374	0.549
$\daleth_i$	-0.526	-0.561	0.274

(f) Table of Spearman coefficient for EnsemDT

	Spearman (rho) coefficient		
	$P(\gamma_{rl})$	$P(\vartheta_{rl})$	$P(\varphi_{rl})$
$\mathfrak{C}_i$	0.215	-0.025	-0.315
$\mathfrak{I}_i$	0.415	0.739	0.067
$\mathfrak{R}_i$	0.823	0.911	0.733
$\mathfrak{Z}_i$	0.524	0.425	0.418
$\mathcal{L}_i$	0.562	0.426	0.527
λ_i	0.624	0.627	0.649
$\mathcal{O}_i$	-0.449	-0.533	-0.562
$\daleth_i$	-0.526	-0.484	-0.417

(g) Table of Spearman coefficient for SLP

	Spearman (rho) coefficient		
	$P(\gamma_{rl})$	$P(\vartheta_{rl})$	$P(\varphi_{rl})$
$\mathfrak{C}_i$	0.076	-0.077	-0.078
$\mathfrak{I}_i$	-0.805	0.764	0.002
$\mathfrak{R}_i$	0.844	0.837	0.703
$\mathfrak{Z}_i$	0.901	0.872	0.871
$\mathcal{L}_i$	0.877	0.913	0.953
λ_i	0.788	0.827	0.891
$\mathcal{O}_i$	-0.303	0.297	0.295
$\daleth_i$	-0.271	-0.302	-0.312

(h) Table of Spearman coefficient for MVVA

	Spearman (rho) coefficient		
	$P(\gamma_{rl})$	$P(\vartheta_{rl})$	$P(\varphi_{rl})$
$\mathfrak{C}_i$	0.115	-0.271	-0.361
$\mathfrak{I}_i$	-0.472	0.824	0.042
$\mathfrak{R}_i$	0.635	0.441	0.561
$\mathfrak{Z}_i$	0.726	0.512	0.572
$\mathcal{L}_i$	0.811	0.755	0.617
λ_i	0.724	0.536	0.772
$\mathcal{O}_i$	-0.403	-0.351	-0.425
$\daleth_i$	-0.361	-0.472	-0.382

shows a lower (positive) correlation between expected and predicted outputs across the test data set for SLP models and much higher (positive) association for other baselines and RFT. Furthermore, from the results of the Spearman (rho coefficient) test done on the independent and dependent variables, it can be seen from Tables 3c–h that the spearman coefficient indicates strongly positive monotonic correlations between turbulence measures (γ_{rl}, ϑ_{rl} and φ_{rl}) and sentiment scores [$(\mathfrak{S}_i, \beth_i), (\mathcal{L}_i, \lambda_i), (\mathcal{O}_i, \daleth_i)$] and moderately positive correlations between the same turbulence measures (γ_{rl}, ϑ_{rl} and φ_{rl}) to both category confidence and entity salience ($\mathfrak{C}_i$, $\mathcal{J}_i$).

Additionally, from Table 3a, it is observed that across all models, strength of associations between predicted and expected outputs tend to be weaker in specifically directed communications. This is observed in Enron's email datasets as opposed to Twitter and Google results. It is analysed that this is due to high relational interference scores which tend to correlate fairly well to entity salience scores. In this scenario, entity salience plays an important function in determining relational turbulence - as opposed to contexts over which the sentiments were expressed. This means that an actor with a higher social status of influence may more readily interfere with other relationships in directed communications. Generally however, it can be observed from Table 3b that as the number of subsample windows increases over the dataset, the MAPE over DCN, EnsemDT and RFT decreases. Whereas MAPE for SLP, IMPALA and MVVA tend to fluctuate about a fixed error. This behavior is attributed to overfitting and gradient saddle points from poor initializations. RFT remains the clear winner across the measured baselines in all k-fold validation experiments.

6   Conclusion

In conclusion, it has been shown that RFT is capable of predicting relational turbulence profiles between actors within a given OSN acquired from anytime data. The results show superior accuracies and performance of the FNN model in comparison to well known baseline models. The feasibility of the learning model has been demonstrated through the implementation on three large scale networks: Twitter, Google Plus and Enron emails. The study uncovers three pivotal long-term objectives from a relational perspective. Firstly, relational features can be used to strengthen medical, cyber security and social applications where the constant challenges between detection, recommendation, prediction, data utility and privacy are being continually addressed. Secondly, in fintech applications, relational predicates (e.g. turbulence) are determinants to market movements - closely modeled after a system of constant shocks. Thirdly, in artificial intelligence applications like computer cognition and robotics, learning relational features between social actors enables machines to recognize and evolve.

References

1. Amari, S.I., Nagaoka, H.: Methods of information geometry, vol. 191. American Mathematical Society (2007)
2. Backstrom, L., Leskovec, J.: Link prediction in social networks using computationally efficient topological features. In: 2011 IEEE Third International Conference on Privacy, Security, Risk and Trust (PASSAT) and 2011 IEEE Third International Conference on Social Computing (SocialCom), pp. 73–80. IEEE (2011)

3. Church, K.W., Hanks, P.: Word association norms, mutual information, and lexicography. Comput. Linguist. **16**(1), 22–29 (1990)
4. De Myttenaere, A., Golden, B., Le Grand, B., Rossi, F.: Mean absolute percentage error for regression models. Neurocomputing **192**, 38–48 (2016)
5. Deng, L., Yu, D., et al.: Deep learning: methods and applications. Found. Trends® Sig. Process. **7**(3–4), 197–387 (2014)
6. Deng, S., Rangwala, H., Ning, Y.: Learning dynamic context graphs for predicting social events. In: Proceedings of the 25th ACM SIGKDD International Conference on Knowledge Discovery & Data Mining, pp. 1007–1016 (2019)
7. Espeholt, L., et al.: IMPALA: scalable distributed deep-RL with importance weighted actor-learner architectures. arXiv preprint arXiv:1802.01561 (2018)
8. Ezzat, A., Wu, M., Li, X., Kwoh, C.-K.: Computational prediction of drug-target interactions via ensemble learning. In: Vanhaelen, Q. (ed.) Computational Methods for Drug Repurposing. MMB, vol. 1903, pp. 239–254. Springer, New York (2019). https://doi.org/10.1007/978-1-4939-8955-3_14
9. Feng, K., Cong, G., Jensen, C.S., Guo, T.: Finding attribute-aware similar regions for data analysis. Proc. VLDB Endow. **12**(11), 1414–1426 (2019)
10. Gan, Y., et al.: A comprehensive evaluation of various sensitivity analysis methods: a case study with a hydrological model. Environ. Model. Softw. **51**, 269–285 (2014)
11. Gao, F., Musial, K., Cooper, C., Tsoka, S.: Link prediction methods and their accuracy for different social networks and network metrics. Sci. Program. **2015**, 1 (2015)
12. Han, Z., Liu, Z., Han, J., Vong, C.M., Bu, S., Chen, C.L.P.: Mesh convolutional restricted Boltzmann machines for unsupervised learning of features with structure preservation on 3-D meshes. IEEE Trans. Neural Netw. Learn. Syst. **28**(10), 2268–2281 (2017)
13. Haunani Solomon, D., Theiss, J.: A longitudinal test of the relational turbulence model of romantic relationship development. Pers. Relationsh. **15**, 339–357 (2008). https://doi.org/10.1111/j.1475-6811.2008.00202.x
14. Huang, G., Sun, Yu., Liu, Z., Sedra, D., Weinberger, K.Q.: Deep networks with stochastic depth. In: Leibe, B., Matas, J., Sebe, N., Welling, M. (eds.) ECCV 2016. LNCS, vol. 9908, pp. 646–661. Springer, Cham (2016). https://doi.org/10.1007/978-3-319-46493-0_39
15. Huang, G.B., Chen, Y.Q., Babri, H.A.: Classification ability of single hidden layer feedforward neural networks. IEEE Trans. Neural Netw. **11**(3), 799–801 (2000)
16. Huang, X., Song, Q., Li, Y., Hu, X.: Graph recurrent networks with attributed random walks. In: Proceedings of the 25th ACM SIGKDD International Conference on Knowledge Discovery & Data Mining, pp. 732–740 (2019)
17. Hutchinson, B., Deng, L., Yu, D.: Tensor deep stacking networks. IEEE Trans. Pattern Anal. Mach. Intell. **35**(8), 1944–1957 (2013)
18. Keshmiri, S., Sumioka, H., Nakanishi, J., Ishiguro, H.: Emotional state estimation using a modified gradient-based neural architecture with weighted estimates. In: 2017 International Joint Conference on Neural Networks (IJCNN), pp. 4371–4378. IEEE (2017)
19. Knobloch, L.K., Theiss, J.A.: Relational turbulence theory applied to the transition from deployment to reintegration. J. Family Theory Rev. **10**(3), 535–549 (2018)
20. Lan, T., Sigal, L., Mori, G.: Social roles in hierarchical models for human activity recognition. In: 2012 IEEE Conference on Computer Vision and Pattern Recognition, pp. 1354–1361. IEEE (2012)
21. Larsson, G., Maire, M., Shakhnarovich, G.: FractalNet: ultra-deep neural networks without residuals. arXiv preprint arXiv:1605.07648 (2016)

22. Li, Y., Zhang, Z.-L., Bao, J.: Mutual or unrequited love: identifying stable clusters in social networks with uni- and bi-directional links. In: Bonato, A., Janssen, J. (eds.) WAW 2012. LNCS, vol. 7323, pp. 113–125. Springer, Heidelberg (2012). https://doi.org/10.1007/978-3-642-30541-2_9

23. Miikkulainen, R., et al.: Evolving deep neural networks. arXiv preprint arXiv:1703.00548 (2017)

24. Nielsen, F., Barbaresco, F.: Geometric Science of Information. Springer, Cham (2015). https://doi.org/10.1007/978-3-030-26980-7

25. Peitgen, H.O., Jürgens, H., Saupe, D.: Chaos and Fractals: New Frontiers of Science. Springer, New York (2006). https://doi.org/10.1007/978-0-387-21823-6

26. Ramanathan, V., Yao, B., Fei-Fei, L.: Social role discovery in human events. In: Proceedings of the IEEE Conference on Computer Vision and Pattern Recognition, pp. 2475–2482 (2013)

27. Simeonova, L.: Gradient emotional analysis (2017)

28. Snijders, T.A.: Markov chain Monte Carlo estimation of exponential random graph models. J. Soc. Struct. **3**(2), 1–40 (2002)

29. Solomon, D.H., Knobloch, L.K., Theiss, J.A., McLaren, R.M.: Relational turbulence theory: variation in subjective experiences and communication within romantic relationships. Hum. Commun. Res. **42**(4), 507–532 (2016)

30. Theiss, J.A., Solomon, D.H.: A relational turbulence model of communication about irritations in romantic relationships. Commun. Res. **33**(5), 391–418 (2006). https://doi.org/10.1177/0093650206291482

31. Wang, P., Xu, B., Wu, Y., Zhou, X.: Link prediction in social networks: the state-of-the-art. Sci. China Inf. Sci. **58**(1), 1–38 (2014). https://doi.org/10.1007/s11432-014-5237-y

32. Watters, N.: Information geometric approaches for neural network algorithms. Ph.D. thesis (2016)

33. Wong, T.T.: Performance evaluation of classification algorithms by k-fold and leave-one-out cross validation. Pattern Recogn. **48**(9), 2839–2846 (2015)

34. Zhang, J., et al.: Detecting relational states in online social networks. In: 2018 5th International Conference on Behavioral, Economic and Socio-Cultural Computing (BESC), pp. 38–43. IEEE (2018)

35. Zhang, J., Tao, X., Tan, L., Lin, J.C.W., Li, H., Chang, L.: On link stability detection for online social networks. In: International Conference on Database and Expert Systems Applications, pp. 320–335. Springer (2018)

Generating Dialogue Sentences to Promote Critical Thinking

Satoshi Yoshida[(✉)] and Qiang Ma[(✉)]

Kyoto University, Yoshidahonmachi, Sakyo, Kyoto, Japan
yoshida.satoshi.62c@st.kyoto-u.ac.jp, qiang@i.kyoto-u.ac.jp

Abstract. Conventional dialogue generating methods focus on generating fluent sentences in context, but insufficient consideration of speaker emotions. They often generate sentences with the same sentimental polarity of the speakers. Such that sentences are hard to change the mood or viewpoints of the speakers, i.e., from a negative mood to positive one, etc. In this paper, we propose a method to generate dialogue sentences to provide different viewpoints. To this end, we propose two novel concepts, polarity co-occurrence (P-Cooc) and modification co-occurrence (M-Cooc). P-Cooc is used to find aspects providing different and comfortable viewpoints, and M-Cooc is used to find proper modification terms to such that aspects. The experimental results demonstrate that our method could provide supplementary viewpoints to promote critical thinking.

Keywords: Dialogue generation · Critical thinking · Polarity co-occurrence · Modification co-occurrence

1 Introduction

Dialogue systems and agents (Siri, Alexa, Cortana, etc.) have become widespread. Existing systems and methods aim to generate natural sentences closed to that written by humans and in the context of dialogue. To the best of our knowledge, there is few work considering the way to change the viewpoints and mood of the speakers. For example, when a speaker is depressed, sentences which can comfort him/her is useful and important. In other words, to provide information from different viewpoints and promote critical thinking is a new challenge in dialogue systems.

Critical thinking is reflective and reasonable thinking and skillfully analyzing [1,2]. When we speak emotionally, the direction of the conversation may be biased and narrow our vision. In these cases, providing information from different aspects to change the viewpoints and mood (sentimental polarity) is important. There are many studies on critical thinking in Pedagogy and media literacy area [3,4]. To the best of our knowledge, there is few work considering critical thinking or providing different viewpoints to enhance the dialogue in the area of chat and dialogue generation systems.

© Springer Nature Switzerland AG 2020
S. Hartmann et al. (Eds.): DEXA 2020, LNCS 12391, pp. 354–368, 2020.
https://doi.org/10.1007/978-3-030-59003-1_23

In this paper, we propose a novel method to generate dialogue sentences, which could provide different aspects and viewpoints on a certain topic to enrich the dialogue and change the mood of the speaker. For example, in response to "Room seemed clean but it smelled very musty." we try to returns positive sentences such as "I want to leave a positive review, because the room had a neat look, and it was a nice quiet stay." Please notice that our method could be used as a dialogue support tool which providing candidate sentences to the responder (a counselor, an advisor, etc.) who wants to comfort the speaker in a negative mood.

In this work, we propose two concepts: polarity co-occurrence (P-Cooc) and modification co-occurrence (M-Cooc). P-Cooc is used to discover aspects that could provide different sentimental polarity, and M-Cooc is used to find the modification terms for such that aspects. For a given sentence (by a speaker), our method acts as the responder to generate a response sentence with different polarities and providing different viewpoints.

The main contributions of this paper are as follows.

- We propose a dialogue generation method to encourage critical thinking. To this end, we propose two novel concept, Polarity-Cooccurrence (P-Cooc) and Modification-Cooccurrence (M-Cooc).
- We performed user evaluation experiments with 20 people and verified the effectiveness of the proposed method.

2   Related Work

In recent years, the field of dialogue sentence generation has been developing remarkably. OpenAI has released GPT-2 (Generate Pre-training 2) [5], which is the state-of-art for dialogue sentence generation. When a certain sentence is input, the subsequent sentence is predicted using a huge amount of learning data. The generated text is very similar to that from a human. GPT-2 was trained on 8 million web pages, which makes it possible to record near-maximum accuracy without learning even for data-set with different domains. The sentence generated by deep learning represented above is a very natural sentence if one sentence is taken. However, sentences that are not natural in dialogue context may be generated.

Research on dialogue systems [6–8] also aims to build systems that can talk like a human. Takayama et al. propose a method to generate chats while preserving the naturalness of responses by predicting characteristic vocabulary that is likely to appear in responses. Words, whose frequency of occurrence is considerably low, are removed from the output using pointwise mutual information (PMI). As a result, it is possible to generate more contextually natural responses sentences.

The Neural Conversation Model (NCM) [9] is a method in which Sequence to Sequence (seq2seq) [10] used in machine translation is applied to response generation in a dialogue system. In such a learning method, the loss does not decrease unless the output word outputs the same word as the reference sentence as a response to the utterance. It outputs many frequently used general words. One of the methods to solve this problem is MMI-bidi [11]. A response that is more co-occurring with the utterance is generated. However, this approach does not maximize mutual information during training. Therefore, the effect cannot be expected if a word having high validity is not output. Takayama et al. increased the effect of applying MMI-bidi by adding an objective function during training to make it easier to generate words that strongly co-occur with the utterance sentence [6].

With recent advancements in neural network research, end-to-end approaches have shown promising results for non-goal oriented dialogue systems [9,12,13]. However, application of this approach towards incorporating emotion in the dialogue is still very lacking. Zhou et al. published their work addressing the emotional factor in neural network response generation [14]. They examined the effect of internal emotional states on the decoder, investigating 6 categories to "emotionally" color the response. However, this study has not yet considered user's emotion in the response generation process. To bridge these gaps, Nurul Lubis et al. exploit emotion appraisal to incorporate emotion into chat-based dialogue systems [15]. They propose a neural chat-oriented dialogue system that captures the user's emotional state and considers it while generating a dialogue response. Objective and subjective evaluations show that the proposed methods result in dialogue responses that are more natural and elicit a more positive emotional response.

To the best of our knowledge, this work is the first attempt to generate dialogue sentences that can promote critical thinking and being natural in context.

3   Proposed Method

3.1   Overview

Figure 1 shows the overview of our method. Two linguistic resources, P-Cooc and M-Cooc, are constructed in advance. P-Cooc (Polarity Co-occurrence) is a dictionary for efficiently searching for aspects that are often mentioned from the opposite polarity. M-Cooc (Modification Co-occurrence) is a dictionary for examining the co-occurrence relation between aspects and its modification terms. Here, an aspect is the subject or object mentioned in a sentence, and its modification term is the adjectives which denotes the sentimental polarity of that aspect. Currently, we apply Standford Sentiment Treebank [16] to extract the aspect, modification terms and the sentiment polarity. The details will be described in Sect. 3.2.

As shown in Fig. 1, suppose the user said "This hotel is expensive." Our method will try to generates a sentence, such as "the food is delicious" to provide different viewpoints as follows.

- Extract the aspect term from the given sentence.
- Search for opposing aspects usually providing different sentiment polarity by using P-Cooc.
- Search for modification terms to the opposing aspects by using M-Cooc.
- Search for candidate sentences from a corpus with the query consisting of the opposing aspect and its modification term.
- Rank the candidate sentences by applying BERT's Next Sentence Prediction [17] to selected the most natural sentences as the output.

3.2 Construction of Co-occurrence Dictionary

P-Cooc. This section describes the way to construct P-Cooc. P-Cooc calculates PMI (Pointwise Mutual Information) to estimate the dependency of two aspects contained in sentences with different polarities within same dialogues. Currently, we use a collection of reviews[1] to construct P-Cooc, and regard one review as one conversation.

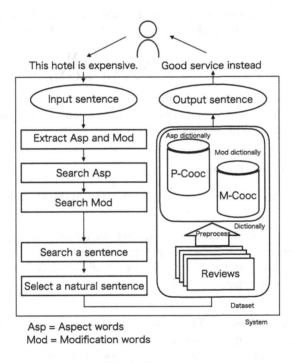

Asp = Aspect words
Mod = Modification words

Fig. 1. Overview

---

[1] In this study, we use the hotel review data-set of Datafinity [18] as the corpus. There are 147,236 sentences in the data-set.

There is an example in Fig. 2 to illustrate the way to calculate P-Cooc score. The P-Cooc of two aspects A and B is defined as follows.

$$P\text{-}Cooc(A, B) = \log_2 \frac{Cnt(A, B) \times N}{Cnt(A) \times Cnt(B)} \tag{1}$$

where $Cnt(A, B)$ denotes the number of conversations (reviews) in such that a dialogue the sentences containing A and B are with different sentimental polarities. $Cnt(A)$ and $Cnt(B)$ are the number of reviews containing A and B, respectively. N is the number of reviews.

One notable point is that we should eliminate rare aspect combinations. If not, it is possible that rare aspect pairs will have a higher score of P-Cooc.

M-Cooc. As mentioned before, M-Cooc (Modification Co-occurrence) is a dictionary for examining the co-occurrence relation between aspects and its modification terms. As same as the construction of P-Cooc, we use a review collection to construct M-Cooc.

Figure 3 shows an example of calculating M-Cooc score. To a given aspect A and its modification term M, the degree of $M - Cooc(A, M)$ is defined as follows based on PMI.

$$M\text{-}Cooc(A, M) = \log_2 \frac{C(A, M)}{C(A) \times C(M)} \tag{2}$$

where $C(A, M)$ denotes the frequency of sentences containing both A and M. $C(A)$ and $C(M)$ represent the number of sentences containing A and M, respectively. N is the number of reviews.

3.3   Sentence Generation

In this section, we describe our sentence generation method in detail.

Aspect and Polarity Extraction. At first, we extract the aspect term from the given sentence and estimate its polarity. In the example of "The hotel is expensive", its aspect and polarity are "hotel" and negative, respectively. The modification term for "hotel" is "expensive".

Opposing Aspect Search. Then, we search the dictionary P-Cooc for opposing aspects, which usually provide different sentiment polarity of "hotel". In our example, the opposing aspects are "staff", "food" and "view".

Modification Term Search. After that, we search for modification terms of these opposing aspects. For each aspect, we pick up modification terms with high probabilities of co-occurrence by using M-Cooc.

In our example, we will have "delicious" or "cheap" for "food", "helpful" or "welcoming" for "staff", "beautiful" or "amazing" for "view".

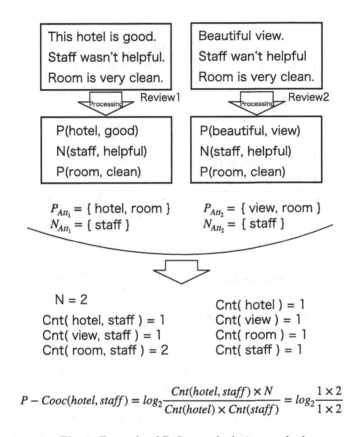

Fig. 2. Example of P-Cooc calculation method

Candidate Sentence Search. Now, we can generate queries to search for candidate sentences from the given corpus. Here, each query is formed as a pair of *(opposing aspect term, modification term)*, such as *(food, delicious)*, *(staff, helpful)*, and so on.

For each searched sentence, we estimate its polarity and remove those with the same sentiment polarity of the input sentence. For example, "The staff was not helpful" will be removed from the candidates because its polarity is negative as same as that of the input sentence.

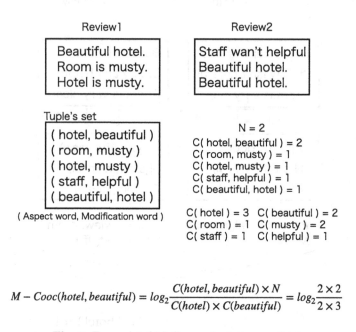

Review1

| Beautiful hotel. |
| Room is musty. |
| Hotel is musty. |

Review2

| Staff wan't helpful |
| Beautiful hotel. |
| Beautiful hotel. |

Tuple's set

(hotel, beautiful)
(room, musty)
(hotel, musty)
(staff, helpful)
(beautiful, hotel)

(Aspect word, Modification word)

N = 2
C(hotel, beautiful) = 2
C(room, musty) = 1
C(hotel, musty) = 1
C(staff, helpful) = 1
C(beautiful, hotel) = 1

C(hotel) = 3   C(beautiful) = 2
C(room) = 1   C(musty) = 2
C(staff) = 1   C(helpful) = 1

$$M - Cooc(hotel, beautiful) = log_2 \frac{C(hotel, beautiful) \times N}{C(hotel) \times C(beautiful)} = log_2 \frac{2 \times 2}{2 \times 3}$$

Fig. 3. Example of M-Cooc calculation method

Ranking. The final step is to rank the candidate sentences by applying BERT's Next Sentence Prediction [17] and selected the most natural sentences as the output.

Intuitively, we use BERT to checks whether two sentences are consecutive, and calculate the degree of continuous sentences. The sentence with the highest value will be returned as the final output.

By using the next sentence prediction of BERT, it is possible to measure which sentence comes after the input sentence. BERT is a language model that has learned the tasks of Masked Language Model and Next Sentence Prediction. This study mainly uses Next Sentence Prediction. This allows the language model to learn the relationship between the two sentences. Learn by inputting two consecutive sentences and two random sentences in the input half each.

By using Next Sentence Prediction, the output sentence guarantees its naturalness and validity. The evaluation value takes a value from 0 to 1, and the closer to 1, the more natural the sentence that comes next. Output the most natural sentence through BERT.

Table 1. Aspect and modification terms

	# of Terms	# of Terms with high frequency
Aspects	9063	2337
Modificaiton	6751	1830

4   Experimental Evaluation

To validate our proposed method, we have conducted two experimental evaluation. One is to evaluate the P-Cooc and M-Cooc; the other is to evaluate the sentences generated by our proposed method based on a user study.

4.1   Data-Set

We choose the hotel reviews from Datafiniti's Business Database [18] as our data-set. This data-set includes review of 1,000 hotels. As mentioned before, we regard a review as one conversation. There are 35912 reviews and 147,236 sentences in our data-set.

To construct P-Cooc and M-Cooc, we have extracted 9063 aspects from 181483 noun terms. Similarly, 6751 modification terms were extracted from 162722 adjective terms. Since we calculate P-Cooc and M-Cooc based on PMI, it is necessary to eliminate rare terms with low frequency. Currently, we removed terms appearing less than five times. As a result, there are 2337 aspects and 1830 modification words. The statistics of aspects and modification terms are summarized in Table 1.

4.2   Evaluation on Sentence Generation

We conduct a user study in a questionnaire manner to evaluate the proposed dialogue sentence generation method. We have cooperated with 20 undergraduate students belonging to the English Study Group. In the questionnaire, we asked the subjects to estimate the generated sentences from three viewpoints as follows.

- Fluent: Whether the conversation is fluent and the context is good.
- Naturalness: Whether the generated sentence is natural like given by a human.
- Critical Thinking: Whether the generated sentence is helpful to promote critical thinking.

We asked the subjects to evaluate the generated sentences with a seven points grading mechanism, where 1 is the worst score and 7 denotes the best one.

In our experiment, we randomly select 7 positive sentences and 6 negative sentences from our review data-set. These sentences are shown in Table 2. For each sentence, we generate a response sentence to construct a conversation as follows.

Table 2. First sentences of conversations used in user study

Room was well outfitted and bed was very comfortable.
The resort was beautiful, well kept, and the room was luxurious without being opulent.
Good experience and very convenient to where I needed to be.
Beautiful hotel.
Staff was great and rooms were clean and very comfortable.
We only stayed one night but this is a good place to stay.
Room seemed clean but it smelled very musty.
Very loud rowdy customers staff would not do anything about the frat like partying.
Room was not clean.
Nice hotel , with very friendly staff and helpful - great choice for breakfast.
The king size bed was the most uncomfortable bed I had slept in for decades.
This hotel is way over rated, it's nothing like the description or the reviews.
The breakfast was adequate and the housekeeping staff were friendly.

Table 3. Result

Method	Fluent	Naturalness	Critical thinking
GPT-2	3.93	4.70	3.27
Proposed	4.07	4.32	3.91

- Mike: This hotel is way over rated, it's nothing like the description or the reviews.
- Jane: This is a good mid priced placed to stay.

In this example, the sentence spoken by Mike is selected from Table 2 and the one by Jane is generated by our method or GPT-2 [5], which is the baseline in our experiment.

Results. The results are shown in Table 3.

We also conduct a paired two-tailed t-test for significance testing. The null hypothesis that there is no difference between the baseline method and the proposed method. The significance level is $\alpha = 0.05$ at the degree of freedom 19^2. The results are shown in Table 4. The null hypothesis was rejected when the p-value was less than 0.05, indicating that there was a significant difference between the proposed method and the baseline method.

There is no significant difference between these two methods from the viewpoint of fluent. The results of the t-test also showed significant differences in naturalness and critical thinking. Our method is superior in critical thinking but inferior in naturalness.

In addition, we investigate the results by distinguish the first sentence into positive and negative ones. The results are shown in Table 5 and 6.

---

[2] We have 20 subjects in our experiment.

Table 4. t-test's p-value

Fluent	Naturalness	Critical thinking
0.433	0.013	0.010

Table 5. Results of first sentence with different polarity

Method	Polarity	Fluent	Naturalness	Critical thinking
GPT-2	Positive	3.86	4.74	3.14
Proposal	Positive	4.13	4.33	3.99
GPT-2	Negative	4.01	4.66	3.43
Proposal	Negative	3.98	4.30	3.83

Table 6. p-value of first sentence with different polarity

Polarity	Fluent	Naturalness	Critical thinking
Positive	0.225	0.043	0.008
Negative	0.561	0.119	0.346

There is a similar difference in the naturalness of sentences. However, about critical thinking, sentences with positive inputs resulted in a large difference. Significant differences were found for positive inputs by the t-test. On the other hand, there was no significant difference for those with negative inputs. One of the considerable reasons is that the first sentence may be erroneously determined to be positive even though it is negative. As a result, the generated sentences may not be considered helpful to critical thinking because the polarity did not change.

Discussion. Although the hotel review data-set need further pre-processing to be suitable for conversation generation, the experimental results reveal that the response sentences generated by our proposed method is superior in promoting critical thinking than that of GPT-2.

From the conversation viewpoint, the limitation of the current method is that it just generate one sentence and change the sentimental polarity. In a real conversation, if we want to persuade someone or comfort someone, at first we may need accept their feeling and then try to provide other viewpoints. We will discuss this issue in our future work.

Fluent. There was no significant difference in fluency. Actually, the candidate sentences searched by the queries consisting of aspect and modification terms were estimated high probabilities of being the second sentences. It reveals that the pair of aspect and modification terms obtained by P-Cooc and M-Cooc is helpful to search for response sentences in context.

Naturalness. All the sentences used in the experiment are written by humans. However, the naturalness of the sentences retrieved by GPT-2 achieved better results in our user study. This may be caused by the characteristics of the review data-set. Usually, sentences are standard sentences such as SVO (Subject, Verb, Object) structure. However, reviews are in free style or in broken language. For example, the sentence with the lowest score was "Hotel was great staff was happy n positive energy everywhere." We will study this issue by utilizing the state of the art NLP tools.

Critical Thinking. The goal of this study was to produce dialogue that encourages critical thinking, and it was better than GPT-2. The generated conversation with the highest score in this experiment is as follows.

- Mike: Room seemed clean but it smelled very musty.
- Jane: I want to leave a positive review, because the room had a neat look, the staff was friendly, and it was a nice quiet stay.

The first sentence is concise, while the sentimental contrast between the sentences is clear. The response sentence provides several aspects with positive polarity.

4.3   Evaluation on P-Cooc and M-Cooc

4.4   P-Cooc

As mentioned before, $P\text{-}Cooc(A, B)$ measures the dependency of two aspects with opposing polarities. Thus, we may have two patterns of the opposing aspect pairs: Negative vs Positive and Positive vs Negative. We compare the results with the two different patterns. The opposing aspects pairs with high scores of P-Cooc are shown in Table 7 and 8. The scores of P-Cooc depend on the order of aspects' polarities. We may need to construct two kinds of P-Cooc dictionaries with the two different patterns. If we want to generate a positive sentence responding to a negative sentence, P-Cooc based on pattern of (Negative, Positive) is preferred, and vice versa.

In order to exclude terms with too low a frequency, term with a frequency less than five were excluded. If we increase the threshold from five to ten and twenty, the number of target terms becomes 1156 and 685, respectively. Table 9 shows the differences when we change the threshold.

As shown in Table 9, increasing the threshold tends to increase the P-Cooc scores. However, the terms used for further processing will be limited. Therefore, to have more aspects for further analysis, the threshold of the appearance frequency was set to 5.

Table 7. P-Cooc based on count of (Negative, Positive) pairs

Negative aspect, Positive aspect	P-Cooc
room, inconveniences	0.6449663977057334
hotel, ports	1.0317250913592815
staff, sabrina	0.6833845190889121
breakfast, shores	1.0783373263749623

Table 8. P-Cooc based on count of (Positive, Negative) pairs

Positive aspect, Negative aspect	P-Cooc
room, controls	0.9080008035395273
hotel, radios	1.7946858940584323
staff, omelet	1.2683470198100684
breakfast, adrenaline	2.6632998270961186

Table 9. P-Cooc with different thresholds of term frequency. B is the aspect having the maximum $P\text{-}Cooc(A, B)$ score with A.

Aspect A	Threshold	Aspect B	P-Cooc (A,B)	Frequency of B	Frequency of (A,B)
room	5	inconveniences	0.64	15	3
room	10	art	−0.09	10	3
room	20	recommendation	−0.36	24	6
hotel	5	ports	1.03	40	22
hotel	10	ports	1.03	40	22
hotel	20	ports	1.03	40	22
staff	5	sabrina	0.68	5	2
staff	10	cake	−0.11	13	3
staff	20	welcome	−0.42	27	5
breakfast	5	shores	1.08	6	2
breakfast	10	innkeepers	0.93	10	3
breakfast	20	love	−0.51	27	3

Table 10. Examples of M-Cooc scores

(Aspect, Modification)	M-Cooc
(room, stunk)	0.6449663977057334
(hotel, weary)	1.0317250913592815
(staff, accommodating)	0.6833845190889121
(breakfast, worldwide)	1.0783373263749623

4.5   M-Cooc

As mentioned before, M-Cooc scores denote dependencies of aspect and its modification terms. Table 9 shows some examples of M-Cooc scores of pairs of aspects and modification terms, in which the aspects appear frequently in our data-set.

These pairs with high M-Cooc scores shown in Table 10 is a bit strange, especially, the pairs of (room, stunk), (hotel, weary) and (breakfast, worldwide). For further analysis, we investigated their frequencies as shown in Table 11. From these tables, it is considerable that some strange modification terms will have high M-Cooc scores due to their low appear frequencies. To confirm this, we investigated the differences when we change the frequency thresholds. The results are shown in Table 12.

Table 11. Frequency of aspect and modification terms

Aspect	Modification	Frequency of modification	Frequency of (aspect, modification)
room	stunk	5	5
hotel	weary	30	22
staff	accommodating	496	301
breakfast	worldwide	6	6

Table 12. M-Cooc scores with different threshold.

Aspect A	Threshold	Modification B	M-Cooc	Frequency of B	Frequency of (A,B)
room	5	stunk	3.32	15	3
room	10	non-smoking	2.92	133	101
room	20	non-smoking	2.92	133	101
hotel	5	weary	1.03	30	22
hotel	10	weary	1.03	30	22
hotel	20	weary	1.03	30	22
staff	5	accommodating	0.68	496	301
staff	10	accommodating	0.68	496	301
staff	20	accommodating	0.68	496	301
breakfast	5	worldwide	4.33	6	6
breakfast	10	fruit	4.23	15	14
breakfast	20	cereal	4.16	26	23

Similar to that of P-Cooc scores, by raising the frequency thresholds it is possible to find more proper modification terms. We will discuss this issue to improve M-Cooc construction method in near future.

We also noticed that some terms may be aspect and modification terms and we need to distinguish their roles to improve the M-Cooc and P-Cooc constructions. For example, "fruit" in Table 12, is a modification term. However, "Fruit" is usually used as noun and in most cases it should be an aspect. This may be caused by the data characteristics: reviews are writing in free style, and there are short sentences without sentence structure. For instance, in the sentence of "Continental breakfast was nice and fresh, fruit and baked goods with.", "fruit" is treated as being parallel to "nice" and "fresh", and has been treated as a modification term. Actually, in our data-set, there are 119 "fruit" appearing as aspects, and 15 are incorrectly extracted as modification terms. As a reasonable solution, it is possible to decide the role of a term based on its role frequencies. For example, as described above, "fruit" is often used as an aspect more than modification terms, we may simply dropout the cases when "fruit" is extracted as modification term. We will discuss these issues in our future work.

5   Conclusion

In this paper, we propose a method to generate dialogue sentences to provide different viewpoints. We propose two novel concepts, polarity co-occurrence (P-Cooc) and modification co-occurrence (M-Cooc) to search for candidate sentences. The experimental results with a real hotel review data-set reveal that our method could provide supplementary viewpoints to promote critical thinking.

In near future, to improve our method, we will conduct further experiments with different data-set. Generate more complex and real conversation with more than two sentences is also necessary. Another important future work is that, we will introduce the insights from psychology to generate sentences Which are useful to promote critical thinking and comfort people.

Acknowledgements. This work is partly supported by MIC SCOPE (172307001, 201607008) and KAKENHI (19H04116).

References

1. Ennis, R.H.: A logical basis for measuring critical thinking skills. Educ. Leadersh. **43**(2), 44–48 (1985)
2. Scriven, M., Paul, R.: Critical thinking. In: The 8th Annual International Conference on Critical Thinking and Education Reform, CA (1987)
3. Hirayama, R., Kusumi, T.: Effect of critical thinking disposition on interpretation of controversial issues: evaluating evidences and drawing conclusions. Jpn. J. Educ. Psychol. **52**(2), 186–198 (2004)
4. Kusumi, T., Matsuda, K.: Structure of media literacy supported by critical thinking attitude. In: The 71st Annual Convention of the Japanese Psychological Association, p. 1PM085. The Japanese Psychological Association (2007). (in Japanese)
5. Radford, A., Jeffrey, W., Child, R., Luan, D., Amodei, D., Sutskever, I.: Language models are unsupervised multitask learners. OpenAI Blog **1**(8), 9 (2019)
6. Takayama, J., Arase, Y.: Chat response generation based on feature vocabulary prediction using pointwise mutual information. In: Japanese NLP Conference 2019, pp. 3–34 (2019)

7. Meguro, T., Sugiyama, H., Higashinaka, R., Minami, Y.: Building a conversational system based on the fusion of rule-based and stochastic utterance generation. In: The 28th Annual Conference of the Japanese Society for Artificial Intelligence (2014), p. 2M5OS20b2. The Japanese Society for Artificial Intelligence (2014). (in Japanese)

8. Rosser, R.J., Sturges, S.B.: Response generator for mimicking human-computer natural language conversation, 24 August 2010. US7783486B2

9. Vinyals, O., Le, Q.: A neural conversational model. arXiv preprint arXiv:1506.05869 (2015)

10. Sutskever, I., Vinyals, O., Le, Q.V.: Sequence to sequence learning with neural networks. In: Advances in Neural Information Processing Systems, pp. 3104–3112 (2014)

11. Li, J., Galley, M., Brockett, C., Gao, J., Dolan, B.: A diversity-promoting objective function for neural conversation models. arXiv preprint arXiv:1510.03055 (2015)

12. Serban, I.V., Sordoni, A., Bengio, Y., Courville, A., Pineau, J.: Building end-to-end dialogue systems using generative hierarchical neural network models. In: Thirtieth AAAI Conference on Artificial Intelligence (2016)

13. Nio, L., Sakti, S., Neubig, G., Yoshino, K., Nakamura, S.: Neural network approaches to dialog response retrieval and generation. IEICE Trans. Inf. Syst. 99(10), 2508–2517 (2016)

14. Zhou, H., Huang, M., Zhang, T., Zhu, X., Liu, B.: Emotional chatting machine: emotional conversation generation with internal and external memory. In: Thirty-Second AAAI Conference on Artificial Intelligence (2018)

15. Lubis, N., Sakti, S., Yoshino, K., Nakamura, S.: Positive emotion elicitation in chat-based dialogue systems. IEEE/ACM Trans. Audio Speech Lang. Process. 27(4), 866–877 (2019)

16. Socher, R., et al.: Recursive deep models for semantic compositionality over a sentiment treebank. In: Proceedings of the 2013 Conference on Empirical Methods in Natural Language Processing, pp. 1631–1642 (2013)

17. Devlin, J., Chang, M.-W., Lee, K., Toutanova, K.: Bert: pre-training of deep bidirectional transformers for language understanding. arXiv preprint arXiv:1810.04805 (2018)

18. Dataset in Datafiniti. https://data.world/datafiniti/hotel-reviews. Accessed 12 2020

Semantic Matching over Matrix-Style Tables in Richly Formatted Documents

Hongwei Li[1,2], Qingping Yang[1,2], Yixuan Cao[1,2], Ganbin Zhou[3],
and Ping Luo[1,2(✉)]

[1] Key Lab of Intelligent Information Processing of Chinese Academy of Sciences
(CAS), Institute of Computing Technology, CAS, Beijing 100190, China
{lihongwei,luop}@ict.ac.cn
[2] University of Chinese Academy of Sciences, Beijing 100049, China
[3] Search Product Center, WeChat Search Application Department, Tencent,
Shenzhen, China

Abstract. Table is an efficient way to represent a huge number of facts
in a compact manner. As practitioners in the vertical domain share lots
of common prior knowledge, they tend to represent facts more concisely
using *matrix-style tables*. However, such tables are originally intended for
human reading, but not machine-readable due to their complex structures including row header, column header, metadata, external context,
and even hierarchies in headers. In order to improve the efficiency of
practitioners in mining and utilizing these matrix-style tables, in this
study we introduce a challenging task to discover *fact-overlapping relations* between matrix-style tables. This relation focuses on fine-grained
local semantics instead of overall relatedness in conventional tasks. We
propose an attention-based model for this task. Experiments reveal that
our model is more capable of discovering the local relatedness, and outperforms four baseline methods. We also conduct an ablation study and
case study to investigate our model in detail.

Keywords: Matrix-style tables · Semantic matching ·
Fact-overlapping relations · Richly formatted documents

1 Introduction

Tables, as a compact representation of data, are widely used on the Web and in
vertical domains. Mining the relationships between tables has its value in a range
of applications, such as table retrieval [22], knowledge base construction, entity
disambiguation, and intelligent reading [1,2,10,12]. Taking intelligent reading
[2,12] as an example, professional documents in vertical domains are usually
hundreds of pages long with dozens of tables. These tables are related to each
other so that they can provide coherent evidence to support the arguments
in the document. However, the related tables might scatter over hundreds of

---

H. Li and Q. Yang—Equal contribution.

© Springer Nature Switzerland AG 2020
S. Hartmann et al. (Eds.): DEXA 2020, LNCS 12391, pp. 369–384, 2020.
https://doi.org/10.1007/978-3-030-59003-1_24

pages, and tracking these linked content requires flipping back and forth in the current reading experiences. It is even more cumbersome for reading on a digital device than using a physical document. Hence, relating tables and displaying them dynamically based on the readers' goal will greatly enhance the reading efficiency.

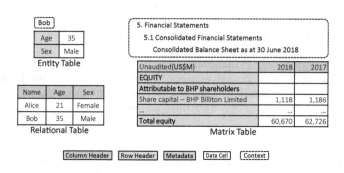

Fig. 1. Examples of entity, relational and matrix tables.

Distinguished by how the information in the table is organized, there are *entity* tables, *relational* tables, and *matrix-style tables*[1] (exemplified in Fig. 1). Matrix tables have a more concise layout than other kinds of tables. Practitioners in vertical domains usually put plenty of data and facts into matrix tables in richly formatted documents [28]. For the public disclosure documents from the financial area, the proportion of matrix tables is as high as 90% based on our empirical study. Therefore, in this paper, we study the relationships over matrix tables within a richly formatted document.

As shown in Fig. 1, a matrix table usually consists of five *components*: *context*, *metadata*, *column headers*, *row headers*, and *data cells* (see the legend for each component in Fig. 1). Here, the *metadata, column headers, row headers* and *data cells* are the areas inside a table [9,23]. And the *context* is the outside-table context, including all the text along the path from the root to this table leaf node in the tree of logical document hierarchy [17]. Since there exist approaches to identify the table components [16,20,23], in this study, we assume that all these table components are extracted in some predecessor steps.

Additionally, each data cell in a matrix table refers to a *fact* whose complete semantics is scattered in multiple table components. For example, in Fig. 2, the fact expressed by the data cell in the dashed box is shown at the bottom. It is a complex composition of several cells from context, metadata, column and row headers. Thus, a matrix table T conveys a set of facts. Their values lie in data cells, while their semantics are presented in the table context, metadata, row and column headers succinctly.

---

[1] *matrix tables* for short in the following of this paper.

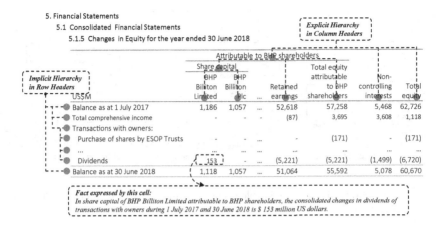

Fig. 2. A table is a set of facts. The fact of a data cell is shown at the bottom, which involves many cells in multiple components, shown in different colors. The row and column headers have hierarchical structures, shown in dotted lines. (Color figure online)

In this paper, we study the semantic matching problem over matrix tables within a document. Our goal is to determine whether two tables have *fact-overlapping relations*. Specifically, if two tables have some facts in common, they have fact-overlapping relations. For example, the two tables from document *BHP Annual Report 2018* are shown in Fig. 3, where the left one is the *consolidated balance sheet*, and the right one is the *consolidated statement of changes in equity*. They are semantically matched as the facts of data cells marked in boxes exist in both tables.

Matching the two tables in Fig. 3 has its practical value. In financial area, people read disclosure documents with different purposes. For investors, they read documents to learn about the operation of the company. When looking at the retained earnings in the left table, readers want to investigate how it changed during the year, which is detailed in the right table. For financial practitioners, they need to ensure the correctness and consistency of the disclosure of the company's financial position. Therefore, they have to cross-check the numbers in the left table with the numbers in the right table. Both groups of people want to link these two tables together to facilitate their reading process and to avoid jumping back and forth to find relating tables in a document of hundreds of pages. To this end, we propose to equip readers with a kind of *table-linking* functionality. When clicking on a table, a list of tables having fact-overlapping relations with that table is shown to the user as a sidebar on the right.

Simply linking two tables with at least one data cell with the same value cannot solve this problem, since the same value might refer to different facts and some mistaken data cells with different values might refer to the same fact [4]. Hence, we need to model the semantics of the facts inside matrix tables. The challenges of this task are summarized in three aspects.

5. Financial Statements

5.1 Consolidated Financial Statements

Consolidated Balance Sheet as at 30 June 2018

Unaudited	2018 US$M	2017 US$M
EQUITY		
Attributable to BHP shareholders		
Share capital – BHP Billiton Limited	1,118	1,186
Share capital – BHP Billiton Plc	1,057	1,057
...		...
Retained earnings	51,064	52,618
Total equity attributable to BHP shareholders	55,592	57,258
Non-controlling interests	5,078	5,468
Total equity	60,670	62,726

Consolidated balance sheet

5. Financial Statements

5.1 Consolidated Financial Statements

Consolidated Statement of Changes in Equity for the year ended 30 June 2018

Unaudited US$M	Attributable to BHP shareholders					Non-controlling interests	Total equity
	Share capital						
	BHP Billiton Limited	BHP Billiton Plc	...	Retained earnings	Total equity attributable to BHP shareholders		
Balance as at 1 July 2017	1,186	1,057	...	52,618	57,258	5,468	62,726
Transactions with owners:							
Purchase of shares by ESOP Trusts	-	-	...	-	(171)	-	(171)
...	...	...	...	...	...	...	...
Dividends	-	-	...	(5,221)	(5,221)	(1,499)	(6,720)
Balance as at 30 June 2018	1,118	1,057	...	51,064	55,592	5,078	60,670

Consolidated Statement of Changes in equity

Fig. 3. Two semantically matched tables. The overlapping facts are shown in the boxes.

First, the same fact can be expressed differently in terms of the table layout and its surface form. For example, for the common fact with value "1,118" in Fig. 3, the date of the fact is contained in its column header and context in the left table, while it is contained in its row header in the right one.

Second, the fact-overlapping relations focus on the semantic similarity at the fact level locally. For example, the two tables shown in Fig. 3 are very different on the whole, but they have overlapping facts. In extreme cases, there could be only one overlapping fact in two semantically matching tables. This requires to retain the detailed local information when modeling the table. However, previous studies in semantic matching usually consider global relatedness in the sense that two objects have similar meanings on the whole, including matching between sentences [15,18,26], documents [29], images [30] and tables [1,10,22].

Third, understanding the exact fact of each data cell is challenging since matrix tables have complex structures so that the meaning of facts scatters over multiple components as we mentioned before. Moreover, row and column headers might have *implicit* and *explicit* hierarchies. Take Fig. 2 as an example. The implicit hierarchy of row headers has three levels, conveyed by their visual cures (e.g. font styles, indentation) and text semantics without a unified standard [5]. Its explicit hierarchy of column headers also has three levels and is presented by the internal table structure with merged cells. With such hierarchies, the integrates of a fact might involve several cells in column and row headers. In Fig. 2, the fact of the data cell in the dashed box involves four cells in the row headers (in orange) and three cells in the column headers (in green).

To preserve the meanings of table facts while tolerating the diverse expressions in tables, we propose an attention-based method for fact-overlapping matching between matrix tables. It employs a deep neural network that consists of two components: table embedding network, and symmetric matching network. The input of this neural network is a pair of tables. First, the embedding network calculates the embedding of each table by considering its 4 components, namely context, metadata, (hierarchical) column headers, (hierarchical) row headers. Then, a symmetric matching network entangles these two embeddings to discover all their local fact-level relatedness and predicts whether these two tables are semantically matched or not. Experiments reveal that our attention-based method is more suitable for this task than four baseline methods with 0.75, 0.77, 0.67 and 0.29 absolute improvement on F_1 respectively. Moreover, our attention-based method has a certain ability to explain why and where two tables are semantically matched, which is illustrated in Sect. 5.3. We also conduct an ablation study to check the importance of each table component and some case studies to investigate our model in detail.

2   Related Work

We introduce the work on semantic matching of two tables as follows. Sarma et al. [22] proposed a table retrieval task that given a query table, retrieved the most similar tables. They took each table as a set of attributes (i.e. column headers in relational table) and linked the attributes of two tables to form a weighted bipartite graph. Then, they used the max-weight matching as matching scores of two tables. We consider this method as a baseline in our experiments. Fetahu et al. [10] proposed a deep learning-based method to recognize two types of table relations: equivalent and subPartOf. But they only focus on the relational tables with column headers, and their method is not suitable for matrix tables. Besides the matching of tables as a whole, some studies on schema matching [3,19] concentrated on the correspondence of columns between two tables [11]. Zhang and Chakrabarti [32] computed semantic matching between columns of web relational tables. If two tables describe different attributes of the same set of entities, they can be used to augment the attributes of entities. The existing studies on the relationships between tables mainly focus on entity and relational tables, since these two types of tables with simple structures are prevalent on the Web (accounting for 98.3% of tables on Web according to Web Data Commons [14]).

Other studies matched tables to other things like query or knowledge base [7,24,33]. Zhang and Balog [33] represent text query and table as a set of vectors of words respectively and compute their similarity. We adopt this method as a baseline in the experiments. Additionally, matching between a table and a knowledge base establishes the mapping between the entities described by them in order to understand the table data [21,34].

There are also some related studies about table classification. Tables have various layouts, and there are many standards to categorize them. For practical purposes, Ahmadov et al. [1] categorized tables to five classes: relational, entity, matrix, layout, and others. In this paper, we adopt this taxonomy and focus on matrix tables as it is prevalent in vertical domains. Also, Wang et al. [27] proposed the taxonomy with three table classes: 1-dimensional tables, 2-dimensional tables, and complex tables. Considering the layout and structure of tables, Crestan and Pantel [6] proposed a more fine-grained classification, which classifies tables into two broad categories: *relational knowledge* and *layout*. Furthermore, based on structural characteristics, Lautert et al. [13] divided relational knowledge tables into concise, nested, multi-valued, and split tables.

3   Network Architecture

As we have discussed in Sect. 1, understanding the exact fact of each data cell is challenging. Therefore, we model whole semantics of tables to determine fact-overlapping relations end to end. Our Semantic Matching (SM) method employs an attention-based deep neural network that consists of two components: table embedding network and symmetric matching network. The input of this neural network is a pair of tables. First, the embedding network calculates the embedding of each table, which preserves its local semantic information. Then, these two table embeddings are fed into the symmetric matching network that predicts whether these two tables are semantically matched or not. Our symmetric matching network leverages an attention encoder to discover the local semantic matching on the fine-grained level between tables, and leverages another attention encoder to aggregate the local matched semantics for classification. We adopt supervised learning to train this deep neural network model with cross-entropy as the loss function.

3.1   Table Embedding

Since the local semantics of tables are essential to the task of determining fact-overlapping relations, we encode one table to a sequence of cell embeddings instead of a fixed-length embedding. We encode four components of a table to form its embedding sequence: row headers, column headers, metadata, and context, as described in Sect. 1. Each component contains some cells (we call each heading in a header cell and also each title in the table context as a "cell" for convenience). We first encode each cell into a vector, add component embeddings, and then serialize vectors of four components into a sequence of vectors as the embedding of the table. Such embeddings preserve the local information in the table.

The first step is cell embedding. For row headers or column headers with explicit hierarchy (like the column headers in Fig. 2), we extend the text of a leaf cell by joining the texts of cells on the path from root to leaf with a special token "&". For example, the extended text of the cell in the second column header in Fig. 2 is

Attributable to BHP shareholders & Share capital & BHP Billiton Plc

As you can see, each joined text contains the complete local and hierarchical information.

As each leaf node corresponds to a column or row in the table, and we have extended its text to incorporate the hierarchy, we only use the leaf cells for table embedding. A transformer [25] encodes the (extended) text of each leaf cell into a text embedding e^t. Moreover, to distinguish cells among different components, we introduce component embeddings e^s for each component (like the segment embedding in [8]). Finally, cell embedding is

$$e = e^t + e^s \in \mathbb{R}^{d_m},$$

where d_m is the dimension of embedding.

The second step is table embedding. We do not compress the table into a fixed-length embedding as it will lose the information of individual cells. Instead, the table embedding is a sequence of cell embeddings. In each component, we stack the leaf cells by order. Then, we stack these four components into an embedding sequence

$$E_T = [e_{c_1}, ..., e_{c_{n_c}}, e_{r_1}, ..., e_{r_{n_r}}, e_{x_1}, ..., e_{x_{n_x}}, e_{m_1}, ..., e_{m_{n_m}}] \in \mathbb{R}^{n \times d_m},$$

where e_{c_i} is the embedding of the i-th cell in column headers (r for row headers, x for context and m for metadata), $n = n_c + n_r + n_x + n_m$ is the number of leaf cells in the four components. Note that we only encode the four components in a table and ignore data cells, since data cells, usually containing digits, are not semantically expressive without the other four table components.

3.2 Symmetric Matching Network

The above table encoding preserves the local semantic information of the table to the maximum extent so that we can explicitly carry out the local semantic interaction between tables in the symmetric matching network. The symmetric matching network takes as inputs two table embeddings, E_{T_1} and E_{T_2}, and outputs the probability that they are matched. First, we entangle the embeddings of these two tables, to get E'_{T_1} and E'_{T_2}, where each cell of T_1 is informed of (by attention on) every cell in T_2 and vice versa. This allows the local semantic matching between two tables. Then, we aggregate E'_{T_1} and E'_{T_2} respectively to get two vectors e_{T_1} and e_{T_2}, which gathers the local matched semantics. Finally, the prediction layer outputs the probabilities. Both entangling and prediction layers are symmetric with regard to T_1 and T_2.

The building block of our network is an attention encoder. In brief, an attention encoder $Q' = Attention(Q, V)$ uses Q, V as input and gets a new embedding Q', where Q and $Q' \in \mathbb{R}^{l \times d_m}$ and $V \in \mathbb{R}^{m \times d_m}$. This attention encoder $Attention(Q, V)$ will be detailed in Sect. 3.3.

In entangling layer, taking E_{T_1} and E_{T_2} as Q and V respectively we get E'_{T_1}:

$$E'_{T_1} = Attention_1(E_{T_1}, E_{T_2}). \tag{1}$$

The attention encoder allows each cell in table T_1 to interact with every cell in table T_2. This is essential to discover the local semantic matching on the fine-grained level between tables. We will show the effectiveness of this encoder in the case study. Symmetrically, we get $E'_{T_2} = Attention_1(E_{T_2}, E_{T_1})$.

In the aggregation layer, a learnable special embedding vector $e_{[TAB]}$ is used to aggregate the information of a table:

$$e^*_{T_1} = Attention_2(e_{[TAB]}, E'_{T_1}). \tag{2}$$

Here, $e^*_{T_1}$ is an aggregated vector representing table T_1 after attention on table T_2. Using the attention allows this aggregation to focus on the local matched semantics flexibly instead of rough semantics such as *mean* or *max*. We also show the effectiveness of this encoder for capturing local matched semantics in the case study. Similarly, we get $e^*_{T_2} = Attention_2(e_{[TAB]}, E'_{T_2})$.

In the prediction layer, we concatenate $e^*_{T_1}$ and $e^*_{T_2}$ in both orders and compute e_O as follows:

$$e_O = \text{ElementwiseMax}(FFN(\text{concat}(e^*_{T_1}, e^*_{T_2})), FFN(\text{concat}(e^*_{T_2}, e^*_{T_1}))) \quad (3)$$

where FFN is a feed-forward network, and the element-wise maximum of the two vectors ensures that this matching network is symmetric. Finally, e_o is used for classification. The symmetric property of our model ensures that no matter the input is (E_{T_1}, E_{T_2}) or (E_{T_2}, E_{T_1}), the result will be the same.

3.3 Attention Encoder

The attention encoder is the building block of the symmetric matching network. It takes as input Q and V, and outputs Q':

$$Q' = Attention(Q, V) \quad (4)$$

where Q and $Q' \in \mathbb{R}^{l \times d_m}$, $V \in \mathbb{R}^{m \times d_m}$. The attention encoder consists of b Layers $(Layer_1, ..., Layer_b)$ in sequence:

$$Q_i = Layer_i(Q_{i-1}, V) \quad (i = 1, \cdots, b) \quad (5)$$

where $Q_0 = Q$ as the initial input and $Q' = Q_b$ as the final output. Each layer consists of Multi-head Attention Layer (MAL) and Feed-Forward Network (FFN). There is Residual Connection (RC) and Layer Normalization (LN) following both MAL and FFN. The detailed structure of $Layer_i$ is as follows:

$$Z_{i,1} = MAL_i(Q_{i-1}, V) \quad (6)$$
$$Z_{i,2} = LN_{i,1}(Q_{i-1} + Z_{i,1}) \quad (7)$$
$$Z_{i,3} = FFN_i(Z_{i,2}) \quad (8)$$
$$Q_i = LN_{i,2}(Z_{i,2} + Z_{i,3}) \quad (9)$$

MAL consists of multiple attention operations. It concatenates the results of these operations to represent a richer attention. The definition of MAL_i is as follows:

$$MAL_i(Q_{i-1}, V) = \text{Concat}(H_{i,1}, \cdots, H_{i,h})W_i^O \quad (10)$$
$$H_{i,j} = W_{i,j}(VW_{i,j}^{V_1}) \quad (11)$$
$$W_{i,j} = \text{Softmax}\left(\frac{(Q_{i-1}W_{i,j}^Q)(VW_{i,j}^{V_2})^T}{\sqrt{d_v}}\right) \quad (12)$$

where $H_{i,j}$ is the j-th head of the multi-head attention in the i-th layer, and h is the number of heads. The projections are parameter matrices $W_{i,j}^Q, W_{i,j}^{V_1}, W_{i,j}^{V_2} \in \mathbb{R}^{d_m \times d_v}$ and $W_i^O \in \mathbb{R}^{(h \cdot d_v) \times d_m}$. And we let $d_v = d_m/h$.

The attention weights can reveal the relations among elements in two inputs, thus give an interpretation of the prediction result. To analyze the effectiveness of our methods in Sect. 5.3, we define the *summary weights* of the attention encoder as:

$$SW = \sum_{j=1,\cdots,h} W_{b,j} \tag{13}$$

where SW is a vector, and its k-th dimension $SW_k \in [0, h]$ indicates the importance of the k-th cell in V to Q. The greater the value of SW_k, the more important the k-th cell in V is to Q. Here, b is the index of the last layer, indicating that we only display the weights of the last attention layer.

4 Experimental Setup

4.1 Dataset

We downloaded public annual reports from CNINFO[2]. They are all long documents with on average 71.12 pages and 100.8 tables per document. The average numbers of cells in row header, column header, context and metadata per table are 7.91, 4.22, 1.84 and 0.18 respectively. We annotated the matching relations among tables in each document, which took three financial practitioners more than a month to complete. In a document, for each table, annotators find out all tables that match with this table, and other table pairs that do not match are set to negative samples. They do not annotate which parts of table facts are overlapping as it is laborious. There are 358,111 table pairs in total in these documents. And the ratio of matching and not matching pairs is 1:66. All the table pairs are randomly divided into a training and test dataset by 9:1.

4.2 Baselines

There are no previous studies that can be directly applied to our task. Thus, we adapt some of them to build the following four baseline methods.

Term-Based Schema Matching (TSM). We represent each table by bag-of-words of its four components[3]. The matching score of two tables is the Jaccard index of their token sets. Two tables are matched if the matching score is greater or equal than a threshold.

---

[2] http://www.cninfo.com.cn. A financial information disclosure website.

[3] We tokenize each cell using Jieba, a popular Chinese word segmentation toolkit.

Embedding-Based Schema Matching (ESM). Using similar methods in TSM, we represent each table as a set of tokens in its four table components. Then, we build a bipartite graph where an edge links each pair of tokens from the two sets. The edge weight is computed as the cosine similarity between the embeddings of the two tokens. Finally, the max-weight matching score of the bipartite is used as the similarity between the two tables [22]. We regard two tables as a positive sample if their similarity is more than a threshold. In detail, we use the method in [31] to train the token embeddings.

Learning-Based Schema Matching (LSM). We perform a learning-based baseline. In this baseline, we convert each table to a binary vector B whose length is equal to vocabulary size. If a word w_i appears in the table and its index in the vocabulary is i, we set B_i to 1, otherwise 0. For a pair of tables, we concatenate the two table vectors as $B_p = [B_1; B_2]$. Then B_p is regarded as the input feature, the relation between the two tables is regarded as the label, which are put into a logistic regression classifier for training.

Semantic Matching Without Attention Encoder (SM$^-$). The above three methods use the bag-of-words model. In this baseline, we first use the *table embedding* component (detailed in Sect. 3.1) to preserve both the term sequence in each table cell and the hierarchies in row and column headers. Then, the *symmetric matching network* is replaced with the following component [33]: four similarity scores (*Early, Late-max, Late-sum, Late-avg*) between two sets of vectors are computed, and used as the features for classification of matching or not.

4.3 Experimental Settings

We use the *Precision, Recall*, and F_1 to evaluate the above methods. Each feedforward network is a fully connected layer with two linear transformations ($d_m \times 2d_m$ and $2d_m \times d_m$) and ReLU in between. We apply Adam for optimization. 10% of the training data are reserved as the validation set to select the best hyper-parameters. As a result, the number of blocks b in each cross encoder is 3 (over 1, 3, 5); the dimension of embeddings d_m is 128 (over 128, 256, 512); the number of heads h in each multi-head attention is 4 (over 4, 8); and the dimension of each head d_v is 32 (over 32, 64); and the learning rate is 10^{-4} (over $10^{-4}, 10^{-3}, 10^{-2}$).

5   Experimental Results

5.1   Results of Different Methods

Table 1. Table matching results on the test set.

Model	Pre.	Rec.	F_1
TSM	0.1016	0.0841	0.0920
ESM [22]	0.0668	0.0682	0.0675
LSM	0.0944	0.7628	0.1680
SM$^-$ [33]	0.6920	0.4559	0.5496
SM	**0.8685**	**0.8090**	**0.8376**

Table 1 shows the results of table matching of each method on the test dataset. The F_1 score of TSM and ESM method peaks at threshold 0.5, 0.8099 in the training set respectively. Their corresponding F_1 on the test set are 9.2% and 6.75%. The learning-based method LSM also has only achieved 16.8% F_1. The poor performance of these three methods indicates that similarity on terms is not adequate for this problem. SM$^-$ method has great improvement comparing with TSM, ESM, and LSM methods. Because it takes into account both the term sequence in each table cell and the hierarchies in row and column headers. This proves the effectiveness of our table embedding network. However, as the SM$^-$ method does not focus on the local semantics, its performance is still poor in this task. SM performs best among these models. The absolute F_1 improvement is 74.56%, 77.01%, 66.96% and 28.80% compared with TSM, ESM, LSM and SM$^-$. Compared with SM$^-$, our SM model takes more attention on local semantics by our symmetric matching network. Supported by the fact of a larger improvement in recall than precision, we argue that SM is more capable of discovering overlapping facts at the local level.

5.2   Ablation Study

To analyze the effectiveness of each component of tables for our model, we ablate each component of tables in the SM method respectively. As shown in Table 2, we list components by F_1 score after ablation in descending order. When we ablate column headers, F_1 only decreases by 2%. By analyzing the dataset, we find most of the column headers are about the time, like 2016, 2017. As one document usually describes financial position within the same period of time, removing column headers has less impact on semantic matching over tables. Metadata also only has little impact on performance, since the metadata appears less frequently in the dataset and usually describes the unit of data cells. However, as removing column headers and metadata mainly deteriorates the precision, we think they provide information to filter out false-positive samples, but less information for

Table 2. Ablation study on components, layout and symmetric model design.

Ablation	Pre.	Rec.	F_1
SM	0.8685	0.8090	0.8376
– Column headers	0.8391	0.7971	0.8176
– Metadata	0.8273	0.8076	0.8173
– Context	0.6574	0.6548	0.6561
– Row headers	0.6540	0.5455	0.5948
– Headers	0.6553	0.7088	0.6810
– Hierarchy	0.8499	0.7984	0.8234
– Component embedding	0.8399	0.8090	0.8242
– Symmetric	0.8630	0.7721	0.8150

matching. When we ablate context, F_1 decreases by 18.15%, this means that it is important information for this task. Also, ablating row headers will decrease F_1 by 24.28% and severely hurt the recall (to 54.55%). So, row headers are very important in discovering matching information.

The result of ablating both column and row headers are shown in the "-Headers" row, which uses only metadata and context (MX). We denote the model ablating only row headers (retaining column headers, metadata, and context) as CMX. An interesting phenomenon is that although with more information (column headers), CMX performs poorer than MX (59.48% vs. 68.10% on F_1). And the difference comes from recall (54.55% vs. 70.88%). We argue that this is because the CMX model overfits the training set on the column information. As discussed above, row headers contain vital information for matching. Thus, both CMX and MX, without row headers, do not have enough information for matching. As the column contains little information for matching, it might become noise for the CMX model. To examine this, we report the result on the training set: the recall of the CMX is higher than (86.11% and 80.56%). It means that the CMX tries to use column headers to improve its recall which did not generalize well. And that results in overfitting and worse results on the test set.

We also study our modeling on the table layout. The first is the hierarchy of headers. For each cell on the tree, we directly use its text (not extended) for text embedding. All cells including non-leaf nodes are used. The result is shown in the "-Hierarchy" row. The F_1 score drops by 1.42% after removing hierarchical information. The second is the component type, which is shown in the "-Component embedding" row. We observed that the recall of the model did not decrease, but the precision drops by 2.86%. The lacking of hierarchical information and component type decreases the effectiveness of the model, which indirectly proves that the table layout affects the fact-overlapping relations. Therefore, it is necessary to keep the layout information of the table in table embedding.

We also test the symmetric design in the model by replacing symmetric E_O with an asymmetric one using

$$e_O = FFN(\text{concat}(e_{T_1}^*, e_{T_2}^*)) \tag{14}$$

instead of Formula (3) in SM. Thus, changing the order of two tables may lead to different results in this model. We select the one with the highest confidence as the result of a table pair. The result is shown in the "-Symmetric" row. The symmetric design outperforms the asymmetric model by 2.26% on F_1.

5.3   Case Study

We show two cases in Fig. 4 to show that the SM can correctly attend on cells related to their overlapping facts. The SM model correctly predicts that both cases have fact-overlapping relations. We show the table pair A and B in each case and visualize the summary weights defined in Formula (13). In table A, we show the summary weight on each cell in $Attention_2$ for $e_{[TAB]}$. In table B, we show the weight on each cell in $Attention_1$ for a specific cell in table A. Cells with darker shading have larger weights.

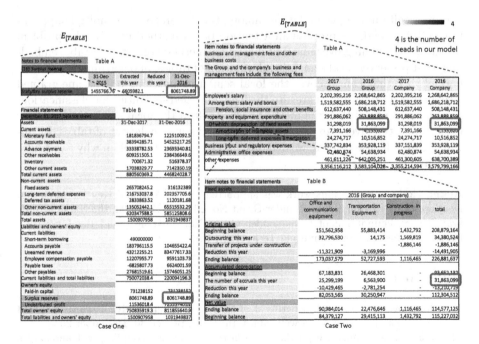

Fig. 4. The illustration of the interpretability of SM method by two cases. The selected data cells have the same fact in each case.

In the first case, Table A is a detailed sheet about "Surplus reserve" with only one row "Statutory surplus reserve". Meanwhile, "Surplus reserve" is one of the row headers in Table B. Weights in Table A indicate that "(16) Surplus reserve" and "Statutory surplus reserve" in Table A are important for classification. So, in table B, we show the weight of each cell corresponding to "(16) Surplus reserve". It demonstrates that "December 31, 2017 balance sheet", "Owner's equity", "Surplus reserves" and "Undistributed profit" are important. That means our model attends to the correct cells.

In the second case, Table B details the "Fixed assets" that contains the changes in depreciation of fixed assets. Meanwhile, "Of which: depreciation of fixed assets" is one of the row headers in Table A. Weights in Table A indicate that "Of which: depreciation of fixed assets", "Amortization of intangible assets" and "Long-term deferred expenses amortization" in Table A is important for classification. Then, in Table B, we show the importance of each cell corresponding to "Of which: depreciation of fixed assets". It demonstrates that the existences of "Fixed assets" and "Accumulated depreciation" in Table B cause the importance of "Of which: depreciation of fixed assets" in Table A for classification.

6   Conclusion

Automatically mining the relations among tables can support multiple applications. To this end, we propose an attention-based method to solve the problem of semantic matching over matrix tables. Our method consists of a table embedding that preserves local semantic information of tables, and a symmetric matching network that can discover the local semantic matching on the fine-grained level between tables and aggregate the local matched semantics for classification. Experiments reveal that the method works well in the local semantic similarity task despite the diverse expressions of facts and complex layout of tables. Meanwhile, through case studies, we demonstrate its ability to explain why two tables are semantically matched.

Acknowledgements. This work was supported by the National Key Research and Development Program of China under Grant No. 2017YFB1002104, the National Natural Science Foundation of China under Grant No. U1811461, and the Innovation Program of Institute of Computing Technology, CAS.

References

1. Ahmadov, A., Thiele, M., Eberius, J., Lehner, W., Wrembel, R.: Towards a hybrid imputation approach using web tables. In: BDC, pp. 21–30, December 2015
2. Badam, S.K., Liu, Z., Elmqvist, N.: Elastic documents: coupling text and tables through contextual visualizations for enhanced document reading. TVCG **25**(1), 661–671 (2019)
3. Bernstein, P.A., Madhavan, J., Rahm, E.: Generic schema matching, ten years later. Proc. VLDB Endow. **4**(11), 695–701 (2011)
4. Cao, Y., Li, H., Luo, P., Yao, J.: Towards automatic numerical cross-checking: extracting formulas from text. In: WWW, pp. 1795–1804 (2018)

5. Chen, X., Chiticariu, L., Danilevsky, M., Evfimievski, A., Sen, P.: A rectangle mining method for understanding the semantics of financial tables. In: ICDAR, pp. 268–273, November 2017
6. Crestan, E., Pantel, P.: Web-scale table census and classification. In: WSDM, pp. 545–554 (2011)
7. Deng, L.: Table2Vec: neural word and entity embeddings for table population and retrieval. Master's thesis, University of Stavanger, Norway (2018)
8. Devlin, J., Chang, M.W., Lee, K., Toutanova, K.: Bert: pre-training of deep bidirectional transformers for language understanding. In: NAACL-HLT, pp. 4171–4186 (2019)
9. Fang, J., Mitra, P., Tang, Z., Giles, C.L.: Table header detection and classification. In: AAAI, pp. 599–605 (2012)
10. Fetahu, B., Anand, A., Koutraki, M.: TableNet: an approach for determining fine-grained relations for wikipedia tables. In: WWW, pp. 2736–2742 (2019)
11. Ju, F., Lu, M., Ooi, B.C., Tan, W.C., Zhang, M.: A hybrid machine-crowdsourcing system for matching web tables. In: ICDE, pp. 976–987 (2014)
12. Kim, D.H., Hoque, E., Kim, J., Agrawala, M.: Facilitating document reading by linking text and tables. In: UIST, pp. 423–434 (2018)
13. Lautert, L.R., Scheidt, M.M., Dorneles, C.F.: Web table taxonomy and formalization. ACM SIGMOD Rec. 42(3), 28–33 (2013)
14. Lehmberg, O., Ritze, D., Meusel, R., Bizer, C.: A large public corpus of web tables containing time and context metadata. In: WWW, pp. 75–76 (2016)
15. Liu, P., Qiu, X., Chen, J., Huang, X.: Deep fusion LSTMs for text semantic matching. In: ACL, pp. 1034–1043 (2016)
16. Liu, Y., Bai, K., Mitra, P., Giles, C.L.: TableSeer: automatic table metadata extraction and searching in digital libraries. In: Proceedings of the 7th ACM/IEEE-CS Joint Conference on Digital Libraries, pp. 91–100. ACM (2007)
17. Mao, S., Rosenfeld, A., Kanungo, T.: Document structure analysis algorithms: a literature survey. In: Document Recognition and Retrieval X, vol. 5010, pp. 197–207, January 2003
18. Pang, L., Lan, Y., Guo, J., Xu, J., Wan, S., Cheng, X.: Text matching as image recognition. In: AAAI, pp. 2793–2799 (2016)
19. Rahm, E., Bernstein, P.A.: A survey of approaches to automatic schema matching. VLDB J. 10(4), 334–350 (2001)
20. Rastan, R., Paik, H.Y., Shepherd, J.: TEXUS: a unified framework for extracting and understanding tables in pdf documents. Inf. Process. Manag. 56(3), 895–918 (2019)
21. Ritze, D., Lehmberg, O., Bizer, C.: Matching HTML tables to DBpedia. In: WIMS, pp. 10:1–10:6 (2015)
22. Sarma, A.D., et al.: Finding related tables. In: SIGMOD, pp. 817–828 (2012)
23. Seth, S., Nagy, G.: Segmenting tables via indexing of value cells by table headers. In: ICDAR, pp. 887–891 (2013)
24. Sun, Y., Yan, Z., Tang, D., Duan, N., Qin, B.: Content-based table retrieval for web queries. Neurocomputing 349, 183–189 (2018)
25. Vaswani, A., et al.: Attention is all you need. In: NIPS, pp. 5998–6008 (2017)
26. Wan, S., Lan, Y., Guo, J., Xu, J., Pang, L., Cheng, X.: A deep architecture for semantic matching with multiple positional sentence representations. In: AAAI, pp. 2835–2841 (2016)
27. Wang, H.L., Wu, S.H., Wang, I., Sung, C.L., Hsu, W.L., Shih, W.K.: Semantic search on internet tabular information extraction for answering queries. In: CIKM, pp. 243–249 (2000)

28. Wu, S., et al.: Fonduer: knowledge base construction from richly formatted data. In: SIGMOD, pp. 1301–1316 (2018)
29. Wu, Z., et al.: An efficient Wikipedia semantic matching approach to text document classification. Inf. Sci. **393**(C), 15–28 (2017)
30. Yu, W., Sun, X., Yang, K., Rui, Y., Yao, H.: Hierarchical semantic image matching using CNN feature pyramid. Comput. Vis. Image Underst. **169**, 40–51 (2018)
31. Zhang, L., Zhang, S., Balog, K.: Table2Vec: neural word and entity embeddings for table population and retrieval. In: SIGIR, pp. 1029–1032 (2019)
32. Zhang, M., Chakrabarti, K.: InfoGather+: semantic matching and annotation of numeric and time-varying attributes in web tables. In: SIGMOD, pp. 145–156 (2013)
33. Zhang, S., Balog, K.: Ad hoc table retrieval using semantic similarity. In: WWW, pp. 1553–1562 (2018)
34. Zhang, Z.: Towards efficient and effective semantic table interpretation. In: ISWC, pp. 487–502 (2014)

On Integrating and Classifying Legal Text Documents

Alexandre Quemy[1,2] and Robert Wrembel[2(✉)]

[1] IBM, Cracow Software Lab, Kraków, Poland
aquemy@pl.ibm.com
[2] Poznan University of Technology, Poznań, Poland
robert.wrembel@cs.put.poznan.pl

Abstract. This paper presents an exhaustive and unified dataset based on the European Court of Human Rights judgments since its creation. The interest of such database is explained through the prism of the researcher, the data scientist, the citizen and the legal practitioner. Contrarily to many datasets, the creation process, from the collection of raw data to the feature transformation, is provided under the form of a collection of fully automated and open-source scripts. It ensures reproducibility and a high level of confidence in the processed data, which is some of the most important issues in data governance nowadays. A first experimental campaign is performed to study some predictability properties and to establish baseline results on popular machine learning algorithms. The results are consistently good across the binary datasets with an accuracy comprised between 75.86% and 98.32% for a micro-average accuracy of 96.44%.

Keywords: Legal text document integration · Text analytics · Text document classification

1 Introduction

Machine learning (ML) algorithms are used in multiple domains (e.g., sales, healthcare, production), as they build prediction models of acceptable quality and yet explainable. However, the application of ML to the legal domain so far has received little attention from research communities [4,17], but the need of ML solutions to support judicial decision is slowly becoming recognized (e.g., study programs combining artificial intelligence and law at Duke University (USA), Swansea University (UK), Maastricht University (NL) [1]).

Applying ML algorithms in the law domain is challenging. First, the legal domain is a messy concept [22] that intrinsically creates some of the most challenging problems for the ML research community including: gray areas of interpretation, many exceptions, non-stationarity, presence of deductive and inductive reasoning, non-classical logic, multiple and complex legal rules, as well as semantic complexity of legal acts. Second, there are few large open repositories

© Springer Nature Switzerland AG 2020
S. Hartmann et al. (Eds.): DEXA 2020, LNCS 12391, pp. 385–399, 2020.
https://doi.org/10.1007/978-3-030-59003-1_25

of legal cases, with clean, adequately structured data. As a consequence, it is challenging to verify ML algorithms on legal data. From the set of ML algorithms [12], classification is a primary technique for building prediction models in the legal domain [15].

There exist few initiatives to provide open data repositories on judicial cases, including the recent one in Australia (AI for Law Enforcement and Community Safety that supports automated classification of online child exploitation material) and Singapore (Intelligent Case Retrieval System that enables retrieval of relevant precedent cases by means of artificial intelligence tools) [24]. From the available judicial repositories, the most known ones include: the *Supreme Court of the United States*[1] and the *European Court of Human Rights*.[2] Even though these corpora of legal cases are available, multiple information are missing and an access interface to these repositories is limited (cf., Sect. 2). Moreover, the content of these repositories has to be pre-processed before ML algorithms are run on them, as incomplete and inadequately prepared data for ML algorithms strongly impact a quality of built prediction models [8,9,18]. Recently, we analyzed and experimentally showed that the way data are pre-processed for classification algorithms impacts the quality of classificators [6,7].

The aforementioned observations motivated us to build and make available an open, exhaustive, and unified data repository, called *ECHR-DB*, about legal cases from the European Court of Human Rights. The repository is accompanied by a comprehensive processing pipeline, neatly documented and supported by rich metadata, to provide reusability, repeatability of experiments, and manageability. In details, the paper **contributes** the following:

1. A **benchmark suit** for ML algorithms in the law domain, based on the European Court of Human Rights. The benchmark is composed of: (1) the *ECHR-DB* repository that stores almost all cases judged by the European Court of Human Rights since its creation, cleaned and transformed to ease the exploration by ML algorithms and (2) 13 standard ML algorithms that can be immediately run on *ECHR-DB*.
2. The whole **extract-transform-load (ETL) and data transformation pipeline** used to generate the benchmark suit, available as an open-source project. As a consequence, the whole data ingestion, transformation, and cleaning processes can be repeated, revised, and extended.
3. A comparison of **13 standard machine learning algorithms** for classification with regards to several performance metrics. These results provide a baseline for future studies and provide some insights about the interest of some types of features to predict justice decisions.

The paper is organized as follows. Section 2 presents related work on analytics in the legal domain. Section 3 outlines the *ECHR-DB* repository. The process of creating the repository is discussed in Sect. 4. Section 5 reports the experiments on the repository. Section 6 summarizes and concludes the paper.

---

[1] http://scdb.wustl.edu/.
[2] https://hudoc.echr.coe.int/eng.

2   Related Work

Predicting the outcome of a justice case is challenging, even for the best legal experts. As shown in [23], 67.4% and 58% accuracy was achieved, respectively for the judges and the whole case decision, using cases from the Supreme Court of the United States and a simple ideology estimation of judges and decisions (liberal versus conservative). Using crowds, the *Fantasy Scotus*[3] project reached 85.20% and 84.85% correct predictions, respectively. In [15], the authors proposed to apply an SVM-based classificator and they were able to correctly predict about 75% of the cases.

A success of research in ML for the legal domain depends on the availability of large datasets of legal cases with judicial decisions. There are a few open data repositories of judicial cases available. The most known ones include: the *SCOTUS* repository[4] of the *Supreme Court of the United States* and the *HUDOC* database[5] of the *European Court of Human Rights*. *SCOTUS* is composed of structured data (in a tabular format) about every case since the creation of the court but it lacks textual information about decisions. *HUDOC* contains all legal cases with judgments. However, its interface has some flaws, e.g., it does not offer any API to allow to access several documents at once and case documents are not unified in the way that they could offer tabular and natural language data. In other words, despite its public availability, the data are hard to retrieve and to work with.

The prediction of the *Supreme Court of the United States* has been widely studied, notably through the *SCOTUS* repository [10,11,14]. To the best of our knowledge, the only predictive models that used the content of *HUDOC* were reported in [2,15]. The data used in [2] are far from being exhaustive: only 3 articles considered (3, 6 and 8) with respectively 250, 80 and 254 cases per article. Using SVM with linear kernel, the authors achieved 79% accuracy to predict the decisions of the European Court of Human Rights. SVM is also used in [15] to reach an overall of 75% accuracy on judgment documents up to September 2017. In [17], the author outlined some practical problems in the field of legal analytics, notably the prediction and the justification problem.

New studies tend to suggest that there will always be a limit in reasoning systems to handle new cases presenting novel situations [5], which emphasize the interest for data-centric methods, hence the need for *large and adequate sets of legal data* (mainly cases and their justifications) available to researchers and practitioners. Such datasets should be equipped with: (1) a user-friendly interface to access and analyze the data and (2) rich metadata to offer means for browsing the content of a repository and to tune ML algorithms. Unfortunately, the aforementioned databases do not fully meet these requirements. This observation motivated us to start the project on building an open European Court of Human Rights repository (*ECHR-DB*).

---

[3]   https://fantasyscotus.lexpredict.com/.

[4]   http://scdb.wustl.edu/.

[5]   https://hudoc.echr.coe.int/eng.

3 ECHR-DB in Brief

The *ECHR-DB* repository aims at providing exhaustive and high-quality database for diverse problems, based on the European Court of Human Rights documents from *HUDOC*. The main objectives of this project are as follows: (1) to draw the attention of researchers on this domain that has important consequences on the society and (2) to provide a similar and more complete database for the European Union as it already exists in the United States, notably because the law systems are different in both sides of the Atlantic.

ECHR-DB is guided by three core values: **reusability**, **quality** and **availability**. To reach those objectives:

- each version of the datasets is carefully versioned and publicly available, including the intermediate files,
- the integrality of the process and files produced are careful documented,
- the scripts to retrieve the raw documents and to build the datasets from scratch are open-source and carefully versioned to maximize reproducibility and trust,
- no data is manipulated by hand at any stage of the creation process to make it fully automatic,
- *ECHR-DB* is augmented with rich metadata that allow to understand and use its content more easily.

The database is available at https://echr-opendata.eu under the **Open Database Licence (ODbL)**. The creation scripts and website sources are provided under **MIT Licence** and they are available on GitHub [20].

We extracted, cleaned, and normalized data including descriptive and textual features gathered from the *HUDOC* database and judgment files. The final data provided are available either in a structured or unstructured format:

- The unstructured format is a JSON file containing a list of all the information available about each case, including a tree-based representation of the judgment document (cf., Sect. 4).
- Structured information files are provided to be directly readable by popular data manipulation libraries, such as *panda* or *numpy*. Thus, they are easy to use with machine learning libraries such as *scikit-learn*. These fields include the description of cases in a flat JSON and the adjacency matrix for some important variables.

4 Database Creation Process

In this section, we outline the process of populating the *ECHR-DB* repository. The process of ingesting data is broken down into the following five tasks discussed in this section: (1) retrieving basic metadata and judgment documents, (2) cleaning cases, (3) pre-processing documents, (4) normalizing documents, and (5) generating the repository.

4.1 Retrieving Basic Metadata and Judgment Documents

Using web scrapping, basic metadata about all entries are retrieved from *HUDOC* and saved in JSON files. Common metadata include among others: case name, the language used, the conclusion in natural language. When available, we also retrieved the judgments in Microsfot Word format.

4.2 Cleaning Cases

HUDOC includes cases in various languages, cases without judgments, cases without or with vague conclusions. For this reason, its content needs to be cleaned before making it available within our project. To clean the content of *HUDOC* we applied a standard ETL process [3]. To ensure a high quality and usability of the datasets, we cleaned and filtered out the cases. As a consequence, *ECHR-DB* includes: (1) only cases in English, (2) only cases accompanied by a judgment document, and (3) only cases with a clear conclusion, i.e., containing at least one occurrence of *violation* or *no violation*.

As part of the ETL process, we also parsed and formatted some raw data: parties are extracted from a case title and many raw strings are broken down into lists. In particular, a string listing articles discussed in a case are transformed into a list and a conclusion string is transformed into a slightly more complex JSON object. For instance, string *Violation of Art. 6-1; No violation of P1-1; Pecuniary damage - claim dismissed; Non-pecuniary damage - financial award* becomes the following list of elements:

```
{"conclusion":
    [
        {   "article": "6",
            "element": "Violation of Art. 6-1",
            "type": "violation"
        },
        {   "article": "p1",
            "element": "No violation of P1-1",
            "type": "no-violation"
        },
        {   "element": "Pecuniary damage - claim dismissed",
            "type": "other"
        },
        {   "element": "Non-pecuniary damage - financial award",
            "type": "other"
        }
    ]
}
```

In general, each item in the conclusion can have the following elements: (1) *article*: a number of the concerned article if applicable, (2) *details*: a list of additional information (paragraph or aspect of the article), (3) *element*: a part of a raw string describing the item, (4) *mentions*: diverse mentions (quantifier, e.g., 'moderate', country. . .), (5) *type*: of value *violation*, *no violation*, or *other*.

4.3 Pre-processing Documents

The pre-processing step consists in parsing an MS Word document to extract additional information and create a tree structure of a judgment file. During this process, we extend the set of features of a legal document with field *decision_body* with the list of persons involved in a decision, including their roles. The most important extension of a case description is the tree representation of the whole judgment document, under the field *content*. The content is described in an ordered list where each element has two fields: (1) *content* to describe the element (paragraph text or title) and (2) *elements* that represents a list of sub-elements. This tree representation eases the identification of some specific sections or paragraphs (e.g., facts or conclusion) or explore judgments with a lower granularity.

Each judgment has the same structure, which includes the following properties: (1) *Procedure*, (2) *Facts*, (3) *Law*, further composed of *Circumstances of the Case* and *Relevant Law*, and (4) *Operative Provision*.

It has been shown in [2] and [15] that each section has a different predictive power. The representation we propose allows to go further to identify each individual paragraph. Each paragraph is an independent statement (e.g., one fact for the *Facts* section, one legal argument for the *Law* section).

4.4 Normalizing Documents

In this task, judgment documents (without the conclusion) are normalized as follows: (1) tokenization, (2) stopwords removal, (3) part-of-speech tagging followed by a lemmatization, and (4) n-gram generation for $n \in \{1, 2, 3, 4\}$.

To construct a dictionary of tokens, we use *Gensim* (an open-source library for unsupervised topic modeling and natural language processing) [21]. The dictionary includes the 5000 most common tokens, based on the normalized documents. The number of tokens to use in the dictionary is a parameter of the script. The judgment documents are thus represented as a Bag-of-Words and TD-IDF matrices on top of the tree representation.

To ease data exploration, notably the connections between cases, we generated adjacency matrices for the following variables: decision body, extracted application, representatives and Strasbourg case law citations.

5 Experiments: Binary Classification

In this section, we perform a first campaign of experiments on *ECHR-DB*. Their goals are twofold. First, to studying the predictability offered by the database. Second, to provide a first baseline by testing the most popular machine learning algorithms for classification. In particular, in this paper we have focused the experiments on **determining if a specific article has been violated** or not, which is an instance of the the **binary** classification problem.

Furthermore, in this experimental evaluation, we are interested in **answering the following four questions**: (1) what is the predictive power of the data in *ECHR-DB*, (2) are all the articles equal w.r.t. predictability, (3) are some methods performing significantly better than others, and (4) are all data types (textual or descriptive) equal w.r.t. predictability?

All the experiments are implemented using *Scikit-Learn* [16]. All the experiments and scripts to analyze the results as well as to generate the plots and tables are open-source and are available on a separated GitHub repository [20] for repeatability and reusability.

5.1 Data Preparation

From *ECHR-DB* we created 11 datasets for the *binary* classification problem mentioned above. Each dataset comes in different flavors, based on: descriptive features, bag-of-words representations. These different representations (listed below) allow to study the respective importance of descriptive and textual features in the predictive models build upon the datasets:

1. *descriptive features*: structured features retrieved from HUDOC or deduced from the judgment document,
2. *bag-of-words* (BoW) representation: based on the top 5000 tokens (normalized n-grams for $n \in \{1, 2, 3, 4\}$),
3. *descriptive features + bag-of-words*: combination of both sets of features.

Each of the 11 datasets corresponds to a specific article. We kept only the articles such that there are at least 100 cases with a clear output, without consideration on the prevalence. Notice that the same case can appear in two datasets if it has in its conclusion two elements about a different article. A label corresponds to a violation or no violation of a specific article. The final datasets have been hot-one encoded. A basic description of these datasets is given in Table 1.

Table 1. Datasets description for binary classification.

	# cases	min #features	max #features	avg #features	prevalence
Article 1	951	131	2834	1183.47	0.93
Article 2	1124	44	3501	2103.45	0.90
Article 3	2573	160	3871	1490.75	0.89
Article 5	2292	200	3656	1479.60	0.91
Article 6	6891	46	3168	1117.66	0.89
Article 8	1289	179	3685	1466.52	0.73
Article 10	560	49	3440	1657.22	0.75
Article 11	213	293	3758	1607.96	0.85
Article 13	1090	44	2908	1309.33	0.91
Article 34	136	490	3168	1726.78	0.64
Article p1	1301	266	2692	1187.96	0.86

Columns min, max, and avg #features indicate the minimal, maximal, and average number of features, respectively, in the dataset cases for the representation *descriptive features + bag-of-words*.

The Bag-of-Words is a rather naive representation that loses a tremendous amount of information. However, we justify this choice by two reasons. First, so far, the studies on predicting the violation of articles for the ECHR cases use only a BoW representation. To be able to compare the interest of the proposed data with the previous studies, we need to use the same semantic representation. Second, from a scientific point of view, it is important to provide baseline results using the most common and established methods in order to be able to quantify the gain of more advanced techniques. This said, future work will consist in investigating advanced embedding techniques that are context aware such as LSTM or BERT-like networks. In particular, we hope not only to improve the prediction accuracy by a richer semantic, but also being able to justify a decision in natural language.

5.2   Protocol

We compared 13 standard classification methods: AdaBoost with Decision Tree, Bagging with Decision Tree, Naive Bayes (Bernoulli and Multinomial), Decision Tree, Ensemble Extra Tree, Extra Tree, Gradient Boosting, K-Neighbors, SVM (Linear SVC, RBF SVC), Neural Network (Multilayer Perceptron), and Random Forest.

For each article, we used three following flavors: (1) descriptive features only, (2) bag-of-words only, and (3) descriptive features combined with bag-of-words. For each method, each article, and each flavor, we performed a tenfold cross-validation with stratified sample, for a total of 429 validation procedures. Due to this important amount of experimental settings, we discarded the TF-IDF representation. For the same reason, we did not perform any hyperparameter tuning at this stage.

To evaluate the performances, we reported some standard performance indicators: accuracy, F_1-score and Matthews correlation coefficient (MCC). Additionally, we report the learning curves to study the limit of the model space. The learning curves are obtained by plotting the accuracy depending on the training set size, for both the training and the test sets. The learning curves help to understand if a model underfits or overfits and thus, shape future axis of improvements to build better classifiers.

To find out what type of features are the most important w.r.t. predictability, we used a Wilcoxon signed-rank test at 5% to compare the accuracy obtained on bag-of-words representation to the one obtained on the bag-of-words combined with the descriptive features. Wilcoxon signed-rank test is a non-parametric paired difference test. Given two paired sampled, the null hypothesis assumes the difference between the pairs follows a symmetric distribution around zero. The test is used to determine if the changes in the accuracy is significant when the descriptive features are added to the textual features.

5.3   Results

Table 2 shows the best accuracy obtained for each article as well as the method and the flavor of the dataset. The detailed results per article are available at [19]. For all articles, the best accuracy obtained is higher than the prevalence. Linear SVC offers the best results on 4, out of 11 articles. Gradient Boosting accounts for 3, out 11 articles and Ensemble Extra Tree accounts for 2 articles.

Table 2. The best accuracy obtained for each article.

Article	Accuracy	Method	Flavor
Article 1	0.9832 (0.01)	Linear SVC	Descriptive features and Bag-of-Words
Article 2	0.9760 (0.02)	Linear SVC	Descriptive features and Bag-of-Words
Article 3	0.9588 (0.01)	BaggingClassifier	Descriptive features and Bag-of-Words
Article 5	0.9651 (0.01)	Gradient Boosting	Descriptive features and Bag-of-Words
Article 6	0.9721 (0.01)	Linear SVC	Descriptive features and Bag-of-Words
Article 8	0.9542 (0.03)	Gradient Boosting	Descriptive features and Bag-of-Words
Article 10	0.9392 (0.04)	Ensemble Extra Tree	Bag-of-Words only
Article 11	0.9671 (0.03)	Ensemble Extra Tree	Descriptive features and Bag-of-Words
Article 13	0.9450 (0.02)	Linear SVC	Descriptive features only
Article 34	0.7586 (0.09)	AdaBoost	Descriptive features only
Article p1	0.9685 (0.02)	Gradient Boosting	Descriptive features and Bag-of-Words
Average	0.9443		
Micro average	0.9644		

The standard deviation is rather low and ranges from 1% up to 4%, at the exception of article 34, for which it is equal to 9%. This indicates a low variance for the best models. The accuracy ranges from 75.86% to 98.32%, with the average of 94.43%. The micro-average that ponders each result by the dataset size is 96.44%. In general, the datasets with higher accuracy are larger and more imbalanced. For the datasets being highly imbalanced, with a prevalence from 0.64 to 0.93, other metrics may be more suitable to appreciate the quality of the results. In particular, the micro-average could simply be higher due to the class imbalance rather than the availability of data.

Regarding the flavor, 8 out 10 best results are obtained on descriptive features combined to bag-of-words. Bag-of-words only is the best flavor for article 10, whereas descriptive features - only for article 13 and article 34. This seems to indicate that combining information from different sources improves the overall results.

Figure 1 displays the normalized confusion matrix for the best methods on article 1 and 13. Similar results are observed for all the other articles. The normalization is done per line and allows to quickly figure out how the true predictions are balanced for both classes. As expected due to the prevalence, true negatives are extremely high, ranging from 0.82 to 1.00, with an average of 97.18. On the contrary, the true positive rate is lower, ranging from 0.47 to 0.91. For most articles, the true positive rate is higher than 80% and it is lower than 50% only for article 34. This indicates that the algorithms are capable of **producing models that are fairly balanced despite the fact that the classes are highly imbalanced.**

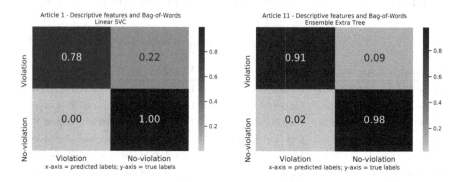

Fig. 1. Normalized confusion matrices for the best methods from Table 2.

Additionally, we provide the Matthew Correlation Coefficient (MCC) in Table 3. The MCC is generally superior to the accuracy because it takes into account the class prevalence. Therefore, it is a much better metric to estimate the model quality. In particular, contrarily to other studies, we decided to not rebalance the dataset because, as shown below, the model underfits. Therefore, the accuracy metric would not allow us to properly compare our approach to the previous studies since we could have a high accuracy but not significantly higher than the prevalence. The MCC ranges from 0.4918 - on article 34 to 0.8829 - on article 10. The best score is not obtained by the same article as for the accuracy (article 10 achieved 93% accuracy, below the average). Interestingly, the MCC reveals that the performances on article 34 are rather poor in comparison to the other articles and close to the performance on article 13. Surprisingly, the best method is not Linear SVC anymore (best on 3 articles) but Gradient Boosting (best on 4 articles). While the descriptive features were returning the best results for two articles, according the MCC, it reaches the best score only for article 34.

Once again, the micro-average is higher than the macro-average. As the MCC takes into account class imbalance, it supports the idea that adding more cases to the training set could still improve the result of these classifiers. This will be confirmed by looking at the learning curves.

Table 4 ranks the methods according to the average accuracy performed on all articles. For each article and method, we kept only the best accuracy among the three dataset flavors.

Surprisingly, neither Linear SVC nor Gradient Boosting are the best methods with a respective rank of 2 and 5, but the best one is Ensemble Extra Tree. Random Forest and Bagging with Decision Tree are the second and third ones,

Table 3. Best Matthews Correlation Coefficient obtained for each article. The flavor and method achieving the best score for both metrics are similar for every article.

Article	MCC	Method	Flavor
Article 1	0.8654	Linear SVC	Descriptive features and Bag-of-Words
Article 2	0.8609	Linear SVC	Descriptive features and Bag-of-Words
Article 3	0.7714	BaggingClassifier	Descriptive features and Bag-of-Words
Article 5	0.7824	Gradient Boosting	Descriptive features and Bag-of-Words
Article 6	0.8488	Linear SVC	Descriptive features and Bag-of-Words
Article 8	0.8829	Gradient Boosting	Descriptive features and Bag-of-Words
Article 10	0.8411	Gradient Boosting	Bag-of-Words only
Article 11	0.8801	Ensemble Extra Tree	Descriptive features and Bag-of-Words
Article 13	0.5770	Ensemble Extra Tree	Bag-of-Words only
Article 34	0.4918	AdaBoost	Descriptive features only
Article p1	0.8656	Gradient Boosting	Descriptive features and Bag-of-Words
Average	0.7879		
Micro average	0.8163		

Table 4. Overall ranking of methods according to the average accuracy obtained for every article.

Method	Accuracy	Micro Accuracy	Rank
Ensemble Extra Tree	0.9420	0.9627	1
Linear SVC	0.9390	0.9618	2
Random Forest	0.9376	0.9618	3
BaggingClassifier	0.9319	0.9599	4
Gradient Boosting	0.9309	0.9609	5
AdaBoost	0.9284	0.9488	6
Neural Net	0.9273	0.9535	7
Decision Tree	0.9181	0.9419	8
Extra Tree	0.8995	0.9275	9
Multinomial Naive Bayes	0.8743	0.8907	10
Bernoulli Naive Bayes	0.8734	0.8891	11
K-Neighbors	0.8670	0.8997	12
RBF SVC	0.8419	0.8778	13
Average	0.9086	0.9335	

respectively, and they never achieved the best result on any article. It simply indicates that these methods are more consistent across the datasets than Linear SVC and Gradient Boosting.

Figure 2 displays the learning curves obtained for the best methods on articles 10 and 11. The training error becomes (near) zero on every instance after only few cases, except for article 13 and 34. The test error converges rather fast and remains relatively far from the training error, synonym of high bias. Those two elements indicate underfitting. Similar results are observed for all methods. Usually, more training examples would help, but since the datasets are exhaustive w.r.t. the European Court of Human Rights cases, this is not possible. As a consequence, we recommend using a more complex model space and hyperparameter tuning. In particular, as mentioned above, the usage of more advanced embedding techniques is an obvious way to explore. Finally, an exploratory analysis of the datasets may also help in removing some noise and finding the best predictors.

If we assume that the process of deciding if there is a violation or not is the same, independently of the article, a solution might be a transfer learning, to leverage what is learnable from the other articles. We let this research axis for future work.

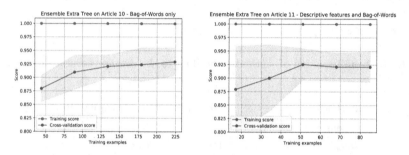

Fig. 2. Learning curves for the best methods as described by Table 2.

Finally, we used a Wilcoxon signed-rank test at 5% to compare the accuracy obtained on the bag-of-words representation to the one obtained on the bag-of-Words combined with the descriptive features. The difference between the samples has been found to be significant only for article 6 and article 8. The best result obtained on bag-of-Words is improved by adding descriptive features for every article. However, statistically, for a given method, adding descriptive features does not improve the result. This could be explained by the fact that descriptive features are rather poor predictors such that adding them to the Bag-of-Words can be considered as some noise that does not help to build a better model. Additionally, we performed the test per method. The result is significant for any method.

In conclusion, the datasets demonstrated a strong predictability power. Apart from article 13 and 34, each article seems to provide similar results, independently of the relatively different prevalence. If the accuracy is rather high, a more informative metric, such as MCC, shows that there are still margins of improvements. Hyperparameter tuning [18] is an obvious way to go, and this preliminary work has shown that good candidates for fine tuning are Ensemble Extra Tree, Linear SVC, and Gradient Boosting.

5.4 Discussion

To sum up, we achieved an average accuracy of 94% which is respectively 15pp and 19pp higher than [2] and [15]. The size of the dataset does matter since we showed that the model underfits. Also, we showed that SVM is far from being the best method for all articles. However, such a huge gap cannot be explained only by those two factors.

In our opinion, the main problem with the previous studies is that the authors rebalanced their datasets. As those datasets were highly imbalanced, they used undersampling, which resulted in a very small training dataset. Most likely, the training dataset was not representative enough of the feature space which leads to underfitting (even more than in our experiments). They justified that rebalancing was necessary to ensure that the classifier was not biased towards a certain class. For this reason, we argue that they modified the label distribution. As some classification methods rely on the label distribution to learn, they introduce themselves a prior shift [13]. In general, rebalancing is necessary only when, indeed, the estimator is badly biased. It is true that the accuracy is meaningless on imbalanced datasets but we can still control the quality of the model using a collection of more robust indicators, including among others: F1-score, MCC, and normalized confusion matrices. In other words, our approach (discussed in this paper) is more neutral in the sense we do not change the label distribution, and it still offers a robust classifier.

This experimental campaign has demonstrated that the textual information provides better results than descriptive features alone, but the addition of the descriptive feature improved in general the result of the **best** method. We emphasise the best method (obtained among all methods) because for a given method adding the descriptive feature are not significantly improving the results.

Another way of improving the results is to tune the different phases of the dataset generations. In particular, our preliminary work reported in [18] has shown that 5000 tokens and 4-grams might not be enough to take the best out of the documents. It might seem surprising, but the justice language is codified and standardized in a way that n-grams for large n might contain better predictors for the outcome.

6 Conclusion

In this paper, we presented an open repository, called *ECHR-DB*, of legal cases and judicial decision justifications. The main purposes of constructing the repository are as follows. First, to provide cleaned and transformed content from the repository of the European Court of Human Rights, that is ready to use by researchers and practitioners. Second, to augment original legal documents with metadata, which will ease the process of analyzing these documents. Third, to provide a benchmark with baseline results for classification models in the legal domain, for other researchers.

Currently, *ECHR-DB* is the largest and most exhaustive repository of legal documents from the European Court of Human Rights. It includes several types of data that can be easily used to reproduce various experiments that have been done so far by other researchers. We argue that providing the final data is not enough to ensure quality and trust. In addition, there are always some opinionated choices in the representation, such as the number of tokens, the value of n for the n-grams calculation or the weighting schema in the TF-IDF transformation. As a remedy, we provide the whole process of dataset construction from scratch. The process is implemented by means of Python scripts and available on GitHub.[6]

The experiments on *ECHR-DB* provide a baseline for future work on classification. The predictability power of each dataset has been tested for the most popular machine learning methods. We achieved the average accuracy of 0.9443. The learning curves have shown that the models are underfitting but, as the datasets are exhaustive, it is not possible to provide more examples. We showed that the textual features help in determining the outcome. Combining descriptive and textual features always help for the best classifier, but overall, the results are not better statistically. Descriptive features surprisingly hold reasonable predictive power.

The preliminary experiments provide several axes of improvements, e.g., better embedding with state of the art encoders, hyperparameter tuning, multi-stage classifier, and transfer learning. From the results, it seems clear that predicting if an article has been violated or not can be handled with the current state of the art in artificial intelligence. However, many interesting questions and problems arise from the proposed repository, e.g. *can we provide legal justification in natural language to a prediction?*, which will be addressed in the future work.

References

1. Maastricht University Law and Tech Lab. https://www.maastrichtuniversity.nl/about-um/faculties/law/research/law-and-tech-lab
2. Aletras, N., Tsarapatsanis, D., Preoţiuc-Pietro, D., Lampos, V.: Predicting judicial decisions of the European Court of Human Rights: a natural language processing perspective. PeerJ. Comput. Sci. **2**, e93 (2016)
3. Ali, S.M.F., Wrembel, R.: From conceptual design to performance optimization of ETL workflows: current state of research and open problems. VLDB J. **26**(6), 777–801 (2017). https://doi.org/10.1007/s00778-017-0477-2
4. Ashley, K.D.: Artificial Intelligence and Legal Analytics: New Tools for Law Practice in the Digital Age. Cambridge University Press (2017)
5. Atkinson, K., Bench-Capon, T.: Reasoning with legal cases: analogy or rule application? In: Proceedings of the International Conference on Artificial Intelligence and Law (ICAIL), pp. 12–21. ACM (2019)
6. Bilalli, B., Abelló, A., Aluja-Banet, T., Wrembel, R.: Intelligent assistance for data pre-processing. Comput. Stand. Interfaces **57**, 101–109 (2018). https://doi.org/10.1016/j.csi.2017.05.004
7. Bilalli, B., Abelló, A., Aluja-Banet, T., Wrembel, R.: PRESISTANT: learning based assistant for data pre-processing. Data Knowl. Eng. **123**, 101727 (2019). https://doi.org/10.1016/j.datak.2019.101727

---

[6] https://github.com/aquemy/ECHR-OD_predictions.

8. Crone, S.F., Lessmann, S., Stahlbock, R.: The impact of preprocessing on data mining: an evaluation of classifier sensitivity in direct marketing. Eur. J. Oper. Res. **173**(3), 781–800 (2006)
9. Dasu, T., Johnson, T.: Exploratory Data Mining and Data Cleaning, vol. 479. Wiley, Hoboken (2003)
10. Guimerà, R., Sales-Pardo, M.: Justice Blocks and Predictability of U.S. Supreme Court Votes. PLoS ONE **6**(11), e27188 (2011)
11. Katz, D.M., Bommarito, M.J., Blackman, J.: A general approach for predicting the behavior of the Supreme Court of the United States. PLoS ONE **12**(4), e0174698 (2017)
12. Kelleher, J.D., Mac Namee, B., D'Arcy, A.: Fundamentals of Machine Learning for Predictive Data Analytics: Algorithms, Worked Examples, and Case Studies. MIT Press, Cambridge (2015)
13. Lemberger, P., Panico, I.: A primer on domain adaptation (2020)
14. Martin, A.D., Quinn, K.M., Ruger, T.W., Kim, P.T.: Competing approaches to predicting supreme court decision making. Perspect. Polit. **2**(4), 761–767 (2004)
15. Medvedeva, M., Vols, M., Wieling, M.: Using machine learning to predict decisions of the European Court of Human Rights. Artif. Intell. Law **28**(2), 237–266 (2019). https://doi.org/10.1007/s10506-019-09255-y
16. Pedregosa, F.: Scikit-learn: machine learning in Python. J. Mach. Learn. Res. **12**, 2825–2830 (2011)
17. Quemy, A.: Data science techniques for law and justice: current state of research and open problems. In: Kirikova, M., et al. (eds.) ADBIS 2017. CCIS, vol. 767, pp. 302–312. Springer, Cham (2017). https://doi.org/10.1007/978-3-319-67162-8_30
18. Quemy, A.: Data pipeline selection and optimization. In: Proceedings of the International Workshop on Design, Optimization, Languages and Analytical Processing of Big Data (DOLAP) (2019)
19. Quemy, A.: ECHR-DB experiments, all detailed results (2019). https://github.com/echr-od/ECHR-OD_project_supplementary_material/blob/master/binary.md
20. Quemy, A.: Predictions of the European Court of Human Rights (2019). https://github.com/aquemy/ECHR-OD_predictions
21. Řehůřek, R., Sojka, P.: Software framework for topic modelling with large corpora. In: Proceedings of the Workshop on New Challenges for NLP Frameworks, pp. 45–50. ELRA (2010)
22. Rissland, E.L.: AI and similarity. IEEE Intell. Syst. **21**(3), 39–49 (2006)
23. Ruger, T.W., Kim, P.T., Martin, A.D., Quinn, K.M.: The supreme court forecasting project: legal and political science approaches to predicting supreme court decisionmaking. Columbia Law Rev. **104**(4), 1150–1210 (2004)
24. Yan, L., Wilson, C.: Developing AI for law enforcement in Singapore and Australia. Commun. ACM **63**(4), 62 (2020)

Prediction and Decision Support

An Explainable Artificial Intelligence Methodology for Hard Disk Fault Prediction

Antonio Galli[1], Vincenzo Moscato[1], Giancarlo Sperlí[1(✉)],
and Aniello De Santo[2]

[1] Department of Electrical and Information Technology, University of Naples
Federico II, via Claudio 21, 80125 Naples, Italy
{antonio.galli,vincenzo.moscato,giancarlo.sperli}@unina.it
[2] Department of Linguistics, Stony Brook University,
Stony Brook, NY 11794-4376, USA
aniello.desanto@stonybrook.edu

Abstract. Failure rates of Hard Disk Drives (HDDs) are high and often
due to a variety of different conditions. Thus, there is increasing demand
for technologies dedicated to anticipating possible causes of failure, so to
allow for preventive maintenance operations. In this paper, we propose a
framework to predict HDD health status according to a long short-term
memory (LSTM) model. We also employ eXplainable Artificial Intelligence (XAI) tools, to provide effective explanations of the model decisions, thus making the final results more useful to human decision-making
processes. We extensively evaluate our approach on standard data-sets,
proving its feasibility for real world applications.

Keywords: HDD maintenance · LSTMs · Explainable AI

1 Introduction

IT infrastructures can be affected by data center's equipment failures, whose
downtime costs have been growing significantly—ranging, for instance, from
5,600/min. in 2010 to 8,851/min. in 2016 [10]. As Hard Disk Drives (HDDs) have
become a primary type of storage in data centers, HDD failure-rate is now one
of the main factor for data center downtime, unavailability, and data loss—with
obvious effects on overall business costs [4,13]. Additionally, HDDs' reliability
is affected by the complex interaction of a variety of factors (i.e., temperatures,
workloads), which are difficult to address directly.

Monitoring HDD's internal status is thus fundamental to reduce overhead
costs due to downtime scheduling maintenance on the basis of self-monitoring,
analysis, and reporting technology (SMART, [1]) for improving its availability and extending its life. While efficient planning of maintenance operations is
clearly valuable, more modern approaches have been focused on *proactive analysis*: predictive strategies to identify HDDs' status in terms of binary classification
(healthy or faulted).

© Springer Nature Switzerland AG 2020
S. Hartmann et al. (Eds.): DEXA 2020, LNCS 12391, pp. 403–413, 2020.
https://doi.org/10.1007/978-3-030-59003-1_26

Obviously, a plethora of approaches have been proposed with this aim, most of them using SMART attributes to predict disk replacement ahead of failure [4,9]. However, health assessment of HDDs based on SMART statistics is not a trivial task, in particular if we are interested in estimating how much functioning time a specific HDD has left (remaining Useful Life, RUL). In particular, the historical data available about HDDs is highly imbalanced, with the majority of information available only describing healthy hard drives. Thus, any efficient solution to health status prediction will have to deal with this issue.

In this paper, we propose a framework that predicts HDD health status, exploiting the peculiarities of LSTMS to take advantage of the variation is SMART attribute values over time. In doing so, we address one of the biggest challenges to the use of deep learning approaches in predicting HDD health status: namely, the imbalanced nature of the available data-sets. Moreover, we show how *explainable artificial intelligence* tools can we used to probe the results of our model, and support practitioners in their decision-making.

The paper is organized as follows. Section 2 briefly summarises state-of-art approaches for HDD health status prediction, and eXplainable Artificial Intelligence (XAI). Section 3 provides details about our approach to the health prediction issue. In Sect. 4, we provide an extensive empirical evaluation of our model, compared so state-of-the-art solutions in the literature. We also show how our model's decisions are made interpretable by the use of XAI tools. Section 6 briefly summarizes our results, and suggests areas for future work.

2   Related Work

The past decade has seen the rise of deep learning techniques applied to prediction tasks in a variety of domains (for a recent review, see [14] a.o.). Generally, it is difficult to cover the entire process in which disk's health deteriorates and forecast when disk drives will fail in the future. Due to a common lack of diagnostic information of disk failures, the majority of approaches relies on the Self-Monitoring, Analysis and Reporting Technology (SMART) data and explore statistical analysis techniques to identify the start of disk degradation. Botezatu et al. [4] propose a disk replacement strategy that predict disk replacement ahead of failure, by using a changepoint-based feature selection strategy, and a compact representation of the time series data for the SMART indicators. Similarly, it has been shown that dynamic tracking methods based on a Rao-Blackwellized particle filter can been also provide online predictions [16]. The core idea being the use a dynamic failure threshold which exploits the statistical property of the tracking residuals. Importantly, the decision process that most of these techniques use to arrive at a specific outcome is often be opaque to human users. However, understanding *why* a certain prediction was made is essential in the context of effective HDD maintenance, as human users have to trust these models enough to use their outcomes to schedule costly preventive strategies. This is the goal of *eXplainable Artificial Intelligence* (XAI) systems (see [6], a.o.), which can be classified, according to the granularity of their analysis, into local and global categories. Local XAI methods aim to explain a deep learning model's outcome on the basis of local information around the prediction. For instance, Baehrens et al. [2], measure local gradients to identify exactly in which ways changing the input affects the prediction. Similarly, Robnik-Sikonja et al. [12] presents a feature importance method which computes the differences between a prediction and the obtained solution. Finally, the model-agnostic (*LIME*) method in [11] is based on an algorithm that faithfully explains the predictions of *any* classifier, by approximating it locally with a fully interpretable model.

The techniques summarized above focus on local explanations to achieve an overall explanation of a deep learning model. However, other techniques aim to build global explanations explicitly. Two popular methods of this latter kind rely on features importance to explain tree models: the Global Mean Decrease in Impurity (MDI, [5]) method—which uses splits' number of samples—and the Mean Decrease in Accuracy (MDA, [7]) method—computing a model's mean increase error on the basis of random permutation of the features. Directly related to the issues discussed in this paper, Xie et al. [18] present a XAI system explicitly designed for disk failure prediction, which is able to infer the prediction rules learned by a model, in order to make the failure prediction process transparent. In what follows, we discuss an approach to estimate the HDD health status based on the analysis of SMART attributes to analyze the Remaining Useful Life (RUL). We also show how XAI tools can be used to probe the rationale behind our model's classification decisions.

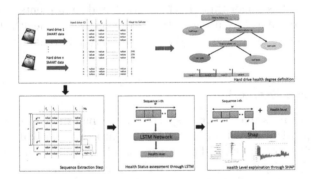

Fig. 1. The proposed schema: Hard drive health degree definition; Sequences extraction; Health Status assessment through LSTM; XAI Explainer of HDD Health Status.

3   Methodology

Since hard drives often deteriorate gradually rather than abruptly, we argue that temporal analysis methods are more appropriate than methods that do not consider time when modeling the sequential nature of the dependencies within SMART attributes. Thus, we suggest an approach to estimate and explain the RUL of a HDD, by automatically identifying specific health conditions on the basis of SMART attributes values. This methodology is grounded in four main steps (Fig. 1): i) *Hard drive health degree definition*; ii) *Sequences extraction*; iii) *Health Status assessment through LSTM*; iv) *Explainer of HDD Health Status via XAI tools*.

3.1   Health Degree Definition

In this step we consider only the hard drives that are going to fail, introducing for each of them an additional feature representing the *time before failure*. Denoting with m_j be the number of samples for the hard disk j, it is possible associate each sample with an index i from 0 to $m_j - 1$, representing the number of samples that follow it in the sequence describing hard disk failure. As a consequence, the sample with index $i = 0$ is the last sample before failure. *Time-to-failure* is the feature representing the time before failure for each hard drive whose meaning depends on sampling period while f_1, f_2, ..., f_n are the SMART attributes. Our idea is to build a Regression Tree (RT) for each SMART attribute f_i with $i = 1, 2 \ldots n$, having the feature representing the time before failure as predictor and f_i as the numeric target value. Among all the resulting trees (one for each SMART attribute f_i), the one with the highest performance is selected, showing the attribute most temporally dependent. Since the selected Regression Tree (RT) presents splits only on the feature *Time-to-failure*, the latter is used to distinguish hard drive health levels according to time before failure. In Fig. 1 is reported an example of hard drive health levels identification by means of the Regression Tree algorithm. Each internal node represents a split on the feature Time-to-failure, resulting in the definition of four health degree levels. The samples belonging to hard drives that will not fail are labelled as *Good* by default.

3.2   Sequence Extraction

We extract feature sequences over specific time windows (TW), to explore the temporal dependencies within the SMART features periodically collected for each hard drive. Let w and a^t be the time window size and the set of SMART features $(f_1, f_2 \ldots f_n)$ at time t, respectively. Our model aims to predict hard drive health status at time $t+1$ ($Hs(t+1)$) considering the sequence $(a^{t-w+1} \ldots, a^{t-1}, a^t)$. For each a^t, the health status $Hs(t)$ is defined, and the feature sequence for each hard drive at time t is extracted considering the $w - 1$ previous samples. Each sequence results in a bi-dimensional array of size $w \times n$, where n is the number of SMART features considered. For each hard drive, sequences are extracted with a stride of one. It follows that $m_j - w + 1$ sequences are extracted for each hard drive, where m_j is the number of samples for the disk j. For each sequence $(a^{t-w+1} \ldots, a^{t-1}, a^t)$, the hard drive's health level is defined by the health level of the set of features a^{t+1}. The result of this step is a sequence-based data-set. More specifically, the data-set consist of bi-dimensional arrays, each associated to a health level representing the hard drive's health condition between two consecutive samples (i.e., a^t and a^{t+1}).

3.3 Health Status Assessment Through LSTMs

This step consists of a multiclass classification task, where each feature sequence is assigned to one of the classes (health levels) introduced in Sect. 3.1. Due to the sequential, gradually changing nature of the SMART features, it is important that our model is able to capture dependencies across features over time. Long Short Term Memory networks (LSTMs) are extension to recurrent neural networks, explicitly designed with the purpose of learning long-term dependencies. They are widely used nowadays, as they work tremendously well on a large variety of problems. In our framework, the input to each LSTM layer is a three-dimensional data structure of size $z \times w \times n$, where: i) z, is the the total number of sequences (or the batch size at each iteration); ii) w is the size of each sequence—i.e., the size of a time window in terms of time steps; iii) n is the total number of features describing each time step. The implemented network has two stacked LSTM layers with 128 units, followed by a single dense layer.

3.4 Explainer of HDD Health Status Through SHAP

Finally, each extracted sequence is explained by means a model-agnostic XAI tool: *SHapley Additive exPlanations* (SHAP, [8]) assigning to each feature of a model, an importance value for a specific prediction. All testing sequences are classified by the LSTM model, delivering a confusion matrix. Denoting by i and j the row and column indexes—$i = 1, 2, \ldots, n$ and $j = 1, 2, \ldots, n$ with n the number of classes—respectively, each element is identified as a_{ij} and represents the number of instances of i class classified as j. Let S_{i_j} be the sequences of the test set belonging to class i but classified as j, our aim is to explain this misclassification, to understand each predicted class characteristic.

4 Evaluation

We test the prediction performance of the model on Backblaze SMART data-set,[1] and then compare its performances against three popular methods in the existing literature: a Classification Tree (CT) model, a Random Forest (RF) model, and a model based on Multiclass Neural Networks (MNN). In particular, we used the Backblaze data-set that contains daily data collected from 50,984 hard disks. We focused on samples belonging to Seagate ST4000DM000, since it is the most populated model in data-set ($29,878$ disks in total; $29,083$ *good* disks and 795 *failed* disks). Among all SMART attributes, the most influential attributes have been selected after a feature selection phase.[2] Finally, the values for every SMART attribute were scaled to the interval $[-1, 1]$. Data pre-processing consisted of two main steps: *Features Selection* and *Health degree computation*. In the first step, the features *Reallocated Sectors Count* and *Current*

---

[1] https://www.backblaze.com/b2/hard-drive-test-data.html.

[2] https://www.backblaze.com/blog/hard-drive-smart-stats/.

Pending, Sector Count were removed in order to preserve their raw values—that is the features *Raw Value of Reallocated Sectors Count* and *Raw Value of Current Pending Sector Count*—since the latter seem more sensitive to the health condition of hard drives. We also excluded the feature representing disk capacity. Importantly, the attributes *failure* and *Serial Number* are necessary in order to distinguish between failed and good hard drives and to create sequences for each hard drive. However, they are not taken into account during sequence classification. For good hard drives, each sample was associated to the health degree level *Good*, while for failed hard drives, their remaining functioning time depends on the number of samples collected for said device. In the Health degree computation step, we focused on the last q samples of each failed hard drives, where q is a *prediction window* that determines the period in which hard drive health status should be assessed. Specifically, our approach is able to predict hard drive health status q days before failure. We explored different values for q, from 15 to 45 days. After choosing the value for q, hard drive health levels are defined according to Sect. 3.1. Since the Backblaze data-set contains daily samples for each hard drive, the feature *Time-to-failure* has been renamed *Day to failure*. We then selected the regression tree built with the feature *Raw value of Current Pending Sector Count*. We then introduced a different level for those hard drives that will not fail. When q is set to 30 or 45, the result is the definition of 4 levels, labelled *Alert, Warning, Very Fair* and *Good*. In turn, if q is set to 15, we define 3 levels, labelled *Alert, Warning* and *Good*. The levels *Good* and *Very Fair* represent HDDs still in good health conditions. Therefore, we classify a hard drive as being in a *Good status*, if its health level is characterized as *Good* or *Very Fair* while a hard drive is classified as being in a *Failed Status*, if its health level is in *Warning* or *Alert*.

4.1   Experimental Setup

We propose an automatic step for hard drive health levels definition, building a Regression Tree (RT) for each SMART attribute f_i, with the feature representing the hours before failure as predictor. The selected tree consider the SMART attribute *Raw Value of Current Pending Sector Count* as numerical target value. The function measuring the quality of a split is the mean squared error (mse). The minimum number of samples required for leaf node in the Regression Tree is 1830, 1380, and 1200 with $q = 45$, $q = 30$ and $q = 15$ respectively. We evaluate our model with respect to three of the *sequence independent* methods most used in the literature: CT, a RF, and a MNN. These models are sequence independent because they generalize over input samples rather than sequences, and thus don't take the temporal dependencies of the SMART attributes into account. We implement the RT, CT and RF models using the Python scikit-learn package, and we use Keras with Tensorflow as the backend for LSTM and Multiclass NN models. As standard for this kind of techniques, the original SMART data-set was divided into training, validation and test sets. More specifically, we take the 70% of the data as training set, the 15% as validation set and the remaining data as test set. Downstream of the parameters optimization, the number of trees for

RF is set to 210 and the minimum number of samples required for leaf node in CT is 20. During the training phase of the LSTM and Multiclass Neural Network models, the maximum number of epochs is set to 10, and the batch size to 500. We use Adam as an optimizer, with learning rate set to 0.001.

The performance of our approach is first evaluated in terms of accuracy, precision, and recall. Since the distinction between good and failed hard drives is preserved in the labelling of the data-set, we express the results in term of accuracy on good sequences (ACC_G) and accuracy on failed sequences (ACC_F)—respectively, the fraction of sequences correctly classified as *Good*, and the fraction of sequence classified as the health levels suggested by the regression trees. We also measure the accuracy of classifying good and failed sequences for a tolerance of misclassification up to one health level (ACC_G^{TOL} and ACC_F^{TOL}). Finally, we evaluate performance in terms of failure prediction, by assessing *failure detection rate* (FDR) and *false alarm rate* (FAR) for each model. This is done by considering the levels *Good, Very Fair* as *Hard drive good statuses*; and the levels *Warning, Alert* as *Hard drive failed statuses*. Intuitively, FDR is the fraction of failed sequences that are correctly classified as failed, while FAR is the fraction of good sequences that are incorrectly classified as failed.

5   Results

Table 1 shows results of our LSTM based approach. Performance is reported for different sizes of the time window (TW) used in the sequence extraction step. We explored time window sizes from 5 to 15 days.

Table 1. Performance values for the LSTM models obtained by varying *prediction window (q)* and TW size on the Backblaze data-set.

q [day]	TW SIZE [day]	Accuracy	Precision	Recall	ACC_G	ACC_F	ACC_G^{TOL}	ACC_F^{TOL}	FDR	FAR
15	5	95.88%	96.90%	95.10%	97.28%	66.56%	97.89%	98.08%	75.53%	2.82%
15	7	95.81%	97.10%	96.00%	97.02%	70.27%	97.93%	**98.45%**	79.34%	2.70%
30	5	94.54%	96.50%	94.60%	96.38%	56.07%	97.68%	88.30%	76.03%	2.73%
30	7	93.93%	96.80%	94.40%	95.59%	59.15%	97.07%	89.37%	80.70%	3.29%
30	10	95.25%	97.40%	96.10%	96.84%	61.84%	97.59%	91.35%	85.48%	2.73%
45	5	94.45%	96.70%	94.93%	95.95%	66.16%	97.80%	90.67%	78.30%	2.50%
45	7	95.82%	97.00%	95.85%	97.28%	68.34%	98.12%	89.37%	77.75%	2.17%
45	10	96.56%	97.72%	96.82%	97.71%	75.08%	98.36%	93.30%	84.18%	1.83%
45	14	**98.45%**	**98.33%**	**98.34%**	**99.21%**	**84.49%**	**99.40%**	96.65%	**91.48%**	**0.72%**

For the latter time-interval, we considered a *prediction window (q)* varying from 15 to 45 days. As expected given the ability of LSTMs to learn long-distance dependencies, the best results are obtained with a time window spanning 15 days. Table 2a reports results for a set of sequence independent models previously explored in the literature, taking hourly samples as input rather than sequences. The best results in terms of accuracy on failed sequences are obtained with MNN. Overall though, these results show that sequence dependent approaches provide higher performance than a sequence independent methodologies.

Table 2. (a) Results of sequence independent models on the Backblaze data-set. (b) Results of best model on the Backblaze data-set detailed by each class.

(a)

Model	Accuracy	ACC_G	ACC_F	ACC_G^{TOL}	ACC_F^{TOL}	FDR	FAR
CT	83.80%	83.87%	56.31%	95.63%	88.46%	63.58%	4.69%
RF	85.77%	85.77%	**71.75%**	93.68%	**93.82%**	**80.66%**	6.49%
MNN	**96.17%**	**99.15%**	39.78%	**99.88%**	69.20%	85.75%	**0.95%**

(b)

Metric	Good	Very Fair	Warning	Alert
Accuracy	99.21%	87.80%	78.10%	84.42%
Precision	99.90%	69.40%	64.70%	73.10%
Recall	98.80%	87.80%	78.10%	84.40%

Finally, Table 2b reports the performance of our best models detailed by each class.

Table 3. Comparison of our best model (LSTM - $TW = 14$ days and $q = 45$ days) on the Backblaze data-set with previously proposed models on the hard drive health status assessment and failure prediction tasks ((a) (b) respectively).

(a)

Author	Methods	Accuracy	Precision	Recall
Zhang et al.[19]	LPAT+All	92.6%	89.3%	88.7%
Basak et al.[3]	LSTM	—	84.35	72.0%
Our Approach	LSTM	**98.45%**	**98.33%**	**98.34%**

(b)

Author	Methods	FDR	FAR
Shen et al.[15]	RF	94.89%	0.44%
Xiao et al.[17]	ORF	98.08%	0.66%
Our Approach	LSTM	**98.20%**	**0.20%**

We also compared our methodology to other sequence dependent approaches, which had been tested on the SMART data-set. Tables 3a and 3b compare our best results with different approaches for hard drive health status assessment and hard drive failure prediction tasks. In particular, Table 3a and 3b compare our best result with some other state-of-the-art methods in the literature: Zhang et al. [19], a method based on adversarial training and layerwise perturbation (LPAT); Basak et al. [3], an LSTM-based prediction model for RUL estimation; Shen et al. [15] and Xiao et al. [17], a prediction model based on part-voting Random Forest and Online Random Forest. Our proposal outperforms all these models in terms of accuracy on failed sequences, FDR, and FAR both for hard drive health status assessment and hard drive failure prediction tasks. Importantly, experimental results demonstrate that our approach is feasible for HDD health status assessment task due to the pre-processing phase and the definition of a specific model (LSTM) relying on temporal sequence. Finally, SHAP tools are used to explain each extracted sequence. To minimize the number of false negative alarms, we are interested in explaining why samples are not placed in the damaged class—that is, Alert. We focused on the damaged class because

HDDs are the main cause of downtime and unavailability for a data center. Since the cost of their replacement has a significant impact on the business continuity and financial resources of a company, we argue that an accurate analysis of these cases is not only desirable, but necessary.

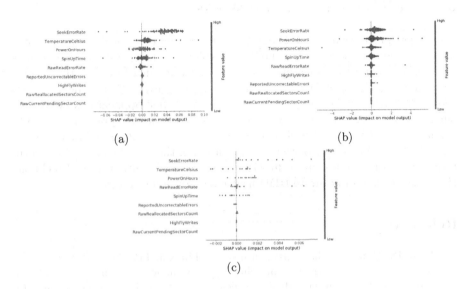

Fig. 2. Summary plot of Good sequences classified as Good.

Figure 2a and Fig. 2b show summary plots combining feature importance with feature impact on model output. Specifically, the y-axes shows features ordered by importance, and and the x-axis shows the related SHAP values. Moreover, each point is characterized by a color representing the feature values from low to high. *Temperature Celsius* (TC), *Seek Error Rate*(SER), *Power on Hours* (PoH) and *Spin Up Time* (SUT) are the most important features. As can be seen from the plots above, the most important features for the sequences classified as part of the Good class are: i) TC, which is almost always low; ii) PoH, with high value; iii) SER, which often assumes low values; iv) SUT, with low values. On the other hand, for the sequences classified as belonging to the Alert class the most important features are: i) SER, which always assumes high values; ii) PoH with low values; iii) TC, which often assumes high values; iv) SUT that takes high values. Finally, Fig. 2c reports an example of a false negative sequence with true class Alert, and classified with Good. This plot highlights the causes that led to an incorrect classification: SER and Raw Read Error Rate contain some outliers (low values) which have a greater impact on the output than the other (high values); moreover, SUT often assumes low values. Overall, these results show the advantages of employing XAI techniques in conjunction with deep learning models. In the case of HDD failure detection, the insights gained from XAI-based analysis helps identify false negatives cases. It should be possible to use this additional information to define better maintenance plans, this allowing companies to optimize operating costs, and increase the reliability of the provided services.

6   Conclusions and Future Works

This paper proposes a methodology to perform hard drive health status assessment, by relying on an LSTM model to exploiting the temporal dependencies of SMART attributes. Our approach is also effective in addressing the issue of highly imbalanced data-sets. An extensive empirical evaluation on big data-sets shows the feasibility of the proposed approach for real applications. Moreover, our model outperforms a variety of existing state-of-the-art approaches. Finally, we have argued how XAI tools can improve HDD preventive maintenance systems, by providing a transparent interpretation of the model's prediction. In the future, it will be interesting to further assess the contribution to XAI technologies to the design of accurate HDD health supervision strategies.

References

1. Allen, B.: Monitoring hard disks with smart. Linux J. **117**, 74–77 (2004)
2. Baehrens, D., Schroeter, T., Harmeling, S., Kawanabe, M., Hansen, K., MÃžller, K.R.: How to explain individual classification decisions. J. Mach. Learn. Res. **11**, 1803–1831 (2010)
3. Basak, S., Sengupta, S., Dubey, A.: Mechanisms for integrated feature normalization and remaining useful life estimation using LSTMs applied to hard-disks. In: 2019 IEEE SMARTCOMP, pp. 208–216 (June 2019)
4. Botezatu, M.M., Giurgiu, I., Bogojeska, J., Wiesmann, D.: Predicting disk replacement towards reliable data centers. In: 22nd ACM SIGKDD, pp. 39–48 (2016)
5. Breiman, L.: Random forests. Mach. Learn. **45**(1), 5–32 (2001)
6. Gunning, D.: Explainable artificial intelligence (xai). DARPA **2** (2017)
7. Louppe, G.: Understanding random forests: From theory to practice. arXiv preprint arXiv:1407.7502 (2014)
8. Lundberg, S.M., Lee, S.I.: A unified approach to interpreting model predictions. NIPS **30**, 4765–4774 (2017)
9. Mahdisoltani, F., Stefanovici, I., Schroeder, B.: Proactive error prediction to improve storage system reliability. In: USENIX ATC 2017, pp. 391–402 (2017)
10. Ponemon, L.: Cost of Data Center Outages. Data Center Performance Benchmark Series (2016)
11. Ribeiro, M.T., Singh, S., Guestrin, C.: "Why should i trust you?" Explaining the predictions of any classifier. In: 22nd ACM SIGKDD, pp. 1135–1144 (2016)
12. Robnik-Šikonja, M., Kononenko, I.: Theoretical and empirical analysis of relieff and rrelieff. Mach. Learn. **53**(1–2), 23–69 (2003)
13. Sankar, S., Shaw, M., Vaid, K., Gurumurthi, S.: Datacenter scale evaluation of the impact of temperature on hard disk drive failures. ACM TOS **9**(2), 1–24 (2013)
14. Sengupta, S., et al.: A review of deep learning with special emphasis on architectures, applications and recent trends. Knowl. Based Syst. **194**, 105596 (2020)

15. Shen, J., Wan, J., Lim, S.J., Yu, L.: Random-forest-based failure prediction for hard disk drives. Int. J. Distrib. Sens. Netw. **14**(11), 1550147718806480 (2018)
16. Wang, Y., Jiang, S., He, L., Peng, Y., Chow, T.W.: Hard disk drives failure detection using a dynamic tracking method. In: IEEE 17th INDIN, vol. 1, pp. 1473–1477 (2019)
17. Xiao, J., Xiong, Z., Wu, S., Yi, Y., Jin, H., Hu, K.: Disk failure prediction in data centers via online learning. In: Proceedings of the 47th ICPP, p. 35. ACM (2018)
18. Xie, Y., Feng, D., Wang, F., Tang, X., Han, J., Zhang, X.: DFPE: explaining predictive models for disk failure prediction. In: 35th MSST, pp. 193–204. IEEE (2019)
19. Zhang, J., Wang, J., He, L., Li, Z., Philip, S.Y.: Layerwise perturbation-based adversarial training for hard drive health degree prediction. In: 2018 IEEE ICDM, pp. 1428–1433. IEEE (2018)

An Improved Software Defect Prediction Algorithm Using Self-organizing Maps Combined with Hierarchical Clustering and Data Preprocessing

Natalya Shakhovska[1] , Vitaliy Yakovyna[2] ,
and Natalia Kryvinska[3(✉)]

[1] Artificial Intelligence Department, Lviv Polytechnic National University,
Lviv, Ukraine
nataliya.b.shakhovska@lpnu.ua
[2] Faculty of Mathematics and Computer Science,
University of Warmia and Mazury in Olsztyn, Olsztyn, Poland
yakovyna@matman.uwm.edu.pl
[3] Department of Information Systems, Faculty of Management,
Comenius University in Bratislava, Bratislava, Slovakia
natalia.kryvinska@fm.uniba.sk

Abstract. An improved software defects prediction algorithm based on com-bination of Kohonen map and hierarchical clustering is presented in this paper. The need for software reliability assessment and analysis growths rapidly due to increasing dependence of our day-to-day life on software-controlled devices and systems. Software reliability prediction is the only tool available at early stage of software development lifecycle when the debugging cost risk of faulty operation is minimal. Artificial intelligence and machine learning in particular are promising techniques to solve this task. Various classification methods have been used previously to build software defect prediction models, ranging from simple, like logistic regression, to advanced methods, e.g. multivariate adaptive regression splicing. However, the available literature still does not allow to make unambiguous conclusion concerning the choice of the best classifier and trying different dimensions to overcome potential bias is suggested. The purpose of the paper is to analyze the software code metrics to find dependences be-tween software module's defect-proneness and its metrics. JM1 public NASA dataset from PROMISE Software Engineering Repository was used in this study. To increase the classification accuracy, we combine self-organizing maps with hierarchical clustering and data preprocessing.

Keywords: Prediction algorithm · Hierarchical clustering · Software defect analysis

© Springer Nature Switzerland AG 2020
S. Hartmann et al. (Eds.): DEXA 2020, LNCS 12391, pp. 414–424, 2020.
https://doi.org/10.1007/978-3-030-59003-1_27

1   Introduction

The rapid development of information technology and artificial intelligence (AI) has led to the fact that we can hardly imagine the daily human activities without computer systems, including embedded and mobile ones. Automated systems manage a variety of technical devices performing safety-critical functions, ranging from nuclear power plants to autonomous vehicles controlled by AI. An integral part of all these systems is software that either performs a control function or implements artificial intelligence algorithms. The failure cost for such software is very high, which can result in events of different severity ranging from economic losses to harm to human life and health. Hence, with increasing responsibility of computer systems functioning the require-ments for their reliability, safety and dependability heavily increase. To meet these requirements, there are two major challenges to be addressed: assessing the reliability of complex technical systems and designing such systems with a given level of reli-ability and safety. However, it is no longer enough to assess the reliability of only the hardware component of such systems, as it was done previously for purely hardware electronic systems. Software has become a crucial part of such devices and systems. There is probably no other human-made material which is more omnipresent than software in our modern society [1]. Therefore, the need for software reliability assessment and analysis become nowadays increasingly relevant.

The cost of bug fixing is increasing rapidly in the later stages of the software life cycle, which causes growing interest in software reliability prediction models having high predictive power. One approach to building such models is to use artificial intelligence technologies. Thus, this paper is devoted to finding patterns and predicting software modules failures using machine learning techniques.

2   Related Works

Software reliability models can be divided into two broad classes – deterministic (static) and probabilistic (dynamic) [2]. Probabilistic models represent failure occur-rence and error correction as random events. Deterministic models use the results of the program source code analysis as input, and do not include any random events or values. The deterministic class includes Halstead Model [3], McCabe Cyclomatic Complexity Model [4], and Complexity Metric Model [5]. In general, these models represent a quantitative approach to measuring computer software. The Halstead Model is used to estimate the number of program defects [2], whereas the McCabe Cyclomatic Com-plexity Model is used to determine the upper limit of the program tests [6]. Note that both models are static in nature, that is, consider processes in the software system unchanged over time, and its reliability is solely a function of software metrics. A substantially different approach to solve problems of the current stage of software reliability theory is the theory of software systems dynamics [7]. The theory software systems dynamics differs from the existing software reliability theory in that it is based on the general theory of system dynamics, not probability theory, and considers the software failures not as a random process but as a result of the influence of determined defect flows. Therefore, this theory also generates deterministic software reliability models, which, however, are not static, unlike the Halstead and McCabe models.

The growth of empirical software engineering techniques has led to increased interest in bug prediction algorithms [8]. These algorithms predict areas of software projects that are likely to be bug-prone: areas where there tend to be a high incidence of bugs appearing. The overall approach is statistical: bug prediction algorithms are based on aspects of how the code was developed and various metrics that the code exhibits, rather than traditional static or dynamic analysis approaches. One common strategy is to use code metrics [9], while another has focused on extracting features from a source file's change history, relying on the intuition that bugs tend to cluster around difficult pieces of code [10]. Another important factor that influences software reliability is the design mechanism, which has a considerable impact on overall quality of the software. A well-designed internal structure of software is a required for ensuring better reliable. There are only few approaches described in literature, which address this question. One of them [11] studies the influence of design metrics on one of the external quality factors, reliability of the software using multivariate regression analysis.

Why are some software modules more defect-prone than others? Unfortunately, now there is no universal answer for this question. However, a lot of research [8, 12] including industry-based, has been devoted to capturing these project-specific properties [13]. The comprehensive benchmarking studies by Lessmann et al. used machine learning classifiers to predict faulty software units [14, 15]. In [9], some project-specific properties (invariants in their history) were discussed, and it was shown that the defect likelihood of some module depends on its history. The machine learning approach in software defect prediction has grown rapidly during the last decade. This approach uses methods adapted from other research fields [16, 17]. However, deep learning models are computationally expensive and hard to train than other classifiers. Hence, while deep learning remains very promising technique to reveal software patterns, which increase it defect-proneness, there is always a trade-off between using complex and simple models. In addition, one should be aware on collecting large enough datasets for deep learning models.

3   Methods and Results

3.1   Dataset

Public dataset from PROMISE Software Engineering Repository [18] was used in this study. The JM1 dataset on software defect prediction was selected. The source of this dataset is NASA and the NASA Metrics Data Program. JM1 is written in "C" and is a real-time predictive ground system, data comes from McCabe and Halstead features extractors of source code. The McCabe and Halstead measures are "module"-based where a "module" is the smallest unit of functionality. The dataset contains 10,885 entries (modules) along with 21 code metrics, used as features and listed in the Table 1, and a "TRUE/FALSE" field indicating either the module contains one or more reported defects, used as a target. Among all modules listed in the dataset, 2,107 have reported defects. Hence, we addressed the data imbalance problem by sampling, randomly choosing the subset containing 50/50 entries with and without reported defects. All calculations were made using RStudio. The randomized 90% of the dataset were used as a training set, and the rest 10% – as a validation set.

Table 1. Software metrics collected in the dataset

Metric	Description
loc	McCabe's line count of code
v(g)	McCabe's cyclomatic complexity
ev(g)	McCabe's essential complexity
iv(g)	McCabe's design complexity
n	Halstead's total operators + operands
v	Halstead's volume
l	Halstead's program length
d	Halstead's difficulty
i	Halstead's intelligence
e	Halstead's effort
b	Halstead's delivered bugs
t	Halstead's time estimator
LOCode	Halstead's line count
LOComment	Halstead's count of lines of comments
LOBlank	Halstead's count of blank lines
LOCodeAnd Comment	Halstead's Count of lines of code that also contain a comment
Uniq_Op	Halstead's Unique operators
Uniq_Opnd	Halstead's Unique operands
Total_Op	Halstead's Total operators
Total_Opnd	Halstead's Total operands
BranchCount	Number of branches in the flow graph

One of the obstacles while building software reliability model based on its code metrics is the fact that a lot of metrics highly correlate. Thus, pairwise Pearson correlation coefficient greater than 0.900 was obtained for 20 metrics pairs from the Table 1.

At the current stage of the study, we have used all features contained in the dataset. The further re-search will include the reduction of the features set to reveal only the most relevant. Such a reduction should be performed considering the possible differences among the programming languages, development methodology and other software engineering processes.

3.2 Research Methods: Classification

The aim is to build a model for software defects classification. The classification task lies in assigning an object to one of the predefined classes based on its formalized features. Each of the classified objects is represented as an N-dimensional vector, each dimension in which corresponds to one of the object's features. The binary classification model should evaluate the influents of each parameter or group of parameters on the defect classification. The analysis process consists of two phases. At the first phase, all parameters were considered. At the second phase, the influence of each parameter was studied.

Phase 1. First, we tried used traditional classification methods, viz. kNN, decision tree, logistic regression, SVM, and Naïve Bayes classifier. k-Nearest Neighbors is one of the simplest non-parametric classification algorithms. To classify each of the objects in the test set, the following operations are performed sequentially: calculate the distance to each of the objects of the training sample; select k-objects of the training sample, the distance to which is the smallest; classify the studied object to the most frequent class among the k-nearest neighbors. The decision tree, in addition to the classification, allows us to determine the weight of the parameters and their effect on the final classification decision. Due to overlearning it usually does not show very good results. Logistic regression also allows us to determine the weight of the parameters. An important feature of the Naïve Bayes classifier is the independence of the parameters of each other. The main task of SVM is to find a hyperplane that divides the data into two classes, viz. "faulty" and "non-faulty". The accuracy was calculated as:

$$ACC = (TP + TN)/N,$$

where TP states for true positive, TN – for true negative, and N is the total size of the validation set.

The accuracy of the used at this phase classification methods varies from 77% for kNN to 81% in case of logistic regression and SVM. As one can see, the classification results are not very good for any of the methods used. Therefore, we try to reduce the dimension of the task by finding the most important features.

Phase 2. To study the influence of each metric to classification model we calculated the attribute weights. The obtained weights are listed in the Table 2.

Table 2. Feature weights by decision tree

Feature (metric)	Relative weight
McCabe's line count of code	100.00
McCabe's design complexity	66.46
McCabe's cyclomatic complexity	65.63
Halstead's line count	62.71
Halstead's count of blank lines	42.87

The random forest consisted of 5 bootstrap-trees has the classification accuracy of 82%. Attribute weights obtained by logistic regression are summarized in the Table 3.

Table 3. Feature weights by logistic regression

Feature	Estimation	Std. error	z-value	Pr(>|z|)
loc	0.0029405	0.0002380	12.354	<2e−16
v(g)	−0.0049244	0.0027161	−1.813	0.06987
ev(g)	−0.0028223	0.0013343	−2.115	0.03445
n	0.0039700	0.0020583	1.929	0.05379
l	−0.0863699	0.0311149	−2.776	0.00552
d	−0.0025469	0.0007939	−3.208	0.00134
BranchCount	0.0041694	0.0015481	2.693	0.00709

In this table, the "std. error" value is the standard deviation of the coefficient point estimate. The "Pr(>\|z\|)" is the so called "p-value" of the test for whether the coefficient point estimate is significantly different from 0. When building a model with only 5 attributes, accuracy increases by 0.3% for logistic regression and by 0.5% for SVM. At the next step, we used backpropagation neural network (NN) to predict software defects. The parameters of the NN were chosen based on the obtained importance of the features. Thus, we constructed NN with two hidden layers, 7 inputs (McCabe's line count of code, McCabe's design complexity, McCabe's cyclomatic complexity, Halstead's line count, Halstead's count of blank lines, Halstead's intelligence, and Halstead's volume) and 1 output referring to target value. The prediction accuracy in this case was 85%.

As one can see, we could not substantially increase the accuracy of prediction nor for any method, neither for ensemble of the classification methods. That is why we choose another strategy: to divide the dataset by different groups (clusters) and make defect-proneness prediction separately for each group. For this task, different clustering methods and their combination were used.

3.3 Research Methods: Clustering

First, we started with classical k-means method. Using gap statistics, the optimal number of clusters was defined as 5. The total sum of square value is 45.2% (Fig. 1).

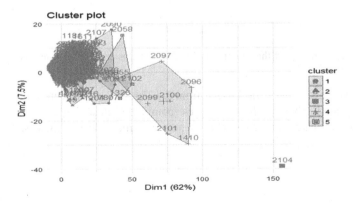

Fig. 1. The results of the whole dataset clustering using k-means algorithm

Next, the Principal component analysis (PCA) for Cluster2 have been made. We used DBSCAN (density-based spatial clustering of applications with noise) for Cluster2. The results show that the density is much higher around PC1 (Fig. 2). That is why we choose examples with characteristics, closed to the centroid of Cluster2 ("loc" value is about 33.5). The accuracy of classical classification methods in this case are varied from 92% for decision tree to 96.8% in case of neural network with two hidden layers, 6 neurons in each, and logistic activation function.

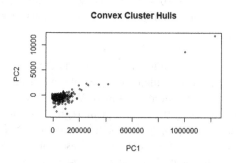

Fig. 2. The DBSCAN plot for the Cluster2

Let us analyze the rest of clusters. We choose subset of examples with "loc" value greater than 400. Directly defect-proneness does not significantly correlate with any feature as it was expected based on available literature analysis (see related works section of the paper). We tried to perform the clustering of the studied subset. The subset was divided into 7 clusters (Fig. 3).

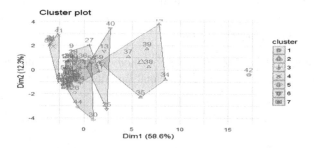

Fig. 3. The results of the subset with LOC > 400 clustering using k-means algorithm

However, as it can be easily seen from the Fig. 3, the intersection of clusters 7, 5 and 3 appeared. That is why we suggest combining clustering and classification to predict the software module defect-proneness classification with increased accuracy.

3.4   Research Methods: The Proposed Algorithm

We use an ensemble of models – the Kohonen map and hierarchical clustering. The heat map shows the weight of the attributes and their grouping, as well as clusters the data. The Kohonen map uses unsupervised learning and the training set consists only of the values of input variables. Kohonen map is trained by sequential approximation. Starting with a randomly selected initial location of the centers, the algorithm is gradually improved to cluster the training data. Kohonen's basic iterative algorithm consistently passes through a number of epochs; one training example is processed for each epoch. Input signals are presented sequentially to the network. The desired output signals are not determined. After processing a sufficient number of input vectors, the synaptic weights of the network are determined by the clusters. In addition, the scales are organized so that the topologically close nodes are sensitive to similar input signals.

To implement the proposed algorithm, it is necessary to determine the degree of neighborhood of neurons (the winner's neuron). For this purpose, we used hierarchical clustering. Some deviations were found, and the corresponding samples were excluded from the further analysis. The proposed algorithm is presented in the Listing 1. In this algorithm the distances d_j from the input signal to each neuron j are determined as:

$$d_j = \text{SUM}\big(x_i(t) \cdot w_{ij}(t)\big)^2,$$

here SUM is summa for all j; $x_i(t)$ is the i-th element of the input signal at time t, $w_{ij}(t)$ is the weight of communication from the i-th element of the input signal to j-th neuron at time t. Adjusted weight values for the neuron j^* and all neurons in its nearest neighborhood are calculated as:

$$w_{ij}(t + 1) = w_{ij}(t) + r(t)\big(x_i(t) - w_{ij}(t)\big),$$

Where $r(t)$ is learning rate, which decreases over time (positive, less than one).

Algorithm 1. An improved classification algorithm
```
Input: Matrix of input signals
Output: Heat map with clustering
1: Initialize the network weights with small random values
2: while (all samples not processed) do
3:    Present new input signal (sample from dataset)
4:    for j = 0...N, j++ do
5:    Calculating the distances dj to all neurons
6:    end for
7:    Choose the neuron-winner //the neuron j*, for which the
distance dj is the smallest one
8:    Adjust the weights of the neuron j* and its neighbors
9: end while
10:  return Heat map with clustering
```

The algorithm uses a gradually decreasing learning rate to fine-tune the new epoch. As a result, the center is established in such a position that satisfactorily clusters the examples for which the given neuron is the winner one. The property of topological ordering is achieved in the algorithm by using the concept of neighborhood. The neighborhood is a series of neurons surrounding the winner neuron. According to the training speed, the size of the neighborhood gradually decreases, so that at first it belongs to a sufficiently large number of neurons (possibly the whole map), at the most recent stages the it consists only of the winner-neuron. In the learning algorithm, the correction is applied not only to the winner neuron, but also to all neutrons from its current neighborhood.

The neighborhood results from agglomerative hierarchical clustering. The idea of agglomerative hierarchical methods is as follows. Initially, each object is treated as a separate cluster. Next, identify the two closest located clusters Qi and Qj and combine them into one cluster $Qi + j$. The merge process continues until all objects form a single cluster. During merging, the distance from the new cluster $Qi + j$ to all other clusters should be calculated. This distance is calculated using Lanes- Williams:

$$d(Qs + j, Qm) = \alpha i \cdot d(Qi, Qm) + \alpha j \cdot d(Qj, Qm) + \beta \cdot d(Qi, Qj)$$
$$+ \gamma \cdot |d(Qi, Qm) - d(Qj, Qm)|,$$

where $d(\bullet, \bullet)$ is a distance measure; Qm ($m \neq i, j$) is current cluster (neighborhood for SOM); αi, αj, β, γ some numerical parameters.

As a result of this change in neighborhoods, the initial rather large sections of the network migrate towards case studies. The network forms a rough structure of the topological order in which similar examples activate groups of neutrons that are closely on the topological map. With each new epoch, the speed of training and the size of the neighborhoods decrease, thus, thinner differences are found inside the map sections, which ultimately leads to a finer tuning of each neuron.

Table 4 presents the classification results obtained using the described algorithm. As it can be seen from the table, the classification using the proposed in this paper algorithm shows somewhat higher accuracy than any single classification method used.

Table 4. The accuracy (%) of the classification for examples near the centroid of Cluster2

Method	Cluster2	The other clusters	The whole dataset
kNN	82	93	77
Decision tree	80	92	75
Logistic regression	83	96	81
Naïve Bayes	81	95	80
SVM	83	96	81
Random forest	83	96	83
Neural network	84	97	85
The proposed algorithm	86	98	86

4 Conclusions and Future Work

An improved software defects prediction algorithm based on combination of Kohonen map and hierarchical clustering was developed in this paper. The algorithm uses a gradually decreasing learning rate to fine-tune the new epoch. As a result, the center is established in such a position that satisfactorily clusters the examples for which the given neuron is the winner one. The property of topological ordering is achieved in the algorithm by using the concept of neighborhood. The neighborhood results from agglomerative hierarchical clustering. The proposed algorithm shows the higher accuracy than other classification algorithm. The data preprocessing allows us to increase the quality of analysis by dividing all data to two clusters.

The further research will include the reduction of the features set to reveal only the most relevant. Such a reduction should be performed considering the possible differences among the programming languages, development method-ology and other software engineering processes.

References

1. Lyu, M.: Software reliability engineering: a roadmap. In: Future of Software Engineering (FoSE'07), Minneapolis, MN, USA, May 2007, pp. 153–170. IEEE (2007)
2. Pham, H.: System Software Reliability. Springer-Verlag London Limited (2006)
3. Halstead, M.H.: Elements of Software Science. Elsevier North-Holland Publishing, New York (1977)
4. McCabe, T.J.: A complexity measure. IEEE Trans. Softw. Eng. SE-2(4), 308–320 (1976)
5. Chen, N., Hoi, S.C.H., Xiao, X.: Software process evaluation: a machine learning framework with application to defect management process. Empirical Softw. Eng. 19(6), 1531–1564 (2014)
6. Rahmani, C., Azadmanesh, A.: Exploitation of Quantitative Approaches to Software Reliability. Survivable Networked Systems (CIST-9900) Report. University of Nebraska at Omaha (2008)
7. Maevsky, D., Kharchenko, V., Kolisnyk, M., Maevskaya, E.: Software reliability models and assessment techniques review: classification issues. In: Proceedings of 9th IEEE International Conference on Intelligent Data Acquisition and Advanced Computing Systems: Technology and Applications, 2017, Bucharest, vol. 2, pp. 894–899 (2017)
8. Lewis, C., Lin, Z., Sadowski, C., Zhu, X., Ou, R., Whitehead, E.J.: Does bug prediction support human developers? Findings from a google case study. In: Proceedings of the 35th International Conference on Software Engineering ICSE'13, San Fran-cisco, CA, USA, September 2013, pp. 372–381. IEEE (2013)
9. Nagappan, N., Ball, T., Zeller, A.: Mining metrics to predict component failures. In: Proceedings of the 28th International Conference on Software Engineering ICSE 06, Shanghai, China, May 2006, p. 452. ACM (2006)
10. Hassan, A.E., Holt, R.C.: The top ten list: dynamic fault prediction. In: 21 IEEE International Conference on Software Maintenance ICSM05, Budapest, 2005, pp. 263–272. IEEE (2005)
11. Selvarani, R., Bharathi, R.: Early detection of software reliability: a design analysis. In: Hosseinian-Far, A., Ramachandran, M., Sarwar, D. (eds.) Strategic Engineering for Cloud

Computing and Big Data Analytics, pp. 83–99. Springer, Cham (2017). https://doi.org/10. 1007/978-3-319-52491-7_5

12. Rahman, F., Posnett, D., Hindle, A., Barr, E., Devanbu, P.: BugCache for inspections: hit or miss? In: Proceedings of the 19th ACM SIGSOFT Symposium on the Foundations of Software Engineering and 13rd European Software Engineering Conference (ESEC/FSE'11), Szeged, Hungary, September 2011. ACM (2011)

13. Zimmermann, T., Nagappan, N., Zeller, A.: Predicting bugs from history. Software Evolution, pp. 69–88. Springer, Heidelberg (2008). https://doi.org/10.1007/978-3-540-76440-3_4

14. Li, L., Lessmann, S., Baesens, B.: Evaluating software defect prediction performance: an updated benchmarking study (2019). arXiv preprint arXiv:1901.01726 [cs.SE]

15. Lessmann, S., Baesens, B., Mues, C., Pietsch, S.: Benchmarking classification models for software defect prediction: a proposed framework and novel findings. IEEE Trans. Softw. Eng. **34**(4), 485–496 (2008)

16. Chen, N., Hoi, S.C.H., Xiao, X.: Software process evaluation: a machine learning framework with application to defect management process. Empirical Softw. Eng. **19**(6), 1531–1564 (2014)

17. Majumder, S., Balaji, N., Brey, K., Fu, W., Menzies, T.: 500+ times faster than deep learning (a case study exploring faster methods for text mining stackoverflow) (2018). arXiv preprint arXiv:1802.05319

18. Shirabad, J.S., Menzies, T.J.: The PROMISE Repository of Software Engineering Databases. School of Information Technology and Engineering, University of Ottawa, Canada (2005). http://promise.site.uottawa.ca/SERepository

A City Adaptive Clustering Framework for Discovering POIs with Different Granularities

Junjie Sun$^{(\boxtimes)}$, Tomoki Kinoue, and Qiang Ma

Kyoto University, Kyoto, Japan
{jj-sun,kinoue}@db.soc.i.kyoto-u.ac.jp, qiang@i.kyoto-u.ac.jp

Abstract. Discovering points of interest (POI) by analyzing social media data (e.g., Flickr) has received a lot of attention in recent years along with the rapid development of social network services. The small-grained POIs started to draw attention from tour recommendation services because more comprehensive information can be recommended when tourists traveling to an unfamiliar destination. To further meet tourists' increasing demands, tour recommendation combining points and areas is an important task, which requires a predefined POI database with different granularities. However, existing POI discovery methods do not well consider the granularity of POI and treat all POIs as the same. To this end, we propose a city adaptive clustering framework for discovering POIs with different granularities in this paper. Our proposed framework takes advantage of two clustering algorithms and is adaptive to different cities. The experiment results demonstrate the effectiveness of our proposed framework.

Keywords: Sightseeing · Points of interest · Location-based social network

1 Introduction

Nowadays, users would like to share their traveling experiences by uploading their photos to social image services such as Flickr and Instagram. The relevant information of these photos (e,g, location, time, and tags) makes these users generated content valuable for various data mining tasks. For instance, the increasing demand for the personalized tour for tourists travels in an urban area motivates more attention on tour recommendation services. These services or applications usually require a predefined POI database as the input data. According to preceding studies on POI identification or discovery [1,2], the POI database can be automatically constructed by applying clustering algorithms over the geotagged photos.

However, the granularity of POI is not well considered in the existing studies. All POIs, whatever a scenic park or a small sculpture, are treated as equal, while

© Springer Nature Switzerland AG 2020
S. Hartmann et al. (Eds.): DEXA 2020, LNCS 12391, pp. 425–434, 2020.
https://doi.org/10.1007/978-3-030-59003-1_28

tourists may want a more detailed recommendation. For instance, Central Park, New York City, is a famous POIs that cover large areas. It would be ambiguous to mark it as single POI in the recommended tours because detailed inner routes are preferred by the tourists. Recently, Wang et al. [4] propose a two-layer tour recommendation framework to provide more detailed information about the recommended tour by introducing the concept of "Super-POIs". Super-POI is a large-scale POI that covers several small-scale POIs, which should be explicitly indicated in the recommended tour rather than showing a unique POI.

Figure 1a shows an example of a recommended tour, and both locations 1 and 2 cover a large area. But tourists would like to know which places are more attractive and take photos on several spots inside the location 1 and lead to different photo quantity shown in Fig. 1b. An inner route to guide tourists with detailed information is helpful when traveling to some places like these. However, manually construct such a small-grained POI database is very time consuming, a method that automatically discovers POIs with different granularities is definitely desired from those tour recommendation services.

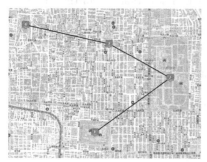

An example of a recommended tour   Photo quantity heat map inside

Fig. 1. Tour recommendation with large-grained POIs

To this end, we introduce the task of automatically detect POIs with different granularities from geo-tagged photos for tour recommendation applications. We consider two levels of granularities in this paper:

– **Areas of interest (AOI):** indicates the large-grained POI, such as a park or temple, is an area that attracts tourists to visit. It can be represented as a polygon or minimum bounding rectangle (MBR) on the map.
– **Sightseeing spot:** is the small-grained POI inside the AOI, such as a sculpture inside a park, which is a specific coordinate that attracts tourists to take photos. It can be represented as a point or coordinate on the map.

Generally speaking, the relationship between these two levels of granularities is that the sightseeing spots are points inside the AOIs. The major contribution of this paper can be summarized as follows:

- We summarize the advantages and disadvantages of applying variant POI discovery techniques to detect sightseeing spots.
- We propose a city adaptive clustering framework to discover POIs with different granularities. The results reveal the effectiveness of our proposed method.

2   Techniques for POI Discovery

Discovering POIs from a collection of geotagged photos can be viewed as a clustering problem on two-dimensional data (i.e., latitude and longitude). Here we summarize the advantages and disadvantages when they are applied to the sightseeing spot discovery task.

2.1 K-means

Kennedy and Naaman [5] use k-means algorithm to find landmarks from geotagged photos. The basic idea behind k-means is to separate a collection of locations into k clusters that make each coordinate belongs to the cluster with the nearest mean. The result of the k-means algorithm will assign a label to each location, and a mean coordinate is assigned to each cluster that can be treated as the detected sightseeing spot. However, the k-means algorithm requires a parameter k to determine the number of clusters, which is very hard to decide the proper number in advance.

2.2   Mean Shift

To address the problem of the fix number of cluster k, mean shift algorithm [7] is applied in [3] to find locations in metropolitan and landmark scales. Instead of fixing the number of clusters, the mean shift requires a *bandwidth* parameter to specify the density radius which can be viewed as the influential area of a location. Same as k-means, the mean shift also is a mean-based clustering method witch will assign a mean location for each cluster. To discover sightseeing spots, one straightforward solution is to globally set a small *bandwidth* value for all locations. However, it is very hard to precisely find a suitable value for different cities, and the unified *bandwidth* value for all locations has also suffered the problem of different scales of sightseeing spots.

2.3   DBSCAN

Density-Based Spatial Clustering of Applications with Noise (DBSCAN) is a very common density-based clustering algorithm for spatial data [6]. The algorithm adopts a range search strategy that requires two parameters. One is the search range radius ϵ and another is *minPts* to determine the minimum number of locations needed to form a cluster. The benefits of DBSCAN that compare to the above clustering algorithms are the high robustness against outliers and there is no constraint on the shape of clusters. However, since the algorithm cannot provide the coordinates of the output clusters, further computation is needed when using it on the sightseeing spot discovery task

3   Sightseeing Spot Discovery Framework

According to the discussion in Sect. 2, we combine two clustering algorithms to deal with POIs of different granularities.

3.1   AOI Discovery

We first detect AOIs according to the photo quantity density. We aim to eliminate low photo quantity density areas that may not worth visiting in the city. This can be done by applying the DBSCAN algorithm which can detect clusters of various sizes and shapes.

DBSCAN takes two parameters, ϵ and $minPts$, and it is difficult to determine appropriate values since different cities have different parameter settings. To automatically determine suitable parameters for various datasets, we propose a self-tuning process for AOI discovery. To be adaptive to different cities, we can use the location and size information of several typical AOIs in the city. We use them as the density criterion by comparing the coverage degree of detected clusters with typical AOIs' area information.

We use the minimum bounding rectangle (MBR) to represents the area of the typical AOI and Dice coefficient to estimate the cover area. The higher the similarity between the cluster detected by DBSCAN clustering and the predetermined AOI rectangle, the higher accuracy of clustering is considered. In this way, the parameter is tuned by the overlapping area of both areas and high photo quantity density AOIs are detected.

Suppose MBRs $s = \{s_1, s_2, ...s_m\}$ are determined m typical AOIs in advance. The area of the rectangle s_i is denoted by a_{s_i}. For each of the n clusters $c = \{c_1, c_2, ..., c_n\}$ obtained by DBSCAN clustering, we can find the smallest rectangle $t = \{t_1, t_2, ..., t_n\}$ covering c. The area of the rectangle t_j is denoted by a_{t_j}. Let $r_{i,j}$ be the overlap rectangle between s_i and t_j, and we define its area as $a_{r_{i,j}}$. Therefore, the Dice coefficient of the two areas is defined as below:

$$d(s_i, t_j) = \frac{2a_{r_{i,j}}}{a_{s_i} + a_{t_j}} \tag{1}$$

Since there are multiple rectangles having an overlapping area with one AOI, we only use the highest one to represent the Dice coefficient of the typical AOI s_i as follow:

$$D(s_i) = \max\{d(s_i, t_j)\}, \forall j = 1, 2, ..., n \tag{2}$$

Finally, the accuracy score for parameter ϵ and $minPts$ are defined as below:

$$DSC(\epsilon, minPts) = \sum_{1 \leq i \leq m} \frac{D(s_i)}{m} \tag{3}$$

We choose the parameters with the highest accuracy score to discover AOIs. The discovered AOIs can be represented by MBRs or polygons by using the points in each cluster. The outline of the tuning process is also showed in Fig. 2.

3.2   Sightseeing Spot Discovery

Mean shift clustering [7] is adopted to discover sightseeing spots from AOIs, which are detected by DBSCAN clustering in the first step. There are two benefits of using mean shift clustering. One is that photos are usually taken around the spot, so we can use the Gaussian kernel for mean shift clustering. The other is that a mean will assign to each cluster, which can be treated as the detected sightseeing spot. We find sightseeing spots in each AOI instead of applying the mean shift to all data points. The benefit of this is that low photo quantity density areas are avoided. Also, the detected sightseeing spots are inside AOIs, which are easier to construct the inner routes.

Accordingly, for data points in each AOI, we mark the clustering mean as the sightseeing spot that is estimated by:

$$m(x) = \frac{\sum_{x_i \in N(x)} K(x_i - x)x}{\sum_{x_i \in N(x)} K(x_i - x)}, \tag{4}$$

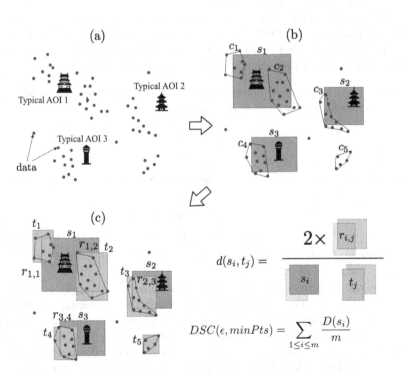

Fig. 2. There are points clustered by DBSCAN and typical AOIs defined in advance (a). The rectangle areas s_i are determined in advance for each typical AOI and n clusters c_j are detected by DBSCAN (b). The smallest rectangle t_j that completely covers each cluster c_j can be defined. Then, the overlap $r_{i,j}$ between s_i and t_j can be computed (c). For each overlapping of these rectangles, a Dice coefficient $d(s_i, t_j)$ is calculated. Finally, we get the accuracy score $DSC(\epsilon, minPts)$ for the parameters of DBSCAN clustering.

where $N(x)$ is the neighborhood of data point x and $K(x_i - x)$ is the kernel function. We use Gaussian kernel (i.e., $K(x_i - x) = \frac{1}{\sqrt{2\pi h^2}} e^{-\frac{\|x_i - x\|^2}{2h^2}}$) as the kernel function that requires the neighborhood of x to satisfy $\|x_i - x\| \leq h$, where h is the bandwidth parameter. The bandwidth parameter can be viewed as the influential area of a sightseeing spot (i.e., clustering center).

However, it is not appropriate to take the same bandwidth for all AOIs since the density of photos and the AOI area vary widely. We propose an area-adaptive parameter setting method to automatically determine the bandwidth for each AOI.

For simplicity, we consider two main factors that affect sightseeing spots discovery in AOIs:

- **1. AOI area:** The bandwidth should be proportional to the AOI area. For instance, the bandwidth of a huge park should be larger than that of a small museum. We use the smallest cover rectangle area a_{t_j} defined in Sect. 3.1 to represent the area of AOI c_j.
- **2. Photo quantity:** The bandwidth should be inversely proportional to photo quantity in the AOI. For instance, the high photo dense area contains much more sightseeing spots and set a smaller bandwidth could find more spots. We use p_{c_j} to represent the number of photos in AOI c_j.

Therefore, we define inverse photo quantity density to represent the relation of the two factors above for each AOI c_j as below:

$$ipd_{c_j} = \frac{a_{t_j}}{p_{c_j}} \tag{5}$$

Finally, we define the bandwidth parameter h_{c_j} for each AOI c_j as below:

$$h_{c_j} = h_m * ipd'_{c_j}, \tag{6}$$

where h_m is used as the hyper parameter to make sure the range of the bandwidth is less than h_m; $ipd'_{c_j} = \frac{ipd_{c_j}}{ipd_{max}}$, which normalize the inverse photo quantity density to the range of $(0, 1]$.

We use the automatically determined bandwidth h_{c_j} to apply to mean shift clustering for each AOI. Consequently, we mark the clustering center as the sightseeing spot. Then, the relevant information for tour recommendation applications can be easily obtained. For example, the visit duration of the spot can be obtained by calculating the average value of the time difference between the last photo and the earliest photo taken by the tourists on the same day.

4   Experiments

We apply our proposed framework on several geo-tagged photo datasets. Quantitative evaluation and case study analysis are presented to demonstrate our findings.

4.1   Experiment Setting

We collect 231,245 geotagged photos taken in Kyoto, Japan and 271,081 photos taken in Paris, France from Flickr API[1] to discover sightseeing spots. We extract 20 MBRs from OpenStreetMap[2], 10 MBRs are used for step one's parameter selection and the rest 10 are used as the validation set, which we manually mark sightseeing spots inside the MBRs as the ground truth (104 spots in Kyoto and 72 spots in Paris). One location is marked as a sightseeing spot based on the symbol on the map (e.g., viewpoint, shrine, etc.).

A grid search strategy is adopted to find the most suitable parameters with the highest dice coefficient for the AOI discovery phase. We set $h_m = 50$ m as the range of the bandwidth hyperparameter for sightseeing spot discovery since we want to find small-grained POIs.

4.2   Results

We compare our framework against the various baselines mentioned in Sect. 2 on two datasets using the following metrics:

- **Number of Spots** The total number of discovered sightseeing spots.
- **Precision** The proportion of discovered sightseeing spots inside the validation MBRs that also exist in the ground truth.
- **Recall** The proportion of sightseeing spots in the ground truth that also exists in the discovered spots inside the validation MBRs.
- **F-score** The harmonic mean of the precision and recall mentioned above.

Considering that the photos may be taken around the spots, we regard the points closest to the ground truth and not more than 10 meters away as the same.

Figure 3 and Fig. 4 report the results of all the baseline methods with different parameter values, in terms of precision and recall on the Kyoto dataset.

The recall of the mean shift with fixed bandwidth parameter decreases as the value of bandwidth increases, while the precision increases with increasing bandwidth, as shown in Fig. 3a and Fig. 4a. The reason is that more spots can be found when setting a smaller bandwidth value, so the recall is higher at first and drops as the value of bandwidth increases.

We vary parameter k in k-means clustering and the result is reported in Fig. 3b and Fig. 4b. It is hard to decide an appropriate value for the parameter k, especially for this sightseeing spot discovery task, the number of spots usually will be large. When k is large enough, the recall is getting better since more spots are marked as sightseeing spots. But this also leads to more non-sightseeing spots are included, which reduce the total quality of the discovered spots.

For DBSCAN, we set $minPts = 2$ for the Kyoto dataset and $minPts = 5$ for Paris dataset due to the difference of photo quantity, only change the search range parameter $epsilon$. The spot is computed by the mean value of the cluster

---

[1] http://www.flickr.com/services/api/.

[2] http://www.openstreetmap.org/.

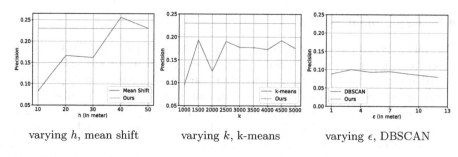

| varying h, mean shift | varying k, k-means | varying ϵ, DBSCAN |

Fig. 3. Comparison on Kyoto dataset in terms of precision

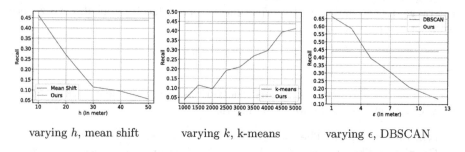

| varying h, mean shift | varying k, k-means | varying ϵ, DBSCAN |

Fig. 4. Comparison on Kyoto dataset in terms of recall

since it does not explicitly return cluster centers. Contrary to mean shift, when the search range ϵ is small, DBSCAN only finds the densest areas, which leads to a high recall at first, as shown in Fig. 3c and Fig. 4b. When the search range gets bigger, the recall quickly drops.

According to the results in Table 1, the mean shift finds the most number of spots since it is applied to all data points with a small influential area. It is hard to choose a proper value of k for the k-means clustering, and the results vary on different datasets. As shown in Fig. 3 and Fig. 4, the results of the baseline methods are greatly affected by the value of parameters. Appropriate parameters also vary depending on the datasets. Our methods rarely depend on parameters and get stable results.

Note that we set a fixed bandwidth of $h = 20$ on the discovered AOIs, and get a better results that compares to mean shift clustering that applied to all data points. It indicates the effectiveness of our idea that apply DBSCAN to remove noises and outliers. Overall, our proposed framework achieves better result in terms of F-score on the average of two datasets, which verifies the effectiveness of the city adaptive parameter tuning strategy in our framework.

Table 1. Comparison in terms of the number of discovered spots and F-score. The bold values indicate the best result for each metric.

	Kyoto		Paris		Average	
Methods	# Spots	F-score	# Spots	F-score	# Spots	F-score
Mean Shift ($h = 20$ m)	7,037	0.21	12,129	0.05	9583	0.13
K-means ($k = 2000$)	2,000	0.11	2,000	0.13	2000	0.12
K-means ($k = 4500$)	4,500	0.26	4,500	0.12	4500	0.19
DBSCAN ($\epsilon = 5$ m)	3,720	0.17	1,175	0.26	2448	0.22
Ours ($h = 20$ m)	1,411	0.25	975	**0.29**	1193	0.27
Ours	1,859	**0.31**	765	0.25	1312	**0.28**

5 Related Work

Landmarks are identified in [3,5] by applying clustering algorithms on geo-tagged photos where the scale of the landmark is subjectively determined. Ying et al. [2] compare variant clustering methods for the POI identification task and propose a self-tuning spectral clustering method. A noise-resistant algorithm is proposed in [1] to improve the quality of detected POIs. POI boundaries detection with given POI relevant information is studied in [9,10]. The authors in [8] study AOI discovery and recommendation by using geotagged photos. However, these works focus on large-grained POIs or AOIs, POIs with different granularities are not well considered.

A semantic location that contains a cluster of stay points with a centroid coordinate and semantic descriptions are mined from GPS data [12]. AOIs and photo shooting spots are detected from geo-tagged photos in [11]. Liu et al. [13] discover AOIs from both social images and check-ins. Hight quality sightseeing spots that are less well-known are discovered social images in [14]. These works focus on detecting spots from social network data, which can be treated as the small-grained POIs. But neither has been applied on tour recommendation applications.

6 Conclusion

In this paper, we summarize various clustering techniques for the POI discovery task and propose a city adaptive clustering framework to discover POIs with different granularities for tour recommendation services. The proposed method combines the advantages of two clustering algorithm and the parameters are automatically determined according to different cities. We compare our proposed framework against various baselines with careful parameter settings. The experimental results on two famous travel destinations reveal the effectiveness of our proposed framework.

Acknowledgement. This work is partly supported by MIC SCOPE(172307001, 201607008).

References

1. Yang, Y., Gong, Z., Li, Q., Leong Hou, U., Cai, R., Hao, Z.: A robust noise resistant algorithm for POI identification from flickr data. In: Proceedings of the 26th International Joint Conference on Artificial Intelligence, pp. 3294–3300 (2017)
2. Yang, Y., Gong, Z., Leong Hou, U.: Identifying points of interest by self-tuning clustering. In: Proceedings of the 34th international ACM SIGIR conference on Research and development in Information Retrieval, pp. 883–892 (2011)
3. Crandall, D.J., Backstrom, L., Huttenlocher, D., Kleinberg, J.: Mapping the world's photos. In: Proceedings of the 18th International Conference on World Wide Web, pp. 761–770 (2009)
4. Wang, C., Gao, Y., Gao, X., Yao, B., Chen, G.: eTOUR: a two-layer framework for tour recommendation with super-POIs. In: Pahl, C., Vukovic, M., Yin, J., Yu, Q. (eds.) ICSOC 2018. LNCS, vol. 11236, pp. 771–778. Springer, Cham (2018). https://doi.org/10.1007/978-3-030-03596-9_55
5. Kennedy, L.S., Naaman, M.: Generating diverse and representative image search results for landmarks. In: Proceedings of the 17th International Conference on World Wide Web, pp. 297–306 (2008)
6. Ester, M., Kriegel, H.P., Sander, J., Xu, X.: A density-based algorithm for discovering clusters in large spatial databases with noise. In: Proceedings of 2nd International Conference on Knowledge Discovery and Data Mining, vol. 96, No. 34, pp. 226–231 (1996)
7. Comaniciu, D., Meer, P.: Mean shift: a robust approach toward feature space analysis. IEEE Trans. Pattern Anal. Mach. Intell. **24**(5), 603–619 (2002)
8. Laptev, D., Tikhonov, A., Serdyukov, P., Gusev, G.: Parameter-free discovery and recommendation of areas-of-interest. In: Proceedings of the 22nd ACM SIGSPATIAL International Conference on Advances in Geographic Information Systems, pp. 113–122 (2014)
9. Bui, T.H., Han, Y.J., Park, S.B., Park, S.Y.: Detection of POI boundaries through geographical topics. In: 2015 International Conference on Big Data and Smart Computing (BIGCOMP), pp. 162–169. IEEE (2015)
10. Vu, D.D., Shin, W.Y.: Low-complexity detection of POI boundaries using geo-tagged tweets: a geographic proximity based approach. In: Proceedings of the 8th ACM SIGSPATIAL International Workshop on Location-Based Social Networks, pp. 1–4 (2015)
11. Shirai, M., Hirota, M., Ishikawa, H., Yokoyama, S.: A method of area of interest and shooting spot detection using geo-tagged photographs. In: COMP@ SIGSPATIAL, pp. 34–41 (November 2013)
12. Cao, X., Cong, G., Jensen, C.S.: Mining significant semantic locations from GPS data. Proc. VLDB Endowment **3**(1–2), 1009–1020 (2010)
13. Liu, J., Huang, Z., Chen, L., Shen, H. T., Yan, Z.: Discovering areas of interest with geo-tagged images and check-ins. In: Proceedings of the 20th ACM International Conference on Multimedia, pp. 589–598 (2012)
14. Zhuang, C., Ma, Q., Liang, X., Yoshikawa, M.: Anaba: an obscure sightseeing spots discovering system. In: 2014 IEEE International Conference on Multimedia and Expo (ICME), pp. 1–6. IEEE (2014)

Deep Discriminative Learning for Autism Spectrum Disorder Classification

Mingli Zhang[1]($\boxtimes$), Xin Zhao[2]($\boxtimes$), Wenbin Zhang[3], Ahmad Chaddad[4],
Alan Evans[1], and Jean Baptiste Poline[1]

[1] Montreal Neurological Institute, McGill University, Montreal, Canada
`mingli.zhang@mcgill.ca`
[2] College of Information Engineering, Dalian University, Dalian, China
`zhaoxin@dlu.edu.cn`
[3] University of Maryland, Baltimore County, MD, USA
[4] Guilin University of Electronic Technology, Guilin, China

Abstract. Autism spectrum disorder (ASD) is a complex neurodevelopmental disorder characterized by deficiencies in social, communication and repetitive behaviors. We propose imaging-based ASD biomarkers to find the neural patterns related ASD as the primary goal of identifying ASD. The secondary goal is to investigate the impact of imaging-patterns for ASD. In this paper, we model and explore the identification of ASD by learning a representation of the T1 MRI and fMRI by fusioning a discriminative learning (DL) approach and deep convolutional neural network. Specifically, a class-wise analysis dictionary to generate nonnegative low-rank encoding coefficients with the multi-model data, and an orthogonal synthesis dictionary to reconstruct the data. Then, we map the reconstructed data with the original multi-modal data as input of the deep learning model. Finally, the learned priors from both model are returned to the fusion framework to perform classification. The effectiveness of the proposed approach was tested on a world-wide cross-site (34) database of 1127 subjects, experiments show competitive results of the proposed approach. Furthermore, we were able to capture the status of brain neural patterns with the known input of the same modality.

1 Introduction

Autism spectrum disorder (ASD) is a structural and functional neurodevelopment disorder, it is also associated with weak communication skills, simple repetitive behavioral pattern and lowered concentration. The common way of diagnosis and treatment of ASD is based on symptoms, and thus, to identify a reliable biomarker is the main challenge [7]. Most diagnosis of ASD is confirmed at around 3 years old in the United States although, it is important to diagnose

---

A. Evans and J. P. Poline—co-last author.

© Springer Nature Switzerland AG 2020
S. Hartmann et al. (Eds.): DEXA 2020, LNCS 12391, pp. 435–443, 2020.
https://doi.org/10.1007/978-3-030-59003-1_29

ASD in the early stage of life for better treatment. Magnetic resonance imaging (MRI) based brain volumetric methods are commonly used to characterize ASD [13]. To better understand the origin of ASD for precise diagnosis, significant progress has been made using neural patterns of functional connectivity of functional magnetic resonance imaging (fMRI) data to caracterize brain changes related to ASD. Identification of Autism Spectrum Disorder from brain imaging provides biomarkers for the mechanisms of the pathology.

In recent years, many representation learning techniques such as discriminative dictionary learning (DDL) [11] and deep neural networks [1] are powerful algorithms to derive high-level latent features from high-dimensional [10] and multi-modal data [4]. DDL has been widely used in resting-state functional connectivity MRI analysis. Wang et al. [7] developed a low rank representation approach for multi-center ASD. Zhao et al. [12] presented an effective 3D convolutional neural network (CNN) based framework to derive discriminative overlap patterns of a spatial brain network that can characterize and identify ASD from healthy controls. However, considering the fact that ASD could be related to subtle feature changes in the brain, it would be difficult to train an end-to-end CNN directly without any pre-determined information, i.e., discriminative features. Most learning based method with extracted dependent or independent features (cortical thickness, cortical volume, connectome of fMRI) may result in a sub-optimal solution.

One of the challenge of ASD identification is to either estimate the corresponding cortical thickness of the subject under the same pre-processing pipeline or to find the correlation of these features for a given cortical area. The trained-rich matrix may be further processed to yield valuable informations that may be more clinically useful by the generation of gray matter thickness with computer-synthesized cortical volume, cortical surface area and thickness relationship.

In this study, we propose a novel multi-modal discriminative subspace learning approach named MMDL for identification of Autism Spectrum Disorder, by fusion of multi-modal brain imaging data. Different from the conventional modeling-based ASD identification methods, we use not only the priors learned by CNN-based learning, but also the priors from discriminative subspace learning. The fusion is performed in two aspects. First, training the dictionary pair learning (DPL) method. Then, the multi-modal features learned by DPL method and the original data as the input of the CNN. The first step can fully utilize the input data by improving the class-specific features of the original data. The CNN can boost the training performance. Capitalizing on the knowledge, the major contributions of this work are as follows:

- In this work, we propose a novel approach (MMDL), which fuses the classifier of discriminative dictionary learning and CNN to identify ASD. In this proposed MMDL method, instead of only using matrix factorization based discriminative dictionary learning, we also apply the CNN based learning to regularize the model. Specifically, during the CNN training, we initialize the reconstructed features from discriminative dictionary learning and the original data as the input of CNN, which boosts the input of CNN training.

Moreover, the trained dictionary pairs are also returned to the classifier fusion section to improve the identification of ASD performance.

- We demonstrate the classification performance of the proposed method on the functional connectivity matrix and gray matter (cortical surface area, cortical thickness and volume) of 1127 subjects from the challenging of predicting autism[1], the data is acquired from multiple sites with different protocols. The proposed model is much more accurate compared to the state-of-the-art. With one of the given features of gray matter, we can estimate the corresponding others of it.

2 Proposed MMDL Approach

The proposed MMDL approach incorporates deep CNN based training into the training framework, and guides the classify work with the learned priors. Figure 1 is an overview of the proposed framework. More details of each step are described as follows.

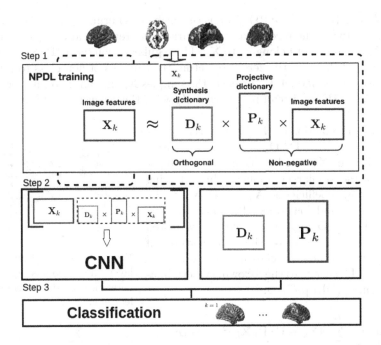

Fig. 1. The Scheme of the proposed MMDL method. The input is tensor format data with T1 MRI volume, surface area, thickness and fMRI connectome.

---

[1] https://paris-saclay-cds.github.io/autism_challenge/.

2.1   Initialize with Discriminative Learning

The modeling of ASD identification is first treated as a 3D tensor based class-wise discriminative dictionary learning problem, we set $\mathcal{X} = fold(\mathbf{X})$ and the operation of *fold* is to fold up each column of the matrix in to the corresponding subject of the tensor, then $\mathbf{X} = [\mathbf{X}_1, \cdots, \mathbf{X}_k, \cdots, \mathbf{X}_K]$ is the data samples and k is the class number but in this case, it is the binary classification for identification of Autism Spectrum Disorder.

Following [2,9], we introduced a linear feature selection dictionary $\mathbf{P}_k \in \mathbb{R}^{M \times S}$ and a reconstruction dictionary $\mathbf{D}_k \in \mathbb{R}^{S \times M}$ for the class k, where M is the number of subject in class k and S is the dimension of the feature for each subject, with $\mathbf{P} = [\mathbf{P}_1, \cdots, \mathbf{P}_k, \cdots, \mathbf{P}_K]$ and $\mathbf{D} = [\mathbf{D}_1, \cdots, \mathbf{D}_k, \cdots, \mathbf{D}_K]$, performing data modeling in two layer fully connected neural network format with low-rank constrain on the selected features of each group:

$$\underset{\mathbf{P},\mathbf{D}}{\arg\min} \quad \sum_{k=1}^{K} \|\mathbf{X}_k - \mathbf{D}_k\mathbf{P}_k\mathbf{X}_k\|_F^2 + \lambda_1\|\mathbf{P}_k\mathbf{X}_k\|_* + \lambda_2\|\mathbf{P}_k\mathbf{X}_{\overline{k}}\|_F^2, \quad (1)$$

where $\overline{k}$ ($\overline{k} \in \{\overline{k} : |k-\overline{k}| \neq 0\}$), $\|\cdot\|_F$ is the Frobenius norm, $\lambda_1, \lambda_2 > 0$ control the trade-off between the reconstruction accuracy and regularization terms, and $\mathbf{X}_{\overline{k}}$ is the data matrix not belonging to $\mathbf{X}_k$. The regularization term $\|\mathbf{P}_k\mathbf{X}_{\overline{k}}\|_F^2$ is used for forcing $\mathbf{P}_k\mathbf{X}_{\overline{k}}$ towards zero, projecting the samples of non-class to a nearly null space. In this model, $\mathbf{P}_k$ projects the samples $\mathbf{X}_k$ into an encoding coefficient matrix $\mathbf{A}_k = \mathbf{P}_k\mathbf{X}_k$, it can reconstruct $\mathbf{X}_k$ with the reconstruct dictionary $\mathbf{D}_k$, such as Fig. 2.

Ideally, the dictionary $\mathbf{D}$ follows orthogonality constraint with $\mathbf{D}_k^\top\mathbf{D}_k = \mathbf{I}$ to avoid overfitting. Hence, $\mathbf{X}_k$ can be taken as a combination of these similar components by enforcing the encoding coefficients $\mathbf{A}_k = \mathbf{P}_k\mathbf{X}_k$ to be non-negative and low rank. To boost the discrimination of $\mathbf{D}$ and $\mathbf{A}$, we explore weighted nuclear norm [3,8] on $\mathbf{A}$, since the features of subjects within the same class have low rank performance. This leads to the following discriminative learning (DL) problem.

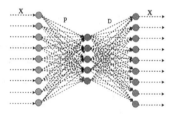

Fig. 2. The flowchart of two layer fully connected neural network based discriminative learning.

$$\underset{\mathbf{P},\mathbf{D}}{\arg\min} \quad \sum_{k=1}^{K} \|\mathbf{X}_k - \mathbf{D}_k\mathbf{P}_k\mathbf{X}_k\|_F^2 + \lambda_1\|\mathbf{A}_k\|_{w,*} + \lambda_2 \sum_{\overline{k} \in \{\overline{k}:|k-\overline{k}|>T\}} \|\mathbf{P}_k\mathbf{X}_{\overline{k}}\|_F^2$$

$$\text{s.t.} \quad \mathbf{D}_k^\top\mathbf{D}_k = \mathbf{I}, \ \mathbf{A}_k = \mathbf{P}_k\mathbf{X}_k, \ \mathbf{A}_k \geq 0, \ k = 1, ..., K. \quad (2)$$

where, the first term is reconstruction error, the second regularization makes the representation low rank, since the components of $\mathbf{X}_k$ are similar and have low-rank performance, $\mathbf{P}_k\mathbf{X}_k \geq 0$ makes representation non-negative and thus, creating sparsity in this way.

To learn dictionary sets $\mathbf{A}$, $\mathbf{D}$ and $\mathbf{P}$, we applied an alternating direction method of multipliers (ADMM) based algorithm as [9].

2.2 Learn the Classification Priors with CNN

Once the $\mathbf{D}$ and $\mathbf{A}$ is acquired, we can get more similar input data with labels as the input of the deep CNN learning. In this sub-section, we will describe the input-output and the architecture of the CNN.

Input-Output: Once the estimated features are achieved, we instead train the CNN with all the multi-model features directly. We use both the estimated multi-model features and the original features $\{\mathbf{DA}\}$ as an input of CNN. This has the advantage of 1) having the estimated features similar to the original multi-model features. 2) with more training data, it can help improve the training accuracy by boosting the learning performance. The intermediate classifiers as the output of CNN. 3) It can work rapidly since the CNN works on features instead of images.

Architecture: We adopt a general CNN architecture. We can apply any CNN models, however here we just adopted the architecture of the CNN has three blocks, which are listed as followed:

- *Conv + BN + ReLU + max pooling*: For the first block, we use 8 filters with size $3 \times 3 \times 8$, the max pooling is done by applying a 2×2 max filter.
- *Conv + BN + ReLU + max pooling*: For the second block, 16 filters with size $3 \times 3 \times 16$.
- *Conv + BN + ReLU + max pooling*: For the third block, 32 filters with size $3 \times 3 \times 32$.
- *AverPooling + FC + Softmax + Classification*

In the CNN, batch normalization (BN) is to accelerate the training, rectified linear units (ReLU) is the activation function, the max pooling layer performs down-sampling and to compute the maximum of each region. The average pooling layer(AverPooling) is for down-sampling and averaging the values of each region. FC is Fully Connected Layer.

2.3 Classification

In the classification process, we input the multi-model testing data into the well trained CNN and the learned $\mathbf{P}$ and $\mathbf{D}$, which can be used to classify samples by measuring the reconstruction error for each class as the approach. Instead of using the classification results via CNN as the final result. It can be as intermediate to further improve the classification performance. Thus the final classification result can be obtained by fusing the results of two classifiers.

To get the intermediate classification results of the initial discriminative learning. Here, we set $\mathbf{x}^i \in \mathbb{R}^{S_i}$ be the features of type i for the subject to classify. We define as $e_k^i = \|\mathbf{x}^i - \mathbf{D}_k^i \mathbf{P}_k^i \mathbf{x}\|_2$ the error of reconstructing $\mathbf{x}^i$ with the dictionaries of class k for feature type i. We then assign the sample to the class whose dictionary gives the lowest error $\widehat{k}_i = \arg\min_k e_k^i$.

To combine the information of the two classifiers to further improve the classification result, the final classification result can be obtained by solving the optimization problem as followed.

$$\arg\min_{\alpha} \left(k_{\text{real}} - \sum_j \alpha_j \widehat{k}_j \right)^2, \quad \text{s.t.} \sum_j \alpha_j = 1, \ \alpha_j \geq 0, \forall j. \tag{3}$$

$\widehat{k}_j$ is the classification result of each classifier j ($j = 2$) and the final output is class label k_{real}. Constraints on regression coefficients α_i enforce the final prediction to be a convex combination of classification results from the classifiers of CNN and discriminative dictionary learning.

3   Experiments

In this section, experiments are conducted on the public set of IMPAC[2] to evaluate the effectiveness of the proposed MMDL approach. We use 1127 subjects with 590 subjects as control and 537 subjects with Autism Spectrum Disorder. The model is evaluated with structural MRI using measures of cortical thickness, surface area and volume and resting state fMRI with 17.01($\pm$10) years old. The structural MRI is preprocessed with FreeSurfer and FSL, then the features are averaged following an adapted Desikan protocol, giving a total of 70 features per type of measure for both brain hemispheres. Connectomes were derived from fMRI using the correlation matrix of each subject, we use the singular values vector of the connectomes of fMRI as the input features. Then, the input is a tensor format data with a subject (subject with the label), volume, surface area, thickness and singular values of fMRI connectome matrix of each subject. For functional MRI in this study, we use the MSDL functional atlas [6], we reconstructed connectivity matrices using 70 brain discriminative regions by applying singular value decomposition (SVD) on these connectivity matrices, the singular values are then rearranged as vectors of 70 features.

For these tasks, We split available 1127 examples into a training set and validation set, the latter containing 10% of examples. The validation set was used to tune the regularization parameters and the size M of synthesis dictionary $\mathbf{D}$. Afterward, the 8-fold cross-validation is applied on these experiments to measure performance in terms of prediction accuracy (ACC), Specificity, Sensitivity, area under the curve (AUC) and root mean square error (RMSE).

3.1   Prediction of Autism Spectrum Disorder

We first demonstrate the proposed framework's performance by predicting the autism spectrum disorder, based on cortical thickness, cortical surface area and

---

[2] https://paris-saclay-cds.github.io/autism_challenge/.

Table 1. Classification results on the database of IMPAC on 8-fold cross-validation.

Method	ACC	Sens.	Spec.	AUC
SVM	0.621 ± 0.027	0.616 ± 0.071	0.629 ± 0.060	0.622 ± 0.028
SVM+CNN	0.664 ± 0.042	0.892 ± 0.030	0.476 ± 0.056	0.684 ± 0.039
RF	0.525 ± 0.034	0.391 ± 0.142	0.628 ± 0.147	0.509 ± 0.022
RF+CNN	0.661 ± 0.019	0.860 ± 0.062	0.481 ± 0.046	0.670 ± 0.030
DL_{In}	0.648 ± 0.040	0.742 ± 0.077	0.543 ± 0.045	0.643 ± 0.041
MMDL	$\mathbf{0.690} \pm 0.055$	0.790 ± 0.049	$\mathbf{0.689} \pm 0.048$	$\mathbf{0.733} \pm 0.051$

cortical volumes of T1 structure MRI and functional connectivity. For functional MRI in this study, we use the MSDL functional atlas [6].

The average ACC, Sensitivity (Sens.), Specificity(Spec.) and AUC of the proposed methods with the comparisons are on 8-fold cross-validation (CV) reported in Table 1, the proposed MMDL method outperforms the SVM and random forest (RF)[5] based methods, as shown in the Table 1, the proposed method has the highest ACC, Specificity and AUC. By fusioning the result of SVM and RF with CNN (i.e., 'SVM+CNN' and 'RF+CNN' in Table 1) separately, the results have improved. Compared to the competed methods in Table 1, our approach yields improvements of about 0.026 in ACC, 0.061 in Specificity (Spec.) and 0.049 in AUC.

In the proposed model, we show the features that are predicted with the discriminative learning model of Eq. (2) in Fig. 3 is an example of with the cortical volume and predicted one, they are quite similar and the RMSE between them is 0.09.

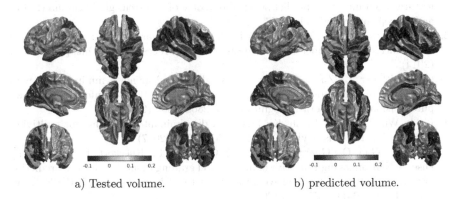

a) Tested volume.   b) predicted volume.

Fig. 3. The tested volume and the predicted one.

4   Conclusion

In this paper, a MMDL method is proposed by fusion discriminative learning and priors of deep CNN to regularize the classification problem. Specifically, in the non-negative discriminative dictionary learning model, this approach learns discriminative features by imposing both orthogonality on the synthesis dictionary, non-negativity low-rank constraints on projective coefficients. We initialize more multi-model data from dictionary learning model as the input of CNN, which can improve the training accuracy. Then, both training priors are returned to the fusion framework to improve the performance. Experiments on the tasks of identifying the ASD showed the benefit of our approach compared to state-of-the-art methods. The proposed method can be used for synthesizing the neural patterns of cortical.

Acknoledgements. This work was supported, in part, by the Fonds de recherche du Quebec (CCC 246110, 271636), National Nature Science Foundation of China (NSFC: 61902220), Shandong Province grant (ZR2018BF009) and the Science and Technology Innovation Fund Project of Dalian, China (No.2019J13SN100).

J.-B.P. was partially funded by National Institutes of Health (NIH) NIH-NIBIB P41 EB019936 (ReproNim) NIH-NIMH R01 MH083320 (CANDIShare) and NIH RF1 MH120021 (NIDM), the National Institute Of Mental Health of the NIH under Award Number R01MH096906 (Neurosynth), as well as the Canada First Research Excellence Fund, awarded to McGill University for the Healthy Brains for Healthy Lives initiative and the Brain Canada Foundation with support from Health Canada.

References

1. Chaddad, A., Rathore, S., Zhang, M., Desrosiers, C., Niazi, T.: Deep radiomic features from mri scans predict survival outcome of recurrent glioblastoma (2019). arXiv preprint arXiv:1911.06687
2. Gu, S., Zhang, L., Zuo, W., Feng, X.: Projective dictionary pair learning for pattern classification. In: Advances in Neural Information Processing Systems, pp. 793–801 (2014)
3. Gu, S., Zhang, L., Zuo, W., Feng, X.: Weighted nuclear norm minimization with application to image denoising. In: Proceedings of the IEEE Conference on Computer Vision and Pattern Recognition, pp. 2862–2869 (2014)
4. Li, J., Wu, Y., Zhao, J., Lu, K.: Low-rank discriminant embedding for multiview learning. IEEE Trans. Cybern. **47**(11), 3516–3529 (2017)
5. Tong, T., Gao, Q., Guerrero, R., Ledig, C., Chen, L., Rueckert, D., Alzheimer's Disease Neuroimaging Initiative et al.: A novel grading biomarker for the prediction of conversion from mild cognitive impairment to Alzheimer's disease. IEEE Trans. Biomed. Eng. **64**(1), 155–165 (2017)
6. Varoquaux, G., Gramfort, A., Pedregosa, F., Michel, V., Thirion, B.: Multi-subject dictionary learning to segment an atlas of brain spontaneous activity. In: Székely, G., Hahn, H.K. (eds.) IPMI 2011. LNCS, vol. 6801, pp. 562–573. Springer, Heidelberg (2011). https://doi.org/10.1007/978-3-642-22092-0_46

7. Wang, M., Zhang, D., Huang, J., Shen, D., Liu, M.: Low-rank representation for multi-center autism spectrum disorder identification. In: Frangi, A.F., Schnabel, J.A., Davatzikos, C., Alberola-López, C., Fichtinger, G. (eds.) MICCAI 2018. LNCS, vol. 11070, pp. 647–654. Springer, Cham (2018). https://doi.org/10.1007/978-3-030-00928-1_73

8. Zhang, M., Desrosiers, C.: High-quality image restoration using low-rank patch regularization and global structure sparsity. IEEE Trans. Image Process. **28**(2), 868–879 (2019)

9. Zhang, M., Desrosiers, C., Guo, Y., Zhang, C., Khundrakpam, B., Evans, A.: Brain status prediction with non-negative projective dictionary learning. In: Shi, Y., Suk, H.-I., Liu, M. (eds.) MLMI 2018. LNCS, vol. 11046, pp. 152–160. Springer, Cham (2018). https://doi.org/10.1007/978-3-030-00919-9_18

10. Zhang, M., Desrosiers, C., Zhang, C.: Atlas-based reconstruction of high performance brain MR data. Pattern Recogn. **76**, 549–559 (2018)

11. Zhang, M., Guo, Y., Zhang, C., Poline, J.-B., Evans, A.: Modeling and analysis brain development via discriminative dictionary learning. In: Knoll, F., Maier, A., Rueckert, D., Ye, J.C. (eds.) MLMIR 2019. LNCS, vol. 11905, pp. 80–88. Springer, Cham (2019). https://doi.org/10.1007/978-3-030-33843-5_8

12. Zhao, Yu., Ge, F., Zhang, S., Liu, T.: 3D deep convolutional neural network revealed the value of brain network overlap in differentiating autism spectrum disorder from healthy controls. In: Frangi, A.F., Schnabel, J.A., Davatzikos, C., Alberola-López, C., Fichtinger, G. (eds.) MICCAI 2018. LNCS, vol. 11072, pp. 172–180. Springer, Cham (2018). https://doi.org/10.1007/978-3-030-00931-1_20

13. Zheng, W., Eilamstock, T., Wu, T., Spagna, A., Chen, C., Hu, B., Fan, J.: Multi-feature based network revealing the structural abnormalities in autism spectrum disorder. IEEE Trans. Affect. Comput. (2019)

From Risks to Opportunities: Real Estate Equity Crowdfunding

Verena Schweder, Andreas Mladenow$^{(\boxtimes)}$ ⓘ, and Christine Strauss ⓘ

University of Vienna, Oskar Morgenstern-Platz 1, 1090 Vienna, Austria
a01452601@unet.univie.ac.at, {andreas.mladenow,
christine.strauss}@univie.ac.at

Abstract. Real Estate Equity Crowdfunding (REECF) enables small investments on big real estate projects through the funding by an online crowd. As there is only limited research on this novel type of financial instrument for the real estate sector, this paper investigates risks and opportunities of REECF in order to establish and improve the understanding of REECF, in particular to identify potential risks and potential opportunities that come along with an engagement in such projects. Based on traditional offline risks of capital investments, three categories of both potential risks and opportunities have been identified: market risks and opportunities, execution risks and opportunities, and agency risks and opportunities. We adopt a holistic approach, as we identified, collected and consolidated existing risks and opportunities in REECF. As a result, we introduce an up-to-date framework on real estate equity crowdfunding.

Keywords: Real estate · Equity crowdfunding · User interface · Information systems · Social web

1 Introduction

The finance industry has often been a driving force in introducing new technologies to decrease risks in commercial transactions and developing new business opportunities. The internet has emerged as an auxiliary platform to promote innovations in banking activities [12]. Examples include online banking, crypto currencies, and internet-based peer-to-peer lending [5]. Hence, it is not surprising that the 2007 housing crisis triggered the implementation of a new internet-based method for financing real estate properties [8]. The latest result of this development is Real Estate Equity Crowdfunding (REECF), which enables real estate investments through the funding of an online crowd. As this method of real estate financing is relatively new, there is only limited research published on this topic. Therefore, this paper provides an in-depth analysis on the risks and on the opportunities associated with REECF. The systematic approach is three-fold as it elaborates on issues relates to market, execution, and agency. The remainder of this paper is structured as follows. Section 2 provides an overview about crowdfunding. Starting with a general summary of the current crowdfunding literature, the focus is gradually moving towards the special type of crowdfunding – REECF. Furthermore, the process of a REECF project is explained. Section 3 presents the

© Springer Nature Switzerland AG 2020
S. Hartmann et al. (Eds.): DEXA 2020, LNCS 12391, pp. 444–454, 2020.
https://doi.org/10.1007/978-3-030-59003-1_30

research framework and the analytical methodology. Section 4 analyses the risks and opportunities of REECF, based on the previously defined methodology. The paper concludes by summarizing the findings of the analysis.

2 Background

2.1 Typology of Crowdfunding

The primary purpose of crowdfunding is to raise capital from the public, i.e. the crowd, for the purpose of financing a specific project [7, 14]. The interaction between the fundraiser(s) and the crowd occurs over crowdfunding platforms, where financial pledges can both be made and collected [7]. There is a variety of crowdfunding models that can be found on crowdfunding platforms, whereas four basic types have emerged: donation-based, reward-based, loan-based and equity-based crowdfunding [3].

Donation-based crowdfunding facilitates individual people in donating money to a specific charitable project in order to meet an overall funding goal. Donators do not receive any financial or material return for their endeavors [7, 13, 14]. Reward-based crowdfunding allows individuals to commit funds to a specific project, without receiving a financial return for their donation. Instead, donators are incentivized by obtaining a non-financial benefit, such as discounts, early delivery of products or tickets to attend a performance [13]. Lending-based crowdfunding (also known as loan-based crowdfunding, crowdlending, peer-to-peer or marketplace lending) refers to the type of crowdfunding, which is available to both individuals and companies. It allows crowdsourcers to borrow funds from individuals through an online lending marketplace in terms of a loan agreement [7, 13]. Equity crowdfunding enables entrepreneurs to allocate funds by issuing any type of securities (e.g. equity, equity-like shares, etc.) that give the holders an ownership stake in the business, in exchange for financial contributions [2, 13]. This type of crowdfunding is the most recent and is typically stipulated as silent partnerships, debt participation rights or subordinate profit-participating loans. In Austria and Germany, profit- participating loans are the most common form of participation on equity platforms [3].

Equity crowdfunding represents a very different approach from other crowdfunding types. There are four main aspects that distinguish this type of crowdfunding from others, particularly from reward-based crowdfunding [13]. First, the key motivation for an investor to engage in equity crowdfunding is generating profit. This motive differs, for example, from the donation-based model, which is driven by philanthropic motives, or from reward-based models that tend to pursue a broader objective which include non-financial benefits [10, 13]. Second, the number of shares to be issued is restricted in order to affect funding dynamics. On the opposite, in the case of reward-based crowdfunding, crowdsourcers usually accept as many requests as possible [10]. Third, the average amount of capital raised and the type of investors differ. Equity crowdfunding projects mostly involve companies and large amounts of capital, whereas in reward-based crowdfunding, individuals tend to invest small amounts [10]. Fourth, equity crowdfunding varies in terms of its risk/reward profile. It usually takes five to eight years for the investors to achieve liquidity; more than half of the investments lead to losses. Thus, the risk is much higher, compared to lending-based crowdfunding, which usually operates on both specified interest rates and time horizons of only three to 36 months [12, 13, 15].

2.2   Real Estate (Equity) Crowdfunding

Real Estate Crowdfunding (RECF) refers to a form of financing, whereby real estate project developers raise and aggregate capital from a wide group of investors through specialized, internet-based platforms [17, 18]. RECF platforms usually operate by offering either debt or equity investments.

RECF emerged as a reaction to the 2007–2008 global financial crisis. The crisis demonstrated that real estate investments were not entirely safe and that housing prices do not always increase [8]. Consequently, investors moved from a high-risk aversion regarding real estate projects towards a preference for low-risk secured lending [17]. Investors were looking for alternative methods of financing, opposite to the traditional investments in the housing market. The concept of crowdfunding, as part of the sharing economy, seemed to be the solution to this need. Thus, RECF has been created to fund real estate developments from several individuals through the Internet [8].

The introduction of relevant laws in the USA, UK, Australia and New Zealand promoted the rapid growth of RECF [17]. These laws permitted RECF as possibility for financing the construction, renovation and ownership of real estate projects [17, 20]. In the early stage, RECF was mainly conducted in the form of loans. Accredited individuals would fund a Limited Liability Company (LLC), which then made a loan to another individual or a small real estate company. These loans were eventually used to renovate and sell properties [19].

As RECF has become more accepted, the focus relocated from small scale lending to raising equity for larger properties [19]. Hence, equity investment, also known as Real Estate Equity Crowdfunding (REECF) has been created. This form usually includes investors buying shares in a limited liability company (LLC) which then invests in a limited partnership holding the actual property. The LLC's daily business is then implemented by a management team, making the investors passive [18].

2.3   Players, Processes and Roles in REECF

REECF involves three parties: the crowdsourcer (also known as project promotor, entrepreneur or creator) [20], i.e. the person/entity, who wishes to purchase or renovate a property; the crowdsourcees (also known as investors, funders, or crowd), who invest capital in order to develop a project, and the platform intermediary, which posts the crowdsourcer's project and provides relevant information to crowdsourcees [8].

The process of a REECF projects involves several activities. In the beginning, a real estate crowdsourcer creates a project on a REECF platform, in order to raise funds for financing the acquisition of a real estate property. As soon as the investors provide the funding objective on the platform, an ad hoc corporation is established, and the funders receive shares in that corporation. The only purpose of this company is to acquire a property, renovate it and to sell it when it seems most profitable. Consequently, the investors receive the returns realized from renting the building and the specified value of the corporation once the property is being sold (cf. Fig. 1).

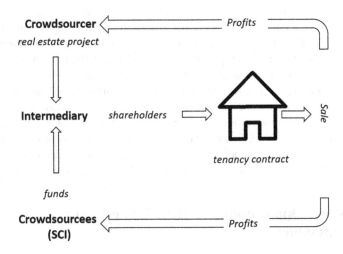

Fig. 1. Process of real estate equity crowdfunding [8]

3 Research Framework and Methodology

The goal of this present research is to understand risks and opportunities regarding the application of REECF. The literature review has shown that REECF is a relatively new and specialized form of crowdfunding and is consequently examined only to a limited extent. To the best knowledge of the authors no prior research has yet provided a systematic overview of potential risks and opportunities of this kind of crowdfunding.

Table 1. Risks and opportunities of real estate equity crowdfunding

	Risks	Opportunities
1. Market	a. Uncertainty of success b. Bankruptcy of the platform/company c. Lack of liquidity of shares d. Governance of the company	a. Fundraising b. Benefits of local investors/social networks c. Community participation d. Self-directed tools
2. Execution	a. Uncertainty of skills b. Cost of capital c. Investor Management	a. Lower fees/cost b. Streamlined process c. Intermediary effects d. Enhanced user experience
3. Agency	a. Information asymmetry b. Fraud c. Lack of regulation clarity	a. Greater transparency b. Enhanced reporting and accountability

Our research approach follows the framework introduced by Mamonov and Malaga (2019), which is based on traditional offline risks of capital investments. Mamonov and Malaga (2019) identified three types of potential risks when investing in a new venture on the success of equity crowdfunding: market risk, execution risk, and agency risk. These three risks form the basis of a framework for identifying potential risks and opportunities of REECF. Additionally, a thorough search on recent literature has been performed to identify relevant papers on crowdfunding, equity crowdfunding and RECF. The results were examined in order to reveal further risks regarding REECF. These additional items were subsequently being consolidated and integrated into the three existing dimensions, based on their compliance. This holistic approach is then transferred to the opportunities of REECF. The three pre-defined categories of investment risks for REECF are adapted to derive and constitute corresponding opportunities, i.e. market opportunities, execution opportunities, and agency opportunities. Table 1 summarizes the ascertained key risks and opportunities of REECF.

4 Risks and Opportunities of Real Estate Equity Crowdfunding

4.1 Risks of REECF

Market Risks. Market risks are external to the business and capture the uncertainty about the venture's market success when introducing a new product or service. These risks mainly comprise parameters that are not controllable by the management team, such as market size, growth trend, accessibility, existing or unforeseen competition, clients, etc. [4, 13].

Uncertainty of Success. REECF includes the risk of not being successful. Especially at the beginning of a REECF project, the market success is highly uncertain. Market risks such as growth trend and market size have been shown to be among the top reasons for the rejection of an investment from professional angel investors [13]. Furthermore, there are overall risks of failing to attract financial resources, which result in an unsuccessful funding attempt. First, the crowdfunder's project might not be accepted in the first place, so the project is not even being posted on a crowdfunding platform. Second, entrepreneurs might simply fail to raise the required capital from the crowd [3]. Third, investment decisions are often not based on solid financial data, but on emotional and other carious decision biases. For example, herding or supporting local projects leads to risks and uncertainty of funding [9]. Finally, REECF does not guarantee profits. Earnings rely on real estate fluctuations. If an acquired property cannot be sold for a price higher than purchased, profit cannot be drawn [8].

Bankruptcy of the Platform/Company. Insolvency of the crowdfunding platform or the company who posted a project are among the most dominant problems of REECF [8, 9]. In the former case, the bankruptcy of the platform has a negative impact on the investors. This was the case with the US-platform iFunding in 2016, when investors could not get information and revenues from the platform, due to the failure of contacting the platform. Of course, this problem also negatively affected developers, as they could not reach out to nor attract potential investors [8].

In the case of the company's bankruptcy, investors are negatively influenced in the subject of the insolvency proceeding. For the most part, investors in REECF projects are determined as having an "ordinary loan" or even a subordinate one. Thus, investors do not have preference regarding the recovery of their contributions of REECF. This is what happened with the German company Zinlands in 2017, where the investors are still involved in protracted insolvency proceedings, aiming to recover the value of their subordinate loans [8].

Lack of Liquidity of Shares. Another main risk of REECF constitutes the lack of liquidity of shares, which is a typical risk in any type of crowdfunding. Investors acquire shares of a business that develops a property. This business is a limited liability company (LLC), which is not allowed to transfer its shares freely.

Subsequently, the company cannot recover its contributions prior to the agreed date, not even for a justified reason [8].

Governance of the Company. Typically, in REECF, investors receive shares of an ad hoc corporation, in exchange for their contributions. The company's only purpose it to acquire a property, to rent it for a specified amount of time (usually between one and two years) and then to sell it when it is suitable. Consequently, investors formally become the owner of the business, however, they do not have any kind of right to actually use the acquired property [8].

Execution Risk. Execution risks are *internal* to the venture and reflect the challenge of executing or implementing a product or service. Furthermore, it also relates to the difficulties regarding the execution of the business model and strategy. Hence, this risk dimension highlights the importance of the business's human resources and factors regarding the execution of the business activities [4, 13].

Uncertainty of Skills. One major factor that is related to the success of a REECF is the equivalent expertise. On the one hand, angel investors and venture capitalists prefer engaging with entrepreneurs with entrepreneurial and industry experience. What is more, those investors also favour entrepreneurial teams over single entrepreneurs, as a team has more potential to possess the set of required skills [3, 13]. On the other hand, investors tend to overestimate their expertise. This phenomenon is based on the idea that people engaging in crowdfunding websites have the knowledge to select good investments. Moreover, they often claim to know specific properties in which to invest with minimum risk [19]. At the same time, crowdfunders are frequently very optimistic about the creator's ability to deliver promises. However, creators often lack experience in building a product or dealing with suppliers and logistics [1].

Cost of Capital. Although the fees and costs of REECF are often lower than traditional financing methods, there are drawbacks regarding the costs of capital of REECF platforms. First, platforms often charge additional fees for due diligence projects or insurances diminishing risk and uncertainty for the funders [9]. Second, on REECF platforms, the information regarding fees are often not provided when signing up for a project. Finally, the amount of fees frequently depends on the type of project [8]. To sum up, costs and fees regarding REECF are frequently untransparent which constitutes a risk for potential investors.

Investor Management. An overall challenge of crowdfunding is managing the investor base [3]. In RECF, however, the average number of investors is relatively low, with an average of seven investors per private real estate partnership. Nevertheless, the key to successful REECF is choosing appropriate partners. As most sponsors do not know their investors, the management of investors is challenging. Conflicts of interests are almost inevitable [19].

Agency Risks. Agency risks occur through information asymmetry, which is a scenario where entrepreneurs have a greater knowledge about the business than the investor. The problem in this case is that managers might pursue their own interest, at the investor's expense [4, 13].

Information Asymmetry. Most of the time, entrepreneurs have more information about the business prospects of their venture than crowdfunders [1, 13]. This issue is known as information asymmetry and potentially leads to market failure, as transactions between creators and funders are not conducted due to the lack of information. Additionally, the imbalance of information is also based on the investor's ex post incapability to initiate effort regarding the entrepreneur [1].

Fraud. The disequilibrium between the two sides of the market might result in opportunistic behaviour, which constitutes a moral hazard. The most extreme version of this manner is outright fraud [1]. This attitude is more common among younger, smaller firms and victimizes inexperienced, overly optimistic investors [1, 16]. Research has shown that it is quite simple to use false or misleading information to fraud funders [1, 3]. As opposed to platforms such as eBay or Airbnb, sellers do not have incentives to build a reputation, since interactions are not constantly repeated. Thus, REECF poses an appealing target for fraudulent behaviour [1].

Lack of Regulation Clarity. In most countries worldwide, there is a lack of clarity on rules and oversight. Commonly, laws regarding crowdfunding and REECF are nascent, vague and complex. What is more, to date, no cross-jurisdictional harmonization in terms of crowdfunding, and certainly not for REECF, has been introduced [17].

4.2   Opportunities of REECF

Market Opportunities. Market opportunities are defined as external to the venture and reflect the advantage regarding the introduction of a new product or service in the market. Thus, such features aim for reducing the uncertainty of the market success.

Broader Investor Base. Traditional real estate projects are mostly financed by investors that are geographically closely located to the property. However, the use of the Web 2.0, allows to bridge these geographical barriers. Subsequently, REEC platforms connect investors and funders from geographically diverse areas, widening the investor base in real estate [1, 3, 17].

Benefits of Local Investors and Social Networks. Despite widening one's investor base through the use of technology, it is nevertheless essential to have investors that are local and/or personally known. Local investors in real estate projects can provide creators with information about what is acceptable in their community. Furthermore, they can appear at community meetings and act as an extra set of eyes to the property [19]. Personal and/or business partners play an essential role during the entire crowdfunding process. Drawing upon social networks is important for entrepreneurs in order to meet their funding objective [17].

Community Participation. From the funder's point of view, investing in a REECF is a social activity. Often, their main motivation is to provide capital in return for recognition form the entrepreneur or the community [1]. Real Estate Crowdfunders also often participate in the development of the venture itself, providing feedback or sharing critical thoughts, which reduces the REECF project's risk of failure [3].

Self-directed Tools. REECF platforms provide efficiency gains as they equip entrepreneurs and funders with self- directed tools that streamline the entire crowdfunding process. In turn, these tools reduce or eliminate the need for professionals such as brokers or advisors. Additionally, platforms allow investors to carry out analyses and risk evaluations as well as building their own portfolios [11, 17].

Execution Opportunities. Execution opportunities are internal to the business and emphasize the importance of the of the creator's team and corporate environment in executing a business strategy. Hence, opportunities regarding the execution deal with the improvements of implementing a REECF project.

Lower Fees/Costs. Due to the structure and layout of online platforms, REECF platforms entail remarkable cost savings [6]. First, REECF platforms remove middlemen of traditional finance methods, who connect borrowers and lenders, resulting in lower costs of capital for entrepreneurs and investors [17]. Second, entrepreneurs might be able to cut costs of capital due to selling properties that are otherwise challenging to sell. Third, the growth of information regarding project information might increase investors' willingness to pay, which in turn lowers the cost of capital [1]. Fourth, REECF platforms in general charge substantially lower fees than traditional real estate financing methods [17].

Streamlined Process. By leveraging modern technology, REECCF platforms enable real estate projects to be significantly quicker, faster, more flexible and more efficient than traditional financing approaches, such as using banks as intermediaries [3, 17].

Intermediary Effects. Prior research suggests that REECF platforms practice intermediary functions similar to banks [6, 17]. These platforms can act as intermediary (broker) or as a principal in investments. Typical functions of REECF platforms include pooling together capital as well as indirectly performing brokerage or appraisal duties by implementing marketing and analyses of projects [6]. REECF platforms providing a large number of services are prone to attract more higher-quality and potentially prosperous projects [17].

Enhanced User Experience. The use of modern technology generates a unique user experience for engaging in real estate project. Features such as information and convenience, efficiency and scale constitute advantages and user friendliness that traditional real estate financing cannot offer [17].

Agency Opportunities. Agency opportunities highlight the potential enhancement between the investor's and the entrepreneur's interests. In conjunction with information transparency, these chances can undermine the investor's ability to capture financial rewards from his/her investment.

Greater Transparency. Arguably, the greatest benefit of REECF is the enhanced transparency for both entrepreneurs and funders. Online technology allows information to flow easily and quickly as well as reporting real time updates [6, 17, 20]. This incomparable level of transparency also fosters crowdfunding success and crowd participation [17].

Enhanced Reporting and Accountability. REECF platforms are mainly focused on providing information, which allows them to present high-quality data in a punctually and easy to understand manner to individuals. As the REECF sector becomes more common, obtaining objective information about the funders and creators should become more uniform and reliable. Analogically to the electronic marketplace website eBay, REECF are likely to integrate accountability mechanisms such as rating and reputation systems, enhancing reporting and accountability [19].

5 Conclusion

This paper aimed at an up-to-date framework on potential risks and potential opportunities in REECF, to support decision-making processes of crowdsourcers, crowdsourcees, and intermediaries engaged in such projects. Based on well-known traditional offline risks of capital investments, three categories of both potential risks and potential opportunities have been identified: market risks and opportunities, execution risks and opportunities, and agency risks and opportunities. Adopting a holistic approach, we identified, collected and consolidated existing risks and opportunities in REECF and derived the envisaged REECF framework. Taking part in REECF poses several risks for the investors and entrepreneur. *Market risks* are external to the business and include the uncertainty of success, the bankruptcy of the platform or the company, the company's interdiction to use the property or the lack of liquidity of shares. *Execution risks* comprise the difficulty of the implementation of REECF, such as lacking skills in engaging in a REECF process, dealing with fees and costs of capital and managing the investor base. *Agency risks* cover potential misalignments between investor and entrepreneur, leading to information asymmetry, fraud and the lack of clarity on rules and oversight. On the other hand, REECF also entails several opportunities. *Market opportunities* reflect external factors that reduce the uncertainty of the market success. In REECF, funders can make use of a broader investor base, local investors and social networks, the participation of the project's community as well as self-directed tools that allow investors to reduce the overall risk of REECF. *Execution opportunities* represent improvements in terms of the implementation of a REECF project. These benefits mainly derive from the use of modern technology, which enables lower fees and costs, a streamlined, a more efficient process, and an enhanced user experience for crowd-funders as well as REECF platforms acting as intermediaries. *Agency opportunities* highlight the potential improvements between the interests of the entrepreneurs and the investors. Advantages entail a greater transparency of information and the potential to enhance reporting and accountability, making REECF more successful and secure.

Future research may focus on the implementation of online tools that support investment decisions of potential crowdsourcees and assist in evaluating risks and opportinities. Such tools shall provide operationalized values for general risks and opportunities in REECF, but should also have options to include local features or/and personal preferences.

References

1. Agrawal, A., Catalini, C., Goldfarb, A.: Some simple economics of crowdfunding. Innov. Policy Econ. **14**(1), 63–97 (2014)
2. Buerger, B., Mladenow, A., Novak, N.M., Strauss, C.: Equity crowdfunding: quality signals for online-platform projects and supporters' motivations. In: Tjoa, A.M., Raffai, M., Doucek, P., Novak, N.M. (eds.) CONFENIS 2018. LNBIP, vol. 327, pp. 109–119. Springer, Cham (2018). https://doi.org/10.1007/978-3-319-99040-8_9
3. Buerger, B., Mladenow, A., Strauss, C.: Equity crowdfunding market: assets and drawbacks. In: 38th International Conference on Information Systems (ICIS), pp. 1–6. Special Interest Group on Big Data Processing, Seoul (2017)
4. Carpentier, C., Suret, J.M.: Angel group members' decision process and rejection criteria: a longitudinal analysis. J. Bus. Ventur. **30**(6), 808–821 (2015)
5. Chen, X., Zhou, L., Wan, D.: Group social capital and lending outcomes in the financial credit market: an empirical study of online peer-to-peer lending. Electron. Commer. Res. Appl. **15**, 1–13 (2016)
6. Cohen, M.L.B.: Crowdfunding as a financing resource for small businesses. Doctoral dissertation, Walden University, Minneapolis, MN (2017)
7. Crowdfunding in the EU capital markets union. https://ec.europa.eu/info/publications/crowdfunding-eu-capital-markets-union_en. Accessed 10 March 2020
8. Garcia-Teruel, R.M.: A legal approach to real estate crowdfunding platforms. Comput. Law Secur. Rev. **35**(3), 281–294 (2019)
9. Gierczak, M.M., Bretschneider, U., Haas, P., Blohm, I., Leimeister, J.M.: Crowdfunding: outlining the new era of fundraising. In: Brüntje, D., Gajda, O. (eds.) Crowdfunding in Europe. FSSBE, pp. 7–23. Springer, Cham (2016). https://doi.org/10.1007/978-3-319-18017-5_2
10. Hervé, F., Manthé, E., Sannajust, A., Schwienbacher, A.: Determinants of individual investment decisions in investment-based crowdfunding. J. Bus. Finance Account. **46**, 762–783 (2017)
11. Hollas, J.: Is crowdfunding now a threat to finance? Corp. Fin. Rev. **18**(1), 17–22 (2013)
12. Mamonov, S., Malaga, R.: Success factors in Title III equity crowdfunding in the United States. Electron. Commer. Res. Appl. **27**, 65–73 (2018)
13. Mamonov, S., Malaga, R.: Success factors in Title II equity crowdfunding in the United States. Venture Capital **21**(2–3), 223–241 (2019)

14. Mamonov, S., Malaga, R., Rosenblum, J.: An exploratory analysis of Title II equity crowdfunding success. Venture Capital **19**(3), 239–256 (2017)
15. Mason, C.M., Harrison, R.T.: Is it worth it? The rates of return from informal venture capital investments. J. Bus. Ventur. **17**(3), 211–236 (2002)
16. Cernicka, R., Mladenow, A., Strauss, C.: The impact of updates in social crowd projects: insights from a german equity crowdfunding platform. In: Proceedings of the 21st International Conference on Information Integration and Web-based Applications & Services, pp. 571–578 (2019)
17. Montgomery, N., Squires, G., Syed, I: Disruptive potential of real estate crowdfunding in the real estate project finance industry: a literature review. Property Manag. **36**(5), 597–619 (2018)
18. Shahrokhi, M., Parhizgari, A.M.: Crowdfunding in real estate: evolutionary and disruptive. Manag. Finance, 597–619 (2019)
19. Vogel, J.H., Moll, B.S.: Crowdfunding for real estate. Real Estate Finance J. Summer/Fall **2014**, 5–16 (2014)
20. Mladenow, A., Bauer, C., Strauss, C.: Crowd logistics: the contribution of social crowds in logistics activities. Int. J. Web Inf. Syst. **12**(3), 379–396 (2016)

Author Index

Printed in the United States
By Bookmasters